TERAHERTZ SENSING TECHNOLOGY

SELECTED TOPICS IN ELECTRONICS AND SYSTEMS

Editor-in-Chief: **M. S. Shur**

Published

Vol. 15: Silicon and Beyond
eds. *M. S. Shur and T. A. Fjeldly*

Vol. 16: Advances in Semiconductor Lasers and Applications to Optoelectronics
eds. *M. Dutta and M. A. Stroscio*

Vol. 17: Frontiers in Electronics: From Materials to Systems
eds. *Y. S. Park, S. Luryi, M. S. Shur, J. M. Xu and A. Zaslavsky*

Vol. 18: Sensitive Skin
eds. *V. Lumelsky, M. S. Shur and S. Wagner*

Vol. 19: Advances in Surface Acoustic Wave Technology, Systems and Applications (Two volumes), volume 1
eds. *C. C. W. Ruppel and T. A. Fjeldly*

Vol. 20: Advances in Surface Acoustic Wave Technology, Systems and Applications (Two volumes), volume 2
eds. *C. C. W. Ruppel and T. A. Fjeldly*

Vol. 21: High Speed Integrated Circuit Technology, Towards 100 GHz Logic
ed. *M. Rodwell*

Vol. 22: Topics in High Field Transport in Semiconductors
eds. *K. F. Brennan and P. P. Ruden*

Vol. 23: Oxide Reliability: A Summary of Silicon Oxide Wearout, Breakdown, and Reliability
ed. *D. J. Dumin*

Vol. 24: CMOS RF Modeling, Characterization and Applications
eds. *M. J. Deen and T. A. Fjeldly*

Vol. 25: Quantum Dots
eds. *E. Borovitskaya and M. S. Shur*

Vol. 26: Frontiers in Electronics: Future Chips
eds. *Y. S. Park, M. S. Shur and W. Tang*

Vol. 27: Intersubband Infrared Photodetectors
ed. *V. Ryzhii*

Vol. 28: Advanced Semiconductor Heterostructures: Novel Devices, Potential Device Applications and Basic Properties
eds. *M. Dutta and M. A. Stroscio*

Vol. 29: Compound Semiconductor Integrated Circuits
ed. *Tho T. Vu*

Vol. 30: Terahertz Sensing Technology — Vol. 1
Electronic Devices and Advanced Systems Technology
eds. *D. L. Woolard, W. R. Loerop and M. S. Shur*

Vol. 31: Advanced Device Modeling and Simulation
ed. *T. Grasser*

Selected Topics in Electronics and Systems – Vol. 32

TERAHERTZ SENSING TECHNOLOGY

Volume 2: Emerging Scientific Applications & Novel Device Concepts

Editors

Dwight L Woolard
US Army Research Laboratory

William R Loerop
US Army Soldier Biological and Chemical Command

Michael S Shur
Rensselaer Polytechnic Institute

World Scientific

NEW JERSEY • LONDON • SINGAPORE • SHANGHAI • HONG KONG • TAIPEI • BANGALORE

Published by

World Scientific Publishing Co. Pte. Ltd.

5 Toh Tuck Link, Singapore 596224

USA office: Suite 202, 1060 Main Street, River Edge, NJ 07661

UK office: 57 Shelton Street, Covent Garden, London WC2H 9HE

British Library Cataloguing-in-Publication Data
A catalogue record for this book is available from the British Library.

TERAHERTZ SENSING TECHNOLOGY
Volume 2: Emerging Scientific Applications and Novel Device Concepts

ISBN 981-238-611-4

Editor: Tjan Kwang Wei

This book is printed on acid-free paper.

Printed in Singapore by Mainland Press

PREFACE

As we begin the new millennium, significant scientific and technical challenges remain within the terahertz (THz) frequency regime and they have recently motivated an array of new research activities. Indeed, the last research frontier in high-frequency electronics now lies in the so-called *terahertz* (or submillimeter-wave) regime between the traditional microwave and the infrared domains. Today, the *terahertz* or *THz* regime has been broadly defined as the portion of the submillimeter-wavelength electromagnetic (EM) spectrum between approximately 1 mm (300 GHz) and 100 μm (3 THz). While the THz frequency regime has long offered many important technical advantages (e.g., wider bandwidth, improved spatial resolution, component compactness) and while significant scientific interest in THz frequency science and technology has existed since the early 1900's, the solid-state electronics capability at THz frequencies today remains extremely limited from a basic signal source and systems perspective (i.e., the highest output power is less than or on the order of milliwatts). The relatively limited development of semiconductor-based electronic circuits within the THz band may be attributed to the confluence of two fundamental factors. First, very challenging development and engineering problems are present in this quasi-optical regime where EM wavelength is on the order of component size. Second, the practical and scientific applications of this shorter-wavelength microwave region, where the atmospheric propagation paths are extremely attenuating, have been previously restricted to a few specialized fields (e.g., molecular spectroscopy for Earth, planetary and space science). Furthermore, engineering efforts to extend conventional three-terminal semiconductor devices upward from millimeter-wave as well as separate efforts to extrapolate traditional solid-state laser technology down from the far infrared have been prohibited due to fundamental physical factors associated with the respective device technologies. Indeed, electron velocities are not high enough to extend the operational of conventional transit-mode transistors (e.g., HFET's and HBT's) into the THz range even for devices with deep submicron dimensions. Alternatively, when the laser end of the spectrum is considered, one finds that the energies of photonic transitions corresponding to the terahertz range are small compared to the thermal energies at room or elevated temperatures (one terahertz corresponds to a 4.14 meV photon energy and to approximately 300 micron wavelength in free space) and effects, such as free carrier generation, plague these devices.

Fortunately, two-terminal semiconductor devices (e.g., Schottky and Heterostructure Barrier Varactor and Schottky mixers) utilize charging effects very near the contact interface and are inherently faster than transistor devices. Hence, two-terminal transport-based semiconductor devices long ago emerged as the key technology for the generation, amplification and detection of electrical signals at submillimeter-wave frequencies. However, the overall performance (e.g., power and efficiency) of even state-of-the-art two-terminal technologies suffers as they are extended for operation high into the THz band. This longstanding limitation in THz electronic technology, along with the excessive cost of instrumentation, have certainly been major stumbling blocks to new scientific inquiries at THz frequencies and has most probably prevented the spread of THz science and technology related issues to the broader scientific and engineering communities. Recent advances in nanotechnology, molecular chemistry and biological science have already begun to chart the course for new and important applications of THz electronics in the coming twenty-first century. In fact, the growing interest in the precise detection, identification and characterization of very small organic and inorganic systems has already begun to emphasize the future value of a robust THz-frequency sensing science and to establish it as an important driver for the rapid advancement of electronics technology at THz frequency. These new and exciting sensing applications only provide added motivation for realizing a practically useful THz electronics technology, which has long been recognized to offer much to conventional electronic application areas such as extended bandwidth for special scenario communications (i.e., short-range, networked and satellite) and significantly enhanced signal processing power.

During the last few years, major research programs have emerged within the U.S. Army and the Department of Defense (DoD) that have been focused on advancing the state-of-the-art in THz-frequency electronic technology and on investigating novel applications of THz-frequency sensing. These basic programs grew out of small seed efforts supported by a number of agencies including the U.S. Army Research Laboratory (ARL), the U.S. Soldier Biological and Chemical Command (SBCCOM) and the Air Force Office of Scientific Research (AFOSR). More recently, these efforts have been intensified and propagated primarily from the support of a Defense Advanced Research Project Agency (DARPA) Program on "Terahertz Technology for Sensing and Satellite Communications" and a Multidisciplinary University Research Initiative (MURI) Program on "Sensing Science and Electronic Technology at THz Frequencies" that is managed out of the Army Research Office (ARO) of ARL. One of the main catalysts for these programs is associated with the idea of using the fundamental interactions of THz radiation at the molecular level for sensing and characterizing chemical and biological (CB) agents. The science and technology emerging from these programs has potentially important ramifications to such areas as CB defense, biomedical applications and molecular science. Historically, the U.S. DoD has been instrumental in establishing the basic foundations for many important endeavors in science and engineering and recent DoD support for THz electronics is once again playing a major role in promoting an interest among the broader international scientific and engineering community. Indeed, there has been a steadily growing interest among these communities in the unique challenges associated with developing a robust electronics technology and with developing a detailed understanding of THz-frequency sensing science.

This growing wave of research and development has already started to shrink the "THz gap" through recent advances in both photonic and electronic technology. On the photonic technology side, the most prominent recent development has been the emergence of relatively high power THz quantum cascade lasers developed by Ruedeger, Köhler and Alessandro Tredicucci of the Scuola Normale Superiore in Pisa and colleagues in Turin and Cambridge [1],[2] (see also earlier work of the Swiss group [3] This device used 1500 alternating layers of gallium arsenide and aluminum gallium arsenide illustrating dramatic improvements in sophistication of modern semiconductor materials growth technology. This device (1.5 mm long and 0.2 mm wide) emitted 2 mW at 4.4 THz (wavelength of 67 μm) in single mode operation at 50 K. On the electronic

---

[1] http://physicsweb.org/article/news/6/5/5
[2] http://optics.org/articles/news/8/5/16/1
[3] Stéphane Blaser, Michel Rochat, Lassaad Ajili, Mattias Beck, Jérôme Faist, H. Beere, A. Davis, E. Linfield, and D. Ritchie, "Terahertz interminiband emission and magneto-transport measurements from a quantum cascade chirped superlattice", Physica E **13**, 854(2002).

technology side, recent progress in planar integration of two-terminal multipliers and in engineering photomixer technology is defining the new state-of-the-art in solid-state electronic sources and new concepts such as ballistic tunneling devices and plasma waves electronics offer promise for considerable potentials improvements in the future.

The recent plethora of new research activities associated with THz-frequency science and technology has motivated the organization of this book. This two-volume book has been structured to stand as an up to date and detailed reference for the new THz-frequency technological advances that are emerging across a wide spectrum of sensing and technology areas. The first volume (see Terahertz Sensing Technology, Vol. 1 Electronic Devices & Advanced Systems Technology) emphasized the ongoing efforts to establish a practical and useful components and system base within the THz regime. This second volume focuses on endeavors that possess higher risk, and much higher payoff, such as innovative applications in sensing, and new emerging device concepts. Hence, the full significance of these research subjects remains to be fully understood and established in many cases.

This second volume of the book presents cutting edge results in two primary areas: (1) research that is attempting to establish THz-frequency sensing as a new characterization tool for chemical, biological and semiconductor materials, and (2) theoretical and experimental efforts to define new device concepts within the "THz gap."

Chapter 1 of this volume begins with an overview of THz-frequency spectroscopic sensing of biological materials by Prof. Tatiana Globus and co-workers. Here, the focus is on spectral measurements of DNA (and related biological materials) that are providing new insight into the microscopic mechanisms associated with the interactions between THz radiation and species-specific phonon modes. This is a very appropriate chapter to begin this book as it provides an interesting and new scientific basis for the development of future bio-sensing technologies. In Chapter 2, Choi *et. al.* report on new efforts to utilize broadband THz-frequency reflection and transmission spectra as a starting point for distinguishing concealed threats in envelops and on personnel. Here, thermal modulation is demonstrated as an effective tool to induce changes to the microwave reflections and points to a new way for gaining qualitative chemical and biological spectra from broadband THz systems. In Chapter 3, Markelz and Whitmire provide results from THz time-domain spectroscopy performed on biomolecules to determine applicability of the technique for chemical and conformational identification. While this research did not reveal distinct normal modes that could be used for identification, it did demonstrate a non-destructive method to rapidly quantify the conformational flexibility of biomolecular species through the FIR response. This research also suggests a potential approach for evaluating cycling times and response time due to mutagenisis, which will be pursued further in future research investigations. Chapter 4 is authored by Federici and Grebel and describes the use of THz, Raman and IR spectroscopy for the indirect assessments of the electronic characteristics of nanocomposites. Here, THz spectroscopy was shown to provide insight into the electron mobility in semiconductive and conductive nanostructures that are suitable for electronic, electro-optic and nonlinear optical applications. Prof. Elliott Brown gives a unique and detailed engineering presentation in Chapter 5 on the subject of Terrestrial Millimeter-wave and THz remote sensing. This excellent treatment of remote sensing addresses all the fundamental principles behind system architectures and overviews all the critical aspects needed for designing a THz-frequency remote sensing system. Indeed, this is an excellent blueprint for anyone who is interested in designing and developing a system for the remote detection of chemical and biological agents and stands as an important and extremely useful technical reference. In Chapter 6, Solomon and coworkers explore the prospects of developing a laser operating in the THz regime using a quantum dot (QD) gain medium. This general device concept holds promise because it overcomes phonon losses in the device active region and is conducive to some THz laser cavity designs. The research presented in this chapter shows that QDs can be selectively placed in whispering-gallery architectures, while maintaining good QD optical quality. This approach suggests an exciting new optical-based source technology. Allen and Scott review the physics of THz-frequency transport in quantum structures in Chapter 7 and evaluate electrically-biased superlattices for their potential as terahertz gain medium. This work demonstrates that embedding superlattice devices in quasi-optical arrays and integrating them into terahertz cavities makes it possible to measure the dynamical conductance of quantum structures. This work sheds new light on fundamental issues related to utilizing gain in superlattice structures. The next two chapters in Volume 2 consider issues related to the theory and modeling of THz-frequency oscillations in quantum structures. In Chapter 8, Woolard and coworkers present an advanced theory of electronic instability in tunneling nanostructures. Here, new theoretical concepts and models are presented for the study of one-dimensional tunneling devices that possess the potential for exhibiting self-oscillations at THz frequencies. This chapter stands as an excellent reference for the study of time-dependent quantum mechanical effects and nonequilibrium phenomena (e.g., multi-band transport). Grubin and Buggeln present results on Wigner-function based simulations in Chapter 9 that can be used to address the effects of device-circuit interactions for quantum devices operating at very fast switching times. This work lays an important foundation for the future simulation of self-sustained THz oscillations in quantum-based circuits. Volume 2 concludes with Chapter 10 where Plusquellic and coworkers present scientific results generated from a continuous-wave linear-absorption spectroscopic system that utilizes solid-state photomixers as its source. Here, the system was applied to the investigation of biomolecules in polyethylene matrices and to line-shape studies of HF for diagnostics of semiconductor etching plasmas. This chapter directly demonstrates the scientific and commercial potential of THz-frequency systems and is a very appropriate conclusion to Volume 2.

The editors would like to thank all the authors for their fine contributions to this outstanding review of "Terahertz Sensing Technology". Both Volume 1 on *Electronic Devices & Advanced Systems Technology*, and Volume 2 on *Emerging Scientific Applications & Novel Device Concepts,* will be useful for technologists, scientists, engineers, and graduate students who are interested in the development of terahertz technology for sensing applications. The book can also be used as a textbook for graduate and senior undergraduate courses on terahertz electronics and as an additional reference text for courses in semiconductor physics, materials science, and electronic device design.

CONTENTS

International Journal of High Speed Electronics and Systems
Vol. 13, No. 4 (2003) 903–936

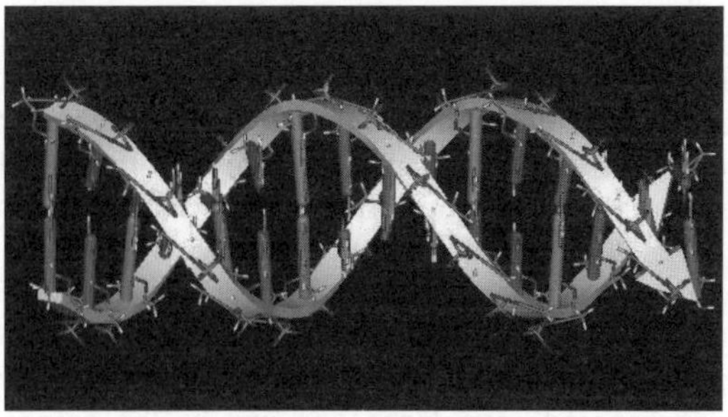

THZ-FREQUENCY SPECTROSCOPIC SENSING OF DNA AND RELATED BIOLOGICAL MATERIALS

T. GLOBUS[a], D. WOOLARD[b], M. BYKHOVSKAIA[c], B. GELMONT[a], L. WERBOS[d] and A. SAMUELS[e]

[a]*Dept. of Electrical and Computer Engineering, UVA, Charlottesville, VA, USA*
[b]*U.S. Army Research Laboratory, ARO, Research Triangle Park, NC, USA*
[c]*Dept. of Biological Sciences, Lehigh University, Bethlehem, PA, USA*
[d]*IntControl, Arlington, VA, USA*
[e]*Edgewood Chemical and Biological Center, Aberdeen Proving Ground, MD, USA*

The terahertz frequency absorption spectra of DNA molecules reflect low-frequency internal helical vibrations involving rigidly bound subgroups that are connected by the weakest bonds, including the hydrogen bonds of the DNA base pairs, and/or non-bonded interactions. Although numerous difficulties make the direct identification of terahertz phonon modes in biological materials very challenging, recent studies have shown that such measurements are both possible and useful. Spectra of different DNA samples reveal a large number of modes and a reasonable level of sequence-specific uniqueness. This chapter utilizes computational methods for normal mode analysis and theoretical spectroscopy to predict the low-frequency vibrational absorption spectra of short artificial DNA and RNA. Here the experimental technique is described in detail, including the procedure for sample preparation. Careful attention was paid to the possibility of interference or etalon effects in the samples, and phenomena were clearly differentiated from the actual phonon modes. The results from Fourier-transform infrared spectroscopy of DNA macromolecules and related biological materials in the terahertz frequency range are presented. In addition, a strong anisotropy of terahertz characteristics is demonstrated. Detailed tests of the ability of normal mode analysis to reproduce RNA vibrational spectra are also conducted. A direct comparison demonstrates a correlation between calculated and experimentally observed spectra of the RNA polymers, thus confirming that the fundamental physical nature of the observed resonance structure is caused by the internal vibration modes in the macromolecules. Application of artificial neural network analysis for recognition and discrimination between different DNA molecules is discussed.

Keywords*: Terahertz, absorption, biological molecules, anisotropy, vibration modes, transmission spectroscopy, normal mode analysis.*

1. Introduction

There exists considerable interest in both the experimental and theoretical investigation of the low-frequency internal vibrations associated with Deoxyribonucleic acid (DNA) polymers. The study of molecular dynamics, achieved via scattering and absorption spectroscopy, is a viable and proven approach that has been applied widely for the general characterization of molecular conformation.[1] Furthermore, there are fundamental physical reasons to expect that an effective application of the spectral data (i.e., especially in the very long-wavelength regime) can yield detailed information about complex biological molecules. Since the submillimeter-wave frequency regime (i.e., ~ 0.1-10 THz) is predicted to be fairly rich with spectral features that are dependent on DNA internal vibrations spread over large portions of the complex molecule,[2] it is reasonable to expect results that are dependent on the primary sequence of the molecule. Hence, the focus of this chapter is to demonstrate the use of submillimeter-wave spectroscopy for the characterization of DNA polymers and to establish a theoretical foundation for the future interpretation of the associated phonon modes.

The submillimeter-wave absorption spectra of DNA reflect low-frequency internal helical vibrations involving rigidly bound subgroups which are connected by the weakest bonds, including the weak hydrogen bonds of the DNA base pairs and/or non-bonded interactions.[2-9] These internal motions are sensitive to DNA composition and topology, have an impact on the main processes related to the transfer of genetic information[10] and eventually can give information regarding the three-dimensional structure and flexibility of the DNA double helix.

Most spectral investigations of DNA have been performed at frequencies above 10 THz or towards the higher-frequency portion of the submillimeter-wave domain. Below this range, a very limited number of spectroscopic studies have been reported until recently [for example, see Refs. 3, 11-20]. Significant progress has been achieved during the last several years with regard to the experimental and theoretical aspects of submillimeter-wave transmission spectroscopy of biological macromolecules. Detailed experimental spectra of DNA macromolecules and complete cellular biological samples over the spectral range 10 cm^{-1} to 500 cm^{-1} have been obtained. The results demonstrated multiple numerical resonances that are associated with low frequency vibrational phonon modes,[21-25] in qualitative agreement with the theoretical prediction.[26,6] Computational methods were also developed to predict the low frequency absorption spectra of short artificial DNA and RNA.[27,28] The preliminary application of Neural Network Analysis for the raw data generalization of very far FTIR spectra of genetic materials from fish indicated a high degree confidence in the recognition of DNA samples.[23] Absorption spectra of polynucleotides with known base-pair sequences (an RNA chain Poly[C]-Poly[G]) (G-Guanine, C-Cytosine) have also been measured in the sub-millimeter-wave range and compared to simulation results.[28,29] The results demonstrate a close correlation between the calculated and experimentally observed spectra of RNA homopolymers confirming the fundamental physical nature of the observed resonance structures that are produced by the internal vibrational modes in macromolecules.

The discovery of submillimeter-wave vibrational spectra of biological molecules is important for a number of reasons. The far-infrared region of the DNA absorption spectra, 3-300 cm^{-1} (or 0.1-10 THz), reflects low-frequency molecular internal motions. In this range, the spectral features from bio-polymer molecules arise out of poorly localized low-frequency molecular internal vibrations such as twisting, bending and stretching of the double helix, sugar pseudorotational vibrations and fluctuations of weakest bonds or non-bonded interactions (van der Waals forces, dispersion forces, and hydrogen bonding).[2-9] The internal motions, dependent on the weak hydrogen bonds of the double-helix base-pairs,[3] are extremely sensitive to DNA composition and topology and have an impact on the main processes related to the transfer of genetic information, such as replication, transcription and viral infection.[10] Hence, phonon modes that arise in this range should reflect features specific to the DNA code. Furthermore, theoretical studies have predicted the occurrence of DNA phonon frequencies throughout this region. [8,30-32] Therefore, the investigation of the far-infrared region of the DNA absorption spectra and the identification of DNA low-frequency internal motions are direct ways to obtain information about the peculiarities of DNA topology and internal motion. Submillimeter wave spectroscopy coupled with theoretical prediction can become a powerful tool for the investigation of the DNA structure and possible biological function.

Detailed insight into the novel physical mechanisms associated with the interactions between radiation and the biological material in the very far-infrared region is also important to biological (bio) detection since this initial work demonstrated that a large number of species-specific phonon modes exist. Thus, phonon modes that arise in the very far-infrared region may serve as a signature of specific DNA-molecules. Data collected by spectroscopic characterization of biological polymer chains have the potential to create a scientific basis for the development of advanced bio-sensing technologies in the future. Here, the sensing methodology will be based upon millimeter-wave transmission spectroscopy that uses terahertz radiation to detect and identify biomolecules. The initial results indicate that the unique portion of the phonon signature occurs at long-wavelengths in the frequency band from 0.1 to 1 THz. This is very important because this corresponds to a regime where it is quite feasible to develop miniaturized, low-cost electronic detectors in the future. Furthermore, the atmosphere is quite transparent below approximately 0.3 THz and this phonon-based approach may offer a means to achieve remote bio-detection. Very recent estimates of sensitivity and discrimination for THz-frequency differential-absorption spectrometers offer the promise of achieving a remote sensing capability for biological spore material.[22, 33]

This chapter will outline recent spectroscopic studies performed on biological materials and present new theoretical models for accurately interpreting phonon mode behavior in spectral signatures.

2. Theory for the Characterization of Bio-Molecules

A. Background

DNA topology and flexibility.

The three-dimensional structure of double-helical DNA is a key factor for its function as a bearer of genetic information. The most commonly found topology of DNA is the B-form,[34] an antiparallel, right-handed double stranded helix. B-form is characterized by the presence of two distinct grooves (major and minor) and by certain helicoidal parameters (10 residues per turn, 34 helix pitch, 20 helix diameter).[10] Theoretical conformational analysis demonstrated that B-DNA is not a single conformation, but rather an ensemble of stable conformations discerned by sugar puckering and by backbone torsion angles (Poncin et al.).[35]

DNA is a highly flexible and polymorphic structure, which can adopt different forms: right-handed A,B,C, D forms and the left-handed Z-form.[36] Left-handed Z-helix DNA is believed to be formed in vivo during DNA supercoiling and to be essential for the processes of replication and transcription.[37] The effect of DNA supercoiling on Z-helix formation[38] suggests that the presence of Z-DNA can serve as a signature of supercoiling. Z-DNA has major structure differences in comparison with B-DNA, such as helicoidal parameters (12 residues per turn, 45 helix pitch, 18 diameter), sugar puckering, and orientation of the base pairs within the helix.[10] Z-form is favored in DNA with altering d(CG) sequences at high salt concentrations.[37,39, 40] Major conformational differences (for example, B-Z transitions) can be detected in spectroscopy or X-ray experiments. However, minor differences between stable conformations belonging to B-family[35] have not been detected experimentally. In addition, experimental studies only yield approximate estimates for the relative percentage of different conformations that are in a solution. Furthermore, information about the quantitative equilibrium between different conformations must be obtained from theoretical calculations of the minima of the potential energy of a macromolecule.[31, 41, 42]

Different aspects of DNA's macromolecular structure affect different modes of DNA vibrational spectra. The range of high absorption frequencies reflects vibrations associated with near-neighbor interactions:[43] stretching vibrations of double bonds (frequencies in the range between 1800 and 1500 cm^{-1}); vibrations of sugar-base glycosidic torsion (frequencies in the range between 1500 and 1250 cm^{-1}); absorptions of phosphate groups and of the sugar (frequencies in the range between 1250 and 1000 cm^{-1}); vibrations of diester chain (frequencies in the range between 1000 and 700 cm^{-1}). The range of low absorption frequencies (below 0.3 cm^{-1}) reflects the dynamics of a polymer as of an elastic continuum.[2] The range of intermediate absorption frequencies reflects internal helical vibrations, such as twisting, bending and stretching, sugar pseudorotational vibrations and vibrations of hydrogen bonds.[5-9]

Calculation of DNA vibrational modes

The vibrational modes of a molecule can be calculated using knowledge about its three-dimensional structure and through the use of an effective harmonic potential (reviewed in Krimm and Bandekar, 1986).[44] The normal modes describing internal vibrations are found by assuming that the molecular potential energy can be approximated as a harmonic function of dynamic variables and then by solving a generalized eigenvalue problem to give an analytical description of vibrations.

Normal-mode calculations of polynucleotides were initially performed under the assumptions that only the interactions of covalent atoms and non-bonded interactions of nearest neighbors contribute significantly to the force field of a macromolecule.[45, 46] This assumption of near-neighbor interactions is plausible for calculations of high frequency modes, however long-range interactions were shown to be non-negligible for calculations of low-frequency vibrations.[5] In the latter study, long-range interatomic interactions were considered using helical symmetry. The DNA homopolymer was modeled as a one-dimensional infinite lattice whose periodic units (or unit cells) are nucleotide pairs or base pairs, and low-frequency acoustic modes corresponding to helical stretchings and twistings were calculated. Pronounced differences in the acoustic modes of A- and B-helixes were demonstrated confirming that low-frequency vibrations may serve as a signature of DNA topology. This lattice dynamics approach (see Prohofsky, 1995)[30] was developed in a number of studies considering the interactions of counterions,[47,48] in studies of a junction of AT/GC helixes,[49] and in studies of the vibrations of hydrogen bonds.[7] A disadvantage of the lattice dynamics method is that it relies on helical symmetry, which makes normal mode analysis of DNA molecules with irregular nucleotide composition and of DNA-drag complexes impossible.

An alternative approach initially developed for low-frequency internal motions of proteins[50-53] is based on the direct calculation of the potential energy of a macromolecule using a semi-empirical force field. This method allows one to calculate molecule vibrations around an energy minimum by making a harmonic approximation to the local energy hypersurface. The potential energy of a molecule is calculated as a sum of energy of 1) the Van der Waals interactions of non-bonded atoms (including hydrogen); 2) torsion rotations; 3) stretching deformations of bond angles and of bond length; 4) electrostatic interactions; 5) deformations of hydrogen bonds.[54] Conformational energy is minimized in the space of atomic coordinates [52] or in the space of torsion and bond angles of a molecule,[50] and normal modes are calculated using the second derivative of the bond's potential energy. This method enabled the calculation of low-frequency (from 0 to 250 cm^{-1}) vibrations of several globular proteins with rather complicated three-dimensional structures.[55] Here, it was demonstrated that anharmonic treatments for potential energy are essential.[53] Recently, low-frequency vibrations of oligonucleotides were calculated using the potential energy function in the space of torsion and bond angles of a molecule.[8, 9] In these studies, it was shown that DNA macromolecule length, nucleotide composition, and topology (A-, B- or Z-form) affect the internal low-frequency (0-250 cm^{-1}) motions corresponding to helical twisting, bending, and stretching, and sugar pseudorotational vibrations.[9] The advantages of normal mode analysis for the study of macromolecules dynamics are that 1) it is a relatively rapid computational technique and 2) the calculated vibrational frequencies can be compared with resonance frequencies found by spectroscopy methods, either Raman or absorption. Thus, theoretical normal mode analysis of macromolecules was developed in the past and some of low-frequency vibrations of polynucleotides have been characterized. However, until recently no detailed test of the ability of normal mode analysis to reproduce DNA vibrational spectra had been demonstrated.

Infrared intensities

In order to analyze the vibrational spectra of DNA, one needs to know what infrared intensities are to be expected for a normal vibration. For infrared absorption there must be a nonzero dipole moment change during the vibration.[56] Intensities of longititude and transverse acoustic modes of DNA were calculated using longitidude and traverse components of the vibrational dipole moment.[4] M. Bykhovskaia and B. Gelmont recently suggested a method for the prediction of far-IR absorption spectra of biomolecules.[27-29] Here, the methods of molecular mechanics and normal mode analysis where employed which made rigorous calculations of biopolymer structure and dynamics possible (reviewed in Ref. 32). Far IR active modes were calculated directly from the base pair sequence and the topology of a molecule with all possible interactions considered. Then, vibrational dipole moments were calculated for each normal mode and spectroscopic results were generated from calculations of the oscillator strengths for these normal modes. This last step made it possible to predict the absorption intensity at each vibrational frequency.

As a first test of this method, normal modes and their oscillator strengths were calculated for two double helical DNA fragments consisting of 12 alternating Thymine - Adenine base pairs: (poly dA) (poly dT) and (Poly dAdT)(Poly dTdA). In addition, absorption spectra were calculated for 12 base pair RNA fragment Poly[C]-Poly[G]. The phonon spectra derived for these molecules were found to be very sensitive to the DNA base pair sequence. A more detailed description of these results follows.

B. Energy Minimization and Normal Mode Analysis

Since vibrations of DNA covalent bonds have frequencies over 750 cm^{-1},[43] it is not necessary to consider vibrations of covalent bonds in order to predict the very far IR absorption spectra. Therefore, normal mode analysis in internal coordinates of a molecule (torsion and bond angles) is the most appropriate method for theoretical study of low-frequency resonance modes of nucleic acids.[8, 9] Normal mode analysis allows calculations of a molecule's vibrations around an energy minimum by making a harmonic approximation to the local energy hypersurface. Initial approximations for the B-helical conformation of the fragment (TA)12 and for the A-helix of double stranded RNA[57] were generated and optimized by the program package JUMNA.[35, 58-60] The conformational energy of each molecule is calculated as a sum of Van der Waals interactions of non-bonded atoms and of electrostatic interactions, torsion rotation potentials, harmonic potentials of bond angle deformations, and the energy of hydrogen bond deformations using FLEX force filed.[61] The energy was minimized in the space of internal helicoidal coordinates of the molecule (torsion and bond angles) using the program package LIGAND[62] and normal modes of the molecule vibrations (eigenfrequencies $\boldsymbol{W}$ and eigenvectors $\boldsymbol{A}$) were calculated as described in[9] as solutions of the equation:

$$\mathbf{HAW} = \mathbf{FA}\,, \tag{1}$$

where $\boldsymbol{W}$ is a diagonal matrix with elements ω_K^2 (vibrational frequencies), $\boldsymbol{F}$ is a matrix of second derivatives of potential energy and $\boldsymbol{H}$ is a matrix of second derivatives of kinetic energy.

C. Oscillator Strengths

Each normal vibration produces the dipole moment of a molecule. Oscillator strengths must be calculated for all the vibrational modes in order to evaluate their weight in the absorption spectrum of a molecule. At room-temperature approximation in the low frequency limit, the normalized dipole moment $\boldsymbol{p}$ can be expressed in terms of the classical model as:

$$\mathbf{p} = \sum_i e_i \mathbf{a}_i / \sqrt{m_i}\,, \tag{2}$$

where index i denotes atoms, e_i is the partial atomic charge, m_i is the atomic mass, and $\mathbf{a}_i$ is an eigenvector in atomic cartesian coordinates (atomic displacements) normalized in such a way that:

$$\sum_i (\mathbf{a}_i)^2 = 1\,. \tag{3}$$

Equation 2 was obtained [63] under the classical mechanical approximation ($h\nu \ll k_B T$). However it is also applicable[27] in ultraquantum approximation ($h\nu \gg k_B T$). Oscillator strength S_k corresponding to k^{th} normal mode can be expressed as a square of normalized amplitude (***p***) of the dipole moment deviation:[63]

$$S_k = (\mathbf{p}^k)^2. \tag{4}$$

Since the normal mode analysis was performed in internal coordinates of a molecule, atomic displacements, $\mathbf{a}_i$, were not readily available from the matrix of eigenvectors ***A*** (solution of Eq. 1), as it would be if eigenmodes were calculated in atomic Cartesian coordinates.[64] Therefore, three dimensional structures of a molecule along normal vibrations were created for each mode and deviations of the molecular dipole moment from the structure in the energy minimum were calculated. The variable dipole moment of a molecule (**P**) is calculated as:

$$\mathbf{P} = \sum_i e_i \mathbf{r}_i \ , \tag{5}$$

where the index i denotes atoms and the vector $\boldsymbol{r}_i$ is a displacement of ith atom from equilibrium. Oscillator strengths S_k are calculated as:

$$S_k = (\mathbf{P}^k)^2_{\max} / \sum_i m_i (\mathbf{r}_i)^2_{\max} \tag{6}$$

where $\mathbf{P}^k$ is produced by k^{th} vibration.

D. Absorption Spectra Modeling and Spectral Line Analysis

The obtained phonon modes can be convoluted to derive the far IR absorption spectrum of a molecule. The dependence of permittivity, ε, on radiation frequency, ω, can be described by the Kramers – Heisenberg dielectric function:

$$\varepsilon(\omega) = \varepsilon_\infty + 4\pi N \sum_k \frac{S_k}{3(\omega_k^2 - \omega^2 - 2\pi i \gamma_k \omega)} \ , \tag{7}$$

where index k denotes normal modes, ω_k are vibration frequencies, S_k are oscillator strengths, γ_k are oscillators dissipations, and N is the number of molecules. The absorption coefficient (α) is proportional to the imaginary part of permittivity:

$$\alpha = (\omega / nc) \operatorname{Im}(\varepsilon) \ , \tag{8}$$

where n is the refractive index. Hence the dependence of α on frequency ν ($\nu = \omega/2\pi$) is determined by:

$$\alpha(\nu) \sim \nu^2 \sum_k S_k \gamma_k / ((\nu^2 - \nu_k^2)^2 + \gamma_k^2 \nu^2) \ . \tag{9}$$

Normal mode analysis enables calculations of eigenfrequencies ν_k. Oscillator strengths S_k can be found from calculations of vibration trajectory along each normal mode. However, oscillator dissipations γ_k cannot be found from harmonic approximation of the potential energy and, hence, remain to be determined from the comparison of theory and experiment. In the first approximation, however, the decay can be considered frequency independent. In this simplified form the absorption will depend on the frequency as:

$$\alpha(\nu) \sim \gamma\nu^2 \sum_k S_k /((\nu^2 - \nu_k^2)^2 + \gamma^2\nu^2). \qquad \textbf{(10)}$$

E. Modelin_ Results for DNA and RNA Fragments

Oligonucleotides (Poly dA)$_{12}$(Poly dT)$_{12}$ and (Poly dAdT)$_6$ (Poly dTdA)$_6$ have 360 degrees of freedom in the space of internal coordinates of a molecule (torsion and bond angles). Respectively, 360 normal modes were found for each sequence with the density higher than 1 mode per cm^{-1} (see Figures 1 and 2). Their frequencies lie below 900 cm^{-1} so there is almost no overlap with vibrations of covalent bonds that have frequencies above 750 cm^{-1}.[65] The two oligonucleotides have similar mode densities in the region above 250 cm^{-1}, which essentially involve vibrations of valence angles.[9] It should be expected that the modes that reflect vibrations of torsion angles would be more sensitive to the conformation and flexibility of the helix. Indeed, the calculated density spectra of the two oligos differ markedly in the region below 250 cm^{-1}. The normal modes of the two examined oligos have strikingly different spectra of oscillator strengths (Figure 2). For the molecule with homogeneous strand's Poly(dA)Poly(dT), the oscillator strengths appeared to be very uniform, while heterogeneity in the strand's composition introduced many-fold variance.

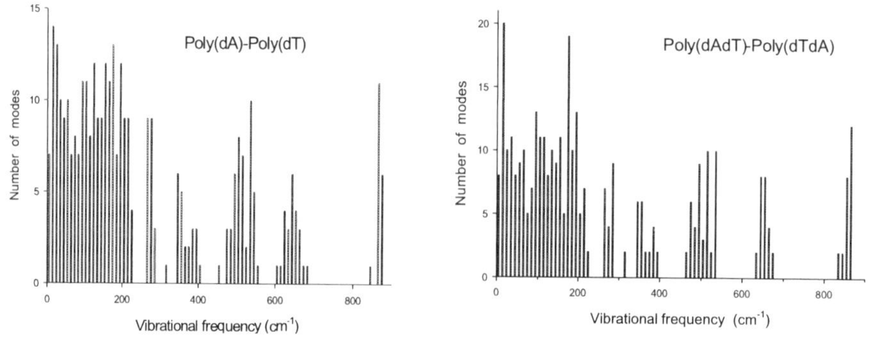

Figure 1. Density histograms of calculated normal modes for two oligonucleotides.

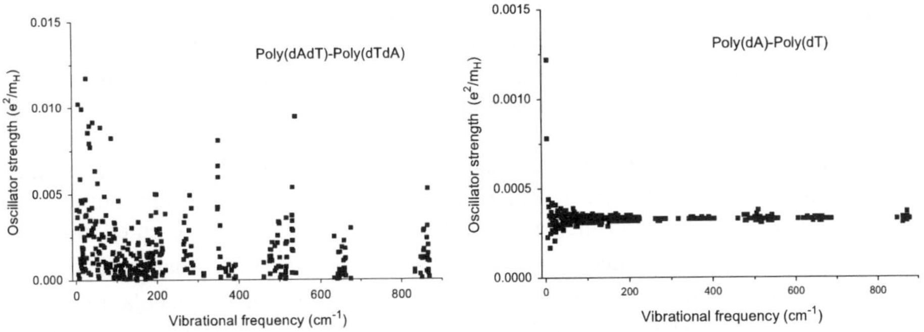

Figure 2. Comparison of oscillator strengths spectra for two double helical DNA fragments.

The absorption spectra of the two molecules shown in Figure 3 were calculated for the damping factor $\gamma = 2\ \mathrm{cm}^{-1}$.[16] Of course, the positions of individual resonance peaks does not depend on the damping factor. However, an experimentally observed resonance peak can be produced by an individual mode (which indicates strong optical activity) or by a large density of weak normal modes (if the damping factor leads to significant overlapping). Hence, non-uniformity in the optical activity of normal modes can alter the optical spectra for molecules with heterogeneous structures, as was observed for poly(dAdT)-poly(dTdA).

This type of analysis makes it possible to understand the physical nature of many spectral features. For example, the decrease in the absorption intensities observed for frequencies above 250 cm^{-1} is obviously due to the fact that most of the low frequency vibrations involve torsion rotations. Hence, the steep fall in absorption intensities as the frequency exceeds 250 cm^{-1} for both sequences is caused by the decrease in normal mode

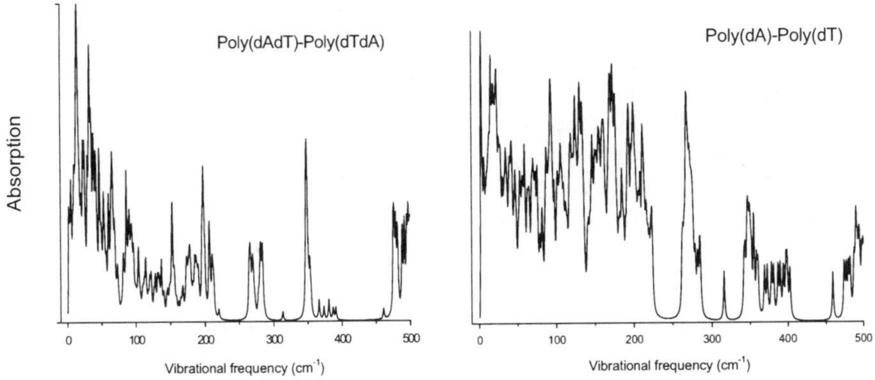

Figure 3. Absorption spectra for two double helical DNA fragments (relative units).

densities and is not related to their optical properties. The two lowest-frequency modes of Poly(dA)Poly(dT) (2.64 and 2.69 cm^{-1}) have strengths 3-4 times greater than the average strength for the other modes (see Figure 2). These two lowest-frequency vibrations have strong **P** components perpendicular to the helical axe, and may reflect helical bending.[9, 45] The strongest modes of the heteropolymer (see Figure 2) also occur in the lowest frequency range (mostly below 50 cm^{-1}). This result is not surprising, for the strongest modes reflect the relative motions of the strands, which involve low-frequency torsion rotations. The fact that the spectrum of optical activities was found to be very sensitive to the DNA base pair sequence is also an important result as it suggests this method can yield microscopic information about the molecules.

Normal modes and absorption spectra were also calculated for a double stranded 12 base pair RNA homopolymer fragment Poly[C]-Poly[G] with the purpose of comparing it directly with experimental data. The calculation results presented in Figure 4 for two values of oscillator decay, γ=0.5 cm^{-1} and γ =1 cm^{-1}, show that the maximum absorption corresponds to the electric field perpendicular to the long molecular axis *z*.

The modeling results demonstrate that the analysis of low-frequency vibrations has the potential to become a clue to deduce bio-molecule structure and topology. These results also confirm that very far IR absorption spectra of biopolymers can be used as fingerprints for some bio-molecules. This suggestion will be further supported when the theoretical predictions are directly compared with experimental data in a later section.

3. Experimental Techniques for the Characterization of Bio-Molecules

A. Review of Techniques

The general need for faster and less expensive techniques that can provide useful structural information has naturally led to the development of spectroscopic methods that utilize the interaction of an applied electromagnetic (EM) field with the phonon (lattice vibration) field of the material (Many references can be found in the 1997 European Conference of Biological Molecules).[66] Indirect techniques such as low-frequency Raman,[3,67] Brillouin,[68, 69] and neutron scattering[70, 71] have clearly

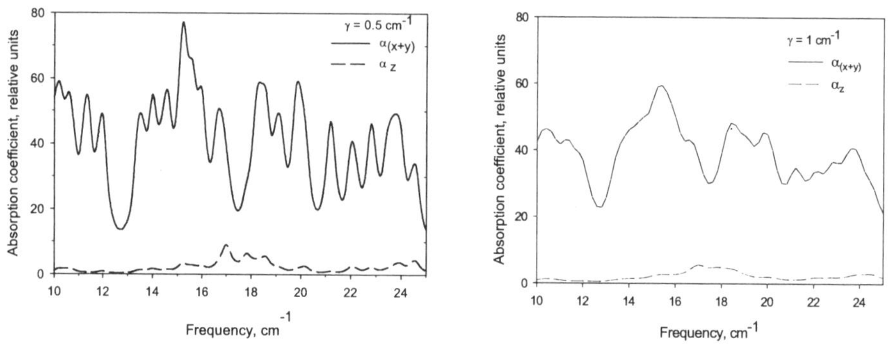

Figure 4. Calculated absorption spectra for electric field E perpendicular to the long axes of a molecule (α_{x+y}) and parallel to the long axes (α_z), and for two oscillator decay values: $\gamma = 0.5$ cm^{-1} and $\gamma = 1$ cm^{-1}.

demonstrated standing wave oscillations in DNA polymer chains in the very-far infrared (IR). From the very beginning, Raman spectroscopy has been widely used to provide insight into the three-dimensional structure and microscopic processes associated with biological molecules (see for example, Liu et all.,[72] Weidlich et all.[3]). Raman scattering produces frequency shifts between the exciting and scattered EM radiation equal to the frequency of the vibrational modes within the material. Hence, the Raman effect arising from nonlinear mixing interactions between the EM field and the naturally occurring phonon modes is a second-order optical process and can give unique information not available from first-order photon absorption. For example, Raman data provided accurate force constants for theoretical refinements in early lattice-dynamical treatments of nucleic acid molecules [73,74] and the supporting experimental evidence for "heteronomous" conformations arising in poly(dA)-poly(dT) homopolymers.[75] DNA in various forms (i.e., thin films, fibers,[76] and solutions[19]) has been studied to determine to the Raman signature dependence on humidity,[76] helix conformation,[3,77] molecular packing (intrahelical and interhelical modes),[78-81] and light polarization.[80] Active Raman absorption modes as low as 25 cm^{-1}, 60 cm^{-1} and 90-100 cm^{-1} have been reported.[72] Although there are some inconsistencies in the identification of the low frequency modes, Raman measurements have demonstrated the existence of phonon modes in DNA and suggested that these modes correlate with DNA sequences.

While Raman is especially effective in the characterization of periodic microstructures (e.g. semiconductors and semiconductor interfaces)[82] it is a very complicated process and correlation between theory and measurement is exceptionally challenging for DNA polymers.[83] In fact, the problem of developing formalisms for the computation of Raman intensities and selection rules in large molecules, such as DNA, has only recently been addressed.[84, 85] Furthermore, experts in the area have noted that while "much experimental efforts has been given to the understanding of the low frequency vibrations of DNA and its constituents, detailed assignment is still lacking." [3] In addition, they point out "even the number of Raman active modes at low frequencies is not known." [3] These remaining challenges and the richness of predicted lines in the low-frequency range have led to an increased interest in the application of submillimeter-wave absorption spectroscopy to biopolymer characterization.

There have been a very limited number of absorption (or transmission) spectroscopic studies in the submillimeter-wave regime[11-20] due to the unique experimental difficulties that are presented there. In this region between the microwave (~ 100 GHz) and the lower end of the far IR (~ 1000 GHz), the output power of the available sources is limited, the absolute absorption of the biological material is relatively weak, and the high absorption of water masks the results from biological materials in solution. The interpretation of results that have been shown to be sensitive to DNA composition, to the environment,[16] and to a sample geometry is not simple. In the THz spectral region where the wavelength of radiation is on the same order of magnitude as the components of the spectrometer, multiple reflections in the system lead to interference or etalon effects and significant care must be used not to misinterpret these type of artifacts as spectral features. These difficulties have severely limited the direct identification of phonon modes in biological materials at submillimeter-wave frequencies in the past. Investigations performed on dried films of DNA in the 3-500 cm^{-1} range have previously revealed only a few modes. Specifically, work by Wittlin *et. al.(*1986)[15] identified modes of 45 cm^{-1} for the Li-DNA and 41 cm^{-1} for the Na-DNA. Most notably are the observation of four sharp bands near 63, 83, 100, and 110 cm^{-1} in polycrystalline poly(dA)-poly(dT) DNA by Powell *et. al.*, 1987.[16] While this work did not report any phonon modes below 63 cm^{-1}, it is important to note that the authors believed that "it is not possible to make meaningful measurements at frequencies lower than 40 cm^{-1} because the polynucleotide samples are too fragile to form the sufficiently thick and large diameter films required to provide adequate absorption below 40 cm^{-1}." More recently, the authors[18] have suggested that the features observed below 3 THz (i.e., which shifted depending on sample thickness[17]) are artifacts due to multiple reflection in samples. Finally, Lindsay and Powell[13] have previously reported the observation of a mode in DNA samples around 12 cm^{-1} and there have been reports of sharp resonances in the microwave range (i.e., < 0.3 cm^{-1}) from measurements on DNA solutions by Edwards *et. al.*[14] However, later investigations were unable to confirm these results.[86] Hence, these early spectroscopic investigations did not definitively established the existence of DNA phonon modes at submillimeter-wave frequencies.

More recently, new studies on dry DNA samples using a W-band network analyzer suggested that microwave-frequency resonant modes exist that are unique to DNA type.[87] Much more recently, pulsed THz time-domain spectroscopy has been applied to examine low-frequency vibrational modes of biomolecules.[20,88] However, the authors basically observed smooth broad absorption over the range 0.1-2 THz, and the advantages of this potentially promising technique, which employs a time-domain methodology,[89] have not been fully realized due to poor spectral resolution.

The latest and most convincing THz-frequency spectroscopic results for biological materials have been obtained from Fourier Transform Infrared (FTIR) studies that were performed during the last few years.[21-24,28,29] Details of these studies will now be presented in the sections that follow.

B. FTIR Transmission Characterization

Sensitivity requirements

FTIR spectroscopic investigations of DNA films and other biological materials were performed over the spectral range from approximately 10 cm^{-1} to 500 cm^{-1} with the most detailed study in the range 10 cm^{-1} -25 cm^{-1}.[21-24] The experiments were performed in transmission mode to search for the occurrence of dielectric resonances induced by interactions of the electromagnetic (EM) field with long-wavelength phonons. Both the material absorption characteristics and the refractive indexes have been determined when possible. The spectral studies utilized a spectrometer Bruker IFS-66 equipped with mercury-lamp and liquid-helium-cooled Si-bolometer ($T = 1.7\ ^{o}K$) for signal detection. All the measurements were made in a vacuum to eliminate any influence of water-absorption lines. The resolution was set between 0.2 and 2 cm^{-1}. Test measurements with a wire polarizer (25 μm wire diameter and 75 μm spacing between wires) were performed indicating 80% polarization of radiation in the FTIR spectrometer in the spectral range of interest and the electric field vector E oriented primarily in the vertical direction. Hence, the measurement apparatus was capable of controlling the angle between the long-axis of the molecules (i.e., for oriented samples) and the electric field vector E of the radiation.

Figure 5 demonstrates the basic sensitivity of the measurement approach. Here, 5 mg of calf thymus (CT) DNA was deposited on a low-loss thin substrate that has an approximately constant dielectric characteristic. The upper curve shows the 100%-line measurement for the FTIR system. This result represents the ratio of two successive background (i.e., empty channel) measurements and demonstrates that the reproducibility is better than 0.3 %. The substrate used was of the type that yielded a transmission characteristic as shown for the polycarbonate (PC) membrane in Figure 5. The variation over the frequency range is minimal (< 0.5 %) with the largest deviation occurring at 18.5 cm^{-1} due to trace water absorption. As these results are statistically typical of the FTIR system used in the studies, the sensitivity can be estimated at greater than 0.5 %. Of course, the accuracy of any measurement depends on the sensitivity, the reproducibility and the final physical interpretation of the data that is received. For

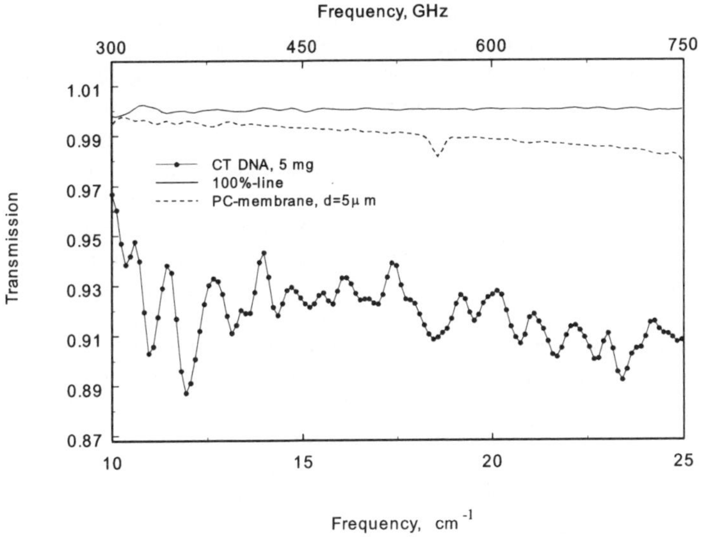

Figure 5. Transmission spectra of calf thymus (CT) DNA. Reproducibility of background spectrum (100% line) and transmission of polycarbonate membrane substrate (PC) are also shown. The structure around 18.5 cm^{-1} is due to traces of water absorption.

example, the spectra for the CT DNA given in Figure 5 show the typical intensity change in the fine resonance structure of the biomolecules considered in these studies. Here, one observes variations between local maximum and minimum values that are on the order of 0.5 – 5.0 %. In addition, it should be noted that any fringes due to multiple reflections in thin film samples (with the thickness less than 1 mm) would be large in wavenumbers (i.e. > 10 cm^{-1}). As will be shown later, this fact was verified by measurements on samples with regular layer thickness. Therefore, these relative differences in extrema define the level of sensitivity required to resolve the phonon modes within the biological material. Certainly, the specific measurement conditions and arrangement of the measurement apparatus can influence the strength of these phonon signatures. For example, depending on the sample and substrate thickness and relative position inside the apparatus it is possible to induce fringing effects. However, such standing wave behavior is always a real result of the specific measurement conditions and the sample geometry. Therefore, if this type of phenomenon is properly understood then physical interpretation can always be used to resolve any influence on the results. However, this may require a significant amount to interrogation of both the operation of the measurement apparatus (e.g., to resolve standing wave patterns along the transmission) and the characteristics of the sample (e.g., to resolve fringing effects inside the material).

Samples preparation

Herring and salmon DNA sodium salts, (Type XIV from herring and salmon testes, with 6 % Na content, obtained from Sigma Chemical Co.), and calf thymus DNA Sodium salts from Fluka were used for this investigation. Data from artificial polynucleotide poly-adenylic acid potassium salt (Poly [A] RNA) and from sporulated Bacillus subtillis variant Niger (BG) are also included.

Far IR spectra show a great sensitivity to factors that are affected by differences in sample preparation, and it was not immediately clear to what extent the fine structures of Figure 5 are reproducible and represent features intrinsic to the DNA. Four different techniques were utilized to prepare samples for this study. This was motivated by the necessity to demonstrate the fundamental character of far infrared resonances in biological molecules and to determine how results depend on the conditions of sample preparation.

Method 1). Free-standing DNA films were obtained using gel material prepared by dissolving herring and salmon DNA sodium salts in glass-distilled water with a concentration ratio between 1:5 and 1:10. This gel was brought to the desired thickness by placing it inside an arbor shim between two Teflon sheets with 50 μm thickness. The assembly was then left to dry at room temperature for several days. After drying, the samples were completely separated from the mold. These free-standing films are extremely fragile and were not oriented (i.e., intentionally) in the first series of measurements.

Method 2). More robust and partially oriented films of herring and salmon DNA sodium salts, and Poly [A] RNA were subsequently deposited on substrates. A gel drier from Labconco Co with controlled temperature and vacuum was used in this preparation. The best results were obtained with porous track-etched polycarbonate (PC) membrane filters (Poretics, Inc.) as a substrate and a cover layer. Membrane thickness is 8μm, and pore sizes ranged from 0.2 μm to 8 μm. This material transmits 92-95% of the radiation in the range 10-50 cm^{-1} and is important for transmission measurements. It also facilitates the process of gel drying. Oriented films were prepared by squeezing lightly heated gel through a syringe-needle onto substrates in one direction. The samples were air dried for 2-8 hours and then were placed in a low vacuum until they were completely dried. The optimal results were received when the gel temperature was about 25-26 C for spreading. Films from materials with short macromolecules were found to exhibit less strain deformation. It was also noted that a gel utilizing 10:1 water to material ratio leads to superior films. Samples with a diameter of 15 to 20 mm were made inside the frame of a steel shim. Films of commercially available DNA and RNA were also prepared using this procedure.

Method 3). The solid calf thymus DNA was dissolved in 0.75 M sodium acetate buffer solution, and an aliquot of this solution was treated in the following manner to prepare the samples. A 2-fold excess of absolute ethanol was added to the solution to precipitate a pellet of DNA. The solution was held at 0 C for 30 minutes to effect precipitation, and the supernatant was discarded. The pellet was rinsed with 70% ethanol in 18 MΩ filtered water, and again refrigerated at 0 C for 30 minutes. The supernatant was discarded and the pellet was again dissolved in 18 MΩ water. This solution was spread onto the polycarbonate membrane and dried at room temperature for a minimum of 24 hours.

Method 4). Bacillus subtillus spores were acquired from a Dugway Proving Ground stock source, originally manufactured by Merc in 1956. The material was subjected to a succession of wash/centrifugation cycles using a saline buffer solution prior to spray-drying. Thin layer films of Bacillus subtillis spores were spread onto Teflon substrate IR-cards (Spectra-Tech, Inc.).

Films of various thicknesses were studied to confirm that the observed resonances were directly attributable to the absorption characteristics of the biological material. The thinnest samples had reasonably good planarity, although they sometimes had an imperfection density (i.e., voids) of about 5-10 %. The drying process typically reduced the sample thickness by 2 to 3 times of its initial thickness. All the samples that were considered in this study possessed a final film thickness between 5 and 300 μm. Film thickness was measured using a Gauge stand ONO SOKKI ST-022. Using the above procedures, samples were produced with randomly oriented and partially aligned DNA chains.

Experimental procedure

In order to successfully resolve phonon resonances in biological materials one must consider the effects of interference and mode coupling strength. Films uniform in thickness are well known to exhibit interference fringes in their reflection and transmission spectra. The interference phenomena can obscure fine resonance features or be misinterpreted as modes themselves.

In the higher-frequency portion of the submillimeter-wave domain, the studied materials possessed sufficient loss and interference effects were not observed experimentally due to variations in the optical thickness across the sample and relatively strong absorption. For this particular condition of quenched fringes, simultaneous measurements of optical transmission through samples of different thickness, prepared under the same conditions, were used to determine the absorption spectra. Data collected for two samples of different average thickness can be used to eliminate the surface-reflectance and define the absorption coefficient as:[90]

$$\alpha = \frac{\ln\left(\langle T_1\rangle / \langle T_2\rangle\right)}{\left(d_2 - d_1\right)}, \qquad \textbf{(11)}$$

where $\langle T_1\rangle$ and $\langle T_2\rangle$ are the average-transmission measurements for samples of thickness d_1 and d_2, respectively.

In the low frequency range, where absorption is low, multiple reflection at the two film boundaries becomes important even in the samples with irregular layer thickness (variation about 5-15%). Since the wavelength of radiation, λ, is much larger than the film thickness, d, standing waves in samples cause the appearance of fringes in the transmission spectra. However, if the periodicity is long in comparison with the signature we are looking for, the transmission can still be used for biological material characterization.

Fringes in transmission spectra can be used to calculate material optical characteristics. In particular, the product of refractive index, n, and the film thickness, d, can be accurately obtained from the position of the transmission extrema, λ_{extr}, through

$$n\,d = \frac{m\,\lambda_{extr}}{4} \qquad \textbf{(12)}$$

where m is an integer number corresponding to the order of extremum, including maxima and minima (m =1 for the first transmission minimum, and m =2 for the first transmission maximum).[90] The order of extremum can be found from the wavelengths of two adjacent extrema in the range, where n does not change appreciably when m is increased by one. Measured transmission spectra were used to extract the absorption coefficient spectra and refractive index of biological material by applying an interference spectroscopy technique (IST) for proper modeling of the multiple reflection behavior.[91,92]

C. Experimental Results

Absorption coefficient: Wide spectral range

An absorption coefficient spectrum of a Herring DNA film over an extended frequency range is depicted in Figure 6. Above about 300 cm^{-1} the absorption spectra consist of numerous sharp, well defined resonances (regions I and II). They represent different covalent short-range, high-energy interactions that have been relatively well studied primarily by Raman spectroscopy. For example, it was demonstrated [93] that in the range 1800-1500 cm^{-1}, the spectrum originates from in-plane double-bond vibration of the bases, and in the range 1500-750 cm^{-1} from backbone-sugar vibration modes. These

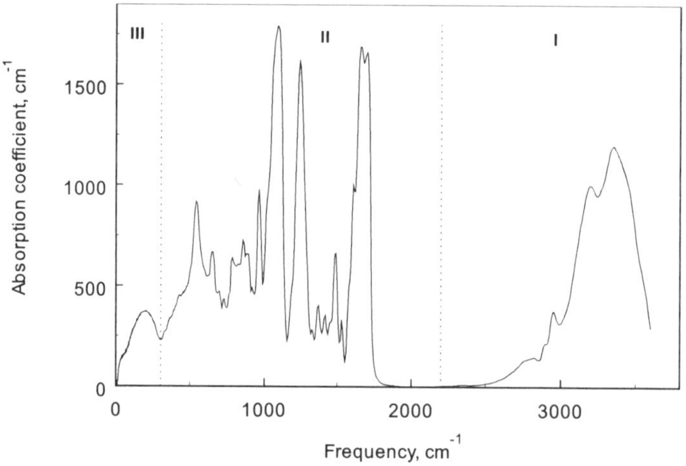

Figure 6. Absorption spectrum of Herring DNA in the extended energy range.

features tend to be independent of the base-pair sequence. The position of these peaks is very well reproducible, does not depend on film thickness, and is the same for both salmon and herring DNA. Since there are no visible interference effects, equation (11) has been used to calculate the absorption coefficient in these regions.

We are primarily interested in region III. The absorption coefficient is significantly weaker in this range and eventually drops dramatically at frequencies below approximately 10 cm^{-1} (see Figure 7). A dip in the absorption coefficient spectrum is observed near 300 cm^{-1}. This is the short wavelength edge of absorption due to low energy vibrations (weak bonds and non-bonded interactions in DNA). This edge was previously predicted by theoretical analysis. [9, 27,47]

In spectral range III, weak features are also observed (i.e., basically shoulders) that are more easily detected in the derivative of transmission spectra (see Figure 8). Earlier, spectral features observed in the spectra of nucleic acids in the region 300-500 cm^{-1} were assigned to the ribose ring vibrations.[11] These features are actually soft phonon modes that are obscured by the absorption roll-off from the very strong resonance at 545 cm^{-1}. The transmission derivative spectra reveal a large number of these weak features over the range 300-500 cm^{-1}. The experimental results show that all clearly resolvable phonon modes are present in both salmon and herring DNA samples. The position of the peaks in the derivative of transmission spectra (i.e., dT/df, where f is a frequency, calculated from smoothing over 9 data points) are the same for herring and salmon DNA

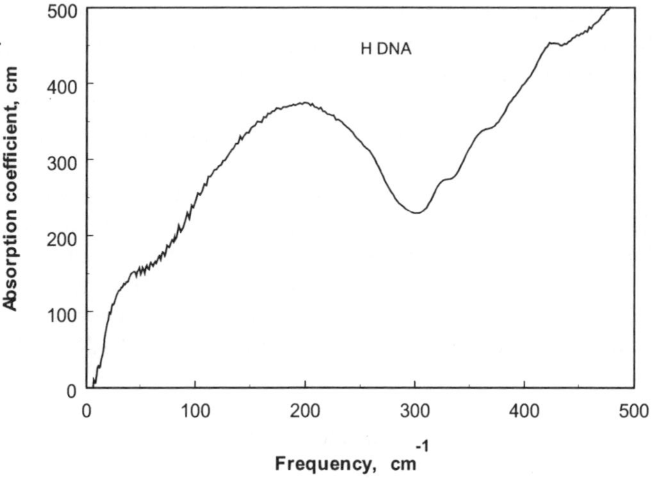

Figure 7. Absorption spectrum of herring DNA in the range 10-500 cm^{-1}.

and does not depend on the thickness. Figure 8 compares the transmission derivative spectra from herring DNA films prepared with 10:1 and 5:1 water-to-DNA concentrations. Clearly, there is a significantly stronger coupling of the electromagnetic energy when the samples are prepared with larger concentrations of water. While the fundamental mechanism responsible has not been identified at this point, the oscillator strength (i.e., either the phonon density or polarizability) is clearly affected by the hydration level. Specifically, the structures disappear when the amount of water is reduced and indicates that the corresponding vibration modes are affected by the

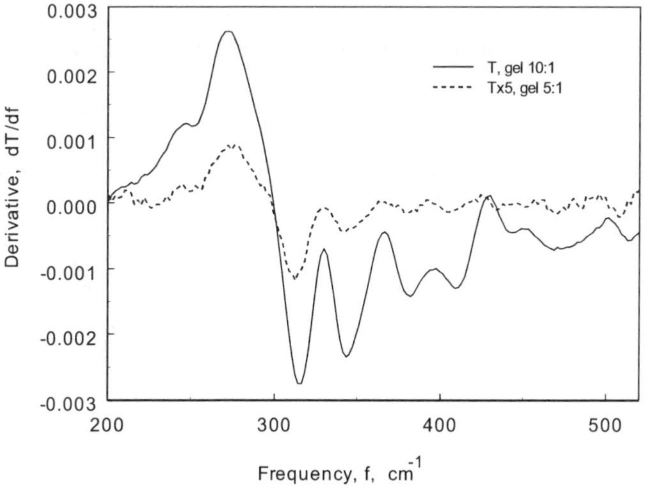

Figure 8. The transmission derivative spectra of herring DNA films with different water content.

conditions of the interface between DNA and the surrounding media. Understanding this phenomenon will have many important ramifications to resolve the dynamics within intrinsic (isolated) DNA macromolecules. For example, if this effect is related to the salinity of the sample (e.g., the DNA samples under study are in fact DNA salts) then the prescription for enhancing the phonon activity is important for interrogating the microscopic physical dynamics.

The broad peak in the long-range absorption spectra as presented in Figure 7 agree qualitatively with earlier studies performed by Powell et. al.[16] on vacuum-dried poly(dG)-poly(dC) DNA (G-guanine, C-cytosine). In fact, the values of extinction coefficient for salmon DNA at the transmission window (i.e. ~0.04 at 300 cm^{-1}) and at the lower-frequency resonant peak (i.e., ~0.105 at 200 cm^{-1}) are in very close agreement with the room temperature measurements on minimal-salt, poly(dG)-poly(dC) films reported earlier.[16]

Resonant modes in transmission: Low frequency regime

An extensive series of measurements carried out in the very far IR frequency region with a higher resolution of 0.2 cm^{-1} revealed fine features in the spectra which can be more or less pronounced depending on the quality of material and sample preparation and measurement conditions.[21,23,24,28,29] The transmission spectra of calf thymus DNA demonstrating this fine structure are given in Figure 5, as do the salmon and herring DNAs in Figure 9, the artificial polynucleotides, poly-adenylic acid potassium salt (Poly [A] RNA) in Figure 10, and of Bacillus subtillis spores in Figure 11. Although the observed amplitude of the resonant modes is not greater than a few percent, the signal-to-noise ratio of the instrument is good enough to detect these features in spectra of free standing films, as well as of films on polycarbonate or Teflon substrates (see discussion in section 3. B). The peak positions in the spectral structure was found to be very independent of the sample-preparation and water content if samples with similar orientations are considered. The fine structure observed here requires spectral resolution greater than 0.5 cm^{-1}, which is difficult to achieve because of the very low source energy in the submillimeter range. This requirement may partially explain

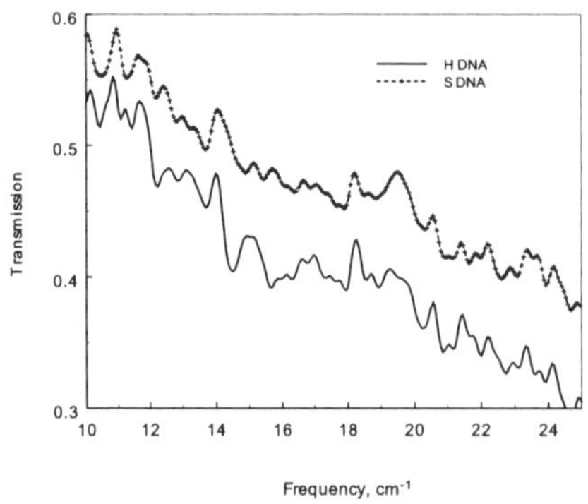

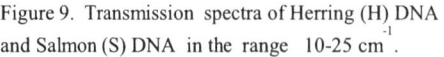

Figure 9. Transmission spectra of Herring (H) DNA and Salmon (S) DNA in the range 10-25 cm^{-1}.

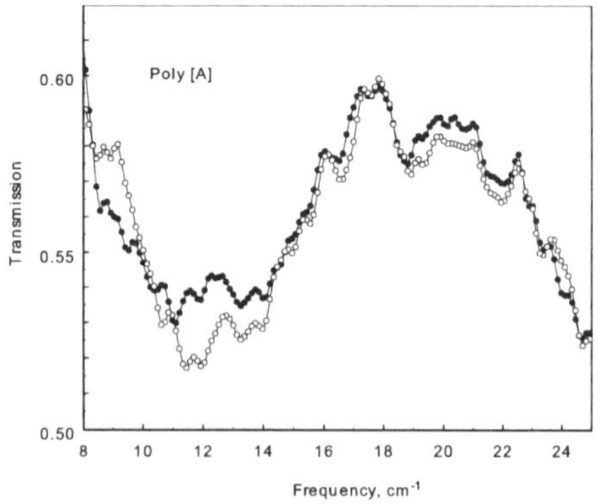

Figure 10. Structure in spectra of Poly [A] RNA. Two independent measurement results are given. Large scale features are due to the multiple reflection in the film.

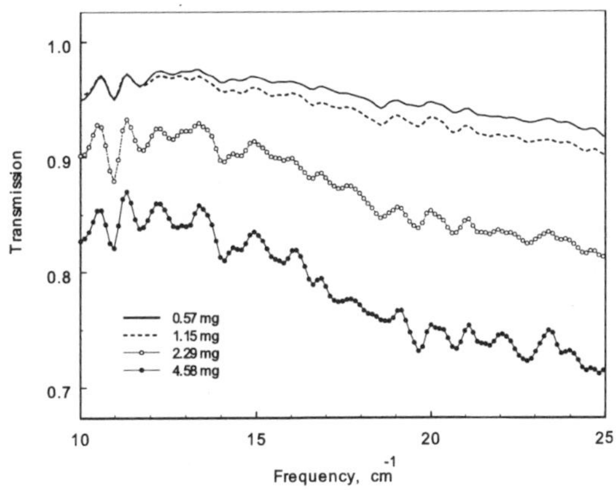

Figure 11. Transmission spectra of Bacillus subtillis [BG] samples made with different amounts of material.

why earlier FTIR investigations were not able to detect phonon resonances. Peaks appear with a density of approximately one per cm^{-1} in the interval between 10 and 200 cm^{-1}. This general density of frequencies in this regime had been predicted earlier in theoretical studies.[9, 27] The intensity of absorption lines gradually drops towards the upper end of the band. Here, the aperture of the spectrometer determined the lower-frequency limit.

Experimental issues

Many previous studies performed on biological materials did not successfully resolve the fine structures in the submillimeter range that are reported here. The sections that follow will consider a number of possible reasons for the poor reproducibility obtained in these earlier spectroscopic studies of biological macromolecules in the far infrared region.

i) Interference Effects in Transmission Measurements

As is shown in Figure 12, the results of transmission measurements performed over broad ranges in frequency are very sensitive to the thickness of the film due to multiple reflection at the film boundaries. Clearly, the position of the transmission extrema depend on d (see also Figure 10). The fine structure is resolved on the envelope of the wide interference pattern, although a roll-off of the transmission in fringes, as shown in Figure 12, can transform resonant extrema into shoulders or inflection points in the spectra. The transmission spectra are consistent with an interference pattern, and equation 12 was used to calculate the refractive index of the bio-materials from the position of transmission maxima and minima of the fringing characteristic. The estimated values of refractive index, n, varied between 1.7 and 2.3 and exhibited a rising trend of about 10 % over the frequency range 10-25 cm^{-1}. These results are agreement qualitatively with the refractive index of natural DNA near 6 cm^{-1} which was measured to be 1.9 using a polarizing millimeter wave interferometer.[94]

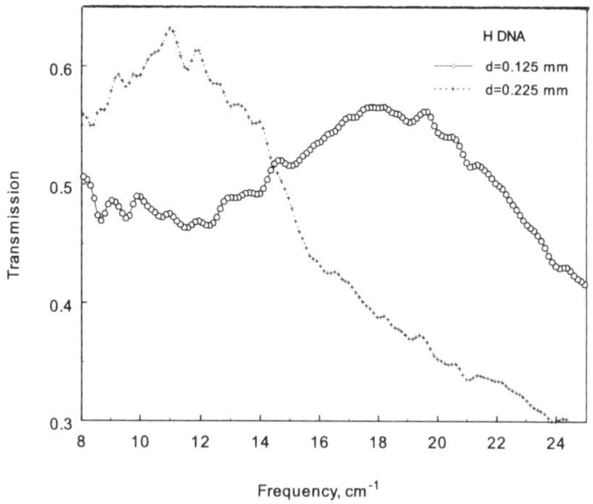

Fig. 12. Transmission spectra of two herring DNA samples with different thickness.

In transmission spectra of very thin films, the interference pattern is not obvious. Figure 13 shows spectra of thin RNA samples with varying thickness between 15 and 70 μm. The increase of transmission at lower frequencies does not necessarily indicate a reduction of the absorption coefficient. Instead, this change of transmission correlates with the first interference fringe between the transmission maximum at zero frequency and the first transmission minimum at higher frequencies. The IST (see section 3. B) was used to extract the absorption coefficient spectra from transmission measured on thin films as well as thick ones.

ii) Polarization Effects

In the very lowest frequency region new factors arise in obtaining reliable and reproducible experimental data. It was found that optical characteristics are dependent on the orientation of the aligned DNA fibers on the film samples in the electromagnetic field of radiation. Specifically, the strength of the spectral features were shown

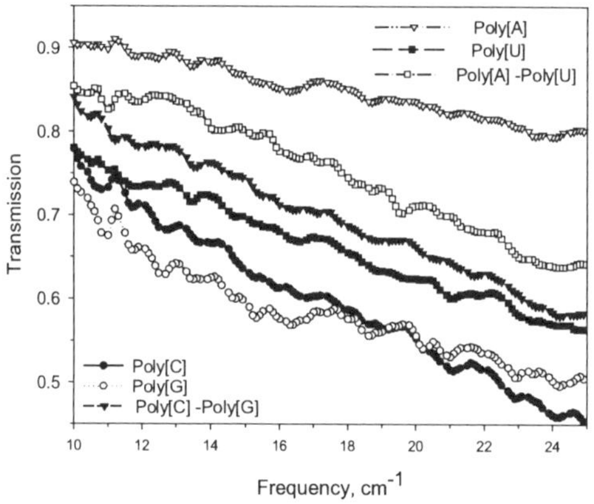

Fig. 13. Transmission spectra of thin samples.

to change with simple sample rotation (see Figure 14) indicating a change of coupling of the incident radiation and the dipole moment of the DNA oscillators (see Section 2. E). It should be noted that some modes are orientation-independent, however, orientation has a profound effect on the observation of some absorption modes. The discovery of this polarization effects is important because this is an exact indication of the phonon's dipole moment within the three-dimensional DNA double helix itself.

iii) Time Instability of Bio-samples

Another reason for poor reproducibility of spectroscopic results is caused by changes in bio-sample characteristics over time. The thickness of the film is usually reduced 2-3 fold upon drying but the final film thickness is also dependent on humidity.

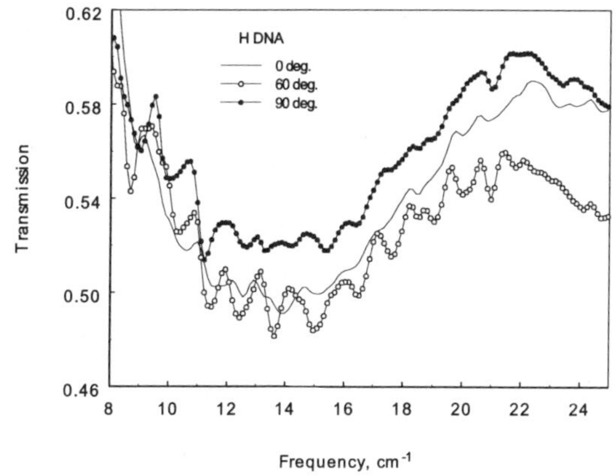

Fig. 14. The dependence of transmission spectra of herring DNA thin film on sample rotation.

Thus, dried films continue to change their optical characteristics including their refractive index. This causes significant changes with time in the long range fringing as illustrated in Figure 15. At the same time, the position of fine features associated with vibrational

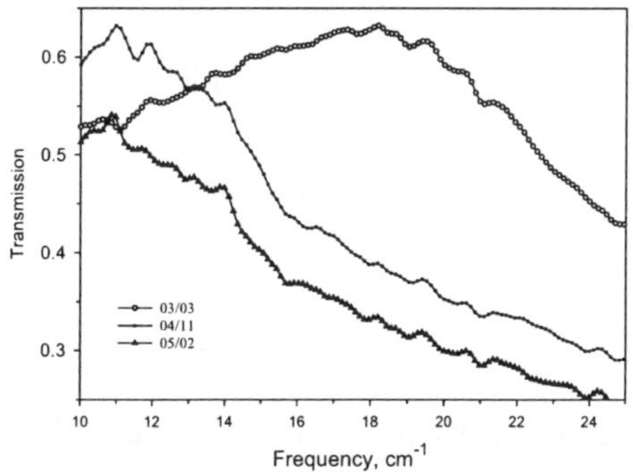

Figure 15. Change in transmission spectra of herring DNA thin film with time due to the thickness and the refractive index change (film "aging").

modes remains practically fixed. While it is more difficult to compare the characteristic frequencies in spectra which are weakened by interference fringes (see transmission spectra of two salmon DNA samples shown in Figure 16), plots of the derivative of transmission with frequency (DTF) for samples with different thickness demonstrate a high degree of consistency, as shown in Figure 17. A direct comparison of DTF plots of aligned herring and salmon DNA samples that were measured at identical orientation shows many common modes. However there are also a number of distinguishing modes (see Figure 18). The high density of spectral features provides a great deal of information for discrimination of the samples even over this limited domain from 10 to 25 cm^{-1}.

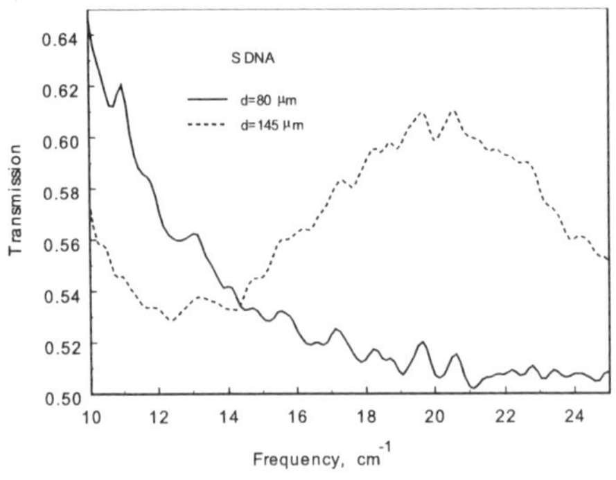

Figure 16. Transmission spectra of two salmon DNA films with different thickness.

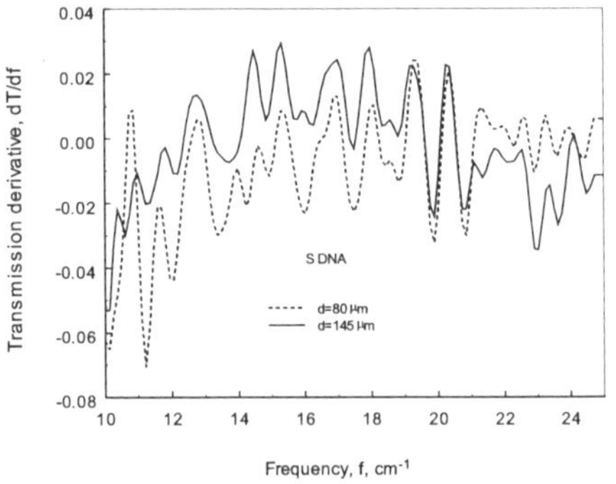

Figure 17. DTF spectra of salmon DNA films with different thickness.

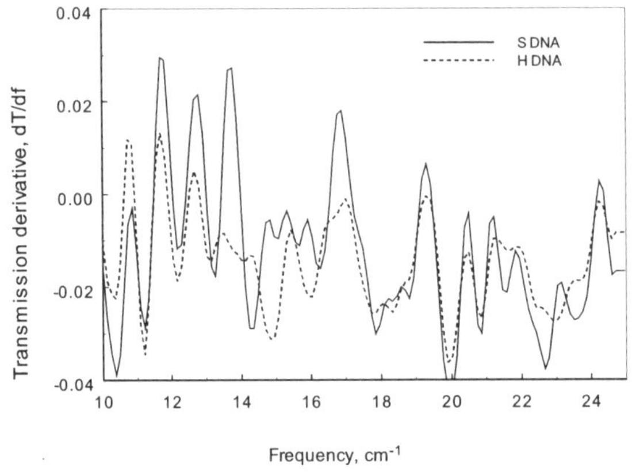

Figure 18. DTF spectra averaged over 20 spectra for herring and salmon DNA samples.

4. Comparison of Experimental Results with Theoretical Prediction

Submillimeter-wave spectroscopy was applied to polynucleotides of known base-pair sequences. Here the primary goal was to demonstrate the fundamental character of the submillimeter-wave spectral absorption, thereby confirming the fact that THz-frequency resonant features result from electromagnetic wave interactions with the DNA and RNA macromolecules. This was achieved by directly comparing the absorption spectra from laboratory measurements with the results of computer modeling as described in the section 2. For this comparison, normal modes and absorption spectra of 12 base pair RNA fragment Poly[C]-Poly[G] were calculated. The second goal was to understand which phonon modes are determined by a double helical topology of nucleic acids, by comparing the spectra of single stranded and double stranded RNA molecules. FTIR spectral measurements were performed on double helical RNA molecules Poly[C]-Poly[G] and Poly[A]-Poly[U] and on single-stranded homopolymers Poly[A], poly[U], poly[C], and poly[G] (Guanine (G), Cytosine (C), Adenine (A), Uracil (U)). This study used solid films of partially oriented macromolecules to derive absorption coefficient spectra from transmission measurements made in the 10 cm^{-1} to 25 cm^{-1} frequency range.

Figure 19 illustrates the experimental absorption spectra of Poly[C], Poly[G] and Poly[C]-Poly[G] macromolecules. The chemical structure of the backbones is similar for these biopolymers, therefore, similar vibration and rotational modes should be expected. A comparison of the experimental spectra in Figure 19 shows a similarity of resonance features indicating that most of the phonons in this spectral range are dependent on the backbone structure. A single stranded RNA macromolecule is generally much more flexible than double helical configuration. Consequently, some of the modes present in single stranded molecules could be absent in the double stranded molecules. For example, the vibration mode at 12.6 cm^{-1} is absent in the absorption spectrum of Poly[C]-Poly[G]. On the other hand, the vibration of the Poly[C]-Poly[G] double helix at 11.2 cm^{-1} is absent in single stranded polymers.

Spectra in Figure 20 obtained from another system (i.e., Poly[A], Poly[U] and Poly[A]-Poly[U]) also indicates the presence of a mode at 11.2 cm^{-1} and the absence of the mode at 12.6 cm^{-1} in the Poly[A]-Poly[U] double-stranded structure. In addition, from comparison of the Poly[C] and Poly[G] spectra in Figure 19, we can deduce that the peak at 14.7 cm^{-1} is an indicator of a guanine base, while the peak at 19.1 cm^{-1} belongs to a cytosine.

Figures 21-23 make a direct comparison of the absorption spectra generated

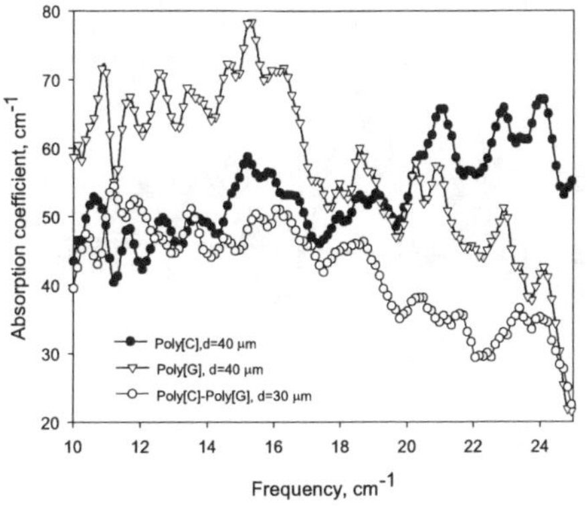

Figure19. Experimental absorption spectra of single stranded polymer films Poly (C), Poly (G) and double stranded polymer Poly[C]-Poly[G].

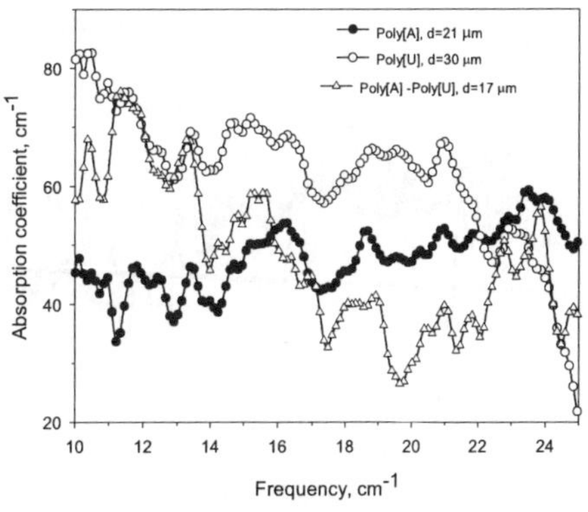

Figure 20. Absorption spectra of Poly[A], Poly[U], and Poly[A]-Poly[U].

by measurements with the results derived from computer simulation. Figure 21 represents an experimental spectrum taken from a relatively thick Poly[C]-Poly[G] sample (~30 μm). Modeling results for two different values of oscillator dissipation parameter, $\gamma = 1\ cm^{-1}$ and $\gamma = 0.5\ cm^{-1}$, are also depicted. For the purpose of comparison, calculated spectra are shown in relative units, and the scaling factor is the only fitting parameter used. The intensity of resonance modes in the thickest samples does not directly correlate with oscillator dissipations γ but more probably with broadening of spectral lines in nonuniform material. The absorption spectrum of a thinner sample (see Figure 22) reveals more spectral details. A comparison of two calculated spectra indicates that the typical oscillator dissipation is less than $1\ cm^{-1}$. It is important to note that the quality of experimental data depends on film thickness and on conditions of film preparation. As a rule, much better resonance structure is observed in thin films. Thin films are normally made from gel with larger water contents and less viscous material. These films are thus more uniform. However, thin films are more difficult to fabricate with areas of sufficient size.

Despite all difficulties, for the thinnest sample (Figure 22, d=6.6 μm) good correlation is observed between almost all measured and calculated resonance frequencies, although several calculated absorption peaks are absent in experimental spectra (for example, peaks at $16.5\ cm^{-1}$ and $19.9\ cm^{-1}$). One of the possible explanations for the lack of peaks in the experimental spectrum is a high sensitivity of peak intensity to the actual value of oscillator dissipation (see Figure 4). Both indicated peaks will not be revealed in calculated spectra if γ is greater than $1\ cm^{-1}$ for these modes. In this thinnest sample, the observed intensity of the resonance structure that occurred around a frequency of $15\ cm^{-1}$ was also very close to the simulation result with $\gamma = 0.5\ cm^{-1}$. It is important to note that these comparisons are made without fitting parameters for individual modes, and from the observed density of peaks one can conclude that the proper value of dissipation for the oscillators is approximately $0.5\ cm^{-1}$.

The above results certainly demonstrate the fundamental character of observed spectral features within submillimeter range and confirm that very long wavelength resonant features are phonon modes resulting from electromagnetic

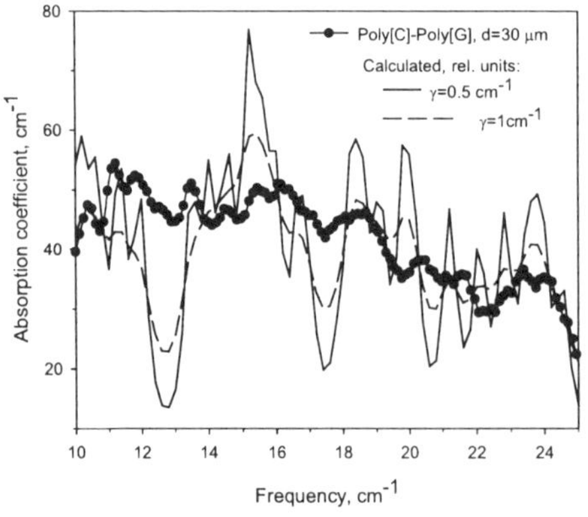

Figure 21. Absorption spectrum of Poly[C]-Poly[G] film (d=30 μm) and modeling results.

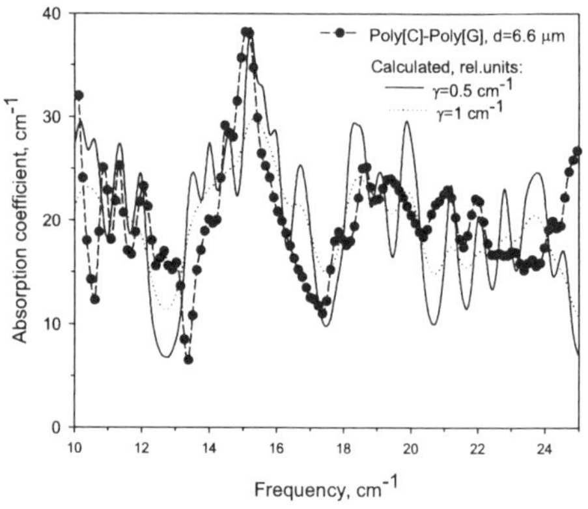

Figure 22. Absorption spectrum of Poly[C]- Poly[G] thin film (d=6.6 μm) and modeling results.

wave interactions with the DNA and RNA macromolecules. The results also show that the mode intensity varies widely over the THz band and that the spectra are sensitive to base pair sequence. Furthermore, experimental results may be combined with theoretical modeling to directly assign vibrational modes to specific structural features and topology of the macromolecules. However, this work is only the beginning.

5. Applications: Artificial Neural Network Analysis

The above results have demonstrated the possibility of exploiting long-wavelength FTIR spectroscopy to characterize films of DNA polymers. While these results are positive and lay an initial foundation for the method, extensive research will be required to completely demonstrate the feasibility of DNA long-wavelength phonon spectroscopy and a new bio-sensing technology. The potential applications include the use of submillimeter-wave spectroscopy as a technique for the detection and identification of biological warfare agents. Important potential bio-medical applications of long-wavelength vibration spectroscopy based on the close relationship between structure and spectrum include the study of protein binding for transport, protein binding with antibodies, the interactions between proteins and nucleic acids (the specificity of which is vital to the functioning of the genetic apparatus), and the study of the binding stability of drug-protein and vitamin-protein systems. In addition, long-wavelength vibrational spectroscopy of biological materials has important environmental applications that cannot be overestimated. Fast, reliable, and inexpensive detection of biological contamination is an ever-present problem. The concerns of food safety, air and water quality, prompt recognition of genetically-altered plants and organisms, DNA analysis of forensic evidence, are all emerging problems, which could become routine procedures if adequate methods based on THz-frequency spectroscopy are perfected. Good experimental techniques for data collection and fast, high-quality data processing may lead to efficient practical implementation.

In order to establish submillimeter-wave spectroscopy as a viable and potentially successful approach for the realization of a biological detection system, a number of scientific and technological demonstrations must be achieved. The existence of sub-millimeter wave resonance phenomena that are both a unique characteristic of the biological agents and yet sufficiently diverse across a class of biological materials must still be demonstrated. It was pointed out in reference [95] that since DNA polymers are of low symmetry, and none of the spectral lines are truly forbidden, this low frequency region should reveal many features related to biopolymer structure. Therefore, one can reasonably expect variations in order, over moderate to short range, within the macromolecules to impact modal frequencies, and absorption intensity. An algorithm has to be created for the recognition and discrimination between different DNA molecules. The detection of DNA phonons via resonant absorption is a difficult process. The weak nature of these phonon modes is certainly consistent with expectations. A DNA molecule represents multiple oscillator structures that are not highly periodic in space. Hence, the oscillators contributing to any individual frequency have a low density per unit volume. The individual features have variations in transmission (typically on the order of several percent) that are above the noise level of the FTIR measurement system (i.e., less than 0.5%). A method of raw data generalization of the very far IR spectra of macromolecules has to be developed to detect regularities and eliminate false structures resulting from system noise, contaminated samples and interference effects. Meaningful information must be extracted from the raw data to confirm and understand the nature of this fine structure, which potentially can be considered a DNA signature. All these issues need to be addressed.

Recently the Artificial Neural Network (ANN) analysis has been applied to experimental data[23] for recognition and discrimination between different DNA molecules. ANNs can be used to memorize characteristics and features of given data and to make associations from new data to the old data. They are based on the ability of the networks to "learn" by example from their environment, and to improve their performance. Therefore, ANN techniques have advantages over statistical methods of data classification because they require no *a priori* knowledge about the statistical distribution of the classes in the data sources in order to classify them. ANNs are well suited both for pattern recognition, classification or clustering and for function approximation, thus allowing quantitative modeling. ANNs are computational systems based on our knowledge about biological neurons in the brain.[96] They represent a complex input-output mapping, and utilize massively parallel structures consisting of non-linear processing nodes (input and output) connected by links with different weights. The Stuttgart Neural Network Simulator (SNNS)[97] was used with a standard error backpropagation algorithm[98] and its modification, known as resilient propagation (RPROP) with modified error function,[99] for supervised learning on a training set (35 samples), which included genetic material from salmon and herring. The goal of modeling was to build input-output relationships between IR spectra of DNAs and the genetic identity of the sample. FTIR transmission derivative spectra were used as the primary source of information.

One of the useful features of the RPROP algorithm is the fact that it minimizes not the simplest error function, like the least sum of squared difference between target and calculated outputs, but the more general error function:

$$E = \sum(t_i - o_i)^2 + 10^{-\alpha}\sum w_{ij}^2, \quad \textbf{(13)}$$

where t_i is the target (desired) output, o_i is the calculated output, w_{ij} is the weight of the link between neurones (nodes) i and j, and α is the weight decay coefficient. The weight decay provided by the modified error function leads to a systematic decrease of the unnecessary weights, thus pruning the network and reducing the number of adjustable parameters. It leads to a more structured network with strong connections between a smaller number of neurons. Modified error functions help to solve one of the most important problems in neural network training: how to avoid overdetermination of the model, when it has too many adjustable parameters (weights). The weight decay procedure consists of a systematic decrease of weights while minimizing the second term in the error function, thus zeroing all but the necessary ones. The "necessary weights" are those that receive reinforcement during backpropagation of error value through the network. So, formally the analysis starts from the full set of random weights, and during the learning process this set is modified: the most important weights, with high values of derivatives of the error function with respect to them, are increased, while the weights that do not significantly change the value of the error function are decreased to zero. It is a local adaptive scheme in which the signs of the partial derivatives of measurement error are used for determining the directions of the parameter weight update. When training is complete, instead of several hundred of almost equivalent weights (as in the case of the simple error) the procedure leads to at most few dozen. It is still more than one would prefer, even with compressed data, however this complexity truly reflects the difficulty of the problem, with almost equivalent spectra for genetically close species.

The structure of ANN with shortcut connections is shown in Figure 23. Although connections between all neurons are formally allowed, most of the links are not visible at the selected level of resolution (i.e., the values of weights for those links are less than 15% of the values for the shown links). Such iterative pruning of weights increases the generalization ability of the neural network. The left part of the Figure 23 represents the examples of spectral intensities vs. measurement range (25 to 10 cm^{-1}). Using this as a scale, we can find out which regions of the spectra make the most informative input.

Data pre-processing for Neural Network Modeling consists of collecting differential transmission spectra of the samples measured in one orientation with the electrical field of radiation perpendicular to the direction of mechanical alignment.

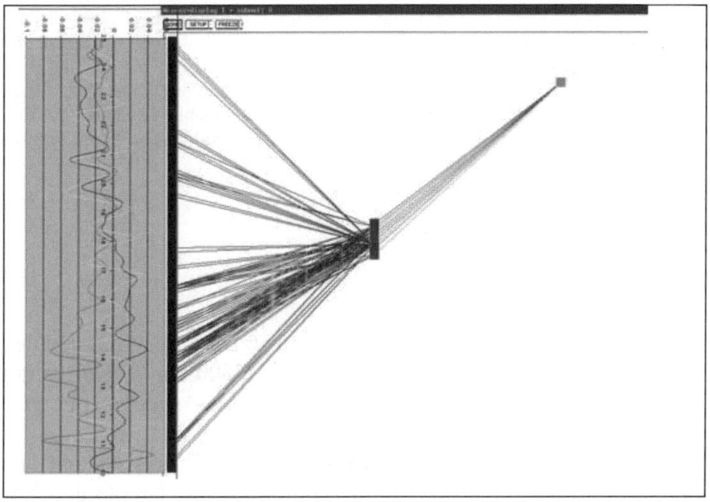

Figure 23. Graphical representation of feedforward Neural network used in the study after training with RPROP algorithm. Left- input layer (125 values for derivative spectra intensities in the range 10-25 cm^{-1}); middle- hidden layer (10 neurons); right-output layer (output value represents the origin – herring or salmon DNA).

This work was performed on 43 DNA samples of different origins : 17 training plus 4 test for herring, and 17 training plus 3 test for salmon. The trained network is then used for prediction. The performance of the trained neural net was tested on the test set (7 samples), which contained patterns not included in the training set (35 samples). Figures 24 &25 are the graphs of the training and test results.

The results of this study present strong evidence that FTIR spectroscopy is a promising new tool in generating spectral data for complex biological objects. The spectra produced with the FTIR technique have a high information content, which is possible to analyze using artificial neural networks (ANNs). Generalization of raw data extracts some regularities unseen before. The preliminary results indicate that recognition of DNA samples can be done with a high degree of confidence using a combination of FTIR spectroscopy and ANNs. Further research is necessary on the influence of data compression (wavelets, Fouirer transform coefficients, Hadamard transform coefficients) on the generalization ability and recognition quality of the network. The described methodology can be generalized to the case of multiple inputs - multiple outputs network, representing the case of multiple classes of species.

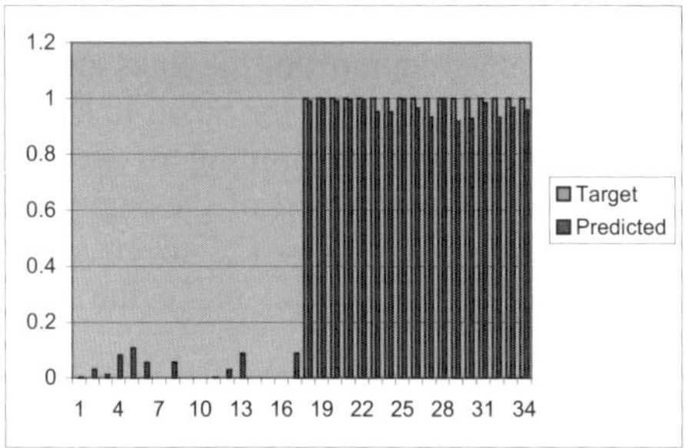

Figure 24. The graph of training results. The value of class variable is shown on Y axis. The experimental value is 0 (salmon) for the first 17 samples of the training set, and 1 (herring) for the last 17 samples. Predicted value of the class variable is shown to the right of the experimental (target) value. The X axis represent the number of samples in the training set.

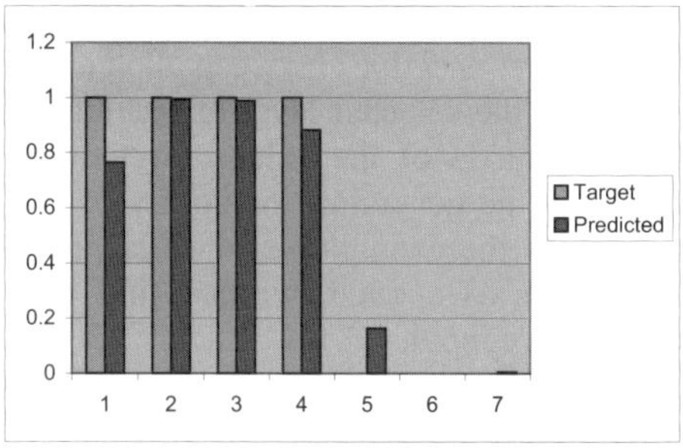

Figure 25. The graph of test results. The value of class variable is shown on Y axis. The experimental value is 1 (herring) for the first 4 samples, and 0 (salmon) for the last 3 samples. Predicted value of the class variable is shown to the right of the experimental value. The X axis represent the number of sample in the test.

6. Conclusions

Highly resolved and reproducible spectra of DNA macromolecules have been recently received in the spectral range from 10 cm^{-1} to 500 cm^{-1} where reliable data was previously absent and where specific features in the spectra of different DNA macromolecules were expected to be found. The potential reasons for the poor reproducibility of results in previous studies have been discussed and analyzed. The dependence of spectra on sample preparation conditions, on sample geometry (multiple reflection at film surfaces), on material aging, and on the orientation of DNA molecules in electromagnetic field of radiation has been studied. The results demonstrate compelling experimental evidence of multiple resonances, due to phonon modes, within the submillimeter wave transmission spectra. Absorption spectra of oligonucleotides with known base-pair sequences have been measured in the submillimeter-wave range. A theoretical study of double helical DNA and RNA fragments has applied normal mode analysis to predict spectra in the far infrared region. Normal modes were calculated and the absorption spectrum was derived for the 12 base pair RNA fragment Poly[G]-Poly[C]. There is a close correlation between calculated and experimentally observed spectra of these RNA homopolymers. These results confirm the fundamental physical nature of the observed resonance structure that is caused by the internal vibrational modes in macromolecules. Furthermore, experimental results may be combined with theoretical modeling to directly assign vibrational modes to specific structural features and topology of the macromolecules.

The long-wavelength vibrational spectra can potentially be used to derive DNA signatures. However, the analysis of FTIR spectra requires the use of advanced non-linear modeling methods, such as Artificial Neural networks, since linear statistical methods are not able to discriminate between different samples. The first applications of Neural Network Modeling for the interpretation of THz-frequency spectra of fish genetic material indicates a high degree confidence in recognition of DNA species.

This work establishes the initial foundation for the future use of submillimeter-wave spectroscopy in the identification and characterization of DNA macromolecules.

This will have wide practical applications in various fields including military defense against biological warfare agents, medical applications, criminology and food safety analysis.

References

1. M. Sarkar, S. Sigurdsson, S. Tomac, S. Sen, E. Rozners, B.-M. Sjoberg, R. Stromberg and A. Graslund, " A synthetic model for triple-helical domains in self-splicing group I introns studied by ultraviolet and circular dichroism spectroscopy", Biochemistry, **35**, 4678-4688 (1996).
2. L. L. Van Zandt and V. K. Saxena, in Structure & Functions, **Volume 1**: Nucleic Acids, Eds. R. H. Sarma and M. H. Sarma (Adenine Press, 1992) and references therein.
3. T. Weidlich, S. M. Lindsay, Q. Rui, A. Rupprecht, W. L. Peticolas and G. A. Thomas, "A Raman Study of Low Frequency Intrahelical Modes in A-, B-, and C-DNA," *J. Biomol. Struct. Dyn.*, **8**, 139 (1990).
4. M. Kohli, W. N. Mei, E. W. Prohofsky, and L. L.Van Zandt, "Calculated Microwave Absorption of Double-Helical B-Conformation Poly(dG)-Poly(dC)," Biopolymers, **20**, 853-864 (1981).
5. W. N. Mei, M. Kohli, E. W. Prohofsky, and L. L. Van Zandt, "Acoustic Modes and Nonbonded Interactions of the Double Helix", Biopolymers, **20**, 833-852 (1981)
6. L. Young, V. V. Prabhu and E. W. Prohovsky, "Calculation of far-infrared absorption in polymer DNA," Phys. Rev. A **39**, 3173 (1989).
7. Y. Feng and E. M. Prohofski, "Vibrational fluctuations of hydrogen bonds in a DNA double helix with nonuniform base pairs", Biophys. J. **57**, 547-553 (1990).
8. D. Lin, A. Matsumoto, and N. Go. "Normal mode analysis of a double-stranded DNA dodecamer d(CGCGAATTCGCG)", J.Chem. Phys. **107**(9), 3684-3690 (1997).
9. T. H. Duong and K. Zakrzewska. "Calculation and analysis of low frequency normal modes for DNA", J.Comp. Chem. **18**(6), 796-811 (1997).
10. R. R. Sinden, DNA structure and Function. Academic press, San Diego (1994).
11. C.P. Beetz and G. Ascarelli, "Far infrared absorption of dNA and polyI-polyC RNA", Biopolymers, **21**, 1569 (1982).
12. R. Giordano, F. Mallamace, N. Micali, F. Wanderlingh, G. Baldini and S. Doglia, , "Light Scattering and Structure in a Deoxyribonucleic Acid Solution", *Phys. Rev. A*, **28**, 3581 (1983).
13. S. M. Lindsay and J. Powell in Structure and Dynamics: Nucleic Acids and Proteins, edited by E. Clementi axnd R. Sarma, p. 241, Adenine, New York (1983).
14. G. S. Edwards, C. C. Davis, J. D. Saffer and M. L. Swiscord, "Resonant microwave absorption of selected DNA molecules," Phys. Rev. Lett., **53**, 1284 (1984).
15. A. Wittlin, L. Genzel, F. Kremer, S. Haseler, A. Poglitsch, and A. Rupprecht, "Far-infrared spectroscopy on oriented films of dry and hydrated DNA", *Phys. Rev.* **A 34**, 493 (1986).
16. J. W. Powell, G. S. Edwards, L. Genzel, F. Kremer, A. Wittlin, W. Kubasek, W. Peticolas, "Investigation of far-infrared vibrational modes in polynucleotides," Phys. Rev., **A 35**, 3929-3939 (1987).
17. T. Weidlich, J. W. Powell, L. Genzel, and A. Rupprecht, "Counterion effects on the far-IR vibrational spectra of poly (rI):poly(rC)", Biopolymers, **30**, 477-480 (1990).
18. J. W. Powell, W. Peticolas and L. Genzel, "Observation of the far-IR spectra of five oligonucleotides", J. Mol. Struct., **247**, 119 (1991).

19. V.Lisy, P. Miskovski, B. Brutovsky, L. Chinsky, “Internal DNA modes below 25 cm-1: A resonance Raman spectroscopy observation”, J. Biomol.Struct. Dyn. **14**, 517 (1997).
20. A .G. Markeltz, A. Roitberg and E. J. Heilweil, “Pulsed terahertz spectroscopy of DNA, bovine serum albumin and collagen between 0.1 and 2.0 THz”, Chem. Phys. Lett. **320**, 42 (2000).
21. T. R.Globus, D.L.Woolard, M. Bykhovskaia, B. Gelmont, J.L.Hesler, T.W.Crowe, A.C.Samuels, *Proc of International Semiconductor Device Researsh Symposium (ISDRS),* p.485, Charlottesville VA (1999).
22. D. L. Woolard, T. R. Globus, A.C. Samuels, B.L. Gelmont, M. Bykhovskaia, J. L. Hesler, T. W. Crowe, J.O. Jensen, J.L. Jensen and W.R Loerop. "The Potential Use of Submillimeter-Wave Spectroscopy as a Technique for Biological Warfare Agent Detection". *Proc to the 22nd Army Science Conference*, Baltimore, MD, Dec.12-13, 2000.
23. T. Globus, L. Dolmatova-Werbos, D. Woolard, A.Samuels, B. Gelmont, and M. Bykhovskaia, "Application of Neural Network Analysis to Submillimeter-Wave Vibrational Spectroscopy of DNA Macromolecules", *International Symposium on Spectral Sensing Research* Proceedings (ISSSR), p.439 (2001), Canada, Quebec, June 2001.
24. T. R. Globus, D. L. Woolard, A. C. Samuels, B. L. Gelmont, J. Hesler, T. W. Crowe and M. Bykhovskaia, “Submillimeter-Wave FTIR Spectroscopy of DNA Macromolecules and Related Materials”, J. Appl. Phys, **91,** 6106-6113 (May 2002).
25.D. L. Woolard, T. R. Globus, B. L. Gelmont, M. Bykhovskaia, A. C. Samuels, D. Cookmeyer, J. L. Hesler, T. W. Crowe, J. O. Jensen, J. L. Jensen and W.R. Loerop, "Submillimeter-Wave Phonon Modes in DNA Macromolecules", Phys. Rev E **65**, 051903 (May 2002)
26. L. L. Van Zandt and V. K. Saxena, “Vibrational Local Modes in DNA Polymer,” J. Biomol. Struct. & Dyn., **11**, 1149-1159 (1994).
27. M. Bykhovskaia, B. Gelmont, T. Globus, D.L.Woolard, A. C. Samuels, T. Ha-Duong, and K. Zakrzewska, "Prediction of DNA Far IR Absorption Spectra Basing on Normal Mode Analysis", Theoretical Chemistry Accounts, **106**, 22-27 (2001).
28. T. Globus, M. Bykhovskaia, B. Gelmont, D.L.Woolard , “Far-infrared phonon modes of selected RNA molecules”, in *Instrumentation for Air Pollution and Global Atmospheric Monitoring*, James O. Jensen, Robert L. Spellicy, Editors, Proceedings of SPIE, **Vol. 4574**, pp.119-128 (2002).
29. T. Globus, M. Bykhovskaia, D. Woolard, B. Gelmont, "Sub-millimeter Wave Absorption Spectra of Artificial RNA Molecules", Advanced Workshop on “Frontiers in Electronics” (WOFE), St. Croix, Virgin Islands, Jan. 6-11 (2002).
30. E. W. Prohofsky, Statistical mechanics and stability of macromolecules. Cambridge University Press (1995)
31. J. A. McCammon and S.C.Harvey. Dynamics of Proteins and Nucleic Acids. Cambridge, Univ. Press (1986).
32. I. Lafontain and R. Lavery, “Collective variable modeling of nucleic acids”, Current Opinion in Struct. Biol. **9**, 170-176 (1999).
33. D. Woolard, et. al., “Sensitivity Limits & Discrimination Capability of THz Transmission Spectroscopy as a Technique for Biological Agent Detection,” in the proceedings to the 5th Joint Conference on Standoff Detection for Chemical and Biological Defense, Williamsburg, VA, 24-28 Sept., 2001.
34. R. Langridge, H.R. Wilson, C.W. Hooper, M.H.F.Wilkins, and L.D.Hamilton. “The molecular configuration of deoxyribonucleic acid", J. Mol. Biol. **2**, 19-27 (1960).

35. M. Poncin., B.Hartmann, and R.Lavery, "Conformational sub-states in B-DNA", J. Mol. Biol. **226**, 775-794 (1992).
36. S.B.Zimmerman, "The three dimensional structure of DNA", Ann. Rev. Biochem., **51**, 395-427 (1982).
37. A. Rich, A.Nordheim, and A.H.Wang, "The chemistry and biology of left-handed DNA", Annu.Rev. Biochem. **53**, 791-846 (1984).
38. J. Kim, C. Yang , S. DasSarma, "Analysis of left-handed Z-DNA formation in short d(CG)n sequences in Escherichia coli and Halobacterium halobium plasmids. Stabilization by increasing repeat length and DNA supercoiling but not salinity", J. Biological Chem. **271**(16), 9340-9346 (1996).
39. A. H. Wang, G.J.Quigley, F.J.Kolpak, J.L.Crawford, J.H. van Boom, G.A. Van der Marel, and A. Rich, "Molecular structure of a left-handed double helical DNA fragment at atomic resolution". Nature (London). **282**, 680-686 (1979).
40. A. H. Wang, T.Hakoshima, G.A. Van der Marel J.H., and J.H. van Boom. "AT base paires are less stable than GC base paires in Z-DNA: the crystal structure of d(m(5)CGTAm(5)CG)", Cell, **37**, 321-331 (1984).
41. R. Lavery "Junctions and bends in nucleic acids: A new theoretical modeling approach". *In Structure and Expression.* Eds, Olson, W.K., Sarma, R.H., Sarna, M.H. & Sundaralingam M. ,**Vol. 3**, pp191-211. Adenine Press, New York (1988).
42. J. S. Weiner, P.A.Kollman, D.A.Case, U.C.Singh, C.Ghio, G.Alogerma, S.Profeta, and P.K.Weiner. "A new force field for molecular mechanical simulation of nucleic acids and proteins", J. Am. Chem. Soc. **106**, 765-784 (1984).
43. E. Tailandier and L.Liquier. "Infrared Spectroscopy of DNA". In *Methods in Enzymology*. ***Vol.1***: *DNA Structure. Part A. Synthesis and Physical Analysis of DNA*. Eds., D.M.J.Lilley and J.E.Dahlberg. Academic Press, pp.307-335 (1992).
44. S. Krimm and J. Bandekar. "Vibrational spectroscopy and conformation of peptides, polypeptides, and proteins", Advances in Protein Chemistry. **38**, 181-365 (1986).
45. J. M. Eyster and E. W. Prohofsky. "Lattice vibrational modes of poly(rU) and poly (rA)", Biopolymers. **13**, 2505-2526 (1974); Eyster J.M. and E.W.Prohofsky. "Lattice vibrational modes of poly(rU) and poly (rA). A coupled single-helical approach", Biopolymers. **13**, 2527-2543 (1974)
46. J. M. Eyster and E W. Prohofsky, "Soft modes and the Structure of the DNA Double Helix", Phys. Rev. Lett., **38**, 371 (1974).
47. V. K .Saxena, L .L. Van Zandt, and W. K. Schroll, "Effective field approach for long-range dissolved DNA polymer dynamics," Phys. Rev A **39**, 1474 (1989).
48. V.K.Saxena, L.L.Van Zandt, "Plasmon Interpretation of 25 cm-1 Mode in DNA", J. Biomol Struct & Dynamics, **10**, 227 (1992).
49. Y. Feng, R. D. Beger, X. Hua, and E. W. Prohofsky, Phys. Rev. A **40**, 4612-4619 (1989).
50. T. Noguti and N. Go, "Collective variable description of small-amplitude conformational fluctuations in a globular protein", Nature, **296**, 776-778 (1982).
51. N. Go, T. Noguti and T. Nishikawa. "Dynamics of a Small Globular Protein in Terms of Low-Frequency Vibrational Modes", Proc. Natl. Acad. Sc. USA. **80,** 3696-3700 (1983).
52. B. Brooks and M. Karplus, "Harmonic Dynamics of Proteins: Normal Modes and Fluctuations in Bovine Pancreatic Trypsin Inhibitor", Proc. Natl. Acad. Sc. USA. **80,** 6571 (1983).
53. A. Roitberg, R. B. Gerber, R. Elber, and M. A. Ratner, "Anharmonic Wave Functions of Proteins: Quantum Self-Consistent Field Calculations of BPTI", Science, New Series, Vol. **268**, No. 5215. pp. 1319-1322 (1995).

54. U. Burkert and N. L. Allinger. *Molecular Mechnics*. Am. Chem. Soc., Washington D.C. (1982).
55. M. Levitt, C.Sander and P.S.Stern, "Protein normal-mode dynamics: trypsin inhibitor, crambin, ribonuclease and lysozyme." J. Mol.Biol. **181**, 423-447 (1985).
56. E. B. Wilson, J. C. Decius, and P.C.Cross. *Molecular vibrations*. McGraw-Hill Book Co.(1955)
57. B. Hartmann and R. Lavery, "DNA structural forms", *Quarterly Reviews of Biophysics* **29**: 309-368 (1996).
58. R. Lavery R. *Structure and Expression, **Vol.3**: DNA Bending and Curvature*. W.K. Olson, R.H.Sarna, M.H.Sarna, and M.Sandaralingam, Eds. Adenine Press, New York. 191-211 (1988).
59. R. Lavery, Zakrzewska, & Sklenar H., "Jumna (junction minimization of nucleic acids) ", Comput. Phys. Commun. **91**, 135-142 (1995).
60. R. Lavery and B. Hartman, "Modelling DNA conformational mechanics", Biophys. Chem. **60**, 33-43 (1994).
61. R. Lavery, H. Sklenar, K. Zakrzewska, and B. Pulman , "The flexibility of the nucleic acids: (II). The calculation of internal energy and applications to mononucleotide repeat DNA", J. Biomol. Struct. Dynam. **3**, 989-1014 (1986).
62. R. Lavery, I. Parker, and J. Kendrick, "A general approach to the optimization of the conformation of ring molecules with an applications to valinomycin", J. Biomol. Struct. Dynam., **4**, 443-462 (1986).
63. M. Born and K. Huang. *The Dynamical Theory of Crystal Lattices*. Clarendon , Oxford (1968).
64. B. Melchers, E.W.Knapp, F.Parak, L.Cordone, A. Cupane, and M. Leone, "Structural fluctuations of myoglobin from normal-modes, Mossbauer, Raman, and absorption spectroscopy", Biophys. J. **70**, 2092- 2099 (1996).
65. F.S. Parker, *Applications of Infrared, Raman, and Resonance Raman Spectroscopy in Biochemistry*. Plenum Press, New York & London, pp. 349-359 (1983).
66. *Nucleic Acids in Spectroscopy of Biological Molecules: Modern Trends, edited by* P. Carmona, R. Navarro and A. Hernanz, Kluwer Academic Publishers, Dordrecht/ Boston/London (1997).
67. Y. Tominaga, M. Shida, K. Kubota, H.Urabe, Y. Nishimura, M. Tsuboi, "Coupled dynamics between DNA double helix and hydrated water by low frequency Raman spectroscopy", J. Chem. Phys., **83**, 5972-5975 (1985).
68. N.J. Tao, S.M. Lindsay, and A. Rupprecht, "The dynamics of the DNA hydration shell at gigahertz frequencies", Biopolymers **26**, 171-188 (1987).
69. N. J. Tao, S. M. Lindsay, and A. Rupprecht, "Dynamic Coupling between DNA and Its Primary Hydration Shell Studied by Brillouin Scattering", Biopolymers **27**, 1655 (1988).
70. H. Grimm, H. Stiller, C.F. Majkrzak, A. Rupprecht, and U. Dahlborg, "Observation of acoustic umklapp-phonons in water-stabilized DNA by neutron scattering", Phys Rev Lett. **59**, 1780-1783 (1987).
71. W.K. Schroll, V.V. Prabhu, E. W. Prohovsky, and L.L. Van Zandt, "Phonon Interpretation of Inelastic Neutron Scattering in DNA crystals" Biopolimers, **28**, 1189 (1989)]
72. C. Liu, G. S. Edwards, S. Morgan and E. Silberman, "Low frequency Raman active vibrational modesof poly(dA).poly(dT)", *Phys. Rev* A **40**, 7394-7397 (1989).
73. G. Maret, R. Oldenbourg, G. Winterling, K. Dransfeld and A. Rupprecht, "Velocity of hgh frequency sound waves in oriented DNA fibres and films determined by Brillouin scattering", *Colloid and Polymer Science*, **257**, 1017-1020 (1979).

74. M. B. Hakim, S. M. Lidsay and J. Powell, "The speed of sound *in DNA", Biopolymers*, **23**, 1185-1192 (1984).
75. G. A. Thomas and W. L. Peticolas, "Fluctuations in nucleic acid conformations. 2. Raman spectroscopic evidence of varying ring pucker in A-T polynucleotides", J. Am. Chem. Soc., **105**, 993-996 (1983).
76. H. Urabe, and Y. Tominaga, "Low lying collective modes in DNA helix by Raman spectroscopy." Biopolymers, **21**, 2477-2481 (1982).
77. H. Urabe, Y. Tominaga, and K. Kubota, "Experimental evidence of collective vibrations in DNA double helix", J. Chem Phys. **78**, 5937-5939 (1983).
78. H. Urabe, H. Hayashi, Y.Tominaga, Y.Nishimura, K. Kubota, and M. Tsuboi, "Collective vibrational modes in molecular assembly of DNA and its application to biological systems – Low frequency Raman spectroscopy", J. Chem. Phys. **82**, 531-535 (1985).
79. C. Demarco, S.M. Lindsay, M. Pokorny, J. Powell, and A. Rupprecht, "Interhelical effects on low-frequency modes and phase transitions of Li- and Na-DNA", Biopolymers, **24**, 2035-2040 (1985).
80. T. Weidlich, S. M. Lindsay, S. A. Lee, N. J. Tao, G. D. Lewen, W. L. Peticolas, G. A. Thomas, and A. Rupprecht, "Low-frequency Raman spectra of DNA: a comparison between two oligonucleotide crystals and highly crystalline films of calf thymus DNA", J. Phys. Chem., **92**, 3315-3317 (1988).
81. T. Weidlich, S. M. Lindsay, "Raman study of the low frequency vibrations of polynucleotides", J.Phys. Chem., **92**, 6479-6482 (1988).
82. L. Bergman, M. Dutta and R. J. Nemanich, in "*Light Scattering in Solids*," **Vol. VIII**, Eds. R. Merlin and W. Weber (Springer, 1999)
83. R. Kutteh and L. L. Van Zandt, "Anharmonic-potential-effective-charge approach for computing Raman cross-section of a gas", *Phys. Rev. A*, **47**, 4046-4060 (1993).
84. R. Kutteh and L. L. Van Zandt, "Dipole Moments and Selection Rles for Raman Scattering from Helical Molecules: The Higgs Rules as a Special Case Subset", J. Mol. Spectrosc., **164**, 1-19 (1994),
85. Y. Z. Chen, A. Szabo, D. F. Schroeter, J. W. Powell, S. A. Lee and E. W. Prohofsky, "Effect of drug-binding-induced deformation on the vibrational spectrum of a DNA.daunomycin complex", Phys. Rev. E **55**, 7414-7423 (1997).
86. C. Gabriel, E.H.Grant, R. Tata, P.R.Brown, B. Gestblom, and E. Noreland, "Microwave absorption in aqueous solutions of DNA", Nature **328**, 145-146 (1987).
87. D. L. Woolard,T. Koscica, D.L. Rhodes, H.L.Cui, R.A.Pastore, J.O.Jensen, J.L.Jensen, W.R.Loerop, R.H. Mittleman and M.C.Nuss, "Millimeter Wave-induced Vibrational Modes in DNA as a Possible Alternative to Animal Tests to Probe for Carcinogenic Mutations," Journal of Applied Toxicology, **17**(4) 243-246 (1997).
88. M. Nagel, P. H. Bolivar, M. Brucherseifer, H. Kurz, "Integrated THz technology for label-free genetic diagnostics", Appl. Phys. Let. **80**, 154 (2002).
89. P.Y.Han, M.Tani, M. Usami, S. Kono, R. Kersting, and X.-C. Zhang, "A direct comparison between terahertz time-domain spectroscopy and far-infrared Fourier transform spectroscopy", J.Appl. Phys, **89**, 2357 (2001).
90. T. S.Moss, G. J. Burrell, B. Ellis. Semiconductor Opto-electronics. (Butterworth & Co. (Publishers) Ltd. (1973).
91. T. Globus, G. Ganguly, P. Roca i Cabarrocas, "Optical characterization of hydrogenated silicon films using interference technique", *J. Appl. Phys.* **88**, 1907 (2000).
92. D.Woolard, T. Globus, E. Brown, L.Werbos, B. Gelmont, A. Samuels, "Sensitivity limits & discrimination capability of thz transmission spectroscopy as a technique

for biological agent detection", *Proc. of 5th Joint Conference on Standoff Detection for Chemical and Biological Defence (5JCSD),* Williamsburg, VA, Sept. 2001.
93. M. Sarkar, U. Dornberger, E. Rozners, H. Fritzsche, R. Stromberg and A. Graslund, " FTIR spectroscopic studies of oligonucleotides that model a triple-helical domain in self-splicing group I introns", Biochemistry, **36**, 15463-15471 (1997).
94. L. Genzel, A. Poglitsh, and S. Haeseler, "Dispersive polarizing mm-wave interferometer", Int. J. Infrared Millimeter Waves, **6**, 741-750 (1985).
95. V.K. Saxena, B. H. Dorfman and L.L.Van Zandt, "Identifying and interpreting spectral features of dissolved poly(dA)-poly(dT) DNA polymer in the high-microwave range", Phys. Rev. **A 43**, 4510-4516 (1991).
96. N.Ampazic, "A Basic Introduction to Neural Networks", http://blizzard.gis.uiuc.edu/htmldocs/Neural/neural.html.
97. A. Zell, G. Mamier, M.Vogt, "Stuttgart Neural Network Simulator In User Manual (Version 4.1)",
Report N. **6/95**, Univ. of Stuttgart, IPVR, 1995.
98. P.J. Werbos, "Beyond regression: new tools for predictions and analysis in the behavioural sciences" in Ph. D. dissertation; Harvard Univ. Press: Cambridge, 1974.
99. M. Riedmiller, "Supervised Learning in Multilayer Perceptions - from Backpropagation to Adaptive Learning Techniques", Intern. J. of Computer Standards and Interfaces, **16**, 265-278 (1994).

International Journal of High Speed Electronics and Systems
Vol. 13, No. 4 (2003) 937–950

SPECTROSCOPY WITH ELECTRONIC TERAHERTZ TECHNIQUES FOR CHEMICAL AND BIOLOGICAL SENSING

MIN K. CHOI
Department of Electrical & Computer Engineering
University of Wisconsin -Madison
1415 Engineering Drive
Madison WI 53706-1691 USA
mchoi@cae.wisc.edu

KIMBERLY TAYLOR
Biophysics Degree Program
University of Wisconsin -Madison
1415 Engineering Drive
Madison WI 53706-1691 USA
kmtaylor@students.wisc.edu

ALAN BETTERMANN
Department of Electrical & Computer Engineering
University of Wisconsin -Madison
1415 Engineering Drive
Madison WI 53706-1691 USA
adb@engr.wisc.edu

DANIEL W. VAN DER WEIDE
Department of Electrical & Computer Engineering
University of Wisconsin -Madison
1415 Engineering Drive
Madison WI 53706-1691 USA
danvdw@engr.wisc.edu

By illuminating the sample with a broadband 10-500 GHz stimulus and coherently detecting the response, we obtain reflection and transmission spectra of common powdered substances, and compare them as a starting point for distinguishing concealed threats in envelopes and on personnel. Because these samples are irregular and their dielectric properties cannot be modulated, the spectral information we obtain is largely qualitative. To show how to gain quantitative information on biological species at micro- and millimeter-wave frequencies, we introduce thermal modulation of a globular protein in solution, and show that changes in microwave reflections coincide with accepted visible absorption spectra, pointing the way toward gaining quantitative chemical and biological spectra from broadband terahertz systems.

Keywords: electronic terahertz techniques, gas spectroscopy, reflection spectroscopy, nonlinear transmission lines, samplers, coherent measurements

1. Introduction

With the advent of short-pulse coherent measurement techniques, based either on optoelectronic or on purely electronic means of generating and detecting broadband radiation, researchers are now equipped with new tools for qualitative sensing and even quantitative measurement of chemical and biological entities. The possibility of using stand-off (i.e. non-contact stimulus/response measurements at a distance) techniques to assess the presence of chemical and biological weapons is of particular interest, and is an application that is well-suited to wavelengths at the millimeter length scale, since these can penetrate clothing and envelopes while still forming images with useful spatial resolution.

Hence, this recent interest in sensing chemical and biological entities with frequencies in the megahertz through terahertz regime has brought about a growing body of experimental results, as well as raising curiosity about the fundamental interactions with, and effects of this broadband radiation on, biological samples.

Such samples can be classified by their phase—vapor, liquid or solid—as well as by their degrees of homogeneity and complexity–purified, admixture, or tissue. Furthermore, the interaction of sample volume and wavelength will set limits on and complicate the interpretation of broadband measurements. Many broadband pulsed systems have $<$ 40 dB dynamic range, which limits the thickness of a highly-absorbing sample, while thin or more transparent samples will require a minimum interaction length with the radiation to produce meaningful contrast above the noise and non-repeatability of broadband systems. Because sample thickness is often commensurate with measurement wavelength, etalon effects of the sample, in which multiple quarter- or half-wave resonances modulate the detected signal, are often compounded with auxiliary standing-wave effects of the measurement apparatus and environment. While radiometric millimeter-wave techniques that rely on the sample's black-body radiation would minimize these standing wave effects, these are not broadband systems, nor do they illuminate the sample.

Thus, in contrast with traditional spectroscopy techniques in the infrared and visible, where sample size is typically much larger than the wavelength, probing chemical and biological samples with broadband terahertz radiation requires reconsideration and careful specification of the sample's physical characteristics. The confounding effects of these characteristics on the raw result usually require a differential (normalized) measurement technique to measure the actual sample behavior, and this can result in magnification of standing wave effects.

While we briefly survey the background of broadband sensing for chemical and biological samples, our principal focus here is on sensing solid-phase, condensed samples that are either pure or admixtures of known substances, typical of what might be expected in a concealed bulk (as distinct from trace) threat.

In addition, to advance an argument for better understanding of sample/radiation interactions, we discuss microwave spectroscopy of purified biomolecules in solution, in this case a globular protein, bovine pancreatic ribonuclease (RNase A), which we thermally denature (unfold) while simultaneously probing the sample with microwave and visible light. In this experiment, we gain two important advantages:

- We can externally modulate the sample (in this case the protein's conformation), and
- We can independently confirm our microwave measurement with an accepted optical technique.

Such modulation and independent, simultaneous confirmation is usually lacking in broadband measurements, making interpretation of the data more challenging.

Since, on the other hand, concealed threats are (almost by definition) not amenable to modulation, single-line or narrowband radiation is probably less useful for qualifying the nature of the threat than is broadband radiation. Although it has more power, and microwave holography has been reported for imaging of concealed weapons,[1] narrowband illumination fundamentally restricts the availability of spectral information in a complex environment. Even with the richer spectrum of broadband radiation, the lack of independent confirmation of contrast mechanisms (such as an optical spectroscopic probe) leaves us with few choices other than to look for characteristic patterns in the broadband reflection or transmission spectra that we measure. The coherence of these synchronous generation/detection spectrometers enables us to measure both magnitude and phase of transmission and reflection. While a characteristic spectral pattern can emerge in the phase spectrum, there is still more progress to be made in quantifying these spectra.

Nonetheless, having millimeter-scale spatial resolution and GHz-level spectral resolution enables qualitative identification of features that, in a pattern-matching algorithm, can be applied to a form of spectroscopic imaging for security screening. Thus the main results presented here are spectra that might be further processed by such an algorithm to result in a false-color image that would indicate the presence of unusual (chemical or biological) substances concealed within envelopes or under clothing.

2. Background

Broadband microwave spectrometers or network analyzers, whether time-[2] or frequency-domain,[3] have been used to measure chemical and biological samples, both in bulk and attached to transducer surfaces, such as planar transmission lines.[4,5] While interpretation of the spectra from all but the simplest samples (e.g. pure water) has been difficult, characteristic signatures of biomolecular conformation have nonetheless been observed.

Biological macromolecules exhibit weak absorptions, if any, in the microwave and terahertz regimes. Proteins are weakly dielectric, and exhibit absorption due to orientational relaxation at radio frequencies.[6] Double-helical DNA does not exhibit such orientational relaxation, since the dipole moments of the two helices nullify each other. Normal mode analysis predicts that both proteins[7] and DNA[8] should exhibit multiple absorptions in the terahertz range due to a variety of collective, vibrational, twisting and librational modes.

All biological macromolecules are surrounded by one or more shells of bound water[6]; molecules in solution are also bounded by bulk solution. Water is a strongly dielectric molecule, with absorptions throughout the microwave and terahertz regimes.[9] Normal mode analysis predicts multiple absorptions for water in the microwave and terahertz ranges.[10] Hence, the primary contributor to the absorption of any solvated biological macromolecule is water, either in bound or bulk form. The observed spectra will be a combination of absorptions from bound and/or bulk water, the biological molecule(s), and interfacial dielectric phenomena. Several authors have described responses of DNA and proteins at microwave and terahertz frequencies.[11-16]

Cells and tissues are complex mixtures of biological macromolecules and water. Spectra from such samples are not amenable to simple analysis, but differences in hydration may be used for imaging and/or identification purposes.[17-19]

3. Broadband stimulus/response

We focus here on coherent broadband stimulus/response measurements, working both in reflection and in transmis sion. While the transmission method can be used for low-loss probing through envelopes, the reflection method can be applied generally from envelopes to the human body, which cannot be imaged in transmission with terahertz systems due to high absorption of water.

To generate energy in the 1-1000 GHz regime, we use nonlinear transmission line (NLTL) pulse generators coupled to wideband planar antennas.[20-25] The GaAs IC NLTLs used in this work consist of series inductors (or sections of high-impedance transmission line) with varactor diodes periodically placed as shunt elements. On this structure at room temperature a fast (~ 0.5-2 ps) voltage step develops from a sinusoidal input because the propagation velocity u is modulated by the diode capacitance, $u(V)=1/\sqrt{LC(V)}$, where L is the line inductance and $C(V)$ is the sum of the diode and parasitic line capacitance.[24, 26] Limitations of the NLTL arise from its periodic cutoff frequency, waveguide dispersion, interconnect metallization losses, and diode resistive losses. Improvements in NLTL design have resulted in sub-picosecond pulses at room temperature.[25]

The first electronic THz systems used resonant or broadband (bowtie) antennas driven directly by NLTLs for generation of freely propagating pulses. These pulses can be focused and collected in a manner exactly analogous to optoelectronic THz systems,[27] using substrate lenses and reflective optics. For coherent detection, the pulse from the receiver's NLTL was used to drive a sampler, essentially a broadband frequency mixer, whose temporal output can be sent to an oscilloscope or Fast-Fourier Transform (FFT) spectrum analyzer for display in the frequency domain.

4. Reflection and transmission spectroscopy with coherent detection

Increasingly sophisticated weapons and explosives require increasingly sophisticated detection technologies. Non-metallic varieties of these threats are especially important because they elude familiar metal-detecting portals, so they have motivated development of a multi-pronged approach to detection, including residue sniffing and computerized tomography. These techniques, however, have significant drawbacks, including invasiveness, slowness, unfamiliarity to the public, and significant potential for false negatives.[28]

Threats like these appear to be readily detectable and even identifiable using a broadband of signals in the sub-THz regime (1-500 GHz), based on experiments reported here. Traditional equipment for generating and detecting these frequencies has been difficult, bulky and expensive. The objectives of this work are to develop and apply all-electronic and monolithically integrated technology for generating and detecting these broadband signals to the problem of imaging the transmission and reflection spectra of unknown chemical or biological powders.

Many of the concepts we employ here are being pursued at lower frequencies for target detection at higher resolutions than traditional narrowband radar allows. This ultrawideband (UWB), carrier-free, impulse, or baseband radar has been rapidly gaining popularity in applications where complex and elusive targets are the norm.[29] UWB radar has benefited from very recent advances in semiconductor technology enabling the production of sub-nanosecond pulses with peak powers of over 1 megawatt but having average powers in the milliwatt regime.

By contrast, the technology we employ—the integrated-circuit nonlinear transmission line (NLTL)—essentially trades power for speed, producing pico- or even sub-picosecond pulses with peak powers less than one watt and average powers in the low microwatt regime. These power levels are non-ionizing and biologically inconsequential, but because we can employ coherent detection, rejecting noise outside the frequencies of interest, we can still use them to measure useful spectra.

Baseband pico- and sub-picosecond pulses of freely propagating radiation, usually generated and coherently detected with photoconductive switches and ultrafast lasers,[30-32] have been useful for broadband coherent spectroscopy of materials, liquids, and gasses in the THz regime. Such systems have even been used for what could be called scale-model UWB radar.[33] These highly versatile beams of ultra-short electromagnetic pulses can be treated quasi-optically: They are diffracted and focused with mirrors and lenses, and the resultant effects can be readily observed in the time-domain waveform at the detector. Consequently, such beams are singularly useful for spectroscopy in a difficult-to-access spectral regime, and recent reports of spectroscopic imaging with these optoelectronic systems have generated much interest.[34] With these systems we have performed a variety of investigations using both transmission and reflection with a completely room-temperature system.

5. Sample preparation

To examine the content of envelopes with an eye toward distinguishing common powders from potentially dangerous ones, we prepared samples of sugar, starch, flour, and talcum powder. These were placed into petri dishes having 90 mm diameter and 20 mm height when covered. The masses of each sample were 56.6, 61.2, 62.0, and 84.6 grams for sugar, starch, flour, and talcum powder, respectively, and they were used in both reflection and transmission. Additional transmission measurements were done through envelopes with the same four powders and with *B. cereus*, a simulant of *B. anthracis*.

Bacillus cereus (from the laboratory of Dr. John Lindquist, University of Wisconsin Department of Bacteriology) was inoculated onto two tryptic soy agar plates (100 x 20 mm) as a lawn and incubated 120 hours at 27 °C. The surfaces of the plates were scraped using a flat spatula to recover spores (and residual of the vegetative cells). A slurry of this material was made by diluting it 1:1 with sterile H_2O. Approximately 200 mg of spore slurry was evenly spread over a 7x10 cm piece of standard office copier paper, which was then dried at room temperature overnight.

6. Reflection

The reflection measurement setup is shown in Fig. 1. First, we used a copper plate with the same size of the samples to obtain a magnitude and phase reference. Each petri dish sample was collected and normalized to the reference values (Fig. 2 and Fig. 3).

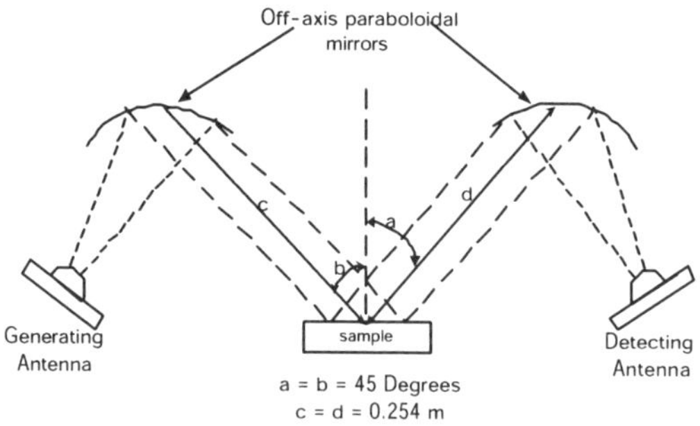

Fig. 1. Reflection measurement setup.

For both lossy and highly conducting materials, reflection measurements are preferable since transmission may not have enough signal-to-noise. As can be seen in figure 2, the reflection magnitude shows some differences in spectral patterns, part of which can be attributed to standing waves when normalized (hence the > 1 reflection coefficient magnitude). The phase information, additionally, can be used to identify the sample powders.

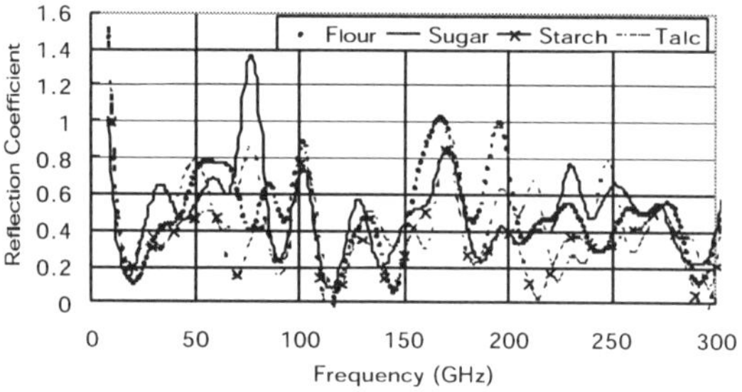

Fig. 2. Normalized reflection magnitude of common powders. Standing-wave effects at ~ 75 GHz cause an anomalous peak in the ratio of the sugar sample to the empty petri dish.

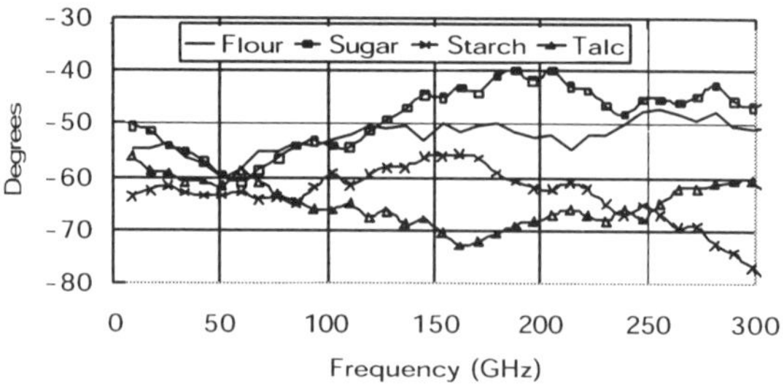

Fig. 3. Normalized, unwrapped reflection phase of common powders.

7. Transmission

The transmission measurement setup is shown in Fig. 4. Transmission through samples was normalized by the detected signal using empty containers in the beam path. We inserted Mylar resistive sheet attenuators into the beam path to check for uniform attenuation across the spectrum

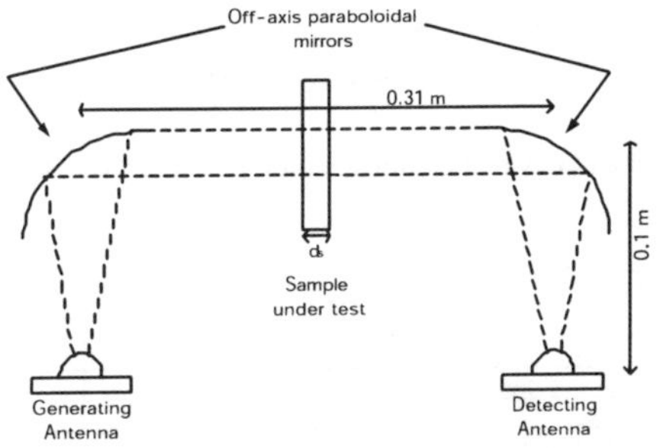

Fig. 4. Transmission measurement setup diagrams: without sample.

With this configuration we examined samples both in petri dishes and in sealed envelopes. The normalization for the petri dishes was performed by using an empty petri dish. For the envelope measurements, the reference values were taken with an empty envelope at the sample position as in Fig. 4.

We expect frequency-dependent attenuation through the samples: Fig. 5 shows more attenuation with increasing frequency than Fig. 7 because the thickness of petri dishes exceeds that of the envelopes. Furthermore, reflection measurements (Fig. 2) do not show this ~1/f type of attenuation. Generally, flour and starch samples have very similar patterns in the magnitude, but they are not identical. A notable result is that for both petri dish and envelope samples, the attenuation for flour and starch is larger than for the others, making identification more feasible.

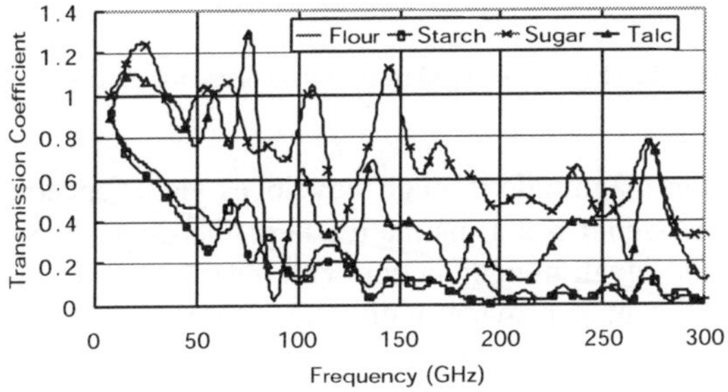

Fig. 5. Normalized transmission magnitude for samples in petri dishes. Note effects of standing waves on ratios plotted.

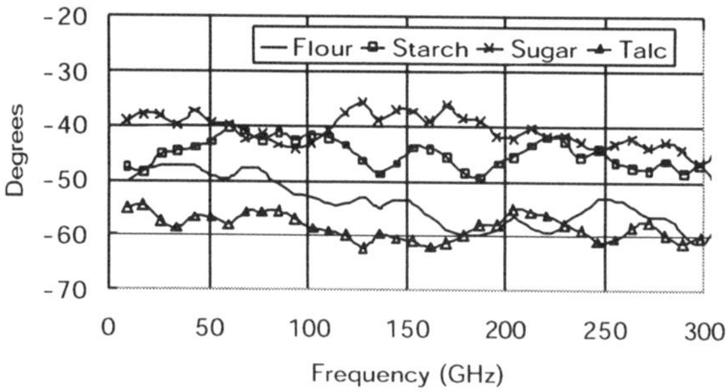

Fig. 6. Normalized and unwrapped transmission phase for samples in petri dishes.

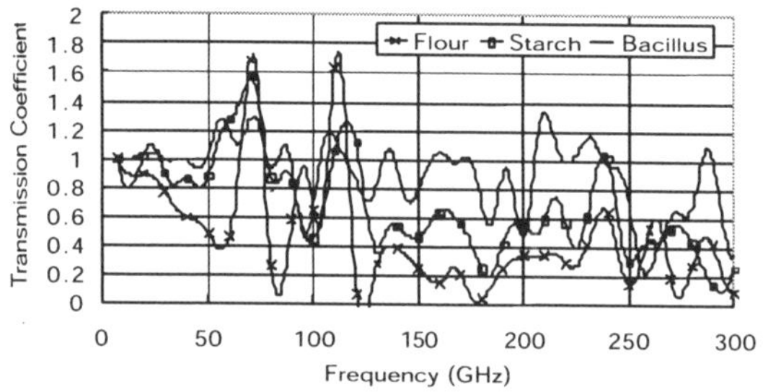

Fig. 7. Normalized transmission magnitude for samples in envelopes, comparing flour, starch, and *B. cereus* spores.

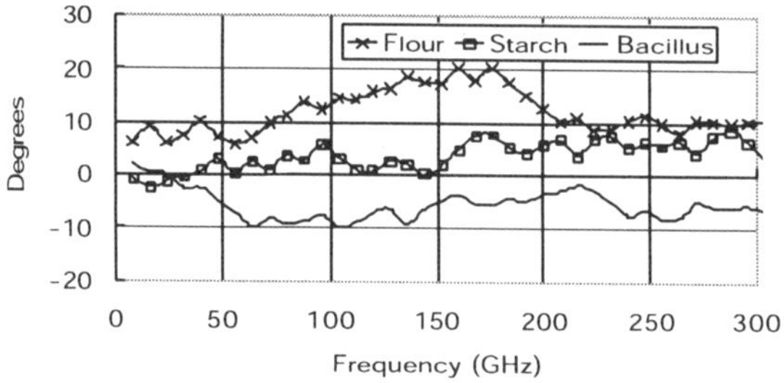

Fig. 8. Normalized and unwrapped transmission phase for samples in envelopes.

While the transmission magnitude through envelopes yields spectra that are difficult to distinguish, the phase signal has a distinctive negative dispersion characteristic of the anthrax simulant. This signal, in conjunction with measuring transmission characteristics of neighboring pixels, would provide additional information about potential threats in envelopes.

Standing-wave effects cause the ratios plotted to have magnitudes > 1 at some frequencies. Inserting attenuators into the beam path could reduce this standing wave problem, at the expense of signal-to-noise. To check for uniform attenuation, we inserted a Mylar resistive sheet (300 ohms per square) in the beam path (Fig. 9). The attenuation was 5 dB at most frequencies, but small variations are clear, and these are magnified when taking the ratio of two measurements.

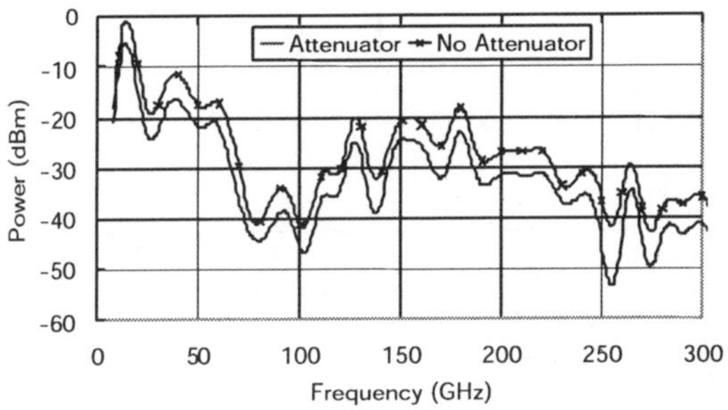

Fig. 9. Transmission with and without a 300 ohm/square attenuating sheet.

8. Reflection from solution proteins

In order to address uncertainties in measuring microwave interactions with biomolecules, we chose a purified protein in solution whose conformation could be thermally modulated and independently verified with optical means. Bovine pancreatic ribonuclease (RNase A), a small globular protein, was used to correlate microwave reflection measurements with changes in protein conformation. RNase A experiences reversible thermal denaturation under a variety of conditions. Conformational changes of RNase A can be detected by monitoring the UV/VIS absorbance at 288 nm.[35, 36] Reflection measurements from 0.5 to 6.5 GHz were performed using a resonant slot antenna attached to a vector network analyzer.[37] The slot antenna was attached to a fused quartz UV/VIS cuvette. A UV/VIS spectrophotometer could be used in conjunction with the antenna/cuvette assembly to obtain simultaneous optical and microwave measurements.

The presence of RNase A in solution could be detected using our slot antenna system. Fig. 10. shows a typical spectrum of RNase A; Fig. 11. shows variation of the position of a single peak with protein concentration. The peak minimum increases approximately linearly with increasing protein concentration. Note that these changes are quite small, less than 0.8 GHz over a concentration range from 0 to 2.97 mg/mL. However, this experiment demonstrates that microwave spectra are sensitive to protein concentration, as expected. This sensitivity would be expected not only for protein, but also for any solute that affects the structure of bulk water.

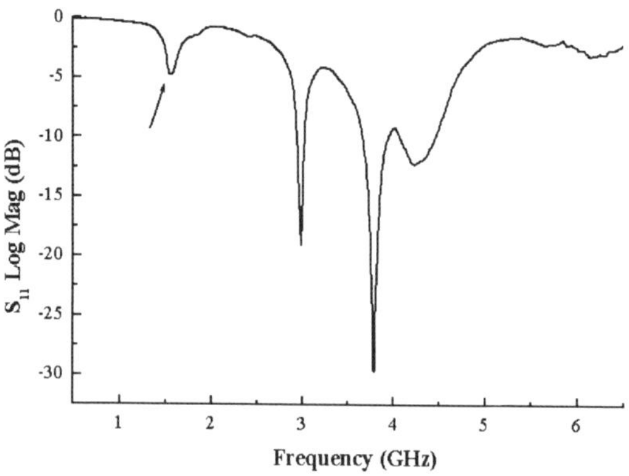

Fig. 10. Reflection spectrum from a slot antenna affixed to a cuvette with RNase-A solution. Arrow indicates a peak that is tracked in t hermal measurements.

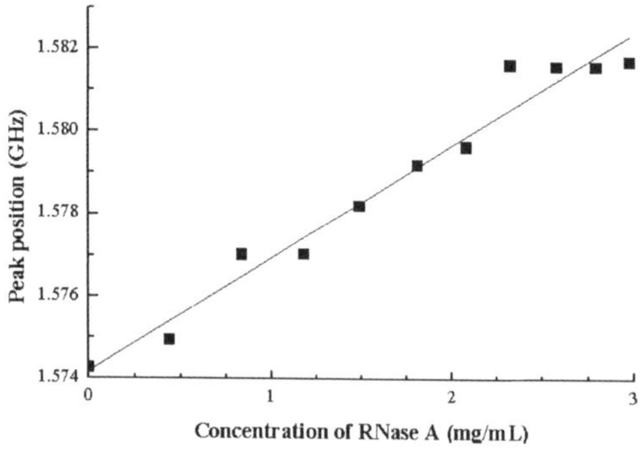

Fig. 11. Concentration dependence of frequency shift for peak noted in Fig. 10.

The response of the spectra of RNase A to increasing temperature is shown in Fig. 12. Note that peaks shift and broaden with increasing temperature. The upper half of Fig. 13. displays the response of a single peak at approximately 3.5 GHz. Note that the position of the peak minimum varies in a sinusoidal manner. Such a sinusoidal response is expected for cooperative phenomena such as protein unfolding[38] and is absent when buffer alone is heated under the same conditions (data not shown). This peak fitted well to a 2-state unfolding model.[35] Results from fitting of the peak at 3.5 GHz, and from fitting to the UV/VIS data alone, are shown in the lower half of Fig. 13. The microwave data predicts a lower midpoint temperature (T_m) and unfolding enthalpy (ΔH_m) than that calculated from UV/VIS data (see Fig. 13.). Since the two data sets were obtained simultaneously, this result is probably due to differences in the phenomena being measured, not to destabilization of the protein by the microwaves. Unfolding results from UV/VIS spectroscopy in the presence and absence of microwaves are identical within experimental error, indicating that the protein is not destabilized under the conditions used (data not shown).

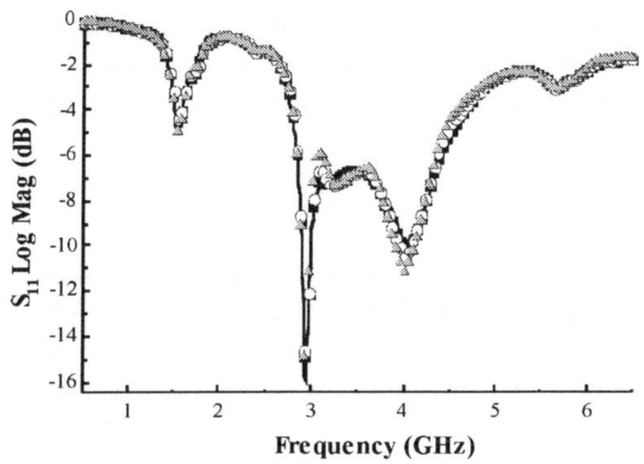

Fig. 12. Spectral response of RNase-A solution to increasing temperature.

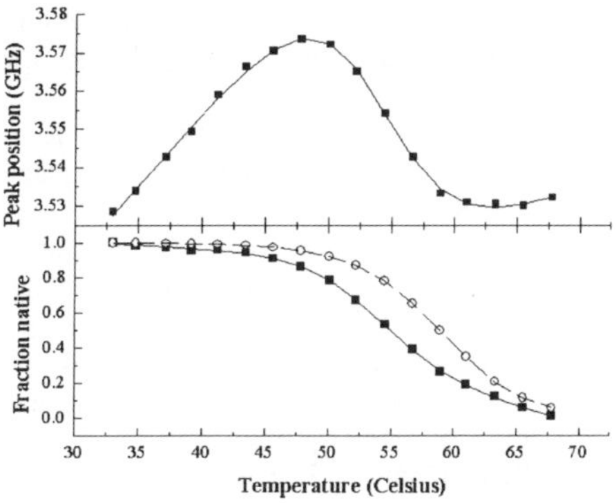

Fig. 13. Raw data from peak shift (above); data fitted to unfolding theory and compared to UV-vis absorption (open circles, below).

9. Future directions

To improve broadband stimulus-response measurements, both higher power systems and those with more array elements will be needed. Video-rate detection and imaging will depend on setting high enough offset frequencies between the coherent source and detector; this requires faster baseband processing.

To extend the protein solution experiments, we will perform additional unfolding at a variety of pH, concentration and microwave power conditions in order to detect limits of system and possible destabilization of the protein. Also needed are enzymology studies to ensure that activity of protein is unaffected. Other macromolecules might also be studied, such as DNA, RNA, carbohydrates and lipids. Finally, we will scale down the size of the experiment to increase the frequency to terahertz or higher.

10. Summary

We have discussed both broadband and resonant probing of biological samples and common powders using microwave and broadband systems. Broadband stimulus-response capabilities enable spectroscopic imaging, and increase the likelihood of sample identification, while experiments on purified and modulated samples demonstrate conclusively the use of water bound to a protein as a label, and that the change of conformation can be observed in the perturbation of electric fields in the near zone of a slot antenna.

Acknowledgments

This work was sponsored by the United States National Science Foundation, Office of Naval Research, Air Force Office of Scientific Research and Army Research Office (Dr. D. Woolard). Kimberly Taylor is supported by an NIH Molecular Biophysics Training Grant. Earlier work on THz imaging was sponsored by the Federal Aviation Administration.

References

1. D. M. Sheen, D. L. McMakin, and T. E. Hall, "Three-dimensional millimeter-wave imaging for concealed weapon detection", *IEEE Trans. Microwave Theory Tech.* **49** (2001) 1581-92.
2. Y. Feldman, A. Andrianov, E. Polygalov, I. Ermolina, G. Romanychev, Y. Zuev, and B. Milgotin, "Time domain dielectric spectroscopy: an advanced measuring system", *Rev. Sci. Instrum.* **67** (1996) 3208-3216.
3. M. A. Hollis, C. F. Blackman, C. M. Weil, J. W. Allis, and D. J. Schaefer, "A swept-frequency magnitude method for the dielectric characterization of chemical and biological systems", *IEEE Trans. Microwave Theory Tech.* **28** (1980) 791-801.
4. J. Hefti, A. Pan, and A. Kumar, "Sensitive detection method of dielectric dispersions in aqueous-based, surface-bound macromolecular structures using microwave spectroscopy", *Appl. Phys. Lett.* **75** (1999) 1802-1804.
5. G. R. Facer, D. A. Notterman, and L. L. Sohn, "Dielectric spectroscopy for bioanalysis: From 40 Hz to 26.5 GHz in a microfabricated wave guide", *Appl. Phys. Lett.* **78** (2001) 996-998.
6. R. Pethig, *Dielectric and electronic properties of biological materials*. John Wiley & Sons, Chichester, 1979.
7. F. Tama, F. X. Gadea, O. Marques, and Y. H. Sanejouand, "Building-block approach for determining low-frequency normal modes of macromolecules", *Proteins-Structure Function and Genetics*. **41** (2000) 1-7.
8. M. Bykhovskaia, B. Gelmont, T. Globus, D. L. Woolard, A. C. Samuels, T. H. Duong, and K. Zakrzewska, "Prediction of DNA far-IR absorption spectra based on normal mode analysis", *Theoretical Chemistry Accounts*. **106** (2001) 22-27.
9. D. J. Segelstein, "The complex refractive index of water." University of Missouri, Kansas City 1981.
10. J. T. Kindt and C. A. Schmuttenmaer, "Far-infrared absorption spectra of water, ammonia, and chloroform calculated from instantaneous normal mode theory", *J. Chem. Phys*. **106** (1997) 4389-4400.
11. R. Pethig, "Protein-Water Interactions Determined by Dielectric Methods", *Annual Review of Physical Chem*. **43** (1992) 177-205.
12. A. G. Markelz, A. Roitberg, and E. J. Heilweil, "Pulsed terahertz spectroscopy of DNA, bovine serum albumin and collagen between 0.1 and 2.0 THz", *Chemical Phys. Lett.* **320** (2000) 42-48.
13. N. Nandi, K. Bhattacharyya, and B. Bagchi, "Dielectric relaxation and solvation dynamics of water in complex chemical and biological systems", *Chemical Reviews*. **100** (2000) 2013-2045.
14. K. Taylor and D. W. van der Weide, "Microwave assay for detecting protein conformation in solution", presented at Photonics Boston, Boston, 2001.

15. K. Taylor and D. W. van der Weide, "Sensing folding of solution proteins with resonant antennas", presented at 9th International Conference on Terahertz Electronics, Charlottesville, 2001.
16. C. Wichaidit, J. R. Peck, Z. Lin, R. J. Hamers, S. C. Hagness, and D. W. van der Weide, "Resonant slot antennas as transducers of DNA hybridization: A computational feasibility study", *IEEE MTT-S Int. Microwave Symp. Dig.* **1** (2001) 163-166.
17. P. Y. Han and X. C. Zhang, "Time-domain spectroscopy targets the far-infrared", *Laser Focus World.* **36** (2000) 117-+.
18. S. W. Smye, J. M. Chamberlain, A. J. Fitzgerald, and E. Berry, "The interaction between Terahertz radiation and biological tissue", *Phys. Med. Biol.* **46** (2001) R101-R112.
19. A. J. Fitzgerald, E. Berry, N. N. Zinovev, G. C. Walker, M. A. Smith, and J. M. Chamberlain, "An introduction to medical imaging with coherent terahertz frequency radiation", *Phys. Med. Biol.* **47** (2002) R67-R84.
20. D. W. van der Weide, J. S. Bostak, B. A. Auld, and D. M. Bloom, "All-electronic free-space pulse generation and detection", *Electronics Lett.* **27** (1991) 1412-1413.
21. D. W. van der Weide, J. S. Bostak, B. A. Auld, and D. M. Bloom, "All-electronic generation of 880 fs, 3.5 V shockwaves and their application to a 3 THz free-space signal generation system", *Appl. Phys. Lett.* **62** (1993) 22-24.
22. Y. Konishi, M. Kamegawa, M. Case, R. Yu, M. J. W. Rodwell, and R. A. York, "Picosecond electrical spectroscopy using monolithic GaAs circuits", *Appl. Phys. Lett.* **61** (1992) 2829-2831.
23. Y. Konishi, M. Kamegawa, M. Case, Y. Rüai, S. T. Allen, and M. J. W. Rodwell, "A broadband free-space millimeter-wave vector transmission measurement system", *IEEE Trans. Microwave Theory Tech.* **42** (1994) 1131-1139.
24. M. J. W. Rodwell, S. T. Allen, R. Y. Yu, M. G. Case, U. Bhattacharya, M. Reddy, E. Carman, M. Kamegawa, Y. Konishi, J. Pusl, R. Pullela, and J. Esch, "Active and nonlinear wave propagation devices in ultrafast electronics and optoelectronics (and prolog)", *Proc. IEEE.* **82** (1994) 1035-59.
25. D. W. van der Weide, "Delta-doped Schottky diode nonlinear transmission lines for 480-fs, 3.5-V transients", *Appl. Phys. Lett.* **65** (1994) 881-883.
26. M. J. W. Rodwell, M. Kamegawa, R. Yu, M. Case, E. Carman, and K. S. Giboney, "GaAs nonlinear transmission lines for picosecond pulse generation and millimeter-wave sampling", *IEEE Trans. Microwave Theory Tech.* **39** (1991) 1194-204.
27. M. C. Nuss and J. Orenstein, "Terahertz time-domain spectroscopy", in *Millimeter and Submillimeter Wave Spectroscopy of Solids*, vol. 74, *TOPICS IN CURRENT CHEMISTRY*, 1998, pp. 7-50.
28. P. Mann, "TWA disaster reopens tough security issues", in *Aviation Week & Space Technology*, vol. 145, 1996, pp. 23-27
29. D. Herskovitz, "Wide, Wider, Widest", *Microwave Journal.* **38** (1995) 26-40.
30. M. v. Exeter, "Terahertz time-domain spectroscopy of water vapor", *Optics Lett.* **14** (1989) 1128-1130.
31. D. Grischkowsky, S. Keiding, M. v. Exeter, and C. Fattinger, "Far-infrared time-domain spectroscopy with terahertz beams of dielectrics and semiconductors", *Journal of the Optical Society of America B.* **7** (1990) 2006-2015.
32. M. C. Nuss, K. W. Goossen, J. P. Gordon, P. M. Mankiewich, M. L. O'Malley, and M. Bhusan, "Terahertz time-domain measurement of the conductivity and superconducting band gap in niobium", *J. Appl. Phys.* **70** (1991) 2238-2241.
33. R. A. Cheville and D. Grischkowsky, "Time domain terahertz impulse ranging studies", *Appl. Phys. Lett.* **67** (1995) 1960-1962.
34. B. B. Hu and M. C. Nuss, "Imaging with terahertz waves", *Optics Lett.* **20** (1995) 1716-1718.
35. T. A. Klink, K. J. Woycechowsky, K. M. Taylor, and R. T. Raines, "Contribution of disulfide bonds to the conformational stability and catalytic activity of ribonuclease A", *European Journal of Biochemistry.* **267** (2000) 566-572.

36. C. N. Pace, G. R. Grimsley, S. T. Thomas, and G. I. Makhatadze, "Heat capacity change for ribonuclease A folding", *Protein Science.* **8** (1999) 1500-1504.
37. H. G. Akhavan and D. Mirshekar-Syahkal, "Slot antennas for measurement of properties of dielectrics at microwave frequencies", presented at National Conference on Antennas and Propagation, 1999.
38. T. E. Creighton, *Proteins: structures and molecular properties*, Second ed. New York: W.H. Freeman and Company, 1993.

International Journal of High Speed Electronics and Systems
Vol. 13, No. 4 (2003) 951–967

World Scientific
www.worldscientific.com

Terahertz Applications to Biomolecular Sensing

Andrea G. Markelz and Scott E. Whitmire
Physics Department, University at Buffalo, USA

THz time domain spectroscopy of biomolecules was performed to determine applicability of the technique for chemical and conformational identification of biomolecules. Measurements were performed on samples of DNA, bovine serum albumin, collagen, hen egg white lysozyme, myoglobin, and bacteriorhodopsin as a function of temperature, hydration and photoexcitation. The results are compared to normal mode calculations. We demonstrate a clear resemblance of the observed broad and near featureless THz absorbance spectrum to the calculated density of normal modes. While the magnitude of the absorbance and the center of the broad response depends on biomolecular species, unique chemical identification would appear challenging. The observed dependence on hydration is in agreement with mass and dielectric loading, and the dependence on temperature is in agreement with decreasing conformational flexibility with reduced temperature. Finally THz absorbance dependence on the biomolecular conformation and mutation is demonstrated for bacteriorhodopsin.

Keywords: terahertz, proteins, biomolecules, conformation, collective excitation.

I. Introduction

Among the reasons for using THz radiation to probe biological systems are: sensitivity to water content, nondestructive and nonionizing radiation technique and frequency range overlap with important biomolecular vibrational modes. A number of authors have pursued terahertz imaging for diagnosis of tissue damage and dental decay and terahertz spectroscopy for DNA identification, protein identification and DNA and protein conformation determination.[1-9] Here we discuss the applications of terahertz spectroscopy to biomolecular conformation and conformational dynamics and will mainly focus on conformational dynamics of proteins. Proteins undergo large-scale conformational change during biochemical processes. Examples of which include the hinging motion of lysozyme when it clamps down on the bacterial wall sugars in the act of destroying the cell wall. The ability for the molecule to access these different conformations is critical to their function, as has been demonstrated by mutagenisis, temperature dependence and ligand binding causing steric hindrance of the necessary motions. Thus current medical chemistry is developing therapies that both cause chemical modification and conformational constraints. For example, a therapies based on introducing a ligand which binds to a virus in such a way as to limit the necessary conformations required for replication.[10] However the apriori understanding of conformational flexibility and conformational change is just beginning. For any given protein, mutagenisis to change its conformation and dynamical conformational properties is mainly combinatorial. Molecular modeling while far advanced from even 10 years ago still is based on often incomplete and unreliable interatomic potentials applied to > 1000 atom systems. Thus there is some desire to develop techniques that can characterize a protein's conformation and conformational flexibility, both statically and dynamically during a reaction. As will be discussed there is a strong relationship between conformation and conformational flexibility and low frequency vibrational modes of the protein. These vibrations lay in the 0.03 – 3 THz range. Until recently the number of THz spectroscopy measurements on proteins has been limited and inconclusive due to instrumental concerns. However with the advent of terahertz time domain spectroscopy (TTDS) the measurements themselves are straight forward, and the main challenge is the fabrication and control of the samples. We have performed a number of measurements on biomolecular samples using TTDS, addressing the spectroscopy and sample concerns. These measurements have thus far demonstrated that the absorbance strongly resembles the normal mode density as calculated by molecular mechanics, the absorbance is broad, and shifts with temperature in accord with shifts in the force fields rather than narrowing of overlapping lines. The response is hydration dependent with increasing absorbance and dielectric index with increasing hydration. Finally we find that the THz response is sensitive to conformational change and mutation. We would suggest that this sensitivity is not due to a redistribution of available vibrational modes, but rather a reflection of the change in net dipole strength as a function of conformation and mutation.

II. Background

A. Conformation and conformational change in biomolecules

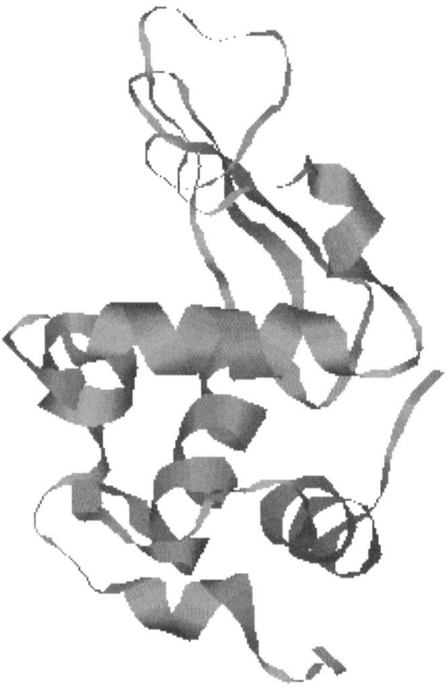

Figure 1. Hen egg white lysozyme. PDB 1BWH.[11,12] Rendered with RasMol

We will first discuss some fundamentals of protein structure. Other authors within this book discuss nucleic acids and we refer to their contributions for background concerning these molecules. All proteins consist of polypeptide chains of amino acids. The 20 naturally occurring amino acids consist of a common unit consisting of an amino group (except for proline which has an imino group) and a carboxyl group. Attached to the common unit is a side group called the residue. There are 20 unique residues constituting the 20 unique amino acids. The polypeptide chain consists of a sequence of amino acids with peptide bonds between the common units and the residues forming side chains. The sequence of residues or amino acids is called the primary structure. A given amino acid chain will fold into a three dimensional structure consisting of secondary structural units such as α helices and beta sheets and these secondary structures are then organized relative one another forming the final tertiary structure. In Figure 1 is shown the structure of hen egg white lysozyme (HEWL) from the Protein Data Bank (PDB 1BWH) as determined by X-ray crystallography.[11, 12] The diagram shows only the polypeptide backbone and does not show the residue side chains, which actually determine the tertiary structure, both through electrostatic interactions as well as steric effects. A given sequence will uniquely fold into a global minimum state, which is referred to as the native state, however current research has demonstrated that a given tertiary structure does not uniquely determine the sequence. As an example, the 7-α helix structure of bacteriorhodposin is common to many transmembrane proteins in particular the G protein coupled receptors (GPCRs).[13]

The ability of biomolecules to deform is critical to their participation in biochemical activity: the binding of the antibiotic calicheamicin to the minor groove of DNA by inducing a conformational change inhibits the binding of transcription factors to DNA;[14] transmembrane receptors, such as the aspartate receptor in E.coli, undergo conformational change upon attractant binding, thereby regulating cellular processes;[15] and the respiration process in myoglobin requires large scale movement to provide oxygen and CO access to the heme pockets.[16] Indeed, bioactive molecules are unstable to conformational changes and when minimal modifications are made to increase stability, the activity is subsequently decreased.[17] Such effects were demonstrated by mutation of functionally important residues in the active site of T4 lysozyme and then subsequent measurement of the melting temperature and chemical activity.[17] Similarly for chemically minor changes in base pair substitutions near the active site for λ phage binding showed a substantial change in the affinity. It has been suggested that this change in affinity results from conformational stiffening due to the base pair exchange.[18] Thus to understand how a protein functions one must understand its conformational dynamics: 1.) how fast does conformational change occur; 2.) What routes does energy take throughout the molecule so that such large scale motions can occur; 3.) what is the new conformation as function of environmental parameters and ligand binding? With a general understanding of these mechanisms one may better understand protein dynamics and how to design and control protein function.

B. Established structural determination techniques

A variety of techniques are employed to study biomolecular structure. The techniques tend to either probe the entire structure statically, or to indirectly probe the large-scale structure with the possibility of dynamical measurements. Traditional techniques for measuring conformational *dynamics* generally monitor local probes and thus the interpretation of the data is reliant on the fortuitous placement of probes or the engineering of probes to be located at the regions of interest.[19] Full three dimensional static structural determination can be accomplished with x-ray crystallography, NMR spectroscopy and electron beam crystallography. When the protein can be crystallized, X-ray crystallography is the preferred measurement. Changes in structure as a function of ligand binding can also be performed if the ligand can be introduced to the crystalline protein and binding can occur or if the ligated protein can be crystallized from solution. If the protein cannot be crystallized then multidimensional NMR can be readily used to determine the structure for proteins smaller than 50 kDa. For larger macromolecules the resolution requirements and spectral modeling become more difficult. Membrane proteins pose a special challenge for these techniques in that they are difficult to crystallize and their secondary and tertiary structure are dependent on being in the bilipid environment. However researchers successfully determined the structure of tubulin, for example, by creating a 2D crystalline phase in a membrane and then performing electron beam crystallography.[20] In general these techniques study conformational change by inducing the change in the molecule either in solution or in some cases directly in the crystal, reaching a new equilibrium state and then acquiring the new structure data.[21, 22]

In dynamical NMR, "tag" nuclei are followed in ~ 1 μs measurements and some qualitative information can be determined. Another dynamical NMR technique is hydrogen exchange measurements. Here the dynamics are determined indirectly by measuring the spatially resolved concentration of deuterium after a conformational change, such as folding, has occurred. For example, as the protein folds the folded domains become protected from the deuterated solution, limiting exchange. Neither the rate nor the pathway of the conformational transition is clear from these data and this method cannot be considered a true dynamical measurement.

Dynamical techniques where conformational change can be monitored in time include circular dichroism (CD) spectroscopy, IR spectroscopy (Raman and absorption) and fluorescence resolved energy transfer (FRET) spectroscopy. These methods allow synchronizing the measurement with a conformational change initiation step such as stop flow mixing with complexing molecules, photolysis of the ligand, or temperature jump of the protein. Both IR spectroscopy and FRET are indirect in that they monitor the conformational change by examining how it effects the local environment of the probe.

Circular dichroism measurements are highly effective at measuring the secondary structure content, where the inherent chirality of the structure necessarily will result in absorption spectra dependent on probe polarization. However, for many systems the important functional conformational changes occur without change to the secondary structure contents. Examples of this are the hinging movement in lysozyme, the pocket opening in myoglobin and the pore opening/closing for the potassium ion channel, KcsA.[23]

IR absorbance and Raman spectroscopy have been extremely effective in studying conformational change through measurements of the vibrational modes of the polypeptide chain and side groups.[24] The interatomic normal modes or local modes in general involving mainly the polypeptide chain backbone bonds such as the carboxyl bond, and the NH. These modes in general have not more than 5 atoms participating in the coherent motion. Clearly in a long polypeptide chain, the modes will have overlap for similar species, however distinct structures, such as α helices or β sheets do have distinct IR signatures. These measurements cannot give quantitative secondary structure content, but can be used to monitor changes in secondary structure. In addition the vibrations of some residues (e.g. tryptophan and cysteine) are distinct from the overlapping regions or regions of solvent absorption. If these residues are appropriately placed at the point of structural movement they can be used to monitor conformational change. Typical concentrations used for these measurements is high, on the order of 10-100 mg/ml.[25]

Fluorescence resolved energy transfer (FRET) is sensitive to movements in tertiary structure by measuring the distance between two chromophores and how that distance changes as a function of folding/unfolding or ligand binding. In this measurement a primary chromophore (donor), often a tryptophan residue is photo-excited. If a neighboring chromophore (acceptor) has an absorption spectrum overlapping the emission spectrum of the donor, then energy may be transferred from the donor to acceptor and thereby quench the fluorescence. The rate of this energy transfer is proportional to $1/r^6$, where r is the distance between the donor and acceptor. Thus any movement that would change r sufficiently will change the integrated fluorescence and the time decay of the fluorescence of the donor.[26, 27] This gives limited information of the actual protein movement since only the relative distance of the two residues in two equilibrium states (e.g. folded versus unfolded) is measured. To measure the dynamics of a transition requires the monitoring the relative distances of the chromophores on the time scale of the dynamic modes which are on the order of picoseconds i.e. multiple fluorescence decay measurements would be measured as the conformational change occurs. The minimum time separation for these repeated measurements would be set by the intrinsic decay time. Thus if the decay time was sufficiently short, < 1ps, one could consider using a time resolved FRET measurement to measure the interatomic beating of these motions. In general fluorescence decay times are on the order of ns, unless there is a fast energy transfer component such as in heme proteins (where decay times as fast as 30 ps have been measured).[27, 28] Thus FRET can only be used for conformational dynamics if the energetically compatible chromophores are properly placed to sense the conformational change and the fluorescence decay time is sufficiently short.

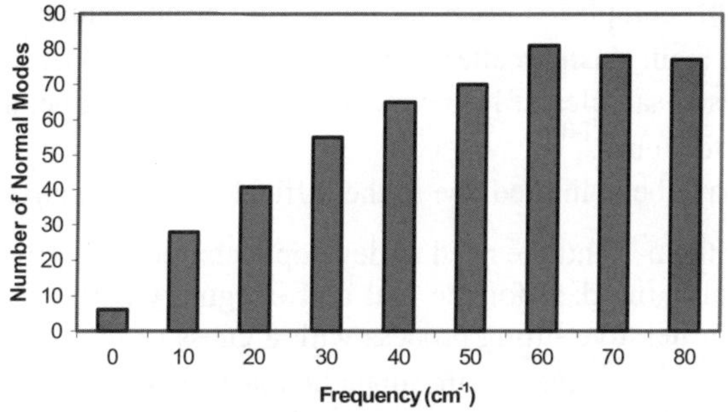

Figure 2. Normal mode density for hen egg white lysozyme (PDB 1BWH) calculated with CHARMm.

C. Collective vibrational modes

The structure of hen egg white lysozyme was shown in Figure 1. Lysozymes contribute to immune response by cleaving polysaccharide chains in the cell walls of bacteria. The active site for interaction with the sugars is in the cleft formed between the top and bottom of the protein. Upon ligand binding in the cleft, lysozyme clamps down on the ligand in a hinging motion which buries the sugar molecule in the enzyme to promote interaction and cleavage. This hinge bending motion corresponds to a low frequency vibrational mode of the macromolecule, calculated to lie at 90 GHz by normal mode molecular mechanics calculations.[29] For such macromolecules with > 1000 atoms the total number of modes ~3000 with a large density of these modes in the terahertz frequency range. In Figure 2 we show our calculation of the density of normal modes for HEWL calculated using CHARMm[30] and the atomic coordinates downloaded from the Protein Data Bank,[12] the details of these calculations will be discussed in Section III C. These modes are due to collective motion of the tertiary subunits beating against one another, or coherent movement of a portion of a subunit. The optical activity of these modes has not been determined. Such calculations of the transition dipole moments are formidable requiring a quantum mechanical treatment with a determination of the wavefunctions for the ground state and the various excited states for > 1000 atoms systems. Few ab initio quantum calculations have been attempted on proteins.[31] However due to the polar nature of most proteins and the high asymmetry of their structures, it would seem likely that many of these modes would be optically active. This is indeed what we find, a strong similarity between the measured spectra and the density of normal modes. By measuring these modes directly, one can attain information of the overall conformation of the protein. If these same measurements can be time-resolved as is the case with time domain terahertz spectroscopy, one can directly follow the dynamics of the tertiary structure.

Attempts to measure these low frequency modes by the diffuse x-ray scattering, inelastic neutron scattering, Raman scattering or FIR absorbance measurements have had limited success.[2, 32-44] While monitoring diffuse x-ray scattering gives some indication of the amplitude of intramolecular motions, it can not say which modes dominate and how the modes evolve as the protein changes state.[34] Interpretation of diffuse x-ray scattering requires the reliable determination of the collective modes by either normal mode analysis or molecular dynamics simulations. Inelastic neutron scattering measurements have been made of dehydrated noncrystalline samples of myoglobin and lysozyme with some success.[33] These measurements are not time resolved and the facilities available for such measurements are limited.

Spectroscopic measurements of these vibrational modes have been studied with Raman and traditional FTIR transmission techniques.[36-39] While Raman studies have been extensive, there is some disagreement in band identification from various groups, possibly due to the application of complex Lorentzian line fitting procedures for extracting weak contributions to the main elastic scattering peak.[37, 42] Recently however Raman measurements on single crystal samples of lysozyme and DNA show distinct and hydration dependent features down to 5 cm^{-1}.[40]

FIR transmission studies have been limited due to the difficulty in accessing this frequency range, especially below 50 cm^{-1} and the need to develop techniques to remove interference effects and to deconvolve the data for the real and imaginary parts of the dielectric function which requires an iterative fitting process with a guess of the number of oscillators to include in the fit.[39, 43, 44] Part of the impediment to performing FIR spectroscopy of biomolecules is the limited dynamic range and S/N for standard FIR sources. The development of terahertz time domain spectroscopy has now lifted that barrier to these biomolecular studies.

III. Terahertz Time Domain Spectroscopy of Biomolecular Conformation

A. Experimental setup

Terahertz time domain spectroscopy (TTDS) systems are based on ultrafast optical or near-infrared (NIR) solid state lasers to both generate and detect the terahertz light. Broadband (2-200 cm^{-1}) short pulsed ($\leq$ 1 ps) terahertz output is generated either by difference frequency mixing techniques [45, 46] or current transient generation in semi-insulating semiconductors.[47] Our TTDS systems are based on photoconductive switches fabricated on semiconductor substrates.[48] The TTDS systems used were enclosed in a dry nitrogen purged boxes to diminish THz absorption due to ambient humidity. In all cases the samples and references were mounted behind metal apertures (diameters = 4 mm) at the focus of the THz beam, (diameter < 2 mm, Rayleigh range > 4 mm) and the transmission was measured through repeated toggling between the reference and the sample to reduce effects due to laser and optics drift. All data shown here is of the *field* absorbance as defined by:

$$\text{Absorbance} = -\log |E_{sample}/E_{reference}| \quad (1)$$

where E_{sample} ($E_{reference}$) is the field transmitted through the sample (reference). We do not discuss the phase measurements here.

The main technical challenges for THz measurements are a bright source, sensitive detection, interference fringe artifacts, low loss optics, and removal of atmospheric absorptions. Our bright source and detection are provided by the TTDS system. We remove interference fringe effects by a.) using either very thin samples or very thick samples and b.) difference spectroscopy, where we reference to the same system as a function of temperature or other environmental parameters. Our window materials are ultra low loss polyethylene and low loss infrasil quartz windows. The window material is sufficiently thick to remove any additional interference effects. We note that wedging of the windows is not viable since this would sufficiently distort the wavefront of the TTDS system to remove the integrity of the coherent field detection.

The first application of TTDS to biomolecules was made by Woolard et al. in which they studied the spectral differences between DNA from different species of fish.[2] Our previous studies have focused on whether narrow THz absorptions corresponding to individual collective vibrational modes of biomolecules could be observed, and how the THz dielectric response changes as a function of hydration, conformation and mutation. These measurements took advantage of the fact that TTDS instrumentation allows for high sensitivity and rapid turn around characterization, thus allowing us to focus on developing protocols for control of hydration and conformation of the samples. We will now review our previous measurements. The first measurements on DNA, BSA and collagen established a protocol to nearly eliminate etalon and extract reliable absorption coefficients following Beer's law as well as to determine the hydration dependence of the dielectric constants for these biomolecules. We next examine the question of the overall broad shape of the observed absorbances by performing measurements on small molecular weight proteins for which we could easily calculate the normal mode spectra. Finally we will discuss the sensitivity of TTDS absorption spectra to conformational change and mutation for bacteriorhodopsin.

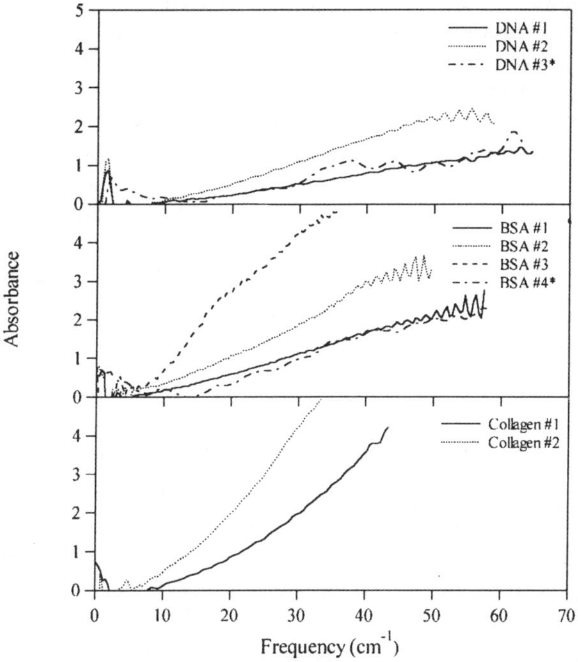

Figure 3. THz spectra of DNA, bovine serum albumin and collagen demonstrating Beer's law dependence on path length and the broad absorption for all biomolecules measured.

B. Etalon removal, and hydration dependence

In our first series of measurements, we attempted to resolve several conflicting results concerning DNA terahertz absorbance and also begin chemical and conformational specificity measurements using TTDS up to 70 cm^{-1}.[9] In these measurements care was taken to remove artifacts due to multiple reflections, etalon. Such artifacts were falsely identified as collective modes by earlier investigators measuring DNA thin films.[39, 43] In addition to these artifacts seen in earlier THz measurements the absolute absorptions varied for different groups and no scaling with concentration was demonstrated.

For these studies we used lyophilized powders from SIGMA. The DNA samples were from calf thymus DNA with average molecular weight, MW, of ~ 12 MDa (10^6 gm/mole). Bovine serum albumin (BSA) is a transporter blood protein of 550 residues with MW = 66 kDa. Its crystallographic structure consists of nine α-helices with a heart shaped tertiary conformation.[49] Type I Collagen consists of a three polypeptide superhelix. Samples had average MW = 0.36 MDa.

The lyophilized powders of the biomolecule under study were mixed with polyethylene powder and pressed into thick free standing pellets, (7 mm diameter, 7 mm long). Polyethylene is transparent at THz frequencies. The procedure insured that etalon effects were below the resolution of the measurement, 0.5 cm^{-1}. Transmissions were normalized to pure polyethylene pellets of the same size. Several concentration pellets were made for each of the biomolecules and all measurements were made at room temperature. The results are shown in Figure 3. The samples labeled with asterisks are thin pellets consisting of only the biomolecular lyophilized powder. As seen in the figure, the results for the two types of pellets are quantitatively similar, however the thin pellets have many dips and features arising from etalon effects. The measured absorption increases with optical pathlength. The results gave good agreement with Beer's law. We used the thin pellets to determine the hydration dependence of the real part of the index. See Figure 4. For DNA samples the index increased with hydration, reaching an asymptotic limit, whereas the index for BSA was nearly independent of hydration. This difference we attributed to the number of sites available for water binding on the phosphate backbone of DNA versus the lower hydration surface for the globulin BSA.

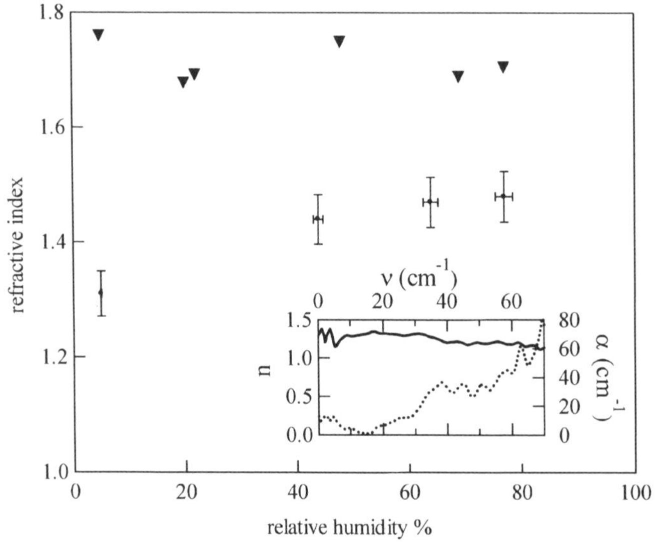

Figure 4. Hydration dependence of the real part of the refractive index at ν = 25 cm^{-1}. DNA (•) and BSA (▼) samples. Inset: the frequency dependent real part of the index, (n, solid line) and absorption coefficient (α, dashed line) for DNA #3 shown in Fig. 3.

In these measurements there is some concern as to whether the lyophilized protein samples represent the fully hydrated native state response. It would be preferable to measure samples in solution. However even for high concentration solutions of BSA and collagen we found that referencing against pure solvent was difficult since the solvent (buffered H_20) is a strong absorber and the net transmission actually increases with the high protein concentration due to solvent exclusion by the protein.

In all samples discussed the THz absorption was broad band with no prominent features. We will now discuss a systematic comparison between THz absorbance measurements and normal mode calculations. These results for demonstrate that the THz absorption strongly resembles the *full* normal mode spectrum.

C. Spectral and chemical sensitivity: comparison with normal mode density

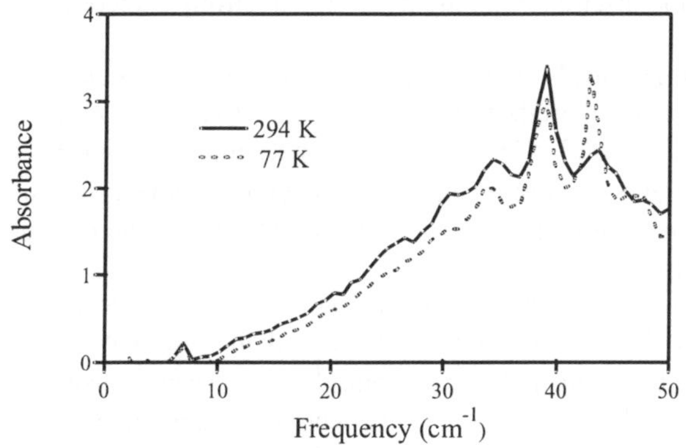

Figure 5. TTDS THz spectrum of hen egg white lysozyme. Density of normal modes shown in Fig. 2.

Among the questions raised by the previous results is why the THz absorbance is so broad. In the case of DNA and collagen, the samples were not monodisperse so that sample nonuniformity may be the cause of the broad response. To examine the question of broadening we measured myoglobin and lysozyme as a function of temperature and compared the results to normal mode calculations. Samples of lyophilized sperm whale myoglobin and HEWL were purchased from SIGMA and were used without further purification. Samples of 30 mg of lyophilized powder 200 mg of polyethylene powder were mixed and then pressed into 7 mm diameter pellets. The resultant pellets were 7 mm long. A pure 200 mg polyethylene pellet was used as reference sample. The samples were mounted in a cold finger cryostat. 4 mm metal apertures with black nonreflective paint were used to eliminate spurious reflections.

In Figure 2 we showed our calculations of the normal mode spectrum for hen egg white lysozyme (HEWL) using the software CHARMm and with structure downloaded from the Protein Data Base (1BWH).[11, 12, 29] The calculation is for a full energy minimized molecule with no cutoffs for nearest neighbor interactions. It does not include any additional waters beyond those in the downloaded structure. The figure is the full normal mode spectrum, rather than an absorption spectrum which would require calculation of the dipole transition elements for each vibrational mode. The total number of modes calculated for the 0 – 80 cm^{-1} spectrum shown is 501 with an average spacing between modes of 0.200 cm^{-1}, and no temperature dependence is included. In Figure 5 we show THz absorption measurements for a thick pressed pellet of HEWL in a matrix with polyethylene. The sample was held in an evacuated cryostat and the measurements were made with a terahertz time domain spectroscopy system with 0.5 cm^{-1} resolution. The

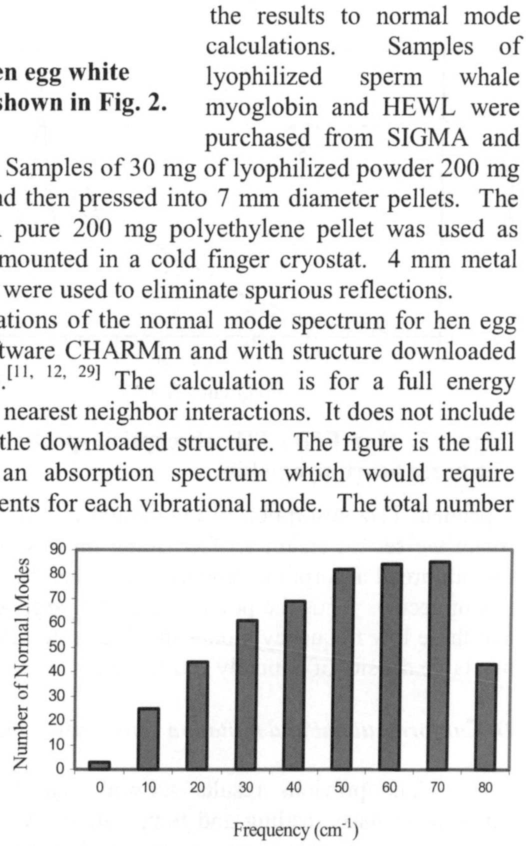

Figure 6. Normal mode density for horse heart myoglobin using CHARMm. See text.

regularly spaced peaks overlaying the smooth curve are due to a remaining interference effects in the experimental set up. The measurement is for both room temperature and 77K. The overall absorbance reduces slightly with cooling as expected. The smooth curve of the experimental data looks very similar to the calculated normal mode spectrum except that the experimental peak is at 40 cm^{-1} for the room temperature data, versus the calculated the peak is at 60 cm^{-1}. The blue shift with cooling suggests that better agreement with the calculated spectra may be attained at lower temperatures. The fact that there is such close agreement with the calculation suggests that all modes are optically active and interestingly the oscillator strength in not a strong function of frequency.

We demonstrate this same result with another important biomolecule, myoglobin. Myoglobin (Mb) consists of a single polypeptide chain linked with a heme group. It has 153 residues, 1217 atoms, 17.2 kDa molecular weight, and size 45 x 35 x 25 Å.[50]. Myoglobin is normally found in muscle cells where it serves as a reservoir of oxygen via the attached prosthetic heme group, where the O_2 or CO bind at the central iron atom of the protoporphyrin group.

As is the case for HEWL, the small size of myoglobin makes molecular modeling feasible. Both normal mode and molecular dynamic calculations have been performed for Mb and it's various liganded and deoxy states.[51] A large density of collective modes below 100 cm^{-1} have been predicted by normal mode analysis.

In Figure 6 we show our calculations for the normal mode spectrum of horse heart oxy-myoglobin (PDB 1A6M) calculated using CHARMm using no cutoffs, only waters in the PDB structure and again no temperature dependence. The total number of modes calculated for the 0.0-2.4 THz range is 496 with average spacing of 0.157 cm^{-1}. Our measured temperature dependent THz absorption of mygolobin in polyethylene matrices is shown in Figure 7. Again we see an etalon artifact riding on top of a smooth broad absorption peak. The smooth broad absorption strongly resembles the normal mode spectrum we calculate for this molecule. Thus the polar nature and large asymmetry of biomolecules and the fact that these low frequency modes involve large scale motion of the entire molecule results in a large density of optically active modes at these frequencies.

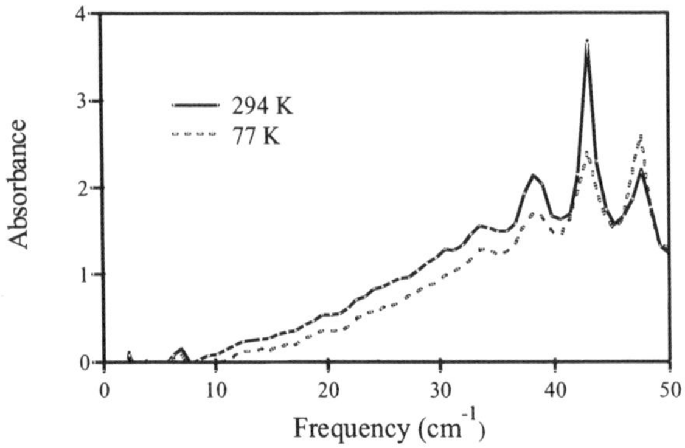

Figure 7. Our TTDS THz absorption spectrum for horse heart myoglobin.

D. Conformational and mutation sensitivity: bacteriorhodopsin

The previous results showed that the THz absorption of biomolecules is sensitive to mass loading and temperature. We now discuss the THz absorption as a function of conformation and mutation. For these studies we use the photoactive protein bacteriorhodopsin (bR) which is ideal for these measurements in that the conformation can be optically controlled, and because of the photoactivity of bR, many groups are exploring applications to electro optic devices, thus a number of mutants are available with well characterized photocycles.

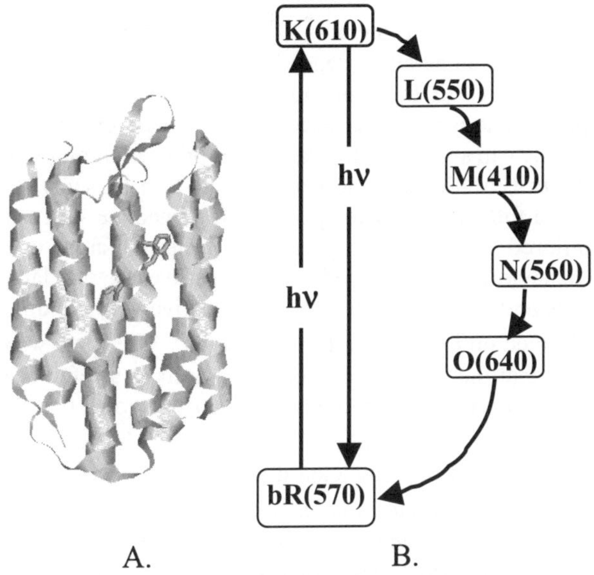

Figure 8. A.) ribbon diagram of the bacteriorhodopsin (bR). The retinal chromophore is shown using a stick representation. (PDB 1FBB) B.) The photocycle for bR.

Bacteriorhodopsin (BR) is a photoreceptor protein found in the purple membrane of the salt marsh bacterium called *halobacterium salinarum.*[52] It is a 7 α-helix transmembrane protein with a retinal chromophore attached at the 216 Lysine, which in the ground state is all trans. Upon illumination at 570 nm, a 13-cis isomerization occurs in the retinal and resultant conformational changes in the surrounding protein form the photocycle that serves to pump protons from the intracellular side of the membrane to the extracellular side as part of ATP synthesis. The photocycle consists of a series of metastable conformational states defined by their visible absorption bands. In Fig. 8 A.) we show the ribbon diagram of BR with the retinal chromophore shown in blue. In Fig. 8 B.) the different stages of the photocycle with the intermediate states labeled by a letter and in parenthesis the peak of the UV/Vis absorption defining the intermediate state is shown. As mentioned earlier bacteriorhodposin has structure and function closely related to a family of signal transduction proteins, the G protein coupled receptors (GPCRs).[53] These proteins are responsible for nearly all signal transduction such as odor, light and electrical current detection and response. GPCRs are transmembrane proteins with an exposed surface on the extracellular side and a surface coupled to a guaninine triphosphate (GTP) protein on the intracellular side. The signal transduction occurs through the GPCR receiving a stimulus either by chemical binding to their extracellular surface or by light absorption by a chromophore buried within the molecule. Once the stimulus is received the protein undergoes a large-scale structural reorganization, or conformational change sufficient to effect the interaction with the intracellular GTP causing a cascade of biochemical events within the cell.[13] Thus it is the geometry of the macromolecular system as well as its ability to access other geometries which determines the biochemical activity. Bacteriorhodopsin's close relationship to the GPCRs has made it a test bed for signal transduction research.

In addition the optical properties of BR and the ease with which the material can be produced has spurred development of all optical data storage system based on BR.[52] Another possible application of this material is an all optical switching device based on the large index change occurring in the early intermediate M state.[52] The lifetime of the

M state is critical to device applications. By mutagenisis this lifetime can extend from 10 ms to 10^3 s. The question arises what controls that lifetime? What parameters control the transition time between configurations? Earlier measurements by neutron inelastic scattering suggest that conformational flexibility or the ability to access low frequency motions is critical to the photocycling of the protein.[54-56] In these measurements, investigators found a cross over in the temperature dependence of the mean square displacement and ascribed this cross over to the transition from a low temperature harmonic regime, where the full photocycle is no longer accessible to a high temperature anharmonic regime where the molecule can switch conformation and proceed through the photocycle. We have used terahertz time domain spectroscopy (TTDS) to study the conformational vibrational modes as a function of the temperature, intermediate state and mutation to establish if the far infrared active modes spectrum reflects this cross over from harmonic to anharmonic behavior, if these modes are sensitive to the large conformational change of the M state and if the THz absorption is sensitive to mutation.

Thick films of BR were formed on infrasil quartz windows.[57] The films were confined to half the window area so that the clear portion could be used as a reference. Etalon artifacts were avoided by using 4 mm thick windows. It has been shown that the stability of the M state can be prolonged indefinitely at 233 K,[56] however for temperatures < 200 K the photocycle can not be completed due to limited access to the necessary anharmonic modes.[52] It was found that samples under either vacuum or dry N_2 purged conditions become severely dehydrated with functionally important water molecules being removed from the protein. The protein color and THz spectrum change accordingly and the samples no longer photocycle, thus humidity control in necessary to perform these measurements. The samples were cooled to 233 K at a controlled hydration by mounting the samples on a peltier temperature control plate, which was inside a humidity control cell. The humidity was controlled by exposure to a saturated salt solution (r.h. = 80%). In all cases, the samples are referenced a bare substrate. Etalon contamination is minimal.

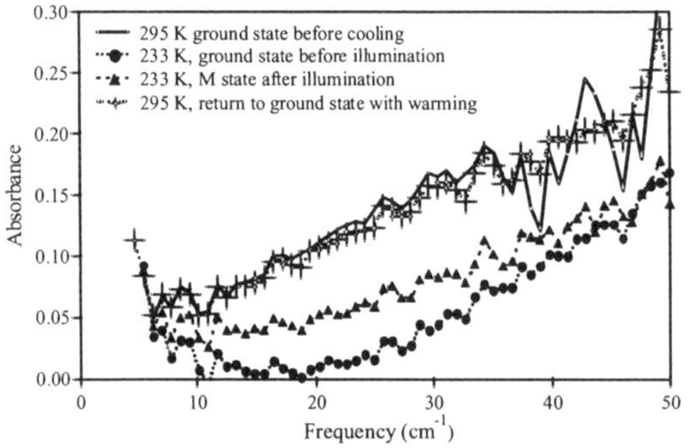

Figure 9. TTDS THz absorption for wild type bacteriorhodopsin as a function of conform-ation. Illumination initiates transfer to the M inter-mediate which is stable at 233K.

THz absorbance measurements were performed at room temperature, then the sample was cooled to 233 K in the dark. THz absorbance measurements were repeated at 233K. The sample was then illuminated with a white light source that was low pass filtered with a cut off at 640 nm for 45 minutes. The conformational switch is verified by monitoring the UV/Vis absorbance during the THz measurements. We note that these secondary measurements are critical, due to incomplete switching of the thick film samples. The data shown in Figure 9 is for a the sample which had complete switching to the M state and the color change from purple to yellow was obvious by visual inspection, however other samples due to thermal and hydration cycling only achieved switching of less than 50%, requiring verification by by UV/Vis absorption. THz absorbance was measured for the M intermediate and then the sample was heated to room temperature and final absorbance measurements were collected to establish the reproducibility of the THz response.

A representative set of THz absorbance data for one of the wild type BR samples is shown in Figure 9. All samples show the same strong FIR absorbance from 10 – 50 cm^{-1} and the relative values of the room temperature THz absorbance directly scale with the BR 568 nm optical densities measured. For all samples the net FIR absorption decreases and appears to shift to higher frequencies as the temperature is lowered. This behavior is in good agreement with temperature dependent neutron inelastic measurements.[54, 55] Upon illumination with the λ < 640 nm light the bR undergoes a conformational change to the M state. We see an increase in the absorption especially at the low frequencies. This increase in absorption with the M state change is exciting in that it is the *first observation of a conformational change through the THz vibrational mode spectrum.* We note that a net dipole moment can be assigned to the molecule and that electrical measurements on oriented films of BR show an increase in the dipole moment as the molecule goes from the ground state to the M intermediate state, in agreement with the observed increase in low frequency absorbance we observe. We hope to extend these measurements to time resolved studies of the THz response as a function photo-initiated conformational change.

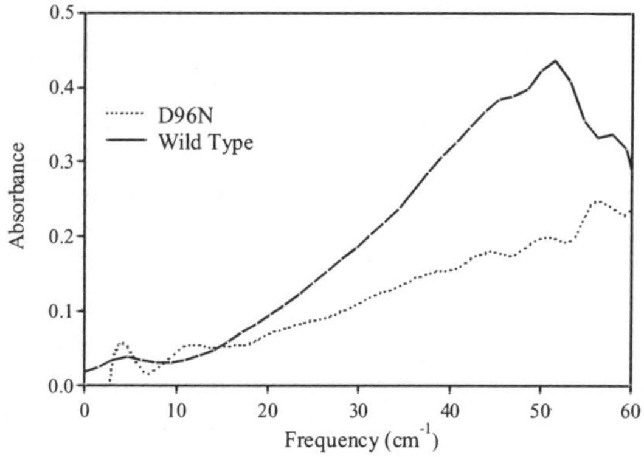

Figure 10. TTDS THz field absorbance of wild type bacteriorhodopsin and the D96N mutant.

Finally because of the technological interest in bR, many mutants for the protein have been developed and characterized enabling a determination of the THz sensitivity to mutation of proteins. We compared the TTDS THz spectrum for thin films of wild type bR and the D96N mutant. This mutant has the 96th aspartic acid residue of the 248-residue polypeptide chain changed to an asparagine residue. This single residue mutation results in a >10^3 increase in the photocycling time compared to the wild type. It has been speculated that the mutation results in a reduction of the conformational flexibility of the protein and this then results in the slowing of transitions between the different conformations of the photocycle. Our measurements of the THz absorption for a mutant film versus the wild type are shown in Figure 10. Films were prepared from wild type and D96N solutions with the thicknesses determined by the UV/Vis optical density measurements. For the samples shown in Fig. 10, both wild type and mutant films had OD = 0.6 at 570 nm, thus the comparison of the spectra is straightforward. As seen in the figure, there is a decrease in THz absorption for the mutant versus the wild type. This lower absorption agrees with a decrease in conformational flexibility as suggest by the increased photocycling time. The behavior is typical of the several mutant samples that we measured. This result is not surprising in that Thus we have demonstrated TTDS is sensitive to a single residue mutation of a biomolecule.

IV. Conclusion

We have used TTDS to examine several biomolecules and find that terahertz absorbance strongly reflects the density of low frequency vibrational modes as calculated using molecular mechanics. The normal mode distributions are similar for various biomolecules suggesting that care must be taken in calibration in order to use TTDS as a method for biomolecular identification. If the samples are not homogenous, spectroscopic determination of the different biomolecules present using TTDS may be challenging. However with regard to studies of conformational flexibility comparisons for homogeneous samples, between mutants of a single biomolecular species or conformational state, TTDS offers a non-destructive method to rapidly quantify the conformational flexibility through the FIR response and perhaps find a direct non-destructive method to evaluate cycling times and response time due to this mutatgenisis. In these studies we have demonstrated that the magnitude of the transmission for proteins strongly resembles the density of normal modes with little information concerning distinct modes. However we see a strong effect of mass loading on both the real and imaginary part of the index and see a strong effect of large scale conformational change and mutation for a single protein. It is not clear from this work that chemical identification through the FIR spectrum will be possible, however the FIR may be a probe of the condition of example an analyte for a particular protein, in that the mass loading due to binding of the target will cause a considerable shift in the FIR absorption. In addition it would appear that the FIR measurements are strongly correlated to the redistribution of charge during a conformational change. In future measurements concerning the possible THz emission due to this charge redistribution.

Acknowledgments

The authors would like to thank our collaborators: R. R. Birge, J. Hillebrecht, A, Roitberg, E.J. Heilweil. We would like to especially thank Y. Zhao for his invaluable assistance with the CHARMm calculations as well as the Center for Computational Research at the University at Buffalo for providing computing resources. We would also like to acknowledge funding through University at Buffalo Pilot Project Program and Army Research Office grant DAAD19-02-1-0271.

References

1. J.E. Boyd, A. Briskman, V.L. Colvin, et al., *Direct Observation of Terahertz Surface Modes in Nanometer-Sized Liquid Water Pools.* Phys. Rev. Lett., 2001. **87**(147401).
2. D.L. Woolard, T. Koscica, D.L. Rhodes, et al., *Millimeter wave-induced vibrational modes in DNA as a possible alternative to animal tests to probe for carcinogenic mutations.* Journal of Applied Toxicology, 1997. **17**(4): p. 243-246.
3. D.M. Mittleman, R.H. Jacobsen, and M.C. Nuss, *T-ray imaging.* Ieee Journal Of Selected Topics In Quantum Electronics, 1996. **2**(3): p. 679-692.
4. P.Y. Han, G.C. Cho, and X.-C. Zhang, *Time-domain tran-sillumination of biological tissues using THz pulses.* Opt. Lett., 2000. **25**(242-244).
5. M. Walther, B. Fischer, M. Schall, et al., *Far-infrared vibrational spectra of all-trans, 9-cis and 13-cis retinal measured by THz time-domain spectroscopy.* Chemical Physics Letters, 2000. **332**(389-395).

6. M. Brucherseifer, M. Nagel, P.H. Bolivar, et al., *Label-free probing of the binding state of DNA by time-domain terahertz sensing.* Appl. Phys. Lett., 2000. **77**(24): p. 4049-4051.
7. M. Nagel, P.H. Bolivar, M. Brucherseifer, et al., *Integrated THz technology for label-free genetic diagnostics.* Appl. Phys. Lett., 2001. **80**(1): p. 154-156.
8. C. Longbottom, D.A. Crawley, B.E. Cole, et al. *Potential uses of terahertz pulse imaging in dentistry: caries and erosion detection.* in *SPIE: BiOS 2002.* 2002. San Jose, CA: SPIE Press.
9. A.G. Markelz, A. Roitberg, and E.J. Heilweil, *"Pulsed Terahertz Spectroscopy of DNA, Bovine Serum Albumin and Collagen between 0.06 to 2.00 THz,".* Chem. Phys. Lett., 2000. **320**: p. 42 - 48.
10. O.M. Becker, Y. Levy, and O. Ravitz, *Flexibility, Conformation Spaces, and Bioactivity.* J. Phys. Chem. B, 2000. **104**: p. 2123-2135.
11. J. Dong, T.J. Boggon, N.E. Chayen, et al., *Bound-solvent structures for microgravity-, ground control-, gel- and microbatch-grown hen egg-white lysozyme crystals at 1.8 A resolution.* Acta Crystallogr D, 1999. **55**: p. 745-752.
12. H.M. Berman, J. Westbrook, Z. Feng, et al., *The Protein Data Bank.* Nucleic Acids Research, 2000. **28**: p. 235-242.
13. D.L. Nelson and M.M. Cox, *Lehninger Principles of Biochemistry.* Third ed. 2000, New York: Worth Publishers.
14. C. Liu, B.M. Smith, K. Ajito, et al., *Sequence-selective carbohydrate DNA interaction: Dimeric and monomeric forms of the calicheamicin oligosaccharide interfere with transcription factor function.* Proc. Natl. Acad. Sci., 1996. **93**: p. 940-944.
15. S.A. Chervitz and J.J. Falke, Proc. Natl. Acad. Sci., 1996. **93**: p. 2545.
16. R.H. Austin, K.W. Beeson, L. Eisenstein, et al., *Dynamics of Ligand Binding to Myoglobin.* Biochemistry, 1975. **14**(24): p. 5355-5373.
17. B.K. Shoicet, W.A. Baase, R. Kuroki, et al., *A relationship between protein stability and protein function.* Proc. Natl. Acad. Sci. U. S. A., 1995. **92**(2): p. 452.
18. M. Ciubotaru, F.V. Bright, C.M. Ingersoll, et al., *DNA-induced Conformational Changes in Bacteriophage 434 Repressor.* J. Mol. Biol., 1999. **294**: p. 859-873.
19. J.A. Glasel and M.P. Deutscher, eds. *Introduction to Biophysical Methods for Protein and Nucleic Acid Research.* First ed. 1995, Academic Press: San Diego.
20. E. Nogales, S.G. Wolf, and K.H. Downing, *Structure of the ab tubulin dimer by electron crystallography.* Nature, 1998. **391**: p. 199=203.
21. K. Edman, P. Nollert, A. Royant, et al., *High resolution X-ray structure of an early inermediate in the bacteriorhodopsin photocycle.* Nature, 1999. **401**: p. 822-826.
22. N.I. Burzlaff, P.J. Rutledge, I.J. Clifton, et al., *The reaction cycle of isopenicillin N synthase observed by x-ray diffraction.* Nature, 1999. **401**: p. 721-724.
23. B. Roux and R. MacKinnon, Science, 1999. **285**: p. 100-102.
24. B.R. Cohen and R. Hochstrasser, *Ultrafast Infrared Spectroscopy of Biomolecules*, in *Infrared Spectroscopy of Biomolecules*, M.H. H. and C. D., Editors. 1996, Wiley-Liss: New York. p. 107-130.
25. T. Miura and G.J.T. Jr., *Optical and Vibrational Spectroscopic Techniques*, in *Introduction to Biophysical Methods for Protein and Nucleic Acid Research*, J.A. Glasel and M.P. Deutscher, Editors. 1995, Academic Press: San Diego. p. 261-313.

26. M.R. Eftink, *The Use of Fluorescence Methods to Monitor Unfolding Transitions in Proteins.* Biophysical Journal, 1994. **66**: p. 482-501.
27. E. Haas, *The Problem of Protein Folding and Dynamics: Time-Resolved Dynamic Nonradiative Excitation Energy Transfer Measurements.* IEEE Journal of Selected Topics in Quantum Electronics, 1996. **2**(4): p. 1088-1106.
28. R.M. Hochstrasser and D.K. Negus, *Picosecond Fluorescence decay of tryptophans in myoglobin.* Proc. Natl. Acad. Sci. USA, 1984. **81**: p. 4399-4403.
29. C.L. Brooks, M. Karplus, and B.M. Pettitt, *Proteins: A theoretical perspective of dynamics, structure, and thermodynamics.* 1988, New York: John Wiley & Sons.
30. B.R. Brooks, R.E. Bruccoleri, B.D. Olafson, et al., *CHARMM: a program for macromolecular energy, minimization, and dynamics calculations.* J. Comput. Chem., 1983. **4**(2): p. 187-217.
31. A. Roitberg, R.B. Gerber, R. Elber, et al., *Anharmonic Wave functions of Proteins: Quantum Self-consistent Field Calculations of BPTI.* Science, 1995. **268**: p. 1319.
32. W. Doster, S. Cuszck, and W. Petry, *Dynamical transition of myoglobin revealed by inelastic neutron scattering.* Nature, 1989. **337**: p. 754-756.
33. M. Diehl, W. Doster, W. Petry, et al., *Water-coupled low-frequency modes of myoglobin and lysozyme observed by inelastic neutron scattering.* Biophys. J., 1997. **73**(5): p. 2726.
34. P. Faure, A. Micu, D. Perahia, et al., *Correlated intramolecular motions and diffuse x-ray scatttering in lysozyme.* Nature Structural Biology, 1994. **1**(2): p. 124.
35. L. Genzel, L. Santo, and S.C. Shen, *Far-Infrared Spectroscopy of Biomolecules*, in *Spectroscopy of Biological Molecules*, C. Sandory and T. Theophanides, Editors. 1984, D. Reidel: Boston. p. 609-619.
36. S.M. Lindsay, S.A. Lee, J.W. Powell, et al., *The Origin of the A to B Transition in Dna Fibers and Films.* Biopolymers, 1988. **27**: p. 1015-1043.
37. V. Lisy, P. Miskovsky, B. Brutovsky, et al., *Internal DNA Modes Below 25 cm-1: A Resonance Raman Spectroscopy Observation.* J. Biomol. Struct. Dyn., 1997. **14**(4): p. 517.
38. J.W. Powell, G.S. Edwards, L. Genzel, et al., *Investigation of far-infrared vibrational modes in polynucleotides.* Phys. Rev. A, 1987. **35**(9): p. 3929.
39. J.W. Powell, W.L. Peticolas, and L. Genzel, *Observation of the far-infrared spectrum of five oligonucleotides.* J. Mol. Struct., 1991. **247**: p. 107-118.
40. H. Urabe, Y. Sugawara, M. Ataka, et al., *Low-frequency Raman spectra of lysozyme crystals and orientedDNA films: Dynamics of crystal water.* Biophys. J., 1998. **74**(3): p. 1533.
41. T. Weidlich, S.M. Lindsay, and A. Rupprecht, *The Optical Properties of Li- and Na-DNA Films.* Biopolymers, 1987. **26**: p. 439.
42. T. Weidlich, S.M. Lindsay, Q. Rui, et al., *A Raman-Study of Low-Frequency Intrahelical Modes in A-Dna, B- Dna, and C-Dna.* J. Biomol. Struct. and Dyn., 1990. **8**(1): p. 139.
43. T. Weidlich, J.W. Powell, L. Genzel, et al., *Counterion Effects On the Far-Ir Vibrational-Spectra of Poly(Ri).Poly(Rc).* Biopolymers, 1990. **30**: p. 477-480.
44. A. Wittlin, L. Genzel, F. Kremer, et al., *Far-infrared spectrsocopy on oriented films of dry and hydrated DNA.* Phys. Rev. A, 1986. **34**(1): p. 493.

45. A. Nahata, A.S. Weling, and T.F. Heinz, *A wideband coherent terahertz spectroscopy system using optical rectification and electro-optic sampling.* Appl. Phys. Lett., 1996. **69**(14): p. 2321.
46. X.-C. Zhang, Y. Jun, K. Ware, et al., *Difference-frequency generation and sum-frequency generation near the band gap of zincblende crystals.* Appl. Phys. Lett., 1994. **64**(5): p. 622.
47. N. Katzenellenbogen and D. Grischkowsky, *An Ultra-wide Band Optoelectronic THz Beam System*, in *Ultra-Wideband, Short-Pulse Electromagnetics*, Betroni, Editor. 1992, Plenum Press: New York.
48. A.G. Markelz and E.J. Heilweil, *Temperature Dependent Terahertz Output from Semi-Insulating GaAs Photoconductive Switches.* Appl. Phys. Lett., 1998. **72**(18): p. 2229-2232.
49. D.C. Carter and J.X. Ho, *Structure of Serum Albumin.* Adv. Protein Chem., 1994. **45**: p. 153.
50. L. Stryer, *Biochemistry.* Fourth ed. 1995, New York: W. H. Freeman and Co.
51. Y. Seno and N. Go, *Deoxymyoglobin Studied by the Conformational Normal Mode Analysis.* J. Mol. Biol., 1990. **216**: p. 111-126.
52. R.R. Birge, N.B. Gillespie, E.W. Izaguirre, et al., *Biomolecular Electronics: Protein-Based Associative Processors and Volumetric Memories.* J. Phys. Chem. B, 1999. **103**: p. 10746.
53. H.E. Hamm, *The Many Faces of G Protein Signaling.* The Journal of Biological Chemistry, 1998. **273**(2): p. 669-672.
54. M. Ferrand, A.J. Dianoux, W. Petry, et al., *Thermal motions and function of bacteriorhodopsin in purple membranes: Effects of temperature and hydration studies by neutron scattering.* Proc. Natl. Acad. Sci USA, 1993. **90**: p. 9668-9676.
55. G. Zaccai, *How Soft is a Protein? A Protein Dynamics Force Constant Measured by Neutron Scattering.* Science, 2000. **288**: p. 1604-1607.
56. H.J. Sass, I.W. Schachowa, G. Rapp, et al., *The tertiary structural changes in bacteriorhodopsin occur between M states: X-ray diffraction and Fourier transform infrared spectroscopy.* EMBO J. ., 1997. **16**: p. 1484.
57. S. Whitmire, A.G. Markelz, J.R. Hillebrecht, et al. *Terahertz Time domain spectroscopy of the M intermediate state of Bacteriorhodopsin.* in *26 International Conference on Infrared and Millimeter Waves.* 2001. Toulouse, France.

International Journal of High Speed Electronics and Systems
Vol. 13, No. 4 (2003) 969–993

World Scientific
www.worldscientific.com

CHARACTERISTICS OF NANO-SCALE COMPOSITES AT THz AND IR SPECTRAL REGIONS

JOHN F. FEDERICI
Department of Physics, New Jersey Institute of Technology, Newark, NJ 07102, USA

HAIM GREBEL
Department of Electrical and Computer Engineering, New Jersey Institute of Technology, Newark, NJ 07102, USA

Electronic characteristics are difficult to monitor in nanocomposites. Here we describe indirect assessments of these characteristics using THz, Raman and IR spectroscopy. Specifically we seek to gain understanding of the electron mobility in semiconductive and conductive nanostructures for electronic, electrooptic and nonlinear optical purposes.

Keywords: Terahertz Spectroscopy, Near-Field Imaging, Nanomaterials, Carbon Nanotubes, Electronic Transport.

1. Introduction

Spectroscopy and imaging technology has progressed rapidly into the THz region of the electromagnetic spectrum during the last few years.[1] This advance is mostly due to development of the THz time-domain (or THz time-resolved) spectroscopy (THz-TDS) technique.[2,3] This method covers a wide spectral window from 0.1 THz to 40 THz, which is rich in electromagnetic phenomena. The THz-TDS system has a small power in the THz beam, but exceptional sensitivity. This combination makes the system a powerful tool for far-infrared imaging[4,5,6,7] and spectroscopy.[8,9]

Here we demonstrate the use of THz and IR spectroscopy (including Raman and Surface Enhanced Raman) in determining the linear and non-linear (non-equilibrium) electrical transport properties in nanostructures. Through the use of visible pump/ THz probe spectroscopy, the nonlinear opto-electronic properties of the nanostructures can be measured. In the longer term, these investigations of nanoscale structures could lead to all optical high-speed communication switches and novel optically controlled THz/millimeter wave devices.

The quest for new dielectric materials in the past decade has turned the attention of researchers to manmade materials, so called, "smart materials", that can be tailored to specific needs. In general, artificial dielectrics (ADs) are composite materials consisting of a dielectric matrix (host) containing clusters of another material (guest). Effective dielectric properties of the composite, which are different than those of either component, can be tailored to achieve desired characteristics. Conditional artificial dielectrics (CADs) are composites whose dielectric properties can be conditioned by light. For example, a system of semiconductor clusters embedded in a glass matrix may be activated by photons having energy above the semiconductor bandgap. In particular, optically activated dipoles in the semiconductor clusters can interact with each other leading to strong, efficient nonlinear interactions among the clusters, which could be exploited for novel device applications. Finally, since the properties of nanocomposites are strongly correlated with their size and dimension, basic research was directed to find the 'transition' between bulk properties and nanosize properties.

Artificial dielectrics were first proposed for construction of microwave lenses.[10] In this case metallic spheres, disks or wires, with dimensions of the order of millimeters, were embedded in a dielectric, such as polyfoam. The index of refraction of the composite was defined by the electric or magnetic dipoles induced in the conducting particles. This concept has been extended later by others and us to the optical frequency domain where sub micron size particles are required.[11,12,13,14,15,16] In general the embedded obstacles have to be smaller than the electromagnetic wavelength of interest. When the particles are made of a semiconductor they may behave as conductor obstacles or semi-insulator obstacles, depending on the presence of photo-induced carriers. Thus, the dielectric properties of such Conditional Artificial Dielectrics (CAD) can be controlled by external illumination. Considering the linear properties of conductive embedded clusters, most information about the ADs can be retrieved from the permittivity function. The absorption properties of Artificial Dielectric Materials depend on the cluster's size and conductivity.[11] In general, the presence of photo-carriers increases the refractive index of the host medium.[10] This determines the dielectric properties at other frequency regions through Kramers-Kronig's relation. At the same time, the conduction energy bands are no longer continuous. The separation of these bands into discrete states is an indication of quantum confinement. The states' separation is in the THz frequency region for clusters sizes on the order of tens of nanometers. Thus by probing the sample with a THz source one may study the underlying physical processes of carrier transport within these structures for potential device design and applications. In studying the fundamental carrier transport properties of clusters, there has been little experimental data, which directly measure the ultrafast (picosecond and shorter time scale) carrier dynamics. The mobility, inter-valley scattering, carrier relaxation times, and mean-free path are important parameters for assessing the electronic and optical properties and potential device applications.

Nanocomposites can be made by many methods. Here we describe three major techniques. These are: laser ablation, ion implantation and Chemical Vapor Deposition (CVD). While these growth techniques result in different film morphologies the common factor is that they are all semiconductor materials. Much knowledge has been accumulated on the non-linear properties of both laser ablated or ion implanted Si films.[17,18,19,20,21,22] Their non-linear values are large $\mathrm{Re}\{\chi^{(3)}\}\sim 10^{-3}$ esu for laser ablation and $\mathrm{Re}\{\chi^{(3)}\}\sim 10^{-8}$ esu for ion implanted films compared to other semiconductor dots such as, CdS or CdS_xSe_{1-x} ($\mathrm{Re}\{\chi^{(3)}\}\sim 10^{-9}$-$10^{-10}$ esu).[23,24] Even their time constants seem unusually short (about 5 nsec limited by the laser pulse duration) when considering their large non-linear values. On the other hand, the origin of the non-linear behavior is not well understood. One can make the case that it may arise from electronic excitations[18] and thus, may be very suitable, as an example, for optical switching purposes. Therefore, studying the electronic mobility within the clusters may clarify the role of electronic contribution to the non-linear characteristics of nano-size clusters. The change of the electronic mobility in aggregates of clusters versus well-separated ones will clarify the role of the cluster-cluster interactions. The insight gained from studying the transport and control of charge carriers in quantum confined nanocluster semiconductors could then be exploited for novel devices based on the underlying physical principles developed.

2. THz spectroscopy

In this section, a few configurations of far- and near-field THz spectroscopy are described. Emphasis is placed on carrier mobility measurements using this non-invasive, all-optical technique. A major advantage of the all-optical pulsed Terahertz technique is that it can be applied to as-grown material without any need for electrical or other physical contacts to the material. Furthermore, patterning of the material is not required as with other ultrafast electrical measurements. These considerations are particularly relevant when studying clustered material for which electrical properties of the host material and the quality of the electrical contacts is critical to reliable electrical measurements. Finally, we comment on the relationship using these measurements and possible nonlinear optical properites at other wavelenths, especially in the near-infrared spectral region.

2.1. *Far-field THz spectroscopy*

Pulsed Terahertz spectroscopy has been used to measure the linear far-infrared transmission of materials[25,26,28,29] as well as the nonlinear transmission.[8,9,30,31] For our purposes, we define "linear" to denote transmission or reflection measurements for which the sample under test is not subject to photoexcitation. In addition, the THz beam is of low enough intensity such that it does not perturb the sample's properties. Typically for nonlinear measurements, the change in reflection or transmission of a THz probe beam is measured subsequent to photoexcitation of the sample.

For analyzing linear THz transmission through semiconducting wafers, generally it is assumed that the complex index of refraction of semiconductor can be approximated using the Drude model:[27]

$$\tilde{n}^2 = (n_r + i\, n_i)^2 = \varepsilon(\omega) = \varepsilon_\infty - \frac{\omega_p^2}{\omega(\omega + i\nu_c)} \tag{1}$$

where $\tilde{n}$ is the complex index of refraction, ε_∞ is the contribution of the dielectric at infinite frequency, and ν_c is the average scattering frequency. The plasma frequency is defined as $\omega_p^2 = Ne^2 / \varepsilon_o m$ where N is the number density of charge carriers, e is the electronic charge of the carriers, m is the effective mass of the carriers, and ε_o is the permittivity of free space. Within the Drude model, the dc conductivity σ_{dc} can be determined by $\sigma_{dc} = e\mu N$ where $\mu = e / m\nu_c$ is the carrier mobility.

In analyzing the THz transmission data for a homogenous sample of thickness L, the intensity absorption coefficient can be related to the imaginary index of refraction as a function of frequency via[37]

$$\ln(I/I_o) = -\alpha L = -(4\pi\, n_i / \lambda_o)L \tag{2}$$

where λ_o is the vacuum wavelength of the THz. The data for a single layer can be easily compensated for reflections from the front and back interface.[26] The result can be extended to multiple layers with corresponding reflections from the interfaces between each layer.[26,28]

In nonlinear transmission measurements, the sample to be studied is typically photoexcited with visible or near visible light while the change in THz transmission is measured. When the THz pulse and pumping visible pulse are short pulses (<1ps), it is then possible to investigate time-resolved ultrafast carrier dynamics in materials. The visible pump/ Terahertz probe spectroscopic technique was first used to investigate ultrafast carrier dynamics in InP and GaAs semiconductors[9] and subpicosecond quasiparticle dynamics in superconductors[8] using nonlinear transmission measurements. In both Refs 8 and 9, the change in THz transmission is measured subsequent to visible excitation of the samples. More recently, the same technique has been applied to studying subpicosecond carrier dynamics in low-temperature grown GaAs,[29] ultrafast carrier trapping in microcrystalline silicon,[30] and a more detailed study of transient photoconductivity in GaAs.[31] One advantage of the visible pump/ THz probe technique is that carriers can be photoexcited to different levels in an energy band which allows one, for example, to selectively occupy different mobility valleys in GaAs.[9,31]

In analyzing the data of photoexited samples compared to linear THz transmission, several factors need to be considered. Typically, a layered model is appropriate for which the photoexcited layer is treated as a separate layer from the bulk substrate. The thickness of this layer is approximately that of the penetration depth of the visible light: ie. the region over which visible light is absorbed and electron-hole pairs are created. For semiconductors such as GaAs and silicon, this photoexcited layer is typically thin enough that thin film transmission formula apply.[9,31] Unlike linear THz transmission measurements on doped semiconductor wafers[26,28,29] for which the charge carriers are predominately electrons (n-doped) or holes (p-doped), nonlinear photoexcitation measurements create equal numbers of electrons and holes. While in general the transport of both charge carriers need to be considered, it may be possible to ignore the role of holes, for example, based on differences in effective mass.[9] Other potential complications include diffusion of carriers such that the photogenerated carrier density and spatial extent of the photogenerated carriers is time-dependant.[30,31]

The 'standard' visible pump/ Terahertz probe spectroscopic technique for measuring ultrafast carrier transport utilizes a subpicosecond visible light pulse (photon energy above semiconductor bandgap) to photoexcite carriers while a subpicosecond, broadband far-infrared pulse probes the photoexcited carriers' transient *electrical* response to the far-infrared electric field in the THz range. As an example, our typical experimental setup is shown in Fig. 1 and Fig. 2. The THz light is be generated using a mode-locked Ti:Sapphire laser and a biased photoconductive (PC) switch.[9] The THz is detected using another photoconductive antenna. Alternatively, the THz could be detected electro-optically using a ZnTe crystal or interferometrically[32] using a Michelson interferometer and liquid He cooled bolometer. While antenna detection is more sensitive, electro optic detection offers a larger bandwidth above 3THz.[33] The instantaneous THz electric field is determined by chopping the applied bias to the transmitting antenna structure and using a lock-in amplifier and phase sensitive detection to measure the resulting current at the receiving antenna. The THz electric field is detected as it interacts with the receiving antenna and polarizes the charge carriers across the photoconductive gap. The gating optical pulse causes the gap to conduct and the resulting current surge is detected. The THz field is mapped out in time by scanning a delay line. A second translation stage is used to vary the arrival time of the pumping optical pulse relative to the arrival of the probing THz pulse on the sample. A second-harmonic generator (SHG) may be necessary to convert some of the Ti:Sapphire laser power at 800nm to blue light to pump the nanomaterials. The nonlinear THz transmission consists of measuring the THz waveform at different time delays of the visible pump beam. From the change in the sample's far-infrared transmission subsequent to photoexcitation, carrier dynamics such as inter-valley scattering, mobility, carrier relaxation times, and mean-free-path can be deduced.

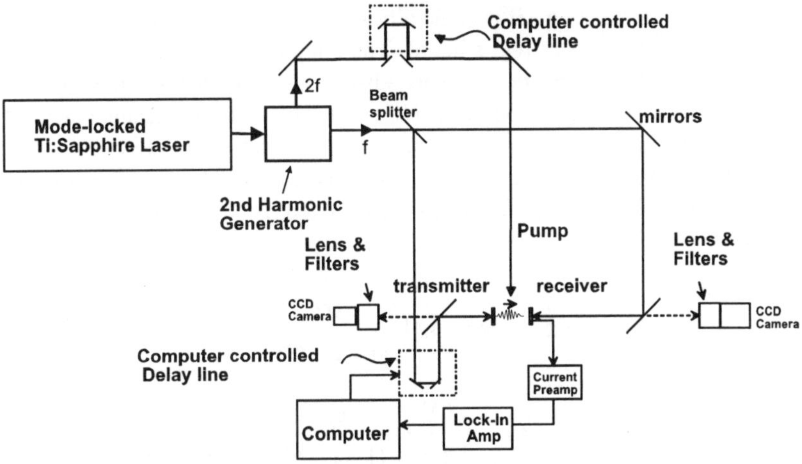

Fig. 1. Visible Pump/ THz Probe Spectroscopic Configuration. A second harmonic generation module is required to generate blue light for photoexcitation of nanocluster films.

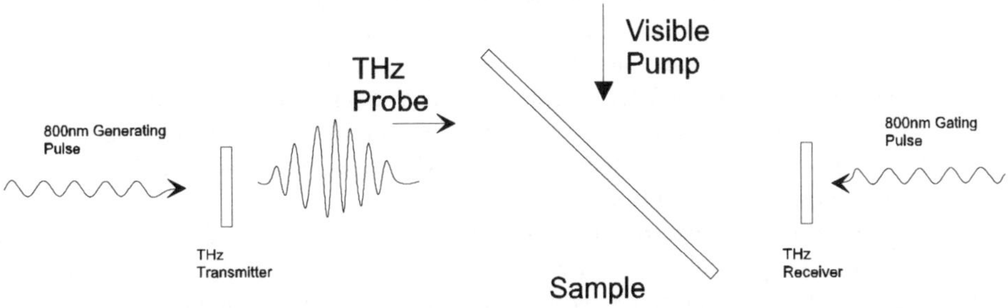

Fig. 2. Sample configuration in standard visible pump/ THz probe configuration. The change in THz transmission is measured as a function of visible excitation pump power and time delay between pump and probe.

In analyzing the THz transmission data of the nanostructure sample, the propagation of an electromagnetic wave is affected by the electric and magnetic dipoles induced in the sample. For the case of an artificial dielectric with clusters embedded in an insulating matrix, dipoles are induced in the clusters. If the size of particles (e.g.. clusters) embedded in the dielectric is much smaller than the optical wavelength, their effect may be described qualitatively by a quasi-static (Lorentz) theory. The electric field, which interacts with the particles, is composed of two components: the external electromagnetic electric field, $\mathbf{E_0}$, and the interaction field, $\mathbf{E_i}$. The interaction field is due to the presence of nearby clusters. Using the effective dielectric constant, $\varepsilon_{eff,}$ and the electrical polarizability, α_e, of each individual cluster, we may write,[34]

$$\varepsilon_{eff} = \varepsilon \cdot [1 + 3\alpha_e N/(1-\alpha_e C)]. \tag{3}$$

Here ε is the permittivity of the dielectric matrix, N is the density of the electrical dipoles formed by the clusters and $\alpha_e=3\varepsilon(\varepsilon_s-\varepsilon)(\varepsilon_s+2\varepsilon)(4\pi a^3/3)$. Here we defined ε_s as the dielectric constant of the scattering sphere of radius 'a'. The electric dipoles can be intrinsic to the cluster (e.g., if it is metallic) or photoinduced by the absorption of visible light by a dielectric cluster (eg. silicon). The coupling constant, C, relates the interaction field (field due to all the neighboring obstacles) $\mathbf{E_i}$ to the polarization, P, such that, $\mathbf{E_i}=C\mathbf{P}/\varepsilon$. In the Lorentz model, $C=N/3\varepsilon$ for an isotropic arrangement of scatterers. However, this model does not account for the scattering effect of an ensemble of scatterers and thus we find that if α_e is real, so is ε_{eff}. In order to account for the scattering effect (the existence of an imaginary part even in the absence of absorption), we approximate,[35]

$$\begin{aligned}\varepsilon_{eff} = \varepsilon_r + i\varepsilon_i = \varepsilon\{[\frac{(1+2\alpha_e C)}{(1-\alpha_e C)}] \\ + ik^3(\frac{3N}{4\pi})|\frac{\alpha_e C}{(1-\alpha_e C)}|^2[\frac{(1-f)^4}{(1-2f)^2}]\}\end{aligned} \tag{4}$$

Here we defined $f=N4\pi a^3/3$ as the volume fraction of the sphere's ensemble. If $f\to 0$ $\varepsilon_{eff}\to\varepsilon$. If $f\to 1$, $\varepsilon_{eff}\to\varepsilon_s$. Therefore, on measuring the real and imaginary parts of the dielectric constant of the nanoclustered media the coupling between clusters can be assessed. This may be obtained for either materials at equilibrium or at non-equilibrium states (namely, under intense optical pumping) permitting direct measurement of the dielectric characteristics of nonlinear materials. Moreover, saturation of the dielectric constant may be easily deduced from Eq. (4).

In the case of cluster absorption, electron-hole pairs are generated (e.g., Conditional Artificial Dielectric material). At saturation, the system may be viewed as N interacting metallic clusters. In this saturation condition, the polarizability α_e of the individual cluster is effectively that of a metal due to copious quantities of photogenerated electron-hole pairs. Thus, if the transition is above an excitonic resonance, $\alpha_e^{(metal)}=4\pi l^2$ is the upper limit on the polarizability of the clusters with, l, being the mean free path within the cluster. In the absence of carrier tunneling, the mean free path of the carriers will determine the optimal cluster size for non-linear applications.

Is the large optical non-linearity in semiconductor clusters due to cluster-cluster coupling (eg. the constant C in Eq. (3)? Using pump/probe THz spectroscopy, one can probe cluster-cluster coupling. For this test, the respective pump and probe beam sizes will be kept fixed while measuring samples with different cluster densities (but same cluster size). If no coupling is observed, the ultrafast spectroscopic signatures will vary proportional to the density of clusters. If cluster-cluster interactions are important, the dependence on cluster density should be nonlinear. As the clusters get closer together, the propensity for interaction should increase. In metals, the average mean free path is approximately 10 nm. Strong coupling effects start at distances of 20 nm between clusters. If the mobility is wavelength dependent, it is indicative of quantum size effects resulting from transitions between sub-bands in the nanoclusters.

The carrier mobility and lifetime can be extracted from the THz transmission data by slightly modifying the data analysis for homogeneous multilayer samples. The nanocluster material is modeled as shown in Fig. 3 as a thin film surrounded by air on one side and the substrate material on the other. The transmission of light through this configuration has been solved in textbooks.[36,26,28] Given the index of the refraction of air and the substrate and the thickness of the thin film under study, the phase and amplitude of the transmitted electromagnetic wave depends on the real and imaginary indices of refraction of the thin film. The transport properties are determined by relating the complex conductivity of the thin film to its complex index of refraction.

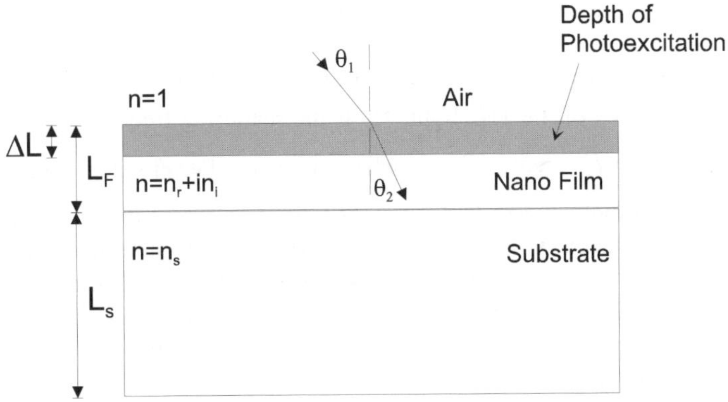

Fig. 3. Schematic of optical analysis of nanocluster film on insulating substrate. The shaded area represents the volume of the nanocluster film that is photoexcited. In the case of photoexcitation, the optically illuminated and unilluminated regions of the film each have their own complex index of refraction. In analyzing the data, reflections of THz from the various interfaces must be taken into account.

The real and imaginary indices of refraction can be related to the complex conductivity (cgs units) by

$$\tilde{n}^2 = (n_r + in_i)^2 = \varepsilon_r + i\varepsilon_i = \varepsilon \tag{5}$$

$$n_r^2 - n_i^2 = \varepsilon_\infty - \frac{4\pi\sigma_i(\omega)}{\omega} = \varepsilon_r \tag{6}$$

$$n_r n_i = \frac{2\pi\sigma_r(\omega)}{\omega} = \frac{\varepsilon_i}{2} \tag{7}$$

Here n_r and n_i are the real and imaginary indices of refraction and ε_∞ is the real part of the dielectric constant due to bound charges. σ_r and σ_i are the real and imaginary components of the electrical conductivity.[37] For our applications, ω is the frequency of the THz wave. In practice, one must make some assumptions concerning the mathematical form of the electrical conductivity in order to infer the transport properties such as DC mobility from the THz transmission.

As with the analysis of homogeneous layers, if a Drude-type free carrier absorption is assumed, then the complex dielectric constant $\varepsilon = \varepsilon_r + i\varepsilon_i$ is given by the expression $\varepsilon = \varepsilon_\infty + 4\pi\sigma\, i/\omega$ where the frequency dependant electrical conductivity $\sigma(\omega)$ is given by

$$\sigma(\omega) = \frac{Ne^2}{m(v_c - i\omega)} = \frac{\sigma_{dc}}{1 - i\omega/v_c} \tag{8}$$

where N is the number density of free electrons, m is the effective mass of the electrons, ω is the angular frequency of the THz light, v_c is the scattering frequency. v_c is related to the mean-free-path by $l_{mfp} \approx v_F / 2\pi v_c$ where v_F is the Fermi velocity of the electrons and σ_{dc} is the conductivity at zero frequency. For the simple case that the holes contribute negligibly to the conductivity (as is the case for GaAs), the DC electron mobility is then given by $\mu=\sigma_{dc}/eN=e/mv_c$. For linear transmission measurements, the number density of free electrons N is determined by the doping while for the visible pump/ THz probe measurements, the number density is determined by the number of photogenerated carriers and the intrinsic free carriers.

However, nanocluster films are not uniform, homogeneous films. In this view, applying the previous analysis to nanocluster films is inferring the *effective conductivity* of the nanocluster thin film. The effective conductivity may be derived by averaging over all the quantum states[38] of the cluster. Then, by using the complex dielectric constant form, Eqs. (4) and (5), the measured phase and amplitude of the transmitted THz wave, the coupling constant and electron damping frequency v_c of the metallic or semiconductive clusters can be determined. In the latter case, the N_e of Eqs. (6), (7), and (8) depends on the optical intensity of the pumping visible light and the absorption cross section of the cluster. The relation between N_e and the intensity are easily incorporated through microscopic models[39,40] or phenomenologically, through a three-level rate equations.

In an alternative configuration to determine time-resolved carrier transport properties, Hu et. al.,[41] used a 10fs pulsed Ti:Sapphire laser to resolve the dynamics of photogenerated carriers in the first 10-1000fs after generation. In this technique, the generation of THz radiation is used to infer the carrier transport. The output of the 10fs Ti:Sapphire laser is passed through a Michelson interferometer to generate two pumping pulses with a variable time delay between them. The pulses are focused onto either biased or unbiased semiconductors. The first pulse generates photo-carriers as well as THz radiation. Subsequent to illumination, the photogenerated carriers move in the presence of a DC electric field leading to a polarization of carriers and partial screening of the DC field. (The DC field may be either an externally applied field or an intrinsic built-in surface field.) The second ultrafast pulse generates additional THz radiation. However, the intensity is dependent on both the DC field and the polarization field of the photoinjected carriers. The magnitude of the polarization field as a function of time is determined by the transport properties of the carriers. A schematic of the experiment is shown in Fig. 4. Using Terahertz generation as a diagnostic of carrier transport, Hu et. al identified the four distinct phases of carrier transport in GaAs and Si with 10 fs resolution. In the first stage, the photogenerated electron-hole pair instantaneously polarizes when the pair is created. Over the next ~70fs, the carriers are accelerated ballistically. In the third stage, scattering processes dominate during which the carriers attain a maximum drift velocity. For materials such as GaAs, hot electrons are scattered into satellite valleys with lower mobility leading to velocity overshoot. Lastly, the carrier velocity equilibrates at a lower level in the steady state (~300fs).

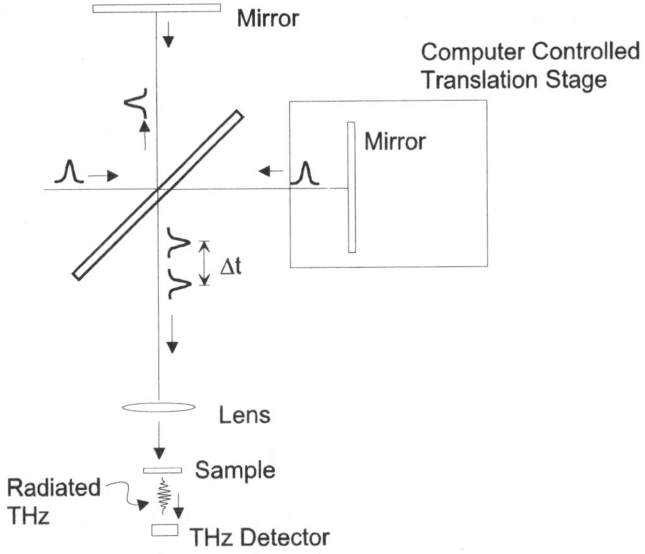

Fig. 4. Schematic of carrier transport measurements using THz generation from sample under study.

In Hu et. al's[41] configuration (Fig. 4), the change in THz energy is related to the motion of photogenerated carriers by

$$\frac{\Delta W(t)}{W} = \frac{2\Delta E_{dc}(t)}{E_{dc}} = -\frac{2}{E_{dc}\varepsilon}\left[Nex(t)\right] \tag{9}$$

Here $\Delta W(t)$ is the change of THz energy from the second pulse at time t after the excitation pulse, W is the total THz pulse energy, E_{dc} is the unscreened bias field, ΔE_{dc} is the change is the bias electric field due to screening by the photogenerated carriers, N is the number density of photogenerated carriers, e is the electron charge, x(t) is the average displacement between electrons and holes at time t after excitation, and ε is the dielectric constant of the material. The above equation, which is valid for t>40fs, neglects the instantaneous polarization term. From the measured change in THz energy ΔW(t), the displacement of the electron-hole pairs as a function of time can inferred. Previous measurements by Hu et. al show that the measured displacement versus time for silicon is linear with time indicating a constant drift velocity as one would expect for steady state transport. The mobility μ is then calculated from $v_d = \mu E_{dc}$ where v_d is the experimentally measured drift velocity (slope of displacement versus time data) and E_{dc} is the applied DC electric field.

By varying the density of clusters (but keeping their size constant), the dependence of cluster-cluster coupling on the carrier transport properties can be investigated. Assuming that cluster-cluster coupling does not change with carrier density, than the change in THz generation (Eq. 9) should only depend on the density of photogenerated charges since the transport properties represented by x(t) remain the same. In this situation, the density of photogenerated charges should be proportional to the density of semiconductor clusters assuming a transparent host material. This implies that the change in THz generation should vary linearly with the density of nanoclusters. However, if cluster-cluster coupling is important, than the change in THz generation [Eq. (9)] should exhibit nonlinear behavior with the density of nanoclusters. Therefore, based on this simple theoretical prediction, it should be experimentally possible to determine if cluster-cluster coupling influences the charge carrier dynamics in nanocluster materials.

A final consideration in applying either linear or nonlinear THz techniques to nanostructures is the calibration of pump/probe and THz techniques such that meaningful values for mobility, mean-free path, trapping rates, etc. can be extracted from the data. In order to provide quantitative values for the carrier transport properties, typically one can use bulk samples of GaAs and Si with known transport properties.

2.2. *Near-field THz spectroscopy*

In this section, we summarize our development of an imaging method that provides a very high spatial resolution and has all the advantages of the THz-TDS technique.[47] Furthermore, we discuss its applications to THz measurements of nanostructures.

The major limitation of THz imaging is poor spatial resolution due to the long THz wavelength. The resolution can be significantly improved by implementing the concept of near-field scanning optical microscopy. Various methods based on this approach have been demonstrated, pushing the resolution limit to a few tens of microns.[42,43,44,45,46,47,49] Among them is a dynamic aperture approach that potentially can improve resolution to a few microns.[46,48] However, application of this method is limited to semiconductor surfaces and images are related to the concentration of photogenerated carriers. In alternative approach, a micromachined near-field probe was fabricated.[49] This device, with a spatial resolution of a few tens of microns, is capable of mapping the propagation of THz pulses on coplanar transmission lines.

The resolution capabilities of our THz near-field method[47] lie in the range of a few microns, which is considerably smaller than the wavelengths of the employed THz radiation (250 μm -1500 μm). Furthermore, the resolution is independent of the wavelength. The combination of the near field microscopy concept with the THz-TDS technique allows for studying the temporal evolution of the electromagnetic field in the near field of objects. The broadband coherent THz source potentially provides the possibility of spectroscopy on a micrometer scale.

High spatial resolution in imaging can be achieved if the evanescent components of the field scattered by the object are detected. The evanescent field exists only at the object and decays very fast with increasing distance from it. Detection of the evanescent field is possible by introducing an aperture-type probe into the near field region of the object. Fields in front of the aperture determine waves that couple into the probe. These waves carry information about the point of the object, where the probe is placed. By scanning the object in front of the probe one constructs a near field image. The spatial resolution of this method is defined by the aperture size and is not limited by diffraction.

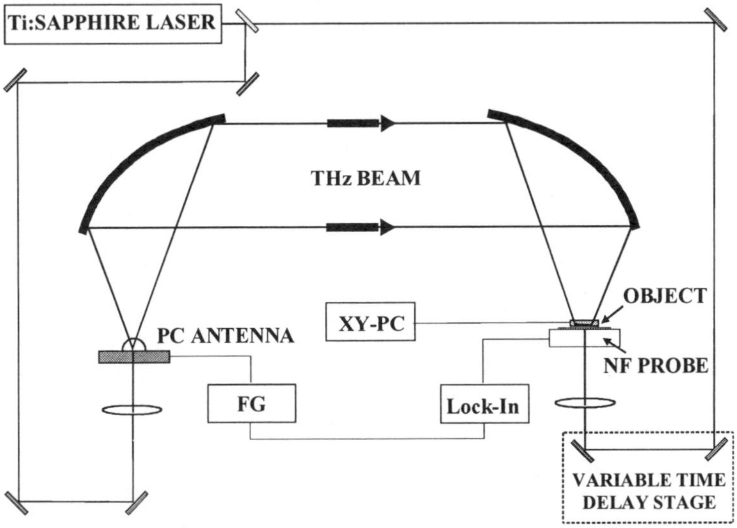

Fig. 5. Schematic diagram of the THz near field imaging setup (XY-PC denotes *xy*-position control equipment for scanning and FG - a function generator, which applies the alternating bias to the PC antenna).

The near field probe is an essential element of the system. The probe makes use of an efficient design that allows the detection of the electric field coupled through an aperture as small as $\lambda/300$. The THz near field imaging setup is presented in Fig. 5. THz pulses are generated by the transient current in a photoconducting (PC) switch excited by optical pulses from a mode-locked Ti-sapphire laser (λ_c=800 nm, τ_{FWHM}=150 fs). The repetition rate of the laser system is 100 MHz. The THz beam is focused on the object through a transparent substrate by means of two off-axis parabolic mirrors. The beam waist in the object plane is ~2 mm (FWHM), which is usually much larger than the object, therefore the illumination can be considered uniform. The near-field probe is located behind the sample almost in contact with the object. The probe consists of a small aperture in a metallic screen and a PC antenna that detects THz pulses. Generation of the THz pulses is slowly modulated applying a square wave alternating bias to the emitting PC switch. The detecting antenna is gated by optical pulses from the same laser. Current induced in the antenna is proportional to the THz field and is measured using a lock-in amplifier.

An automated *xy*-translation stage scans an object perpendicular to the optical axis. A variable time delay stage allows for time domain sampling of the THz pulse. The image is constructed using the THz signal collected either at a fixed time delay or in the time domain for every position.

A schematic diagram of the near-field probe is presented in Fig. 6. An entrance subwavelength aperture of size d (5 μm - 50 μm) is lithographically defined on a surface of the probe in a 600 nm gold film evaporated on a thinned GaAs layer. A GaAs protrusion through the aperture enhances field coupling into the probe. The PC planar antenna is embedded between a thin layer of GaAs (3-10 μm, n~3.6) and a sapphire substrate (n~3.1). Note that the space behind the aperture is filled with a high refractive index material that reduces the effective wavelength. The antenna is fabricated on a 1μm thick low temperature grown GaAs epilayer. Details of the probe fabrication are described elsewhere.[50] The sapphire substrate supports the structure and allows the optical gating pulses access the antenna from the substrate side.

Most of the incident THz power is reflected from the metallic screen, and transmission through a subwavelength aperture is extremely small.[51] The electric field that exists behind the illuminated subwavelength aperture can be divided into modes with real and imaginary longitudinal k-vectors.[52] The latter are usually referred to as evanescent modes. Electric field amplitude of the evanescent modes is significantly larger than that of the propagating modes at distances from the aperture $z<d/2$.[53] At a distance approximately equal to the aperture size their contribution is comparable. As distance z increases the amplitude of both mode types decreases, but decay is much more rapid in the case of the evanescent modes, which do not transfer energy into the far field region. Only the modes with real wavenumbers can propagate to distances $z>>\lambda$. Therefore evanescent modes are not detected in conventional collection mode near field microscopy. An important feature of our probe design is that the electric field that couples through the aperture is detected inside the probe in the near field zone of the aperture ($z<d/2$). Therefore, not only propagating but also evanescent modes of the radiation transmitted through the aperture contribute to the signal. Detection of the evanescent modes of the aperture results in a higher sensitivity of the near-field probe.[50]

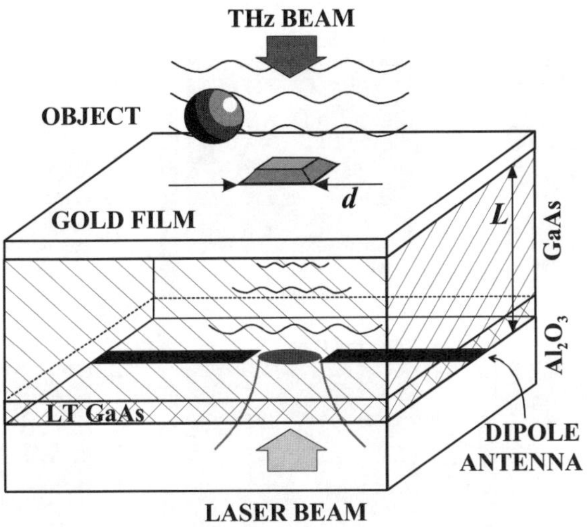

Fig. 6. Schematic diagram of the near field probe.

The spatial resolution of the near-field probe is defined by the aperture size. To demonstrate it, we performed an edge resolution test on the probes with different aperture sizes. Boundary conditions for the electric field at a metallic edge are different for the two principal polarizations (parallel and perpendicular to the edge). If the edge is oriented parallel to the polarization of the incident THz pulse, then the electric field in the plane of the object exhibits a sharp contrast between the metallic and the open areas.[54] These tests reveal a 7μm spatial resolution for a 5 μm aperture probe (L=4 μm) when the edge is scanned over the probe at a distance h~2 μm. The resolution test on the probes with larger apertures showed that spatial resolution scales with the aperture size and is independent of wavelength.[47]

Electric fields with high spatial frequency only exist in the proximity of the object (evanescent fields) and decay over distances comparable to the size of object features. In order to detect these fields the near-field probe must be placed very close to the object. The fast decay of the high spatial frequency fields is observed when performing an edge test for various separations between the probe and the object h. The sharp edge profile smears as h increases. In practical THz near-field imaging, the probe-sample separation is less than several microns. Waveform distortion due to interference is negligible at this range, however the variation of the amplitude of the detected THz field can create an uneven background in the image, if the separation is not maintained constant during the scan.

It should be emphasized that the near-field image of an object is not a direct replica of the instantaneous electric field scattered by the object. The probe aperture alters the detected waveform. In principle the original waveform can be extracted if the transfer function of the aperture is known. Analysis of THz near-field image formation appears as the essential task in order to apply this technique. In this respect we would like to mention the finite-difference time-domain numerical method[50,55,56] which can be used to simulate THz near-field images.

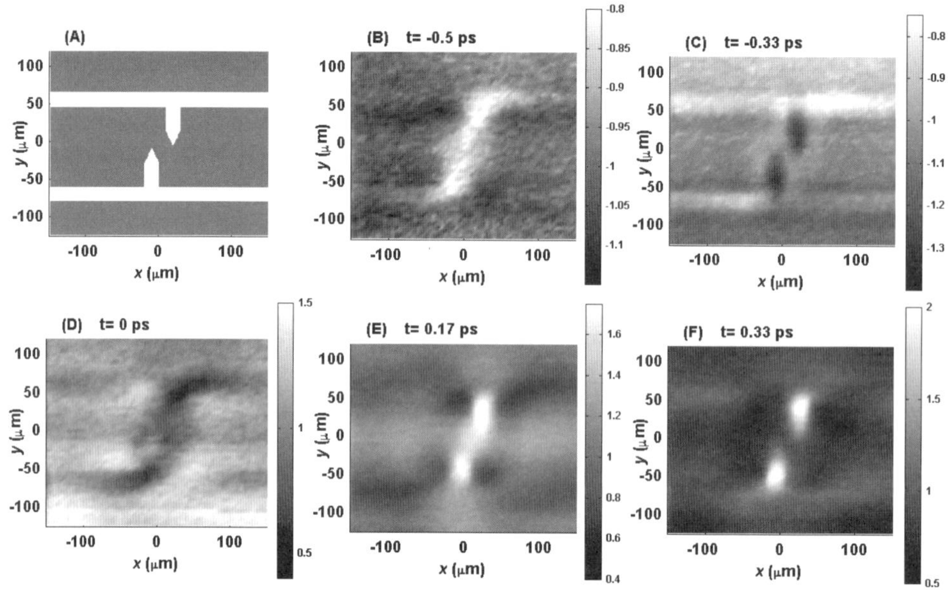

Fig. 7. (a) Schematic diagram of a planar antenna on sapphire. (b-f) Series of near field images taken at different time delays of the gating optical pulse. The gray level corresponds to the measured electric field.

The THz-TDS technique allows studying objects in time domain. These images provide a wealth of spatial and spectral information if full THz waveforms are recorded at each spatial position. As an example of the near-field probe's imaging ability, the series of images in

Fig. 7 demonstrates changes of the image pattern with time. At various time slices, the interaction between the near-field THz and the object vary. The object in this experiment is a gold dipole antenna embedded in a planar transmission line. The antenna is lithographcally printed on a sapphire substrate and consists of two sharp ended 50 μm long arms, slightly shifted with respect to each other and two 20 μm wide striplines separated by 105 μm. A schematic diagram of the antenna is shown in

Fig. 7a. The object contains only gold features oriented either parallel to the direction of polarization (dipole arms) or perpendicular to it (strip lines). The images are obtained at consequent moments in time using the 10 μm aperture probe (L=4 μm).

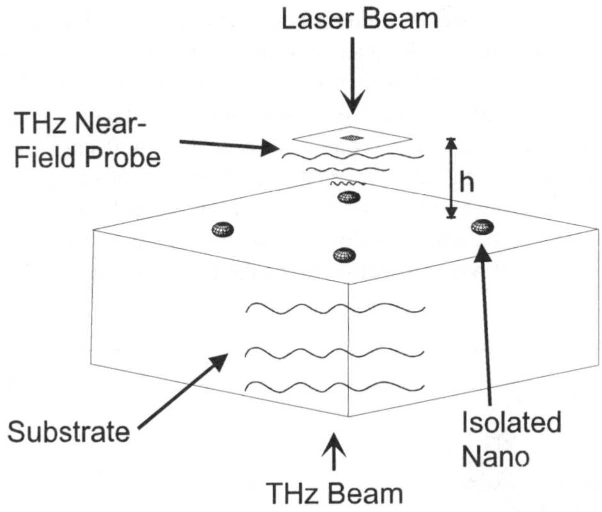

Fig. 8. Schematic illustration of using THz near-field probe to measure THz image of an isolated nano structure.

Clearly, the 7μm spatial resolution of near-field THz is much larger than an individual nanoparticle. In order to measure a individual nanoparticle, it would have to be physically isolated from other dots as shown in Fig. 8. In this configuration, the advantage of near-field THz is that an individual nanoparticle could be measured with more THz radiation interacting with the nanoparticle as compared to the far-field. An alternative configuration would be to directly fabricate nanoparticles on top of the near-field probe. The near-field image of

Fig. 7 suggests that near-field THz imaging might be useful for studying near-field THz images of collective modes in nanostructures for which the THz electric field behavior is determined by a collective response of nanostructures.

2.3. *IR and Raman spectroscopy*

Raman spectroscopy is a well-known technique.[57] It probes the energy exchange between a probe laser and phonon energy levels. Raman spectroscopy proves very useful when non-destructive interrogation of small nanoclusters is required. In addition, Raman spectroscopy is a proven tool to evaluate the crystallinity of small clusters.[58] Finally, Raman spectroscopy in concurrence with THz spectroscopy may reveal electron-phonon coupling mechanisms in nanocomposites. A μ-Raman system is a Raman system equipped with a translation stage and a microscope through which the laser light is admitted. In that way we achieve a signature of small areas of the composite film. This system is especially useful for films deposited by laser ablation. Since the laser plume is mostly concentrated in a normal direction to the target, the clusters' morphology may depend on its relative position of formation. This effect is easily deduced from the Raman spectroscopy of the clusters as a function of clusters' position on the film. The question still remains whether Raman spectroscopy is a true measure of the cluster's size or is it influenced by strain factors. The answer is usually determined between the Raman and the TEM measurements. The signature of the Raman shift frequency and its width depend on the morphology of the clusters as well as their sizes. As seen in Fig. 9, the Raman signature of the bulk Si-Si bond is at 520 cm^{-1}. For very small silicon clusters (3-4 nm of sample S1 in the figure) the curve becomes very broad with a peak at 480 cm^{-1}. As the cluster size increases to 5-6 nm (sample S2) the curve has a shifted peak of –2 cm^{-1} compared to bulk silicon.[22] The curve is also asymmetric. Finally, laser ablated silicon films (sample S3) portray a symmetric curve, which is down shifted by 4 cm^{-1}. Such behavior has been attributed to the hexagonal wurtzite symmetry of the laser-ablated crystallites.

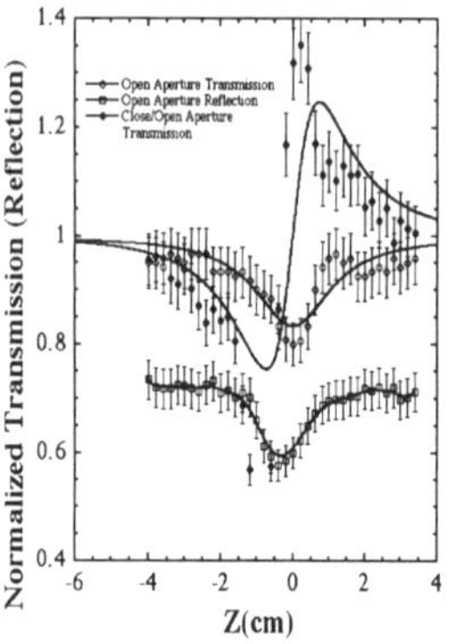

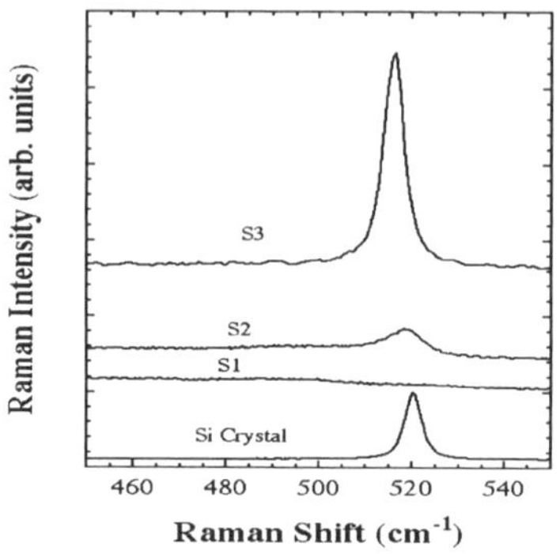

Fig. 9. (a) Z-scan measurement of 5-6nm ion implanted clusters (b) Raman Spectroscopy of 3-4 nm Si clusters (sample S1), 5-6 nm Si clusters (S2) and laser ablated Si (S3).

Surface Enhanced Raman Spectroscopy (SERS)[59] is a modified Raman spectroscopy: The typical weak signals of the spontaneous Raman signature are amplified by interaction with surface plasmons. Coupling to surface plasmons is typically made by use of gratings. Surface plasmons are also easily excited by use of a typical THz or microwave waveguide setting such as, co-planar waveguides. Thus, one may combine the THz spectroscopy and SERS together.

It is widely known that whenever a strong absorption is measured at Infrared (IR) spectral range the corresponding Raman signal is weak. Thus a Fourier Transform InfraRed (FTIR) spectrometer is a complementary system to the Raman spectrometer. The system is made of an infrared source and a controllable etalon, which help analyzing the corresponding IR wavelength. For example, laser ablated silicon films have been characterized using FTIR to find special IR band, which are different than those of cubic silicon wafer.[17] Such studies are helpful in determining the vibrational states of the film under test and eventually, electron-phonon (polariton) interactions.

3. Nano-materials: fabrication and properties

Small clusters have been achieved by use of a laser assisted deposition technique, ion implantation and Chemical Vapor Deposition. The first two growth methods result in clusters, which are similar in size yet, are distributed differently. Laser ablation of dielectric, conductive, semiconductor and superconductor materials is a proven method of obtaining thin films of the host and guest materials.[17] It results in an agglomeration of nano-clusters in between micron-size droplets. The system involves a high power Excimer laser and vacuum system. Full control of the distance between clusters on adjacent layers and within layer is achieved by varying the pressure, deposition times and substrate temperature. Ion implantation is a proven method of producing clusters as well [Ref 21 and references therein]. It results in well-separated nano-clusters within the matrix material. Clusters ranging from 3nm to 6nm have been produced in a large range of crystalline and amorphous matrices. These clusters are distributed quite uniformly throughout the film. The size distribution and uniformity are particularly important when studying the effect of individual clusters in a given matrix (Artificial Dielectric).

A third and most useful technique to produce nanoclusters has been the Chemical Vapor Deposition (CVD) method. Specifically we have grown Single-Wall Carbon Nanotubes (SWCNT). A precursor, (CO gas in our case) is heated up and dissociate in the presence of a catalyst. The result is a uniform growth of SWCNTs. Growth of these in a matrix made of an ordered array of silica nanospheres (opal) is shown in Fig. 10. Such growth technique gives the hope that nano-size organic conductors will make massive electrical interconnects a reality.

Fig. 10. SWCNT inside a matrix made of an ordered array of 275 nm silica spheres. The nanotubes are the small 'wirelike' structures which are extended from one silica sphere to the other.

4. THz spectroscopy of nanocomposites

In this section, we present some preliminary results of THz spectroscopy of various nanomaterials. The experimental configuration is shown in Fig. 1. One modification in applying the Terahertz technique to semiconductor nanoclusters is to frequency double the Ti:Sapphire laser wavelength (tunable from about 700 to 1000nm) in order to generate light at the 355 or 500nm pumping wavelength suggested by our previous non-linear Z-scan measurements for semiconductor clusters.[17,18] Unlike our earlier measurements of visible pump/ THz probe spectroscopy[9] for which an amplified short-pulse laser was used to photogenerate carrier densities of roughly 10^{18} cm^{-3}, our recent improvements in THz radiators and detectors permit photoexcitation measurements of fairly dilute carrier concentrations (10^{12} cm^{-3})[60] without the need for an amplified short-pulse Ti:Sapphire laser.

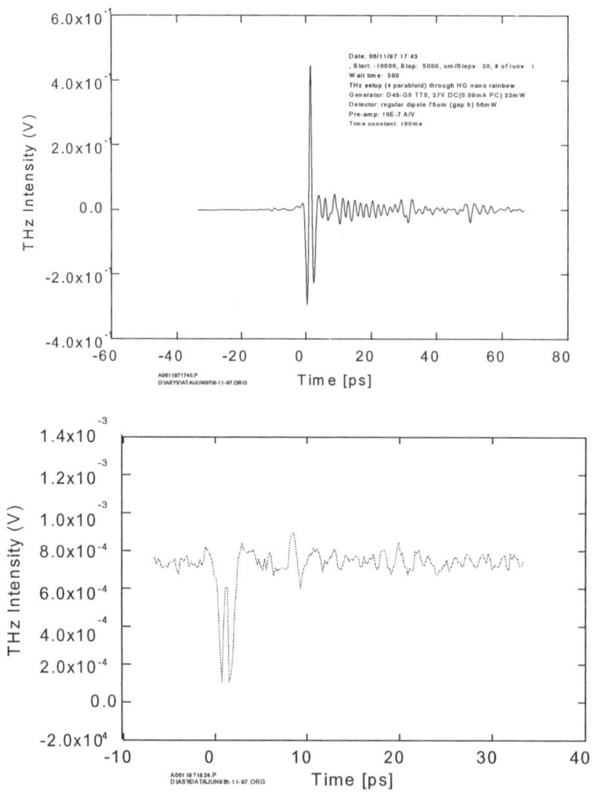

(a)

(b)

Fig. 11. (a) Time domain THz transmission through Si nanocluster film. The unilluminated film is essentially transparent to THz radiation. (b) Visible Pump (514nm, 0.53W CW) and THz probe measurements of Si nanocluster film. The two large negative going spikes near t=0ps is indicative of a photogenerated carrier density.

Our preliminary measurements of linear THz transmission (Fig. 11a) show that the Si nano clusters are transparent. Upon illumination with a visible pumping source (Ar laser) the THz transmission is reduced as shown in Fig. 11b. For this visible pump/THz probe measurement, the visible pump beam is mechanically chopped. While this configuration of visible pumping (essentially CW) will not allow us to investigate the time-resolved carrier dynamics, it does allow us to demonstrate a time-averaged optically induced change in the transport properties of the nanoclusters. Time-resolved measurements are in progress and will be reported elsewhere.

The resultant detected THz radiation is measured using a lock-in amplifier that is locked to the chopping frequency of the visible beam. Using this configuration, the modulation of the THz waveform due to the pump beam excitation is directly measured. This method of detection measures the differential THz transmission as a function of time - essentially transmission with excitation minus transmission with no excitation. We hypothesize that the change in THz transmission is due to the photogenerated carrier density. If NO photogenerated carriers were generated, the measured waveform would be a flat line. The measured change in THz transmission (normalized to the unilluminated transmission) through the illuminated nanoclusters is roughly 10^{-4}. Clearly, the visible pump beam is creating changes in the nanocluster film's THz transmission.

Similar results are obtained for Single-Wall Carbon Nanotubes (SWCNTs). Without optical excitation, the samples are essentially transparent in the THz region. Upon visible excitation, the free carriers that are generated respond to the E-field of the THz probing wave and alter the THz transmission through the sample. The time domain signals are similar to that measured for silicon nanoclusters. A Fourier transform of the time domain data gives the power and phase spectrum. The figure below shows the differential transmission of SWCNT normalized to the transmission with no excitation ($\Delta T/T$). Frequency components below 0.1THz and above 0.8 THz should be ignored because most of the power of the THz pulse is between 0.1 and 0.8 THz. Strong differential absorption near 0.4 THz is observed. A similar differential transmission spectra with a broad reduction in transmission near 0.4THz is observed in Si nanoclusters.

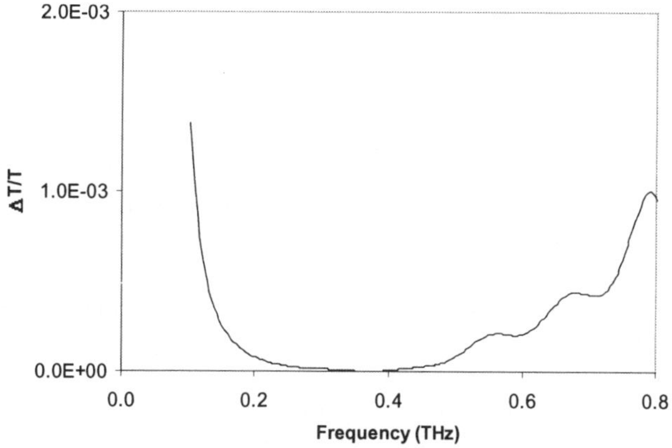

Fig. 12. Normalized differential transmission of SWCNT samples. Note the preferential absorption near 0.4 THz.

Calibration of the system is done using semi-insulating silicon as a test sample. From our previous measurements, it is known that photoexcited Si acts as a Drude-type (free carrier) absorbing medium. In essence this means that the phase of the THz probe is essentially unchanged by the photogenerated carriers, but the amplitude of the THz transmission is attenuated. The differential spectra is shown in Fig. 13. Note that unlike the SWCNT and silicon nanocluster samples, the differential transmission of a Drude-type photoconductor (bulk silicon) is essentially constant at low frequencies and increases as the frequency approaches 0.7THz. Please note that the magnitude of the differential transmission will depend on the optical penetration depth of the pump compared to the film thickness. The SWCNT film has a thickness of 10μm while the semi-insulating silicon is 500μm thick.

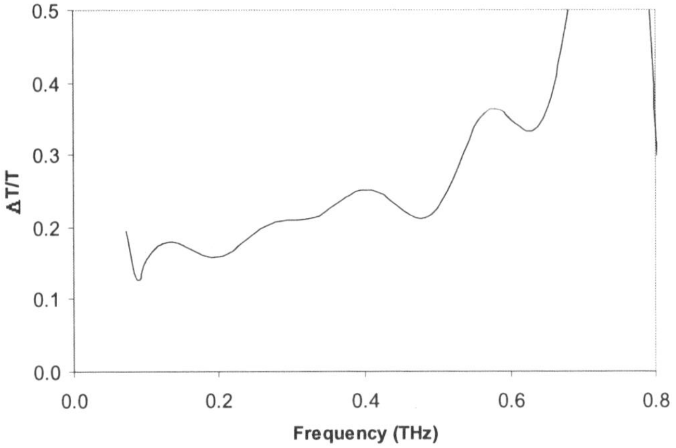

Fig. 13. Normalized Differential Absorption of semi-insulating Si wafer.

5. IR and Raman spectroscopy of nanocomposites

In our laboratory, we have fabricated optically controlled optical waveguides and optically controlled microwave transmission lines by embedding semiconductor clusters in the dielectric of the waveguide.[16,61,62,63] We have found that the phase modulation index was about 20 times larger than the amplitude modulation index; the attenuation of a propagating beam was substantially smaller than the change in the propagating constant.

The non-linear properties of compact cluster structure and isolated clusters have been conducted in the visible and IR.[17,18,19,20] Typical results of Z-scan measurements for ion-implanted Si in silica matrix are shown in Fig. 9a. The nonlinear absorption (open aperture) and nonlinear refraction (close aperture) indicate a positive nonlinear value at λ=532 nm. The existence of nano-size particulate is exhibited by the Raman shift with respect to bulk silicon. Samples S1 and S2 contain 3-4 nm and 5-6 nm clusters formed by ion implantation and followed by annealing. At a wavelength of λ=532 nm the 3-4 nm silicon clusters do not exhibit any nonlinearity. This fact together with the amorphous-like Raman signature supports the assumption that the nonlinear process at this wavelength is phonon mediated. Sample S3 is formed by laser ablation. To the best of our knowledge, the non-linear parameters measured in these references are the largest demonstrated for Si nano-clusters. The non-linear properties of ion-implanted Si clusters in SiO_2 matrix have been measured as well.[21,22] The commonality of these experiments is that at nanosecond pulse durations the nonlinear is advanced via loss mechanism, namely, excitation of carriers in the clusters. These are referred to as energy driven mechanisms. As the pulse duration of the excitation laser becomes shorter, intensity driven nonlinear mechanism take place. In principal, one can distinguish between these two nonlinear mechanisms by use of THz spectroscopy: The former will exhibit a large signal in a pump/probe type experiment while the latter will exhibit a THz radiation.

The Raman signals from CVD grown SWCNT have been measured as well. In Fig. 14 we show a rich Raman spectra, which may be associated with various diameter SWCNT of chiral structure. The Raman signature of SWCNT has two major features: in the high frequency range (around 1600 cm^{-1}) it portrays vibrations along the tube axis. In the low frequency range (around 300 cm^{-1}) it exhibits radial vibrations. The tubes are excited by an Ar laser at =514.5 nm. The line is close to a resonance (Van Hove singularity which is a singularity in the density of states) and helps detection of the typically weak signals. In nonlinear pump/THz probe measurements on GaAs, the THz signal was correlated to the density of states.[9] Similar arguments may apply to SWCNT. Based on TEM measurements we concluded that the diameter of the SWCNT varies between 0.7 to 1.1 nm and that they are well separated or in very small bundles.

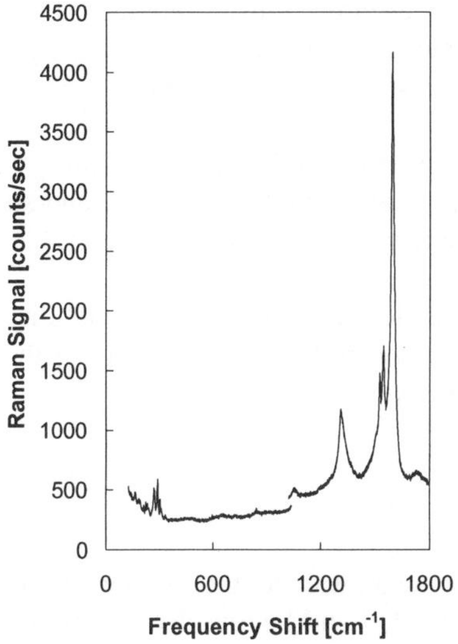

Fig. 14. Raman spectra of SWCNT within and ordered array of 275 nm size silica sphere

6. Conclusion

The dielectric properties of passive thin films, such as glasses or polymers can be tailored by embedding nano-scale crystallites of conductor, semiconductors or even superconductors. Nano-clusters possess properties, which are unique and are not seen in bulk material. They are in the gray region between the quantum limit and bulk properties.

The ability to tailor the properties of both individual clusters and the coherent effect of many interacting clusters will undoubtedly lead to very efficient nonlinear optical devices. For example, by mixing well-known distributions of nanoclusters we will be able to understand the correlation between clusters of various sizes. It is already known that the nonlinear effect in nanoclusters is inhomogeneously broadened, namely, the nonlinear effect triggers one size of clusters with little impact on the others. However, strong coupling may occur at proximity below the Bohr radius size. In fact, there are already theoretical models, which predict that there is an optimal distance for which the coupling is the strongest.

Both linear THz spectroscopy and non-linear visible pump/ THz probe spectroscopy are non-invasive, non-contact techniques to characterize charge carrier transport in semiconductors and nanostructures. The major advantage of these techniques is that transport properties can be measured without electrical contacts to the samples. By measuring THz transmission or reflection one can deduce the complex dielectric constant of the nanolayer. By assuming a Drude model response of the charge carriers, the frequency dependant dielectric, carrier density, and scattering rates can be determined as well as the phenomenological coupling between nanoclusters.

The work presented here focuses on the basic science of charge mobility in nano-size clusters. However, this basic science potentially impacts development of THz and nanotechnology:

- Novel THz sources and detectors: Carrier mobility is related to the cluster media as well as the interface between cluster and its surrounding matrix. Thus, understanding the charge mobility will result in better THz sources and nano antennas.

- THz spectroscopic tools: the carrier mobility and the interplay between THz frequencies and optical pumping affects other nano-size systems such as, nano-electronics, optoelectronic gates and memory devices.

- THz and IR shield: By understanding the carrier mobility in nanoclusters one is able to deduce the absorption and reflection coefficient of the media. Novel thermal and THz screens can then be realized.

Acknowledgments

The authors gratefully acknowledge the support of the Army Research Office, Contract No. DAAD19-01-1-0009 and helpful discussions with O. Mitrofanov and J. M. Joseph.

References

[1] D.M. Mittleman, M. Gupta, R. Neelamani, R.G. Baraniuk, J.V. Rudd, and M. Koch, "Recent advances in terahertz imaging," *Appl. Phys. B* vol. 68 (1999), 1085-1094.

[2] M.C. Nuss and J. Orenstein, "Terahertz time-domain spectroscopy," in: G. Gruner (Ed.), "Millimeter-wave spectroscopy of solid," Springer Topics in Applied Physics, vol. 74, Springer, Berlin, 1998.

[3] M. van Exter and D.R. Grischkowsky, "Characterization of an optoelectronic terahertz beam system," *IEEE Trans. Microwave Theor. and Tech.* vol. 38 (1990) 1684-1691.
[4] B.B. Hu and M.C. Nuss, "Imaging with terahertz waves," *Opt. Lett.* vol. 20 (1995) 1716-1719.
[5] P.Y. Han, G.C. Cho, and X.-C. Zhang, "Time-domain transillumination of biological tissues with terahertz pulses," *Opt. Lett.* vol. 25 (2000) 242-244.
[6] S. Mickan, D. Abbott, J. Munch, X.-C. Zhang, and T. Van Doorn, "Analysis of system trade-offs for terahertz imaging," *Microelectronics J.* vol. 31 (2000) 503-514.
[7] P.Y. Han, G.C. Cho, and X.-C. Zhang, "Time-domain transillumination of biological tissues with terahertz pulses," *Opt. Lett.* vol. 25 (2000) 242-244.
[8] J. F. Federici, B. I. Greene, D. R. Dykaar, F. Sharifi, and R. C. Dynes, "Direct picosecond measurement of photoinduced Cooper pair breaking in lead", *Phys. Rev. B* **46**, (1992) 11153
[9] P. Saeta, J. F. Federici, B. I. Greene, and D. R. Dykaar, "Intervalley scattering in GaAs and InP probed by far infrared absorption spectroscopy", *Appl. Phys. Lett.* **60** (1992) , 1477.
[10] R. E. Collin, Field Theory of Guided Waves, McGraw Hill, New York, 1993; (b) J. Kong, "Electromagnetic Field Theory", Wiley & Sons, 1990.
[11] S. D. Stooky and R. J. Araujo, "Selective polarization at light due to absorption by small elongated silver particles in glass", *Appl. Opt*, **71** (1968) 777-779.
[12] N. Borrelli, J.B. Chodak and G.B. Hares, "Optically induced anisotropy in photochromic glasses", *J. App. Phys.*, **50** (1979) , 5978.
[13] W.J. Kaiser, E.M. Logothetis and L.E. Wegner, "Dielectric properties of small metal particle composites", *J. Phys. C*, **181**, (1985) L837-L842.
[14] S. Lee. T.W. Noh, J.R. Gaines, Y. Ko and E.R. Kreidler, "Optical studies of porous glass media containing silver particles", *Phys. Rev. B*, **37** (1988) , 2918.
[15] R.G. Barrera, G. Monsivais and L.W. Mochan, "Renormalized polarizability in the Maxwell Garnett theory", *Phys. Rev. B,* **38** (1988) , 5371.
[16] H. Grebel and P. Chen, "Artificial dielectric polymeric waveguides: metallic embedded films", *J. Opt. Soc. Am.* **8** (1991) , 615.
[17] S. Vijayalakshmi, M. George, J. Sturmann and H. Grebel, "Pulse laser deposition of Si nanoclusters", *Appl. Surface Science*, **127-129** (1998) , 378.
[18] S. Vijayalakshmi, F. Chen and H. Grebel, "Artificial Dielectrics: Non-linear optical properties of silicon nanoclusters at λ=532nm", *Appl. Phys. Letts.*, **71** (1997) , 3332
[19] S. Vijayalakshmi, M. George and H. Grebel, "Non-linear optical properties of silicon nanoclusters", *Appl. Phys. Letts.*, **70**(6) (1997) , 708.
[20] S. Vijayalakshmi, M. GeorgeJ. Federici, Z. Iqbal and H. Grebel, "Non-linear optical properties of silicon nanoclusters", *Thin Solid Films* **339** (1999) , 102-108.
[21] S. Vijayalakshmi, H. Grebel, Z. Iqbal and C. W. White, "Artificial dielectrics; non-linear properties of Si nano clusters formed by ion implantion in SiO2 glassy matrix", *J. Appl. Phys.* **84** (1998) , 6502.
[22] Z. Iqbal, S. Vijayalakshmi, H. Grebel and C. White, "Microstructure and optical properties of nanostructured silicon thin films and artificial dielectrics", *Nanostructured Materials*, **12**, (1999) 271.
[23] G. R. Olbright and N. Peyghambarian, "Interferometric measurement of the nonlinear index of refraction, n_2, of CdS_xSe_{1-x} doped glasses", *App. Phys. Letts.* **48** (1986) , 1184.
[24] R. K. Jain and R. C. Lind, "Degenerate four-wave mixing in semiconductor-doped glasses", *J. Opt. Soc Am.,* **73** (1983) , 647.
[25] M. Herrmann, M. Tani, K. Sakai, and R. Fukasawa, "Terahertz imaging of silicon wafers", *J. Appl. Phys.* **91** (2002) , 1247.
[26] M. van Exter and D. Grischkowsky, "Carrier dynamics of elctrons and holes in moderately doped silicon", Phys. Rev. B 41 (1990) , 12140.
[27] N. W. Ashcroft and N. D. Mermin, Solid State Physics, Saunders College Publishing (1976).

[28] M. Li, J. Fortin, J. Y. Kim, G. Fox, F. Chu, T. Davenport, T.M. Lu, and X. C. Zhang, "Dielectric constant measurement of thin films using goniometric terahertz time-domain spectroscopy", *IEEE J selected topics Quant. Electron.* **7** (2001) 624
[29] M. C. Bear, G. M. Turner, and C. A. Schmuttenmaer, "Subpicosecond carrier dynamics in low-temperature grown GaAs as measured by time-resolved terahertz spectroscopy", *J. Appl. Phys* **90**, (2001) 5915.
[30] P. Uhd Jepsen, W. Schairer, I. H. Libon, U. Lemmer, N. E. Hecker, M. Birkholz, K. Lips, and M. Schall, "Ultrafast carrier trapping in microcrystalline silicon observed in optical pump-terahertz probe measurements", *Appl. Phys. Lett* **79** (2001) , 1291.
[31] M. C. Bear, G. M. Turner, and C. A. Schmuttenmaer, "Transient photoconductivity in GaAs as measured by time-resolved terahertz spectroscopy", *Phys. Rev. B* **62** (2000) , 15764.
[32] B.I. Greene, J.F. Federici, D.R. Dykaar, R.R. Jones, and P.H. Bucksbaum, `Interferometric characterization of 160 fs far-infrared light pulses', *Appl. Phys. Lett.* **59** (1991) , 893,.
[33] Y. Cai, I. Brener, J. Lopata, J. Wynn, L. Pfeiffer, J. B. Stark, Q. Wu, X. C. Zhang, and J. F. Federici, "Coherent Terahertz Radiation Detection: Direct Comparison Between Free-Space Electro-optic Sampling and Antenna Detection", *Appl. Phys. Lett.* **73** (1998) , 444.
[34] R. E. Collin, Field Theory of Guided Waves, McGraw Hill, New York, 1993.
[35] J. Kong, "Electromagnetic Field Theory", Wiley & Sons, 1990.
[36] M. Born and E. Wolf, *Principles of Optics,* Pergamon Press (1980).
[37] J. I. Pankove, *Optical Processes in Semiconductors*, Dover (1975).
[38] F. Hache, D. Ricard, and C. Flytzanis, "Optical nonlinearities of small metal particles: surface-mediated resonance and quantum size effects", *J. Opt. Soc. Am. B*, **3** (1986), 1647
[39] E. Hanamura, "Very large optical nonlinearity of semiconductor microcrystallites", *Phys. Rev. B* **37** (1988) , 1273.
[40] L. Belleguie and S. Mukamel, "Nonlocal electrodynamics of arrays of quantum dots", *Phys. Rev. B* **52** (1995) , 1936.
[41] B. Hu, E. A. de Souza, W. H. Knox, J. E. Cunningham, M. C. Nuss, A. V. Kuznetsov, and S. L. Chang, "Identifying the distinct phases of carrier transport in semiconductors with 10 fs resolution", *Phys. Rev. Lett.* **74** (1995). , 1689,
[42] S. Hunsche, M. Koch, I. Brener, and M.C. Nuss, "THz near-field imaging," *Opt. Comm.* vol. 150 (1998) 22-26,.
[43] K. Wynne and D.A. Jaroszynski, "Superluminal terahertz pulses," *Opt. Lett.* vol. 24 (1998) 25-27.
[44] O. Mitrofanov, I. Brener, R. Harel, J.D. Wynn, L.N. Pfeiffer, K.W. West, and J. Federici, "Terahertz near-field microscopy based on a collection mode detector," *Appl. Phys. Lett.* vol. 77, (2000) 3496-3498.
[45] O. Mitrofanov, I. Brener, M.C. Wanke, R.R. Ruel, J.D. Wynn, A.J. Bruce, and J. Federici, "Near-field microscope probe for far infrared time domain measurements," *Appl. Phys. Lett.* vol. 77 (2000) 591-593.
[46] Q. Chen, Z. Jiang, G.X. Xu, and X.-C. Zhang, "Near-field terahertz imaging with a dynamic aperture," *Opt. Lett.* vol. 25 (2000) 1122-1124.
[47] O. Mitrofanov, M. Lee, J. W. P. Hsu, I. Brener, R. Harel, J. Federici, J. D. Wynn, L. N. Pfeiffer, and K. W. West, 'Collection mode near-field imaging with 0.5 THz pulses', *IEEE Journal of Selected Topics in Quantum Electronics* **7** (2001) , 600.
[48] Q. Chen and Z. C. Zhang, "Semiconductor Dynamic Aperture for Near-Field Terahertz Wave Imaging", *Sel. Topics in Quantum. Elecron.* **7** (2001) , 608
[49] H. Lee, J. Lee, and J. Kim, "A micromachined Photoconductive Near-Field Probe for picosecond pulse propagation measurement on coplanar transmission lines", *IEEE J. Selected Topics Quant. Electron.* **7** (2001) , 674.
[50] O. Mitrofanov, R. Harel, M. Lee, L.N. Pfeiffer, K. West, J.D. Wynn, and J. Federici, "Study of single-cycle pulse propagation inside a THz near-field probe," *Appl. Phys. Lett.* 78 (2001) , 252

[51] K. Wynne, J.J. Carey, J. Zawadzka, and D.A. Jaroszynski, "Tunneling of single-cycle terahertz pulses through waveguides," *Opt. Comm.* vol. 176 (2000) 429-435,
[52] R. D. Grober, T. Rutherford, and T.D. Harris, "A Modal Approximation for the electromagnetic field of a near-field optical probe," *Appl. Opt.* vol. 35, (1996) 3488-3495.
[53] H. A. Bethe, " Theory of diffraction by small holes", Phys. Rev 66 (1944), 163.
[54] I. Brener, S. Hunsche, Y. Cai, M.C. Nuss, J. Wynn, J. Lopata, and L. Pfeiffer, "Time resolved near field imaging and diffraction with subwavelength far-infrared dipole sources," Ultrafast Phenomena XI, (1998) 171-172.
[55] A. Taflove and S. Hagness, "Computational Electrodynamics: The Finite-Difference Time-Domain Method," Artech House, 2000.
[56] J. Bromage, S. Radic, G.P. Agrawal, C.R. Stroud, Jr, P.M. Fauchet, and R. Sobolevski, "Spatiotemporal shaping of half-cycle terahertz pulses by diffraction through conductive apertures of finite thickness," J. Opt. Soc. Am. B vol. 15 (1998) 1399-1405.
[57] A. Mooradian, *Raman Spectroscopy of Solids, Laser Handbook*, Vol 2, North Holland, Amsterdam, 1972.
[58] S. Veprek, F. Sarott and Z. Iqbal, "Effect of grain boundaries on the Raman spectra, optical absorption, and elastic light scattering in nanometer-sized crystalline silicon", *Phys. Rev. B.* **36** (1987) , 3344.
[59] H. Grebel, Z. Iqbal and A. Lan, "Detecting single wall nanotubes with surface enhanced Raman scattering from metal coated periodic structures", *Chem. Phys. Letts.*, **348**, (2001)203 ; H. Grebel, Z. Iqbal and A. Lan, "Detection of C_{60} using surface enhanced Raman scattering from metal coated periodic structures", *Appl. Phys. Letts.*, **79** (2001) , 3194.
[60] Y. Cai, Ph.D. Thesis, New Jersey Institute of Technology (1998).
[61] H. Grebel P. Chen, "Artificial dielectric polymeric waveguides: semiconductor-embedded films", *Opt. Lett.* **15** (1990) , 667.
[62] H. Grebel and M. Jimenez, "Conditional artificial dielectric transmission lines; Optically controlled microstrip phase modulators", *IEE Proc.-J*, **140** (1993) 232-236.
[63] S. C. Wu and H. Grebel, "Phase shifts in coplanar waveguides with patterned conductive top cover", *J. Physics D: Appl. Phys.*, **28** (1995) 437-441.

International Journal of High Speed Electronics and Systems
Vol. 13, No. 4 (2003) 995–1097

World Scientific
www.worldscientific.com

Fundamentals of Terrestrial Millimeter-Wave and THz Remote Sensing

E.R. Brown

Professor of Electrical Engineering
University of California, Los Angeles
Los Angeles, CA 90095, USA
Email: erbrown@ee.ucla.edu ; drerbrown@earthlink.net

Having long been the realm of molecular chemistry, astronomy, and plasma diagnostics, the upper millimeter-wave band (~100 to 300 GHz) and the THz region above it have recently become the subject of heightened activity in the engineering community because of exciting new technology (e.g., sub-picosecond optoelectronics) and promising new "terrestrial" applications (e.g., counter-terrorism and medical imaging). The most challenging of these applications are arguably those that demand remote sensing at a stand-off of roughly 10 m or more between the target and the sensor system. As in any other spectral region, remote sensing in the THz region brings up the complex issues of sensor modality and architecture, free-space electromagnetic effects and components, transmit and receive electronics, signal processing, and atmospheric propagation. Unlike other spectral regions, there is not much literature that addresses these issues from a conceptual or system-engineering viewpoint. So a key theme of this chapter is to review or derive the essential engineering concepts in a comprehensive fashion, starting with fundamental principles of electromagnetics, quantum mechanics, and signal processing, and building up to trade-off formulations using system-level metrics such as noise-equivalent power and receiver operating characteristics. A secondary theme is to elucidate aspects of the THz region and its incumbent technology that are unique, whether advantageous or disadvantageous, relative to other spectral regions. The end goal is to provide a useful tutorial for graduate students or practicing engineers considering the upper mm-wave or THz regions for system research or development.

Keywords

Active and passive sensor, direct and coherent detection, heterodyne and homodyne receiver; blackbody and thermal radiation, Planck distribution, Rayleigh-Jeans limit, brightness temperature; atmospheric propagation, water vapor absorption, PCLnWin, HITRAN96 database; waveguide, feedhorn, planar antenna, parabolic dish, lenses, hyperhemisphere, antenna theorem, antenna directivity and gain, effective aperture, diffraction limit; Gaussian beams, beam waist, Rayleigh length, ABCD matrix; radiation noise, shot and thermal noise, quantum noise; classical and quantum mixer, classical and quantum square-law detection, bolometer, Golay cell, Schottky diode, low-noise amplifier; coupling and quantum efficiency, responsivity, conversion gain, noise figure, RF, IF, and post-detection bandwidth; Gaussian, Boltzmann, Poisson, Rayleigh, and Rician statistics; signal-to-noise ratio (SNR), noise-equivalent power (NEP), noise-equivalent delta temperature (NEΔT), probability of detection, probability of false alarm, receiver operating characteristics (ROC); Poynting's theorem, Friis' transmission formula, Johnson-Nyquist theorem, Nyquist's sampling theorem, North's theorem, Whittaker-Shannon theorem.

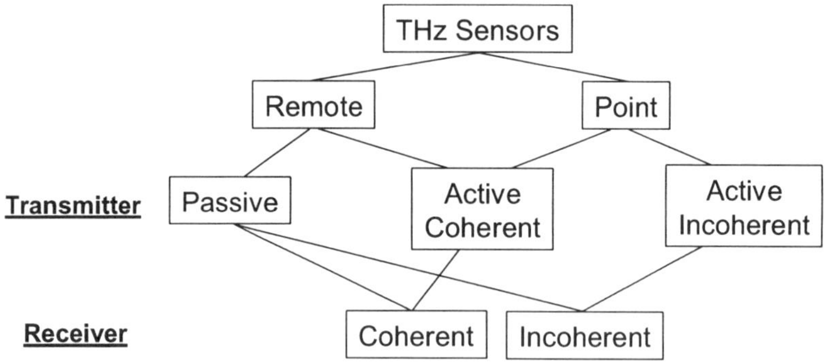

Fig. 1. Heirarchy of sensor technology in the mm-wave and THz regions

I. Introduction

A. Background and Purpose

The author has been in and around the field of remote sensing for over 20 years and over this period has observed a significant decline in the education and understanding of this topic, particularly since the end of the U.S./Soviet Cold War around 1990. There was little challenge finding interest and support for remote sensing when the biggest threat to U.S. National Security was intercontinental and theater ballistic missiles. This threat alone created entire industries to produce high-performance RF electronics, electromagnetic components, digital signal processing electronics, and signal-processing software, to name a few. Now, with the threats of terrorism, global warming, and bio-catastrophe, the field of remote sensing is experiencing a resurgence. And even though the threat ranges have shrunk from thousands of kilometers to thousands of meters and the threat agents have shrunk from missiles to molecules, many of the principles and methodologies developed during the Cold-War period should be applicable. A key motivation for this chapter is to provide a primer for students or young engineers or scientists considering remote sensing as a research topic or as a career. In the limited space of one book chapter, it is impossible to cover this topic in a comprehensive fashion. Therefore, it will focus on fundamental principles behind system architectures and remote sensing in the millimeter-wave and THz regions, which many consider to be at the upper end of the useful RF spectrum.

A secondary purpose for this chapter is to provide some quantitative contrast of the various remote-sensor architectures. There have been several new approaches demonstrated during the past decade that fall outside the traditional types of mm-wave and THz remote sensors. For example, an active sensor has been developed to detect concealed weapons, and passive focal plane arrays have been developed for all-weather imaging. The present chapter will contrast these primarily within a sensor system context, not at the device or component performance. In spite of its limited popularity, the THz region offers several new sensor applications such as the remote detection of airborne toxic bioparticles, the identification of concealed weapons and contraband, and all-weather aircraft landing.

By focusing on the sensor principles and architectures, a sacrifice will be made in terms of technological detail. There is simply not enough space to cover all the electromagnetics, solid-state physics, and signal processing associated with the myriad of components and devices being used today in THz sensors. There is also not enough space to give proper credit to some of the emerging THz technologies that have shown great results in the laboratory but have not yet been used for remote sensing. Principal among these are the optoelectronic technologies, such as THz generation by ultrafast photoconductive switches and mixers using solid-state and semiconductor lasers operating mode-locked or cw at near-infrared wavelengths. One of these techniques, called "T-ray imaging," has advanced to the point of commercial production. Fortunately, several excellent review articles have already been published on both the traditional[1,2,3] and optoelectronic technologies,[4,5,6] and the reader is referred to one of these articles for more information.

B. Sensor Systems

The THz region of the electromagnetic spectrum has long been the realm of basic sciences, such as molecular chemistry and astrophysics, but has not been broadly utilized for commercial or military systems because of the lack of transceiver technology and because of the strong attenuation of THz radition through the terrestrial atmosphere. Historically, the most common THz remote sensors have been passive radiometers utilizing coherent (i.e., heterodyne) detection. The most common systems for point sensors have been Fourier transform spectrometers utilizing incoherent transmitters and receivers.

As in other regions of the electromagnetic spectrum, scientists and engineers have long sought to construct systems in the THz region to *sense* the propagation through or reflection from objects or regions of space having scientific or technical interest. Such *sensor* systems are generally classifed according to three criteria which together define the sensor *modality*. They are: (1) the proximity of the sensor and the object of interest, (2) the source of THz radiation or, equivalently, the design of the *transmitter*, and (3) the sink of the THz radiation or, equivalently, the design of the *receiver*. These three criteria are shown in the sensor hierarchy of Fig. 1. In point sensors, the object or material of interest (henceforth referred to as the *target*) is located in close proximity to the sensor and the THz radiation is provided by a transmitter. The transmitter can be either coherent or incoherent. A good example of a THz point sensor with an incoherent transmitter is a Fourier transform spectrometer having a hot incandenscent or discharge-tube source. A good example of a THz point sensor having a coherent source is a network analyzer, be it vector or scalar. Because of the proximity, practically all of the transmitted power can be directed through or off-of the target, so little transmitted power is required.

In remote, or stand-off sensors, there is a relatively large separation between the target and the sensor. The separation generally means that only a small fraction of the transmitted power can be delivered usefully to the target. If this small fraction can be detected back at the receiver, the modality is called *active* and the resulting sensor is called a *radar* (*ra*dio *d*etection *a*nd *r*anging). It is fascinating that the radar concept was first proposed in the early part of the 20th century, promoted by none other than G. Marconi, and was successfully developed into working systems even more quickly than wireless communications. But in the THz region the relatively low power of sources and the high opacity of the atmosphere has hindered radar development, and system developers have relied on a second modality called *passive*. This generally entails the illumination of the object or region by incoherent radiation from its surroundings, which if intense enough can be detected by the remote receiver. A good example of a passive THz remote sensor is the total power radiometer often used to measure trace gases in the atmosphere or emissions from interstellar media. The advantage of a passive system is clearly that the source radiation comes for free. The downside is that the atmospheric attenuation and fading effects can quickly degrade the sensor performance.

The third defining criterion for sensor modality is the receiver architecture. As in the transmitter there are two types, coherent and incoherent, both discussed later in Sec. IV. Within the coherent category there are two types: (1) heterodyne and homodyne. Within the incoherent category there are also two types: (1) power detection and photon detection. As shown in Fig. 1, a coherent transmitter is generally coupled to a coherent receiver, and an incoherent transmitter is generally coupled to an incoherent receiver. It is surprising to some people that a passive system can have either receiver type. A coherent receiver measuring an incoherent signal (e.g., white noise) essentially filters out that particular Fourier component of the signal that *coheres* to the local oscillator waveform. In other words, it cross-correlates with only one space-time mode of the otherwise random spectrum of incoming radiation.

One might on first glance think that a passive THz sensor operating at temperature T and relying on power received via thermal radiation also of temperature T is a violation of the first law of thermodynamics. After all, this law would require that in equilibrium the power transferred from the object to the sensor should be matched by a power transferred from the sensor to the object. But THz sensors, like RF systems in general, are categorically not in thermodynamic equilibrium ! A good way to see this point is to think about the fact that the first active component (i.e., one capable of doing or receiving electromagnetic *work*) in most RF receivers is a low-noise amplifier (LNA). Modern solid-state LNAs operating up to several GHz have very low noise figures and correspondingly low noise temperatures, typically in the range of 100 K or below. This means that they produce noise equivalent to a device at much lower temperature, made possible by the fact that the LNA is an active device operating well out of thermodynamic equilibrium. This is why active devices are so pervasive in all electromagnetic sensors, both in the transmitter and the receiver.

C. THz Solid-State Devices: A Longstanding Challenge

No summary on the THz region would be complete without some discussion of its technology, ranging from passive quasi-optical components to quantum-effect solid-state detectors and unique coherent sources. Because many articles in recent years have reviewed this technology thoroughly, the discussion here will be kept to a minimum. And in keeping with the author's viewpoint on research, this summary occurs after the discussion of phenomenology and system-level issues. The development of new technology with a system perspective allows the engineer or scientist to design and characterize components and devices with far greater insight and practicality than otherwise.

Because of the relative lack of components in the THz region, some researchers have referred to it as the electromagnetic frontier. Although a bit misleading, this statement is not without justification. The THz region has presented pioneering investigators with one experience common to most "frontiers", namely technical difficulty. The difficulty is based in two well-known but seldom discussed physical characteristics of operating any system in the THz region at room temperature. First, in the THz region the photon energy hν is much less than k_BT around room temperature, where k_B is Boltzman's constant. This is in contrast to systems operating at near-infrared and shorter wavelengths. Then according to the Boltzmann *canonical* distribution, quantized states separated by hν tend to have the same population. Hence, stimulated absorption is nearly as likely as stimulated emission. This affects the ability to measure or utilize atomic and molecular transitions at terrestrial temperatures and pressures.

Second, the THz region tends to be high enough in frequency that the classical behavior (i.e., drift and diffusion) of electrons and holes in semiconductors begins to disappear, and the coupling between electromagnetic waves and semiconductor devices weakens. In short, at frequencies for which $\omega\tau \geq 1$ an electromagnetic wave is oscillating too fast for free carriers to respond. A more precise way to understand this effect is to recall from classical electromagnetics that the energy transfer from an electromagnetic wave to free carriers is described by $\overline{J \cdot E}$ where **J·E** is the Joule term of Poyning's theorem and denotes the time average over many cycles of the electromagnetic fields. When the $\omega\tau > 1$, the free carrier current density J according to (1) approaches a quadrature relationship to the electric field so that $\overline{J \cdot E} \rightarrow \overline{\cos\omega t \cdot \sin\omega t} = 0$. The incoming electromagnetic wave that every device must respond to simply cannot transfer energy into the free carriers, and the device function disappears.

A good example of this is the channel conductance of any field effect transistor (FET). Be it a 3D, 2D, or 1D confinement, the transport of free carriers in the channel is governed by their concentration and their ac conductivity. In the simple model of ac conductivity put forth by Drude, this is given by

$$\sigma = \frac{\sigma_0}{1 + j\omega\tau}$$

where σ_0 is the dc conductivity and τ is the momentum relaxation time. As an example, we consider electrons in GaAs which at room temperature have a mobility of approximately 6000 cm^2/V-s and an effective mass m* of $0.067m_0$. According to kinetic theory, the mobility is related to τ by $\mu = e\tau/m^*$, so that under these conditions $\tau \approx 0.23$ ps. The corresponding 3-dB conductivity frequency is $f = (2\pi\tau)^{-1} = 0.69$ THz. Of course, under device operating conditions the large internal electric fields make the above small-field approximations less accurate.

A second example is the bipolar junction transistor (BJT). Whether constructed with homo- or hetero- p-n junctions, a BJT is also ultimately limited in speed by an RC time constant. A leading contribution to R is the base majority carrier resistance, which depends on the majority carrier concentration and ac resisitivity. Again, in the THz region, the resistivity begins to increase significantly. The second phenomenon that greatly affects electronics is that in solid-state material the THz radiation generally obeys $\omega\tau_c > 1$, where τ_c is the momentum relaxation time, and generates an electrical photocurrent at the difference frequency. The difference-frequency current, in turn, generates THz power by connecting it to a suitable load, such as a planar antenna.

Nevertheless, several technological breakthroughs have occurred during the past decade that make THz sensors more practical than ever before. First there has been a rapid development of THz detector and mixer technology including Schottky varistive mixers, superconducting-insulating-superconductor tunnel junction mixers, and superconducting hot electron bolometers. Second, there has been a parallel rapid growth of THz solid-state source technology including Schottky varactor multipliers. Third, there has been a rapid development of ancillary technologies in the millimeter-wave and near-infrared regions to support the THz sources. In particular, there has been a steady development of monolithic microwave- and millimeter-wave integrated circuits (MMICs) up to around 100 GHz . Specifically, MMICs solid-state power amplifiers are now used to drive Schottky multipliers much more efficiently than ever before. Second: (1) growth and fabrication of semiconducting material having photocarrier lifetime less than 1 ps, (2) modern microfabrication techniques that allow sub-micron electrode features to be patterned on the photoconductor surface, leading to subpicosecond electrical time constants, and (3) integration of photoconductive elements with compact planar antennas, leading to efficient coupling of the THz radiation to free space.

An ancillary breakthrough that strongly supports the optoelectronic THz generation approach has occurred in the field of solid-state and semiconductor lasers. Solid-state materials such as $Ti:Al_2O_3$, have been developed that provide unprecedented values of gain-bandwidth andcan provide high levels of power tunable over 10s of nm. Various techniques such as distributed Bragg reflectors, distributed feedback structures, and external cavities have all been integrated with semiconductor laser diodes to produce sources with useful output power (>1 mW) and high spectral purity. And in the popular fiber-optic telecommunication band around 1550 nm, the erbium-doped fiber amplifier (EDFA) has been developed that can boost the power of spectrally-pure laser-diode sources up to ~1-W level.

A key advantage of THz sensors over infrared and visible ones is ultimate noise limits. As will be shown later in Sec. VI, both direct and coherent receivers operate against fundamental noise limits that depend on the background radiation and the photon frequency. In the coherent case, the limit is simply the photon shot noise that has noise equivalent power equal to $h\nu/\eta$, where η is the receiver coupling efficiency. Fig. 2 shows this limit plotted for $\eta = 1$ in terms of the photon energy $h\nu/e$ and equivalent temperature, $h\nu/k_B$. Of course, this same advantage is shared by RF receivers operating at lower frequencies, which is one of the reasons RF communications, be it wired or wireless, is generally superior to photonic communications in terms of sensitivity.

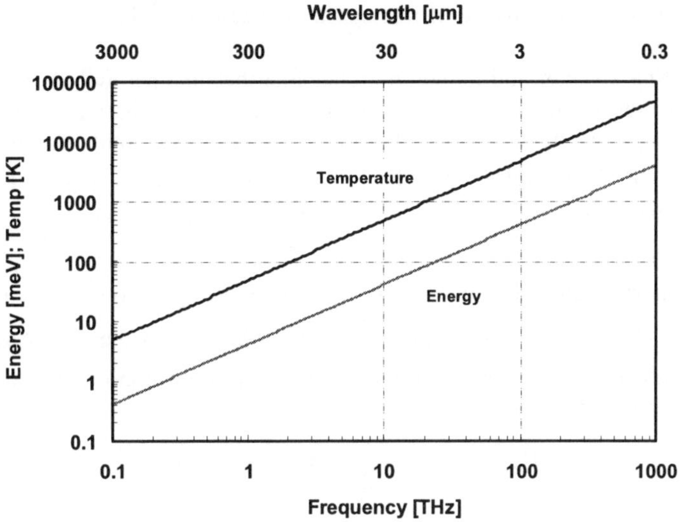

Fig. 2. Quantum-limit defined by the minimum energy – one photon – per spatial mode from the mm-wave region through the visible region of the electromagnetic spectrum.

II. THz Radiation

The THz portion of the electromagnetic spectrum is loosely defined as that region between approximately 300 GHz and 3 THz. As such it overlaps two older regions – the submillimeter-wave region between 300 GHz and 3 THz and the far-infrared region between 3 and 30 THz. Although, lagging behind the RF, infrared, and visible regions in terms of technological maturity, but is still subject to the same laws of electromagnetics (i.e., Maxwell's equation), quantum mechanics (i.e., photon effects), and radiative transport as any other region.

A. Poynting's Theorem: Classical Picture

One of the profound results of electromagnetic theory (i.e., Maxwell's equations) is the existence of traveling or propagating waves, which must always be associated with two quantities, E and H – the electric and magnetic field vectors. An important theorem pertaining to propagating waves is the energy, or Poynting's theorem, which states that the propagating instantaneous power associated with the wave is given by $\vec{S} \equiv \vec{E} \times \vec{H}$. S represents the instantaneous power in the wave, and is applicable no matter what its time or space dependence. Since E and H are always orthogonal in free-space, $H = E/z_0$, and

$$\vec{S} \equiv \frac{E(t)^2}{\eta_0} \hat{s}$$

where z_0 is the intrinsic impedance of free space (377 Ω) and $\hat{s}$ is the unit vector along the direction of S.

Poynting's vector and radiation propagation become particularly simple in the special case of sinusoidal time dependence, sometimes called harmonic waves. The electric and magnetic field vectors, E and H, then have solutions of the form $\vec{E} = \mathrm{Re}\{\tilde{E}e^{j\omega t}\}$ and $\vec{H} = \mathrm{Re}\{\tilde{H}e^{j\omega t}\}$, where the wig over denotes a phasor (complex, time-independent), quantity. And the time-averaged power is given by $\bar{\vec{S}} \cdot = 1/2\,\mathrm{Re}\{\vec{E} \times \vec{H}^*\}$.

An important quantity for sensors is the time-averaged power flowing into the sensor aperture, $P_{inc} = \vec{S} \cdot \vec{A} = (1/2)\,\mathrm{Re}\{\vec{E} \times \vec{H}\} \cdot \vec{A}$, where $\vec{A}$ is the sensor areal vector (pointed perpendicular to the sensor surface). An even more important quantity is the power *usefully* absorbed,

$$P_{abs} \equiv \eta \cdot P_{inc} = \frac{1}{2}\eta \cdot \mathrm{Re}\{\vec{E} \times \vec{H}\} \cdot \vec{A} \tag{1}$$

and η is the power coupling efficiency. Since η is the fraction of incident power *absorbed*, it must account for the effects of reflection at the environment-sensor interface, unabsorbed radiation that passes through the sensor, etc. The majority of sensors couple radiation in from free space propagating perpendicular to the surface. In this case $\vec{H}$ is perpendicular to $\vec{E}$, $|\vec{H}| = |\vec{E}|/z_0$, and

$$P_{inc} = ½(E^2 A)/z_0 = ½(\varepsilon_0 c E^2 A) = cU_E A\,, \tag{2}$$

so that

$$P_{abs} = \eta\, c\, U_E\, A \tag{3}$$

where U_E is the energy density for the electric field and c is the speed of light in vacuum.

B. Harmonic-Oscillator States: Quantum Picture

When the traveling wave phenomenon is addressed in quantum mechanics, the oscillatory nature in the classical picture is captured by the harmonic oscillator model in the quantum picture. The reader is assumed to understand the important derivation, so here we only state the result for the energy eigenvalues for a harmonic oscillator,

$$U_n = (n + 1/2) \cdot h\nu \tag{4}$$

In this expression, n is a positive integer starting at zero, h is Planck's constant, ν is the oscillator natural frequency, and the ½ is the "zero-point" energy, i.e., the residual oscillator energy eigenvalue when n = 0. As taught in basic quantum mechanics courses, the zero-point energy can not do useful work. For example, it can not be used to transmit information in a radar or communications system.

C. Types of Radiation

C.1 Coherent Radiation

The classical sinusoidal waveforms are both replaced by a wave function called the *coherent state* of frequency $\nu = \omega/2\pi$. The amplitude, or occupancy, of this state corresponds to the instantaneous power associated with the classical field amplitudes E or H, and the occupancy number represents the number of *photons* in this state. Each photon has energy $h\nu$, where h is Planck's constant. The coherent state is emitted by oscillators, be they electronic resonator-based sources or atomic lasers.

A necessary aspect of the quantum-mechanical picture is fluctuations associated with the sensor measurement process. If one had a detector fast and sensitive enough to measure the individual photons, one finds that the photon number in the coherent state itself is random and obeys Poisson statistics. In other words, the probability of measuring n photons in the mode in an arbitrary time interval is given by

$$p(n) = \frac{<n>^n}{n!} e^{-<n>}$$

where <n> is the mean number of photons measured in this same interval over many different measurements. It is simple to verify that this distribution is normalized (i.e., $\sum_0^\infty p(n) = 1$) and that $\sum_0^\infty np(n) = <n>$.

This may first appear to be contradictory to what we know about coherent radiation – namely, a sinusoidal dependence of the field quantities (E and H) at frequency ω and a $\sin^2(\omega t)$ dependence of the instantaneous power. This applies to a coherent state as well, but in terms of the mean photon number in the mode ν. In other words, the Poisson distribution gets replaced by

$$p(n) = \frac{[<n>(t)]^n}{n!} e^{-<n>(t)}$$

where $<n>(t) \propto \sin^2(\omega t)$, the same time dependence as the instantaneous intensity. A lucid discussion of the coherent state and its statistical properties is found in Ref. [7]

Given the photon picture, an equivalent way to represent a coherent wave and a sensor is through the average measured photon rate J_P, defined as the average number of photons usefully absorbed by the sensor per unit time. From Eqn 3 This is given simply by

$$J_P = \frac{P_{abs}}{h\nu} = \frac{c \cdot \eta \cdot U_E \cdot A}{h\nu}$$

where h is Planck's constant and ν is the frequency. Even for the relatively weak THz coherent sources, this flux is astronomically high. For example, a source putting out 1 μW at 600 GHz ($h\nu$ = 2.48 meV) is emitting a photon rate of 2.5×10^{15} photon/s !

A practical issue associated with any real sensor in the THz region is that the limited detector bandwidth does not allow for resolution of individual photons at such a high rate. In practice, the average number of photons is given by the sensor resolution time δt, and the mean number of photons measured during this interval is

$$<n> = J_P \cdot \delta t = \frac{P_{abs} \cdot \delta t}{h\nu} \approx \frac{P_{abs}}{\delta f \cdot h\nu}$$

where δf is the measurement bandwidth.

C.2 Thermal Radiation

Thermal radiation results from a volume of electromagnetic modes coming into thermal equilibrium with a heat bath at temperature T. In the terrestrial environment of the Earth, thermal radiation is omnipresent and tends to be very strong in the THz region because of absorption by water vapor. To get a quantitative expression for the average power received by a sensor exposed to thermal radiation, it is best to start with the quantum picture in which the radiation is described not by the electric field and intensity (Poynting) vector, but rather by the occupation of each quantized electromagnetic mode. The occupation probability is then given by the Boltzman distribution p(n) = Cexp[-U_n/kT] where C is a normalization factor , $C=\sum_{n=o}^{\infty} e^{-U_n/kT}$, and k_B is Boltzman's constant [1.38x10^{-23} W/K, MKS]. The mean occupation is then given by the Planck function

$$\langle n(\nu)\rangle = \frac{1}{e^{h\nu/k_BT}-1} \equiv f_P$$

where k_B is the Boltzman constant and T is the temperature of the bath. And the mean energy is given by

$$<U>=(<n>+1/2)\cdot h\nu=(f_P+1/2)\cdot h\nu$$

So if a sensor is pointed at a thermal source, such as a highly attenuating sky, how much of theradiation is incident on the sensor ? The answer lies in electromagnetic modal theory and in thermodynamics. The plane wave modes are the most convenient for decomposing free-space radiation for analytic purposes. However, they are not physically realizable in any receiver because of their infinite lateral spatial extent. More convenient modes are the spatially-orthogonal set defined by the sensor antenna. These so called lateral or "spatial" modes can be used to decompose any radiation transmitted or received by the antenna provided they are defined properly in the antenna coordinate system. A remarkable concept from statistical mechanics is that the Planck function is valid for any orthogonal set of modes, no matter what antenna they apply to. So we arrive at the result that the mean thermal energy *incident* from free space at frequency ν is just the energy quantum, hν, times the mean number of photons in that mode, summed over all spatial modes

$$<U>=\sum_{m=1}^{M} f_P(\nu)\cdot h\nu$$

where m is the spatial mode index, and the zero-point has been excluded because it can not couple energy on the average. Note that the maximum number of spatial modes M is shown explicitly, representative of the fact that many THz sensors accept well more than one spatial mode but rarely enough that the summation can be approximated by a spatial integral as is usually done in the infrared and visible regions.

What happens if we fix the spatial mode and change the frequency ? Each frequency in the Planck function corresponds to a unique harmonic oscillator and, therefore, to a unique mode. In the language of lasers, the different frequencies correspond to different "longitudinal" modes. Therefore, the total energy in a frequency range $\Delta\nu$ is to be thought of as a sum over all the possible longitudinal modes for each lateral mode

$$<U>=\sum_{m=1}^{M}\sum_{n=1}^{N}f_P(\nu)\cdot h\nu$$

where n is the longitudinal-mode index. A simple way to estimate the number of longitudinal modes is to assume that the thermal radiation is separated from the sensor antenna by a distance L and that boundary conditions require that the electromagnetic intensity be a maximum at both the radiator and sensor. Then the lowest frequency longitudinal mode corresponds to a half-wavelength between the two, ν_{min} = c/2L. In open-cavity lasers and Fabry-Perot resonators, this quantity is called the free spectral range. We also assume that the sensor is filtered so that it responds only to radiation lying within a "passband" ν_0 to $\nu_0 + \Delta\nu$, and that the sensor responds only to the half of the longitudinal modes propagating in the direction from the source. The number of longitudinal modes N(ν) that exist at each ν is then given by

$$N(\nu)=\frac{1}{2}\cdot\frac{\nu-\nu_0}{c/2L}$$

And the mean electromagnetic energy is given by

$$<U>=\sum_{\nu_0}^{\nu_0+\Delta\nu}\sum_{m=1}^{M}N(\nu)\cdot h\nu\cdot f_P(\nu)$$

Remote sensors are usually configured so that targets and radiators are both in the far field of the transmit and receive antennas, and hence c/2L << ν_0, where ν_0 is the bottom of signal passband. In this case the sum over n is quasi-continuous and we can approximate the sum by an integral

$$<U>=\sum_{m=1}^{M}\int_{\nu_0}^{\nu_0+\Delta\nu}f_P(\nu)\cdot g(\nu)\cdot h\nu\cdot d\nu=\sum_{m=1}^{M}\frac{L}{c}\int_{\nu_0}^{\nu_0+\Delta\nu}f_P(\nu)\cdot h\nu\cdot d\nu$$

where g(ν) = dN/dν = L/c is the density of forward-traveling states.

We can now estimate the average energy density <U'> just in front of the antenna aperture of area A,

$$<U'>=\frac{1}{AL}<U>=\sum_{m=1}^{M}\frac{1}{cA}\int_{\nu_0}^{\nu_0+\Delta\nu}f_P(\nu)\cdot h\nu\cdot d\nu\ ,$$

and from Eqn 2 the ensemble-averaged power received by the sensor antenna becomes

$$<P_{inc}>=\sum_{m=1}^{M}\int_{\nu_0}^{\nu_0+\Delta\nu}f_P(\nu)\cdot h\nu\cdot d\nu\,.$$

Using the coupling (Eqn 1) coefficient from the classical analysis, we find the power absorbed by the sensor is

$$<P_{abs}>=\sum_{m=1}^{M}\int_{\nu_0}^{\nu_0+\Delta\nu}\eta_m(\nu)f_P(\nu)\cdot h\nu\cdot d\nu \qquad (5)$$

Remarkably, this expression has no dependence on c and no direct dependence on the antenna area, although this is included implicitly through the number of spatial modes contained in the sum as we shall see in Sec. III.C. In addition, it is a straightforward exercise in statistical mechanics to show that it *can be generalized to any form of radiation statistics between the source and the sensor*, so that

$$< P_{inc} >= \sum_{m}^{M} \int_{\nu_0}^{\nu_0+\Delta\nu} h\nu < n_m(\nu) > \cdot d\nu \qquad (6)$$

$$< P_{abs} >= \sum_{m}^{M} \int_{\nu_0}^{\nu_0+\Delta\nu} \eta_m(\nu) \cdot h\nu < n_m(\nu) > \cdot d\nu$$

And we show explicitly the dependence of n on m since other possible radiation statistics, besides the Boltzmann distribution behind the Planck function, may show a dependence of occupancy on mode type.

In the THz region Eqn 5 for thermal radiation commonly gets applied under two limiting cases – high and low background thermal radiation. In the high background case, $T > h\nu/k_B$ over the entire band $\Delta\nu$, so that $e^{h\nu/k_BT} - 1 \approx h\nu/k_BT$. Let's suppose that a sensor is receiving thermal noise power in this limit in M spatial modes over an arbitrary spectral bandwidth $\Delta\nu$. Let's also assume that the electromagnetic waves associated with these modes are perfectly coupled to the sensor (i.e., all incident waves are absorbed without reflection). In this case, the average received power is the product of the average occupancy per mode $<n(\nu)>$ times the energy per mode, $h\nu$, integrated over the longitudinal modes (i.e., the spectral band). So neglecting the zero-point term

$$< P_{abs} >= M \int_{\nu_0}^{\nu_0+\Delta\nu} < n(\nu) > h\nu \cdot d\nu = M \int_{\nu_0}^{\nu_0+\Delta\nu} k_B T \cdot d\nu = M \cdot k_B T \cdot \Delta\nu \qquad (7)$$

This case, called the Rayleigh-Jeans limit, is very common at THz frequencies and below. For example, if $\nu = 1$ THz and $T = 290$ K (room temperature), we have $h\nu = 4.1$ meV, and $k_BT = 25.0$ meV, so that $h\nu/kT = 0.164$ and $e^{h\nu/k_BT} = 1.178$, and then the Rayleigh-Jeans approximation is accurate to about 9%. One interesting aspect of the Rayleigh-Jeans limit is that $\Delta\nu$ can be arbitrarily large if the highest frequency in the spectral band satisfies $h\nu/kT << 1$. So it is not limited to narrow-band sensors as is sometimes stated in the literature.

C.3 The Special Case of Blackbody Radiation

In the special case of uniform, isotropic thermal radiation in an enclosed cavity, the spatial modes can be counted exactly. This is a classic problem in elementary physics and results in the following expression commonly known as blackbody radiation law.

$$\frac{d|S|}{d\nu} \equiv I_\nu = \frac{2\pi h\nu^3 \cdot \varepsilon}{c^2 (e^{h\nu/k_B T_S} - 1)} , \qquad (8)$$

where T_S is the source physical temperature and ε is the source emissivity. This expression, which defines the commonly used optical quantity called the frequency specific irradiance I_ν, gives the randomly polarized power per unit area that crosses a flat surface of area A when illuminated from *one side* by isotropic blackbody radiation. It is very useful when a mm-wave or THz detector is exposed to thermal "background" radiation over a much greater solid angle than the desired incident signal. The Rayleigh-Jeans approximation then applies with the simple result (written in terms of the free-space wavelength, c/ν),

$$I_\nu = \frac{1}{A}\frac{dP_{inc}}{d\nu} = \frac{2\pi \cdot \varepsilon \cdot k_B T_S}{\lambda^2}.$$

A second form of the blackbody law is more useful when one is concerned with calculating the thermal radiation into a mm-wave or THz receiver over a narrow solid angle as defined, for example, by a receiver antenna. In this case, one quantifies the power per unit area emanating from the reference surface of area A in the direction perpendicular to the surface and over a solid angle dΩ. The resulting *brightness* function is given by

$$B_\nu = \frac{2h\nu^3 \cdot \varepsilon}{c^2 (e^{h\nu/k_B T_S} - 1)}$$

Note that the irradiance derives from the brightness by integrating over a hemisphere and weighting by the Lambert factor. And in the Rayleigh-Jeans limit one obtains the following form

$$B_\nu = \frac{2 \cdot \varepsilon \cdot k_B T_S}{\lambda^2} \equiv \frac{2 \cdot k_B T_B}{\lambda^2} \qquad (9)$$

Note the definition given here for the brightness temperature T_B includes the emissivity, which is not known for most sources. This is the temperature quantity most often used in remote sensing, particularly mm-wave and THz astronomy.

Fig. 3 shows brightness curves for three temperatures, 290 K, 120 K, and 40 K. The values T = 120 K and T = 40 K correspond to the typical sky equivalent blackbody temperature looking up from sea level along the zenith at 100 GHz and 70 GHz, respectively.[8] Also shown are the linear curves for the Rayleigh-Jeans approximation. Clearly the latter is a good approximation at 290 K, deviating by at most a factor of 2.6 at the highest plotted frequency of 10 THz. But as the temperature decreases, the Rayleigh Jeans becomes progressively worse, overshooting the peak that occurs in the THz region for temperatures roughly less than 150 K. The displacement of this peak to longer wavelength with decreasing temperature is the famous Wien law,[9]

$$\lambda_{max} T = \text{constant} = 5.098 \text{ [mm-K]}$$

For example, the peak brightness of a source having a frequency-independent ε and a physical temperature of 290 K is 17.6 μm (17.1 THz). This peak shifts to λ_{max} = 127 μm (2.35 THz) for a 40-K source, and 1.89 mm (159 GHz) for a 2.7-K source. The latter temperature corresponds to the cosmic background brightness temperature – a remarkable discovery by radio astronomers and strong evidence for the "big bang" theory of the universe. The 159-GHz peak also suggests that the THz region is very important for cosmic radiative transfer – a key reason why much of the technology development in the THz region during the past 25 years or so has been driven by the astronomy application.

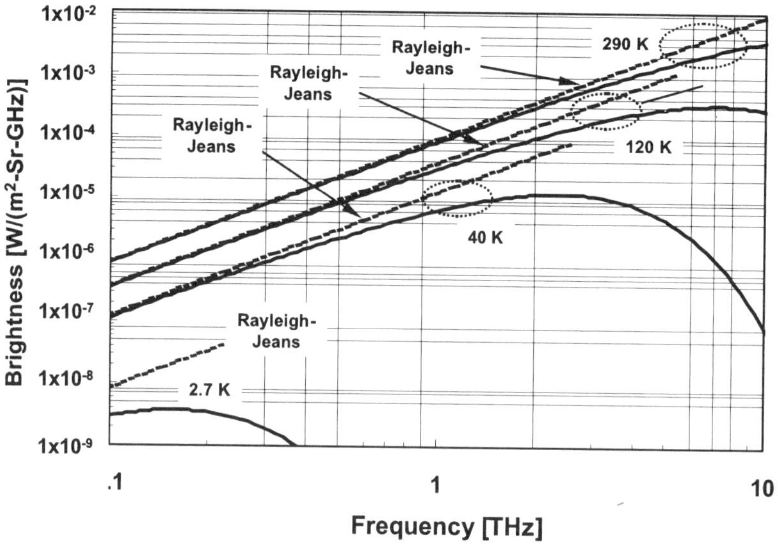

Fig. 3. Blackbody brightness spectra in THz region for four different temperatures.

D. Fluctuations of Radiation

Although most of the principles of electromagnetics necessary to understand THz sensors were understood prior to 1900, a few came later with the advent of quantum mechanics. Perhaps the most fundamental principle is the occurrence of fluctuations in the radiation. Radiation fluctuations associated with the sensing process does depend on the spectral region and has two limiting forms: classical (electric-field thermal-noise) fluctuations in the long-wavelength limit and corpuscular (photon shot noise) fluctuations in the short-wave limit. In the THz region, both types of radiation noise may be exhibited depending on the type of sensor, the type of radiation detected, and other factors. Since a quantitative understanding of radiation noise cannot be made without the tools of statistical mechanics and probability theory, this section starts out with a short review of the basic concepts in these two fields.

D.1 Quantum Derivation

Quantum mechanics teaches us that the act of measuring radiation, like the measurement of any other physical observable, is inherently probabilistic with a metric of uncertainty or deviation given by Planck's constant h. Even when measuring the power from an extremely coherent source, such as a laser or high-Q cavity oscillator, one expects a fluctuation in the measurement results depending on the type of radiation that one is measuring. In general, this is a complicated issue in quantum statistical mechanics or quantum electrodynamics, but the analysis becomes relatively simple in two extremes: (1) coherent radiation, or (2) thermal, or blackbody, radiation. In these cases, we learn that the most common description for radiation fluctuations is the Poisson statistics.

A convenient launch point for the discussion of all radiation noise is Eqn 6, which is generally applicable to any incident radiation of arbitrary photon statistics provided that the fluctuations are perfectly random. The fluctuation in the incident power around the average is then given by

$$\Delta P_{inc} \equiv P_{inc} - < P_{inc} > = \sum_{m}^{M} \int_{\nu_0}^{\nu_0+\Delta\nu} h\nu \cdot \{n_m(\nu) - < n_m(\nu) >\} d\nu \equiv \sum_{m}^{M} \int_{\nu_0}^{\nu_0+\Delta\nu} h\nu \cdot \Delta n_m(\nu) \cdot d\nu$$

where n is now to be regarded as a discrete random variable. Probability theory teaches us that an important measure of the fluctuations is the variance, or "mean-square" fluctuation

$$< (\Delta P_{inc})^2 > = < (\sum_{m}^{M} \int_{\nu_0}^{\nu_0+\Delta\nu} h\nu \cdot \Delta n_m(\nu) \cdot d\nu) \cdot \sum_{m'}^{M'} \int_{\nu_0}^{\nu_0+\Delta\nu} h\nu \cdot \Delta n_{m'}(\nu) \cdot d\nu) >$$

where m' is a dummy summation index. In evaluating this expression we utilize the fact that perfectly *random* fluctuations in different orthogonal modes will be uncorrelated, so that only cross-products having the same mode index will survive the ensemble averaging

$$< (\Delta P_{inc})^2 > = < \sum_{m}^{M} (\int_{\nu_0}^{\nu_0+\Delta\nu} h\nu \cdot \Delta n_m(\nu) \cdot d\nu)^2 >$$

In general, this expression is quite difficult to evaluate because it requires the knowledge of the radiation statistics in all possible modes at all possible frequencies. It is greatly simplified by the "band-limited" assumption that the spectral width Δν is narrow enough that both hν and Δν can be considered constant over the range of the integration. One then gets

$$< (\Delta P_{inc})^2 > \approx < \sum_{m}^{M} (h\nu_0 \Delta n_m(\nu_0) \cdot \Delta\nu)^2 > = \sum_{m}^{M} \cdot (h\nu_0 \Delta\nu)^2 < [\Delta n_m(\nu_0)]^2 > \qquad (10)$$

where the order of mode summation and ensemble averaging have been interchanged since Δn is the only random variable.

To get the mean-square fluctuation of the absorbed power, we weight each term in Eqn 10 by the power coupling coefficient

$$< (\Delta P_{abs})^2 > \approx < \sum_{m}^{M} \eta_m \cdot (h\nu_0 \Delta n_m(\nu_0) \cdot \Delta\nu)^2 > = \sum_{m}^{M} \eta_m \cdot (h\nu_0 \Delta\nu)^2 < [\Delta n_m(\nu_0)]^2 >$$

Note that even after band limiting, the sum must still be carried out over η_m and n_m since the coupling efficiency of different spatial modes in the THz region is highly variant, and the radiation statistics may be different than Boltzmann. In the special case of only one spatial mode, we find

$$< (\Delta P_{abs})^2 > = \eta \cdot (h\nu_0 \Delta\nu)^2 < \Delta n(\nu_0) >^2 \qquad (11)$$

In the literature on remote sensing and RF signal processing this narrow Δν approximation is often called "band-limited". In the author's experience, this is too often treated as an assumption rather than an approximation. Fortunately, it almost always leads to best-case predictions, often called fundamental noise limits. Experimental reality can only be worse, so system engineers need not be concerned about violating the "laws of physics." These fundamental limits will be examined in the subsequent sections in the two special cases of purely coherent and purely incoherent radiation.

D.2. Coherent Radiation: Poissonian Photon Statistics

We start with the case of coherent radiation – so important to the wide array of active sensors (e.g., radars) used in the RF region, and passive coherent sensors that use a local oscillator. When quantized, this form of radiation becomes the coherent state, whose quantum statistics are given by the Poisson distribution.

$$p(n) = \frac{[<n>(t)]^n}{n!} e^{-<n>(t)}$$

where $<n>(t) \propto \sin^2(\omega_c t)$. This state corresponds to only one electromagnetic mode at frequency ν_c which generally has a very narrow bandwidth $\delta\nu$ that is assumed to lie within the sensor passband, $\Delta\nu$. To apply Eqn 11 above, we need to calculate the variance of the Poisson distribution – a standard exercise in probability theory. It is found that

$$<(\Delta n)^2> = \sum_0^\infty (\Delta n)^2 p(n) = <n>$$

In other words, the variance is equal to the mean. Substitution then yields the power variance

$$<(\Delta P_{abs})^2> \approx \sum_m^M \eta_m \cdot (h\nu_C \Delta\nu)^2 <n_m(\nu_C)>$$

To simplify this further, we note that the use of Eqn 11 assumes that the mean population per longitudinal mode is flat with n, which is clearly not the case with a coherent source. But we can replace the coherent source with an effective source, flat across $\Delta\nu$, through the relation $<n_m>_{eff} = P_{inc,m} / (h\nu_c \Delta\nu)$, where $P_{inc,m}$ is the time-averaged incident power in the mth spatial mode. When integrated over the sensor spectral passband, this always gives the correct photon flux, and leads to the power variance

$$<(\Delta P_{abs})^2> \approx \sum_m^M \eta_m \cdot h\nu_C \cdot \Delta\nu \cdot P_{inc,m} \qquad (12)$$

In the special but practical case that the coherent power is contained in only one spatial mode, we can immediately derive the standard deviation and RF power spectral density

$$S_P(\nu) \equiv \frac{\sqrt{<(\Delta P_{abs})^2>}}{\Delta\nu} = \frac{\sqrt{\eta \cdot h\nu_C \cdot P_{inc}}}{\sqrt{\Delta\nu}} \qquad (13)$$

Note that Eqn 12 represents another interesting influence of quantum mechanics on sensor theory. Namely, fluctuations occur in the incident power of a coherent oscillator that, classically, one would expect to be noise-free. These fluctuations occur because of the quantized nature of the photons in the oscillation and the fact that such quantization always brings an uncertainty when "measurements" are made. In this case, measurement is simply the process of absorption.

D.3 Incoherent or Thermal Radiation: Boltzman Photon Statistics

For any remote sensor calculations, particularly in the THz region, one needs to consider the fluctuations of thermal radiation. The terrestrial thermal radiation level is quite high, and even in outer space there is a (cosmic) background radiation corresponding to a 2.7-K thermal source. It is a basic exercise of statistical mechanics to show using the harmonic oscillator energy function and Boltzman statistics that

$$<(\Delta n)^2> \equiv <(n-<n>)^2> = <n^2> - (<n>)^2 = <n>(<n>+1) = f_P(1+f_P) \qquad (14)$$

where f_P is the Planck function. This expression has two interesting limits. In the low-frequency, or Rayleigh-Jeans limit when $h\nu << kT$, $f_P >> 1$ and $<(\Delta n)^2>$ goes to f_P^2, which is approximately $(kT/h\nu)^2$. In the high-frequency limit when $h\nu > kT$, $f_P < < 1$ and $<(\Delta n)^2>$ goes to $<n>$ or f_P , which goes to $e^{-h\nu/kT}$. The first limit is generally valid in the THz region at typical terrestrial temperatures. The second limit is generally anywhere in the near-infrared or shorter wavelengths under the same terrestrial conditions. For example, for T = 300 K, $<(\Delta n)^2>$ aproaches $<n>$ asymptotically just above 10 THz.

From the viewpoint of probability theory, note that the tendency for $<(\Delta n)^2>$ to go to $<n>$ for low photonic state occupancy is exactly the behavior predicted by Poisson statistics, and the tendency to go to $(<n>)^2$ at high occupancy is the behavior expected for binomial statistics, which become Gaussian in the limit of high state density. It is then appropriate to think of thermal radiation of all types as inherently Poissonian. This is fully consistent with the central-limit theorem of probability theory.[10]

Substitution of Eqn 14 into the band-limited expression for absorbed power fluctuation of M spatial modes then yields

$$<(\Delta P_{abs})^2> \approx \sum_m^M \eta_m(\nu_0)\cdot(h\nu_0\Delta\nu)^2 <n_m>(<n_m>+1) = [\sum_m^M \eta_m(\nu_0)]\cdot(h\nu_0\Delta\nu)^2 f_P(f_P+1)$$

where the sum over modes is carried out only over the coupling efficiency. This expression represents a remarkable effect of thermal photons called "bunching." As the occupancy of photon states grows past unity, there is a tendency for the fluctuations to grow in the amplitude much more quickly . This is a ramification of the fact that photons are really massless bosons and, as such, are correlated quantum mechanically such that they tend to condense in a single state at a rate that grows with the population of that state. Remarkably, such "bunching" occurs routinely in THz sensors and RF sensors, in general, as shown in the next section. It has a classical explanation in terms of mixing between the electric fields in the same spatial but different longitudinal modes.[11]

D.4 Low-Frequency Limit: The Johnson-Nyquist Theorem

A very important result for power fluctuations in the low-frequency limit is the Johnson-Nyquist theorem. Since in this limit $h\nu/kT$ is very small, $e^{h\nu/kT} \approx 1 + h\nu/kT$ (by Taylor expansion) so that $f_P \approx kT/h\nu >>1$, and thus $<(\Delta n)^2> \approx (kT/h\nu)^2$, which is a big number at THz frequencies for T near room temperature. As done above, let's suppose that a sensor is receiving thermal noise power in this limit contained in M spatial modes over an arbitrary spectral bandwidth Δν. In this case the mean-square absorbed power from the statistical fluctuations Δn is given by

$$<(\Delta P_{abs})^2>=[\sum_m^M \eta_m](h\nu_0\Delta\nu)^2 f_P(f_P+1)\approx\cdot(h\nu_0\Delta\nu)^2(kT/h\nu_0)^2=[\sum_m^M \eta_m](kT\Delta\nu)^2$$

which can be stated in a more familiar form as

$$P_{RMS}\equiv\sqrt{<(\Delta P_{abs})^2>}=\sqrt{\sum_m^M \eta_m}\cdot k_B T\cdot\Delta\nu=<P_{abs}>\Big/\sqrt{\sum_m^M \eta_m}$$

So by comparing this to the average power of Eqn 7, we observe the remarkable result that for one spatial mode (M=1) and perfect coupling, the rms absorbed power fluctuation is exactly equal to the average power – a sign of Gaussian statistics.

An important consequence of the Johnson-Nyquist theorem follows from the relationship established above between the fluctuations and the power spectral density, which allows us to write

$$S_P(\nu)\approx\frac{P_{RMS}}{\Delta\nu}=\sqrt{\sum_m^M \eta_m}\cdot k_B T$$

a result of such fundamental importance to THz sensor theory, and RF systems in general, that it is found in nearly every calculation of system-level performance.

D.5. High-Frequency Limit: Photon Shot Noise

Although RF systems are traditionally analyzed assuming validity of the Johnson-Nyquist theorem, the THz region is exceptional because for some systems hν can approach kT so that <n> drops to order unity. As done above, let's suppose that a sensor is receiving thermal noise power in M orthogonal spatial modes over a "band-limited" Δν that is narrow enough that <n> can be treated as a constant. Let's also assume that the electromagnetic waves associated with these spatial modes are all equally-well matched to the sensor (i.e., all incident waves experience the same coupling efficiency). With these assumptions, the power from the fluctuations Δn is given by

$$<(\Delta P_{abs})^2>=\sum_m^M \eta_m\cdot(h\nu_0\Delta\nu)^2 f_P(f_P+1)$$

which can be written immediately as an effective power spectral density

$$S_P(\nu)\approx\frac{P_{RMS}}{\Delta\nu}=h\nu_0\sqrt{\sum_m^M \eta_m}\sqrt{f_P(f_P+1)}$$

This can be clarified somewhat by noting that the mean power incident from all spatial modes is simply $\sum_m^M f_P\cdot h\nu\cdot\Delta\nu$, so we can re-write this as

$$S_P(\nu)\approx\frac{P_{RMS}}{\Delta\nu}=\frac{\sqrt{h\nu_0\cdot[\sum_m^M \eta_m]\cdot P_{inc}\cdot(f_P+1)}}{\sqrt{\Delta\nu}}$$

In the high-frequency limit, $h\nu >> k_B T$, we have $f_P << 1$ so that the fluctuations in absorbed power become

$$<(\Delta P_{abs})^2>=\sum_m^M \eta_m\cdot h\nu_0\cdot\Delta\nu\cdot P_{inc}\ ,$$

and the spectral density is given by

$$S_P(\nu) \approx \frac{P_{RMS}}{\Delta\nu} = \frac{\sqrt{h\nu_0 \cdot P_{inc} \cdot \sum_m^M \eta_m}}{\sqrt{\Delta\nu}}$$

Note that in the unimodal case, this is identical in form to the spectral density for a coherent signal, Eqn 13. It is satisfying that such a fundamental result can be derived in two different ways. As will be explained later in Sec. V.B, this result is called "photon shot noise".

E. THz Propagation

E.1 Atmospheric Absorption

More than any other region of the electromagnetic spectrum, the atmospheric transmission in the THz region is dominated by one factor – water vapor. Being a polar molecule with a nonlinear molecular orientation, water displays a strong absorption line for nearly all of its rotational modes. Most of these modes and their optical properties have been conveniently catalogued over the past few decades into a database called HITRAN96. Started by the U.S. Air Force Geophysics Laboratory, the database contains the kinetic coefficients for over 1 million atomic and molecular lines of importance to atmospheric radiative transport between the microwave and visible regions. The set of water vapor lines occurring between 100 and 1000 GHz is shown as a "stick" diagram in Fig. 4. The ordinate is a line-strength coefficient for the Voight model of atomic absorption – a model that combines the essential characteristics of the Lorentzian model of collisional broadening with the Gaussian model of Doppler broadening.

Note that the spectral region in Fig. 4 contains 228 spectral lines. Perhaps the best known of these is the 183 GHz line, which has been used by NASA and other government agencies as the basis for space-based atmospheric-sounding radiometers. Remarkably, many of the water lines that occur at higher frequencies approaching 1 THz and beyond are orders-of-magnitude stronger than the 183-GHz line.

Given the water lines and a model of atmospheric pressure and temperature, one can compute the transmission over chosen paths using a radiative transport code. A popular tool over the past 20 years has been FASCODE, also developed by the U.S. Air Force. In the late 1990s this was converted to a Windows environment.[12] An example transmission spectrum up to 2.0 THz is plotted in Fig. 5. This was computed over a 1 km path length at 1 atm of pressure, 296 K temperature, 60% relative humidity, and a slant angle of 5°. The strongest lines of Fig. 4 are now pressure broadened, and even stronger lines occur above 1.0 THz. In places where neighboring lines are in close proximity, the absorption remains very strong between the features, creating what some have called the "THz pea soup."

An important comment about the HITRAN96 database is that all of its absorption lines apply to monomeric water. At 1 atm pressure and high humidity, it is known that water behaves in a more complicated fashion whereby individual molecules interact to form dimers, trimers, etc. This is, of course, the physical basis for precipitation so it is not surprising that it is occurring at high humidity. The effect of dimerization is an increase in the number of lines in Fig. 4 and a further thickening of the absorption spectrum in Fig. 5.

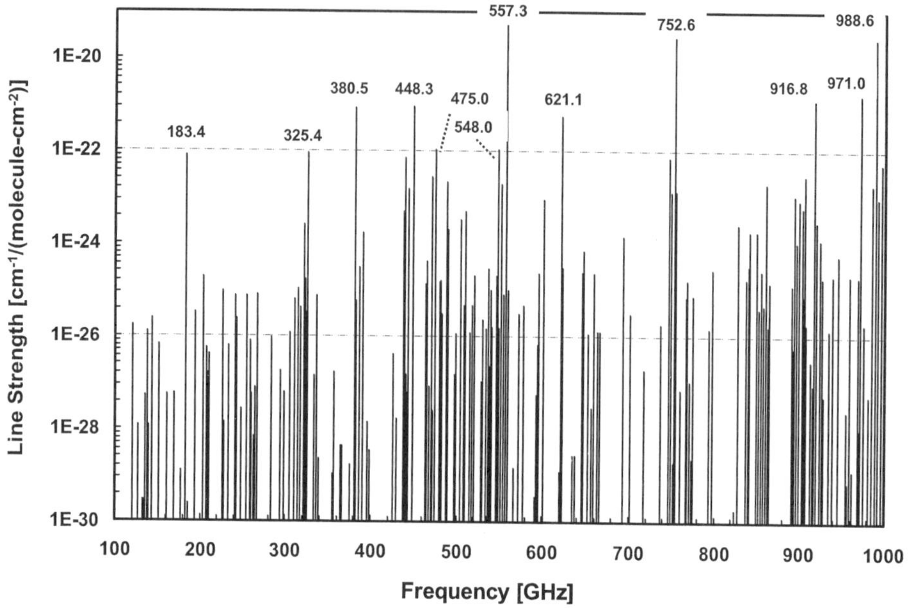

Fig. 4. Stick diagram of water vapor absorption lines between 0.1 and 1.0 THz (from HITRAN96 database). Labels denote the center frequency of 12 strongest lines.

E.2. Effect on THz Signals: Radiative Transfer

Given the existence of a molecular absorption and a pressure at or near 1 Atm, the density of molecules is generally large enough that the interparticle separate is less than a wavelength in the THz region. Hence, the attenuation from absorption can be approximated through the exponential fall in intensity predicted by the Lambert-Beer law

$$\tau \equiv \frac{I_t}{I_i} = \exp(-\rho \cdot \sigma_A \cdot z)$$

where I_i is the incident intensity, I_t is the transmitted intensity, ρ is the particle density, σ_A is the absorption cross section, and z is the path length.

Of course, absorption is concomitant with energy transfer, so we expect an absorbing medium to also contribute radiation according to the principle of detailed balance. The overall radiative transfer has a well-known solution when the source body of thermodynamic temperature T_S subtends a solid angle Ω with respect to the sensor and the intervening medium is much broader than Ω. Because brightness is the quantity conserved over a fixed solid angle, it is the convenient radiative function and is given at the sensor receiver by[13]

$$B_R = \tau(\nu)\, B_S + (1-\tau)\, B_M$$

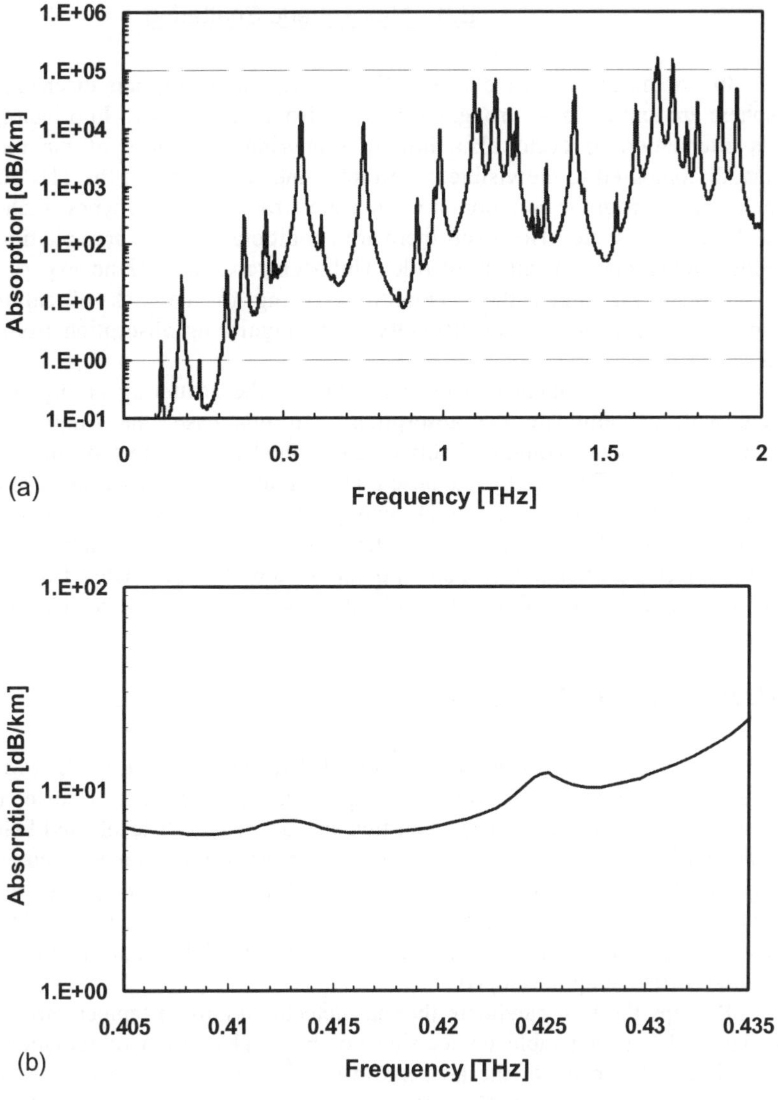

Fig. 5. (a) Transmission of atmosphere through 1 KM of path as computed by PcLnWin. (b) Magnified view of the region around 420 GHz, which will be used later in the chapter for an active sensor simulation.

where B_S and B_M are the brightnesses of the source and medium, respectively. Again, for terrestrial remote sensing the Rayleigh-Jeans limit is quite accurate in the mm-wave and THz regions, so that from the definition of brightness Eqn 9, we can write

$$T_R = \tau(\nu)\, T_S + (1-\tau)\, T_M$$

Note that if the background and medium temperatures are in thermodynamic equilibrium, $T_S = T_M$ and we find $T_R = T_S$.

E.3. Atmospheric Scattering

In the millimeter-wave and THz regions, the absorption mechanisms in the atmosphere are known to be strong, particularly from water vapor. In principle, one also expects attenuation to occur also through scattering. Because of the much longer wavelength compared to the visible or infrared, one would expect the THz scattering to be much weaker from the common, micron-scale airborne particle types, such as fog, ice grains, dust, pollens, etc. However, there are conditions, such as precipitation, in which scattering should play an important role. Unfortunately, little or no experimental data exists on scattering under these conditions owing to the lack of calibrated THz instrumentation and the general difficulty in distinguishing absorption from scattering effects.

A general statement can be made that if the scattering is strong enough to compete with or dominate the absorption. In this case, the simple "rectilinear" expression given above should probably be abandoned and a more sophisticated radiative transfer carried out. To see why, suppose that a ground-based sensor and airborne source are separated by a rain cloud thick enough that the attenuation by scattering is comparable to that by absorption. Under this condition, the raindrops will almost certainly reflect significant radiation into the sensor that originates from the ground below. The reader is referred to two excellent texts that address the scattering problem in great detail.[14,15]

F. Measurement of THz Power

An essential task for the calibration of detectors and qualification of sources is the accurate measurement of power and frequency. Measurement of power in the THz region is a surprisingly difficult task, particularly at microwatt levels and below. As in the lower millimeter-wave region, thermal detectors are often the most accurate detector, although not the most sensitive. This is because of their relative simplicity and ability to efficiently absorb the radiation. The practical requirement on any thermal detector is the sensitivity be good enough to detect the low ($<< 1$ μW) THz output that occurs when a new source is first being developed.

Perhaps the best candidate thermal detector for room temperature operation is the Golay cell – a venerable device that combines absorption of radiation by a thin metallic film and transfer of the heat into a gas.[16] The expansion of the gas moves a membrane on which is mounted an optical grating or similar component. The motion of the grating diffracts light between a source and a photodetector, causing a large change in output voltage of the detector. In the author's experience, the Golay cell is very uniform in spectral response, being limited only on the long-wavelength end by the entrance aperture. For the typical 3-mm-aperture Golay cell, this long-wavelength cutoff is roughly 100 GHz. For frequencies around 1 THz, the NEP is roughly 2×10^{-10} W-Hz$^{-1/2}$. For the typical detector bandwidth of 10 Hz, the minimum detectable power is thus ~1 nW.

The best candidate for cryogenic operation is the composite bolometer operating at 4.2 K (liquid helium).[17] This device consists of a thin absorbing film in an integrating cavity on which is mounted a small Si (or perhaps some other semiconductor) thermistor having a high coefficient of thermal resistance, $(1/R)\, dR/dT$. The key advantage of cryogenic over room-temperature operation is reduction in the thermal noise while maintaining high responsivity. Around 1 THz the 4.2-K Si composite bolometer can provide an optical NEP of roughly 1×10^{-12} W-Hz$^{-1/2}$. So in the typical bandwidth of 10 Hz, the minimum detectable power is ~10 pW. The disadvantage of the composite bolometer compared to the Golay cell is uniformity of response. Because of standing waves that can occur in an integrating cavity, the spectral variation in responsivity can be a factor of ten or more – a fact not appreciated by many people in the field.

III. Coupling of THz Sensors to Free Space

A critical aspect of any remote sensor is "connecting" THz radiation between the various signal processing components and devices, and "coupling" the THz radiation to and/or from the external medium in which the target is embedded. The connecting task, generic to all electromagnetic sensors, is carried out with THz transmission lines. The coupling is carried out with various types of antennas.

A. Routing Between Devices: THz Transmission Lines

THz transmission lines are usually scaled-down versions of microwave and millimeter-wave transmission lines and, as such, are relatively simple to analyze and design. The two most common types are metallic waveguides and coplanar-strip transmission lines. The former type is popular because if fabricated with high conductivity metal (e.g., oxygen-free copper) and smooth inside walls, it can display high bandwidth and low insertion loss. The latter type is popular because if fabricated on a dielectric or a high-resistivity semiconductor (e.g., quartz or semi-insulating GaAs), it can be integrated with THz devices such as superconducting tunnel junctions or Schottky diodes. Various types of scaled-down optical structures have also been investigated as THz transmission lines, such as dielectric waveguides. But because of high insertion loss, inability to integrate with devices, or both, these structures have not gained popularity and will not be addressed in this article.

A.1. Metallic Waveguides

Perhaps the most common THz transmission line of all is the rectangular or circular metallic waveguide. The former is generally made up to about 300 GHz according to the "WR-N (Waveguide Rectangular) standard whereby the width-to-height ratio is approximately 2.0 and N is the width in units of 1/100th of an inch. For example, WR-3 has a width W= 0.03 inch and a height H = 0.015 inch. According to waveguide propagation theory, the cut-off frequency of the fundamental (TE_{10} mode) is then $f_c = c/(2W) = 197$ GHz. WR-3 is about the smallest rectangular waveguide that is available commercially. Smaller sizes are made and follow the WR standard but are generally only made in a custom fashion.

Circular waveguide is also popular, primarily because it is easier to fabricate by simple machine-shop (e.g., milling) techniques. The fundamental mode is the TE_{11} whose cut-off properties are given by $0 = J_0(kr)$ where $k = \omega/c$ is the wavenumber, r is the radius, and J_0 is the Bessel function of the first kind and order zero. The first "zero" of this Bessel function is $J_0(1.841) = 0$, so that the cut-off wavelength in terms of diameter is simply $(\pi/\lambda_C)d = 1.841$. This yields $\lambda_C = \pi d/1.841$, or $f_C \equiv c/\lambda_C = 1.841c/\pi d$. For example, a circular waveguide having a diameter of 1 mm (a convenient size to make by conventional machine-shop techniques) displays a cut-off frequency of 175.8 GHz.

At frequencies well above the cut-off of the fundamental mode in either rectangular or circular waveguide, higher-order modes begin to propagate. For any uniform waveguide, these modes obey the fundamental relation for the cutoff frequency in terms of the width (a) and height (b) of the waveguide:

$$\nu_C^{m,n} = c\sqrt{(m/2a)^2 + (n/2b)^2} \tag{15}$$

An interesting use of this expression is the to plot the number of modes N(ν) that can propagate at any frequency as a function of frequency above cut-off. Such a curve is easily generated by a spreadsheet with sorting capability, as exemplified later in Sec. III.D.

B. Coupling to Free Space: THz Antennas

A critical aspect of any remote sensor is the coupling to the external medium in which the target is embedded. Since the external medium is usually free space, the component that carries out this coupling is traditionally called the "antenna". As in other regions of the electromagnetic spectrum, THz antennas generally fall into one of two categories: (1) wire antennas, such as dipoles, and (2) aperture antennas, such as feedhorns. The distinction between these two is a key topic of many textbooks and will not be addressed here.[18] What distinguishes the THz region from microwave bands and below is that wire-like antennas are generally fabricated on substrates and feedhorns are often operated overmoded, as described further below.

B.1. THz Wire-like Antennas

Perhaps the first wire-like antenna to be used successfully at THz frequencies is the traveling-wave, unbalanced, linear ("cat-whisker") antenna. It consists of a long wire (relative to a free-space wavelength) connected at one end to a ground plane (or a device over a ground plane) and at the other end to a feed port. To force the antenna pattern into one direction of space and thereby improve the directivity, the wire is mounted in a corner-cube reflector.[19] Historically, because of its broadband nature and small reactance, the cat whisker antenna played a key role in the development of THz technology as the demonstration vehicle for Schottky diodes, resonant-tunneling diodes, and other THz semiconductor devices.

The drawback of cat-whisker corner-cube antennas was and is the difficulty in their packaging and incompatibility with integrated circuit techniques. Therefore, during the 1970s and early 1980s, significant research and development was put into printed-circuit antennas, also called planar antennas.[20] A key key driver for this development was the success of one type of PC antenna – the microstrip patch – at microwave frequencies. By this time the utility of the patch was well known, particularly its amenability to two-dimensional arrays. But researchers found that the patch was prone to inefficiency at THz frequencies because of its propensity to launch substrate modes no matter how thin the substrate.[21] This led to the exploration of other antenna types,

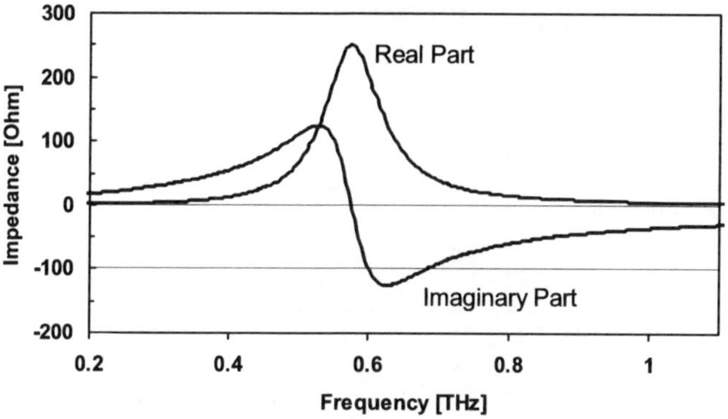

Fig. 6. Real and imaginary parts of the driving point impedance of a planar full-wave dipole antenna

including broad-side dipoles, slots, and various self-complementary antennas. End-fire planar antennas were also investigated, particularly linearly and exponentially-tapered (i.e., Vivaldi) antennas.[22]

Amongst the first planar antennas to be used experimentally were the class of self-complementary antennas, such as the bow-tie and the logarithmic spiral. These are antennas that after some rotation operation less than 2π become their own Babinet complement. For both types the required rotation angle is just $\pi/2$. To achieve this in the log spiral, the edge of each arm emanates out on a locus given by $r_0e^{c\theta}$ and each edge is rotated from its nearest neighbor by 90°. All self-complementary antennas display special electromagnetic characteristics, namely a driving-point impedance that is real and independent of frequency over a wide operational bandwidth (determined by the minimum and maximum lateral extent of the self-complementary structure). For a three-turn spiral, this bandwidth can easily by one decade. The value of the driving-point resistance is then given by

$$R_A = 60\pi/(\varepsilon_{eff})^{1/2}$$

where ε_{eff} is the effective dielectric constant given by $(1+ \varepsilon_r)/2$. For GaAs, $\varepsilon_r \approx 13.0$ so that $R_A = 72\ \Omega$. Another remarkable feature of self-complementary spirals is their insignificant load susceptance B_L over the bandwidth.

For some devices a higher antenna driving-point resistance is desirable. A good choice is a resonant planar antenna, such as a planar dipole or slot. The driving-point impedance of a planar dipole on a semi-infinite GaAs half-space is shown in Fig. 6.[23] Note that its resonant resistance is 250 Ω - over three times greater than that of the self-complementary spiral.

Coupling wire-like antennas to free space

Although planar antennas offer a large variety of pattern and impedance characteristics, they do not necessarily solve the primary problem behind patch antennas – substrate modes. For example, a planar dipole on a GaAs substrate exposed to air on the back side will also support at least one substrate mode and thus will suffer from

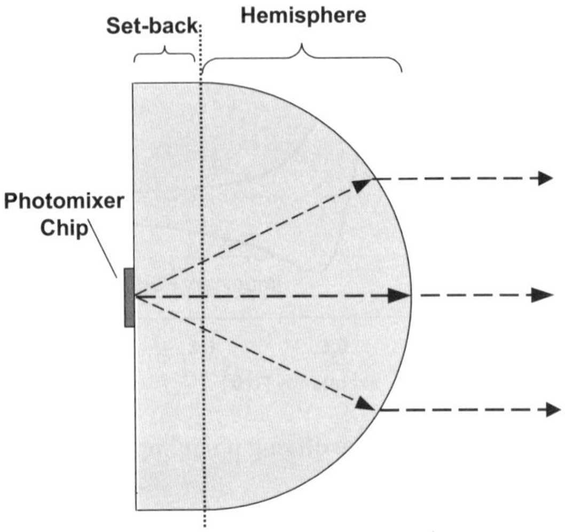

Fig. 7. Cross sectional view of hyperhemispherical optical coupling commonly used for mm-wave and THz direct detectors, mixers, and even some types of sources.

inefficiency. The problem is not so much the antenna type as it is the fundamental cause of substrate modes, which is the total internal reflection at the interfaces of the substrate. Realizing this problem, various techniques were explored to mitigate substrate mode generation and improve efficiency. Probably the most successful is to couple the radiation through the substrate using a back-side hemispheric lens whose dielectric constant is close to that of the substrate. For device integration purposes, planar antennas are often fabricated on quartz or semi-insulating GaAs. Therefore, a good choice for the lens material is quartz or a low-cost, high-resistivity semiconductor, such as float-zone silicon.

The common application of the lens-coupled antenna has been to locate a THz device, such as a mixer or direct detector diode, at driving gap of the planar antenna, and locate the driving gap at the center of curvature of the lens. If the lens is thick enough so that the spherical surface is in the far-field of the antenna, then all the radiation from the antenna reaches the spherical surface at normal incidence and passes in to free space with a reflection coefficient of $(n_L-1)^2/(n_L+1)^2$. This is a great improvement over the slab substrate, but yields a highly diverging beam from planar antennas having large beam solid angle. In such cases, a common practice is to locate the antenna behind the center of curvature at a point where the radiation will be refracted in the forward direction. If the hemispherical surface and set-back are fabricated in the same dielectric material, the resulting optic element is called a *hyperhemispherical* lens.

If n_L is perfectly matched to n_S and the antenna is located behind the center of curvature by a distance r/n, then the lens focusing is *aplanatic*. This is a property from geometric optics which states that all rays from a given point in the object plane (in this case, the planar antenna) are refracted to a parallel bundle after the lens, as shown in Fig. 7. Clearly, most planar antennas can not be accurately analyzed by geometric optics. A treatment of the focusing problem by modal (Gaussian-beam) analysis shows that the optimum set-back is just short of r/n, as might be expected.[24] This point is addressed further in Sec. III.E.

B.2. THz Aperture-like Antennas: Dishes, Lenses, and Feedhorns

Parabolic dishes: the diffraction limit

Perhaps the most universal of the aperture antenna types is the parabolic dish – a two-dimensional surface consisting of a parabola rotated 360^0 about one (optical) axis. The surface of the paraboloid is generally made from a high-conductivity material, such as a good metal. Parabolas are just as effective at THz frequencies as in the RF bands. And like their application in RF systems, parabolic dishes tend to get used as the primary reflector in the front end of THz sensors. This is because they are straightforward to design and manufacture, and display relatively predictable performance in terms of radiation coupled to free space. Their performance is much more easily predicted than practically any wire-like antenna.

To quantify the performance, we imagine a parabolic dish of diameter D lying at the origin of a spherical coordinate system whose origin is located at the center of the dish and whose polar axis coincides with the axis of symmetry. Then, as with any wire-like antenna, the electric field transmitted by the dish at a far distance from the origin will tend to display a modified spherical-wave form

$$E(r,\theta,\phi) \propto \sqrt{F(\theta,\phi)}\,\frac{e^{-jkr}}{r} \tag{16}$$

where k is the free-space propagation constant ($= \omega/c = 2\pi/\lambda$) and F is the (normalized) intensity pattern function, $F \equiv |S(r,\theta,\phi)|/S_{max}$ where S is the Poynting vector and S_{max} is its maximum magnitude, wherever in space that occurs. The range of r for which Eqn 16 is valid is called roughly the "far-field limit." Parabolic dishes, like all wire-like antennas, display a limited region in space where F (θ,ϕ) is large and other regions where it is negligible, in contrast to isotropic (point) sources. Therefore, a useful metric is the directivity

$$D = 4\pi\left(\iint_{4\pi} F(\theta,\phi)d\Omega\right)^{-1} \equiv 4\pi/\Omega_B$$

where Ω_B is the beam solid angle. Conceptually D defines how much greater the intensity is at the peak of F compared to the isotropic radiator emitting the same total power, for which $\Omega_B = 4\pi$ and $D = 1$.

Like other aperture-type antennas used in the mm-wave and THz regions, parabolic dishes generally have a pattern function that displays a predominant, symmetric or quasi-symmetric peak (i.e, "major lobe") in a single direction of space θ_P, ϕ_P. In this case it is useful to approximate $F(\theta,\phi)$ by an equivalent spherical cone or sector having a symmetry axis along θ_P, ϕ_P, and polar angular width (or widths) equal to the full-widths at the half-maximum points $\beta(\phi)$ of the real major lobe. And throughout the cone or sector, $F(\theta,\phi) = 1.0$ If the pattern has perfect conical symmetry (generally true for parabolic dishes and lenses, and often the design goal for feedhorns), then one finds

$$\Omega_B \equiv \iint_{4\pi} F(\theta,\phi)d\Omega \approx \int_0^{2\pi} d\phi \int_0^{\beta/2} \sin\theta \cdot d\theta = 2\pi[1-\cos(\beta/2)]$$

and

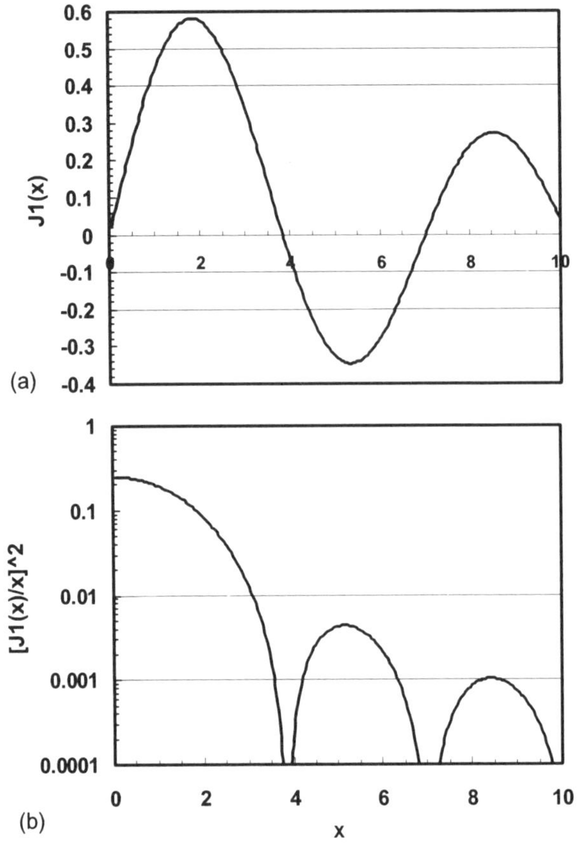

Fig. 8. (a) Bessel's function of the first kind of order 1. (b) Plot of the function $[J_1(x)/x]^2$, which according to scalar diffraction theory, is the angularly dependent part of the radiation pattern from a uniformly illuminated circular aperture.

$$D \approx \frac{2}{1-\cos(\beta/2)} \quad . \tag{17}$$

In the limit of a narrow "pencil" beam where β is small (<< 1 rad), one can Taylor expand the denominator, yielding

$$D \approx \frac{16}{\beta^2} \tag{18}$$

Note that in most books make the simpler approximation $\Omega_B \approx \beta^2$, so that $D \approx 4\pi/\beta^2$ - a less precise expression but one easier to remember.

The characterization of the parabolic dish then reduces to knowing the -3-dB full-width the main lobe. This is where dishes and some other aperture antennas, such as lenses, are simpler to analyze than wire-like antennas because they are often physically large in lateral dimension compared to a free-space wavelength and have simple boundary conditions. In this case, one can approximate the radiation pattern with scalar diffraction theory - a formalism that provides an approximate solution to the vector electromagnetic wave (Helmholtz) equation for radiation passing through the aperture.[25] The scalar formalism results in the famous Kirchoff-Fresnel integral which, in essence, approximates the radiation pattern as the superposition of point sources filling the aperture, each point source radiating a spherical wave. A key issue in using this integral is the amplitude distribution of the point sources inside the aperture. In the special case of *uniform* illumination, scalar diffraction predicts a far-field pattern that goes as[17]

$$|F(\theta,\varphi)|^2 \propto \left[\frac{J_1(ka\theta)}{ka\theta}\right]^2$$

where J_1 is the ordinary Bessel function of 1st order and θ is the angle of the measurement point relative to the optical axis. Historically, this function has generally been avoided or grossly approximated in textbooks because of its lack of tabulation. With modern computation tools (e.g., Excel, Matlab), it is simple to compute and plot, and yields much useful information.

Like the more familiar sinc [sin(x)/x] function, $J_1(x)/x$ peaks at $x = 0$. But its peak value is 0.25, not 1.0 as for the sinc(x). The first null occurs at the first zero of the J_1 function, $x = 3.835$ or $\theta = 3.835\ \lambda/(2\pi a) = 0.610\ \lambda/a$. A secondary peak of magnitude 0.00437 occurs at approximately $x = 5.14$, corresponding to $\theta = 0.818\ \lambda/a$. Note that this secondary peak (first sidelobe) has a value of 0.0175 or –17.6 dB relative to the main lobe. This is to be contrasted to the more familiar value of –13.2 dB for the relative size of the first sidelobe for a square aperture of uniform illumination.

The undulating radiation pattern of Fig. 8 and the associated minima and maxima, are collectively associated with *diffraction* – a phenomena that tends to be as important at millimeter-wave and THz frequencies as it is in the lower RF bands. This is because the size of components, such as parabolic dish antennas, tends to shrink with operating wavelength, making the important ratio λ/a approximately constant. For non-circular antennas or other components, this ratio gets replaced by λ/d where d is the maximum lateral extent of the component.

From an antenna-design standpoint, the most important feature of the diffracted radiation pattern is often the –3-dB beam width. From the uniform-illuminated case of Fig. 8, the –3-dB point is at $x = 1.616$, so that the beam full-width is given by $\beta = 2\theta = 3.232\ \lambda/(2\pi a)$. This can be substituted into Eqn 17 or approximated by Eqn 18 if the dish is large enough that $\beta << 1$ rad. In this case $D \approx 16/\beta^2 = [16(2\pi a)^2]/(3.232\cdot\lambda)^2 = 1.53\cdot[4\pi A/\lambda^2]$ where $A = \pi a^2$ is the circular area . The last quantity arises frequently in the analysis of aperture and wire-like antennas and so has a special place in the electromagnetic field as the maximum or "diffraction-limited" directivity

$$D_{max} = 4\pi\ A/\lambda^2 \qquad (19)$$

The circular-aperture example shows this expression is not precise. But it has long been used as a metric in component and system engineering, and is often assumed to be more precise than it really is. This is because it is so useful, applying also to wire-like antennas when the physical aperture is replaced by the "effective" aperture A_{eff}.

Another useful result that comes out of scalar diffraction theory is an estimate of the far-field-limit criterion. For the generic aperture antenna of uniform illumination, or a wire-like antenna, this is given roughly by the range expression

$$r > 2d^2/\lambda \qquad (20)$$

where d is again the maximum lateral extent. Note, however, that this expression is even rougher than the diffraction-limited directivity, and r might have to be much greater than this, perhaps up to $10d^2/\lambda$, depending on the exact aperture shape and boundary conditions.[26]

Given Eqns 19 and 20 one can quickly estimate the maximum directivity and far-field criterion as a function of frequency. Those accustomed to dealing with antennas in the microwave region are often impressed with how high the directivity can be in the THz region per unit aperture. For example, a round dish of 1-inch diameter operating at 1 THz (λ = 0.3 mm) has a "diffraction-limited" directivity of just over 7×10^4, or 48.4 dB ! By the same token, as the frequency goes up one must go out to an increasingly further range to apply the far-field criterion. For this same dish the far field criterion becomes r > 430 cm = 14.1 ft, which seems surprisingly far considering the size of the aperture.

If a THz antenna is separated from its target by a distance less than indicated by Eqn 20, then the far-field form of the electric field can not be approximated by Eqn 16. An important consequence of this violation is that the magnitude of electric field at a given θ can not be assumed to fall as 1/r or, equivalently, the intensity can not be assumed to fall as $1/r^2$. This is a motivation for the derivation and application of the Gaussian-beam solutions to Maxwell's equations.[27] Such modes are commonly applied to the free-space radiation transmitted or received by THz antennas, and are addressed in Sec. III.E.[28]

An interesting aspect of parabolic dishes in the THz region is the limitations imposed by surface roughness or nonparabolicity. Up to approximately 100 GHz, parabolic dishes can be made very accurately by simple mechanical means such as molding. But in the THz region, a surface error of just a few thousandths of an inch can impact the performance of the dish. So surface machining and polishing becomes necessary to achieve diffraction-limited directivity – a requirement well known to people from the optical end of the spectrum. Surface machining has been scaled up to produce large area (10-m diam) parabolic dishes operable up to at least 600 GHz in astronomical observatories.[29] Even so, a THz parabolic dish is far easier to fabricate than one in visible or IR regions where submicron-scale smoothness is required.

Lenses

Almost since the beginning of the THz field, researchers have recognized that the wavelengths in this region become a small fraction of the diameter of common-sized lenses and other transmissive optics. For example the free space wavelength at 1 THz, l = 333 micron, is 75 times less than the diameter of a 1-inch lens. In fact, lenses often are used as a paradigm for the field of "quasi-optics" – a broad term that means roughly the free-space control of THz (or lower frequency) radiation by conventional optical elements. This does not imply, however, that optical techniques for analyzing the propagation are a good approximation with THz lenses or mirrors. To get acceptable accuracy, diffraction effects usually must be accounted for either by the uniform-illumination or Gaussian-beam approximations (Sec. III.E).

An advantage of transmissive optics in the THz region compared to the visible or infrared is that they can be made from plastics – perhaps the best material choice of all from the perspective of manufacturability and low-cost. Good examples of plastics having low-enough loss and good manufacturability for lenses and windows are Teflon, polyethylene, and TPX. All of these are readily machined, and simple machine-shop surfacing methods leave small enough tool marks and shallow-enough scratches to cause negligible scattering of electromagnetic radiation in the THz region.

Feedhorns

Probably the best performance of all antennas up to about 1 THz is displayed by well-designed feedhorns. Through the microwave and millimeter-wave bands, pyramidal feedhorns are very popular because of their compatibility with rectangular waveguide and good control over the beam pattern. But in the THz region, rectangular-waveguide standards are not prevalent and, because of size scale, it becomes much easier to fabricate conical feedhorns rather than pyramidal ones. Unfortunately, conical feedhorns are not as flexible to design because of their one fewer degrees-of-freedom. They are prone to inefficient and asymmetric patterns that can not be easily designed by analytic means because of their required representation by special (Bessel) functions.

With the advent of full-wave numerical simulation tools, it has become possible to design THz conical feedhorns with excellent antenna characteristics. One of the most popular numerical codes presently available– High Frequency Structure Simlator (HFSS) – is based on the method of finite elements. Given such a simulation capability, there are several design tricks that can produce excellent patterns from conical feedhorns. One such trick is to add a step in the throat region to excite just enough of the first higher-order (TM_{11}) mode to cancel the sidelobes in the fundamental (TE_{11}) mode.[30] A second trick is to fabricate grooves down the length of the conical horn to excite the mode exclusively. The resulting design, called a scaler feedhorn, produces outstanding beam patterns with practically no sidelobes – arguably the best possible pattern from any antenna in the THz region. Unfortunately, scaler feedhorns are difficult to design and even more difficult to fabricate.

As more and more modes begin to propagate in the circular waveguide to which the feedhorn is connected, the radiation pattern begins to resemble the ray transfer function of a condensing cone, sometimes called a light cone. In other words, the feedhorn begins to follow the laws of geometrical optics. In the 1970s infrared researchers began investigating other possible condensing cone tapers besides the simplest (linear) case. For example, Winston discovered that a dual-parabolic taper can produce a transfer function near the ideal (step) function.[31] Not surprisingly, the Winston cone concept works very well at THz frequencies provided that the throat of the horn is heavily overmoded.

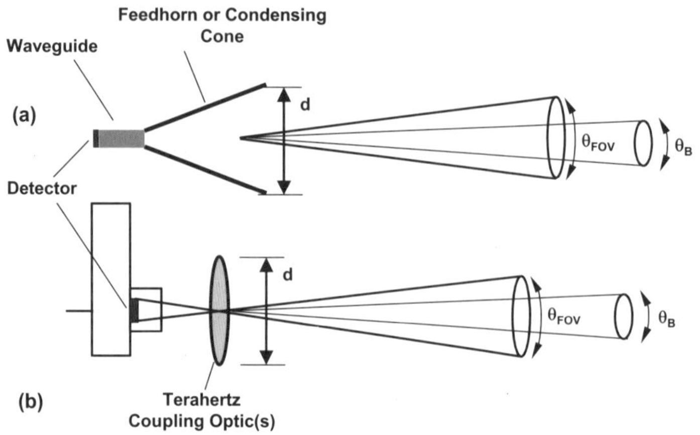

Fig. 9. (a) Feedhorn and (b) Lens coupling of mm-wave and THz detectors and devices

C. The Number of Modes: Antenna Theorem and Interpretation of Planck's Law

Whether the sensor is collecting coherent or thermal radiation, the number of spatial modes collected is an important issue from a system standpoint, is difficult to estimate, and is even more difficult to measure. In the limited space here, we only provide an approximation.

The approximation technique is based on a result from electromagnetic theory called the antenna theorem, which derives from two separate definitions of directivity given above,

$$D_{\max} = \frac{4\pi A}{\lambda^2} = \frac{4\pi}{\Omega_B} \tag{21}$$

Since D_{max} is, in principle, the maximum possible value of the directivity as predicted by diffraction theory, then the corresponding Ω_B is the minimum possible beamwidth and, therefore, corresponds to the *fundamental spatial mode* of the antenna. So if we imagine rotating the Ω_B beam in spherical coordinates to just fill up surface of a sphere, it would take approximately $4\pi/\Omega_B$ rotations to do this. And since Ω_B is the fundamental spatial mode, D_{max} represents the number of spatial modes required to fill the entire sphere. Of course, rarely, if ever, is a THz sensor designed to operate over an entire sphere. More often, they are designed to respond to a much smaller solid angle, called the field-of-view, Ω_{FOV}. And the number of spatial modes then becomes

$$M = \frac{\Omega_{FOV}}{\Omega_B} \tag{22}$$

Note that the correspondence represented by Eqn 21 is often stated as the following "antenna theorem"

$$A \cdot \Omega_B \geq \lambda^2$$

Conceptually, this means that if the antenna is diffraction-limited, the product will reach its minimum value of λ^2. But it reminds us that practical antennas have electromagnetic or mechanical limitations which usually cause the fundamental beamwidth to grow beyond this minimum value.

Illustrations are given in Fig. 9 for the Ω_B and Ω_{FOV} for the two most common millimeter-wave and THz coupling structures: feedhorns (or coupling cones), and quasi-optical components (e.g., lenses). In both cases, one can always define the beam angle Ω_B. For example, for the simple round lens, it is predicted from scalar diffraction theory $\theta_B \approx 2.44\ \lambda/D$, and $\Omega_B \approx (\theta_B)^2$ if $\theta_B << 1$. For the feedhorn or collection cone, there will always be a fundamental electromagnetic mode defined as that mode collected by the sensor at the maximum possible wavelength. Generally, this mode and its cutoff wavelength are defined by the dimensions of the throat of the feedhorn or cone.

An interesting application of the antenna theorem comes in a practical interpretation of Planck's radiation law for RF and THz systems. We can now re-write Eqn 8 ($\varepsilon = 1$) as

$$\frac{dP}{d\nu} = \frac{A \cdot 2\pi h \nu^3}{c^2 (e^{h\nu/k_B T} - 1)} = \frac{2\pi A}{\lambda^2} \frac{h\nu}{(e^{h\nu/k_B T} - 1)} = \frac{2\pi}{\Omega_B} \frac{h\nu}{(e^{h\nu/k_B T} - 1)}$$

where D_{diff} is the diffraction-limited directivity of the element of area A. We now recognize the factor $2\pi/\Omega_B$ as the number of spatial modes coupled into an antenna whose orthogonal-mode beamwidth is Ω_B. So the Planck law is consistent with the antenna theorem . Of course, rarely, if ever, is a THz sensor designed with a hemispherical field-of-view. So a more practical form of Planck's law is simply

$$\frac{dP}{d\nu} = \frac{\Omega_{FOV}}{\Omega_B} \frac{h\nu}{(e^{h\nu/k_B T} - 1)}$$

This is particularly useful for collecting cones, integrating cavities, and other antenna-like structures.

D. Illustrative Examples of Modal Function

For WR-10 rectangular waveguide, we can compute the number of modes at any frequency knowing its dimensions (0.10 inch wide x 0.05 inch high) and Eqn 15. The number of modes is found by incrementing m and n, sorting by the cut-off frequencies and then counting all entries including the degenerate cases. The resulting distribution is plotted in Fig. 10. Clearly, it is approaching a quadratic dependence as the frequency increases. For analytic purposes, it is very useful to fit the curve to the following form,
$M = \mathrm{Int}(\nu/\nu_R)^2 + 1$

where Int is the integer function (rounding down) and ν_R is a fit frequency somewhat greater than the lowest modal cut-off frequency of 59 GHz for the TE_{10} mode. The fit

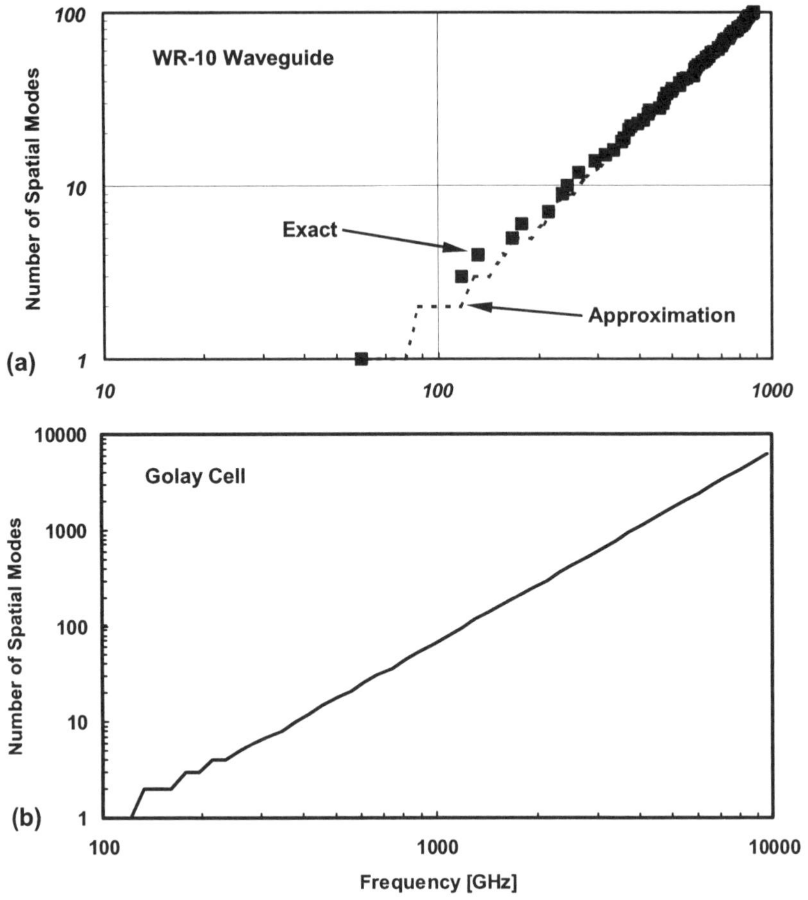

Fig. 10. Density of electromagnetic spatial modes for (a) WR-10 waveguide and (2) Golay cell having a 6-mm diameter (circular) aperture.

function is shown in Fig. 10(a) for $\nu_R = 85$ GHz.

For the lens-coupled Golay cell, we use the definition given in Eqn 22 and assume a cell diameter of 6 mm, and a lens diameter and focal length of 2 inch, respectively. The resulting modal distribution vs frequency is shown in Fig. 10(b). Because the analysis does not account for cut-off effects, the curve is perfectly quadratic and is fit very well by the same fit function as above but with $\nu_R = 0.125$ THz.

E. Gaussian Beams

E.1. Gaussian-beam Methodology

A key assumption behind the circular-aperture diffraction results in Sec. III.B from scalar diffraction theory is that the illumination across the aperture is uniform. This is a good assumption in some circumstances such as predicting the power collected by a receive antenna from a distance source whose pattern beam-width measured at the receive antenna is much larger than the lateral extent of the receive antenna. But there are other times when the uniform-intensity assumption is inaccurate, such as in describing the radiation transmitted or received by a fundamental-mode feedhorn.

One of the useful features of scalar-diffraction theory is its ability to predict what happens to the radiation once the uniform-illumination assumption is violated. For example, if the aperture is circular and if the illumination distribution is a Gaussian in the lateral plane with respect to the axis of symmetry, then the radiation pattern is also Gaussian, at least in the far-field limit. Intuitively, this makes sense since in this limit the Fresnel-Kirchoff integral reduces to a Fourier transform, and the Fourier transform of a Gaussian is always a Gaussian. It turns out that this result, commonly known as the Gaussian beam pattern, also applies to the near-field behavior with increasing accuracy as d/λ increases far beyond unity.

Although the Gaussian-beam result was known early in the history of electromagnetics, it was apparently not fully appreciated until the advent of the laser. The gain media in gas and solid-state lasers typically have very large values of d/λ, and generally emit much greater intensity at the center than at the lateral edges. A useful way to analyze the Gaussian behavior is to model the gain medium with a quadratic complex refractive-index lateral profile.[32] The resulting intensity, described mathematically below, displays a Gaussian dependence on the lateral (r in cylindrical coordinates) dimension, and azimuthal symmetry about the propagating (z) axis.

In the mm-wave and THz region, the applicability of Gaussian beams is less obvious on first glance, but becomes plausible when one considers the coupling between antennas and circuits. MM-wave and THz antennas are often operated in their fundamental spatial modes for which the radiation intensity is maximum but rather slowly varying along the propagation axis. Then at the characteristic angle $\theta \approx \beta/2$ away from the axis the radiation begins falling rapidly in the lateral directions, with some radiation inevitably occurring at larger angles in the sidelobes because of diffraction. All these properties except the undulation of the sidelobes are described rather well by a Gaussian function vs r with perfect azimuthal symmetry about the z axis. This is the so-called fundamental Gaussian mode. Other possible symmetric functions, such as a sech^2 or Lorentzian, are either too steep about the propagation axis or decay too slowly at large angles.

MM-wave and THz antennas are often operated in fundamental mode for practical reasons. One reason is that the fundamental mode is generally the most symmetric and has the smallest beam width of all possible antenna modes. Another reason is that the devices and circuits to which the antenna is coupled are designed for their own fundamental mode, be it in high-frequency transmission line or waveguide. This is usually the easiest and most effective way to design mm-wave and THz active devices and circuits, but it generally makes the coupling between the circuits and antenna efficient only for one antenna mode, usually the fundamental mode. These considerations break down, of course, if none of the electronics or components coupled to the antenna need to process radiation at high frequencies. Such is the case, for example, in mm-wave and THz bolometers which merely rectify any incident power absorbed. Hence, bolometers can be and often are mounted in multimode antenna-like structures, such as integrating cavities, to maximize the sensitivity and spectral bandwidth.

Finally, it is important to realize that any Gaussian beam is just one of an infinite number of modes forming an orthonormal TEM basis set. This basis set is as applicable to representing arbitrary radiation in free space as any other orthonormal basis, and must comply with the antenna theorem defined above in Sec. III. C. Each function of the basis set differs from the others in the degree of azimuthal symmetry about the propagation direction. As alluded to above, the fundamental or TEM_{00} mode is the only one with perfect azimuthal symmetry, and is therefore the most popular and useful in solving free-space propagation problems in mm-wave and THz systems.

E.2. Gaussian-beam Formulation

One benefit of the Gaussian-beam approach is it tends to add improved accuracy in mm-wave and THz design with only a minor increase in difficulty. This is because Gaussian beams, much like Gaussian distributions in probability theory, behave well mathematically under system-level operators. Given a propagation direction along the z axis, the fundamental TEM_{00} Gaussian beam is given in cylindrical coordinates (r,θ,z) by

$$E(r,z) = E_0 \frac{\omega_0}{\omega(z)} \exp[-r^2/\omega^2] \cdot \exp[-j(kz - \phi + kr^2/2R)] \tag{23}$$

where E_0 is the maximum electric field amplitude, η is the intrinsic impedance of the medium of propagation, ϕ is a phase constant, ω is the radius where the intensity drops by e^1 relative to the on-axis intensity (i.e., the "spot size"), or "beam waist," and R is the radius of curvature. Fortunately, all of these quantities are inter-related through simple algebraic expressions:

$$\omega^2(z) = \omega_0^2 \cdot \left[1 + \left(\frac{\lambda \cdot z}{\pi \cdot \omega_0^2}\right)^2\right]$$

$$R(z) = z \cdot \left[1 + \left(\frac{\pi \cdot \omega_0^2}{\lambda \cdot z}\right)^2\right] \tag{24}$$

$$\phi(z) = \tan^{-1}\left(\frac{\lambda z}{\pi \omega_0^2 n}\right)$$

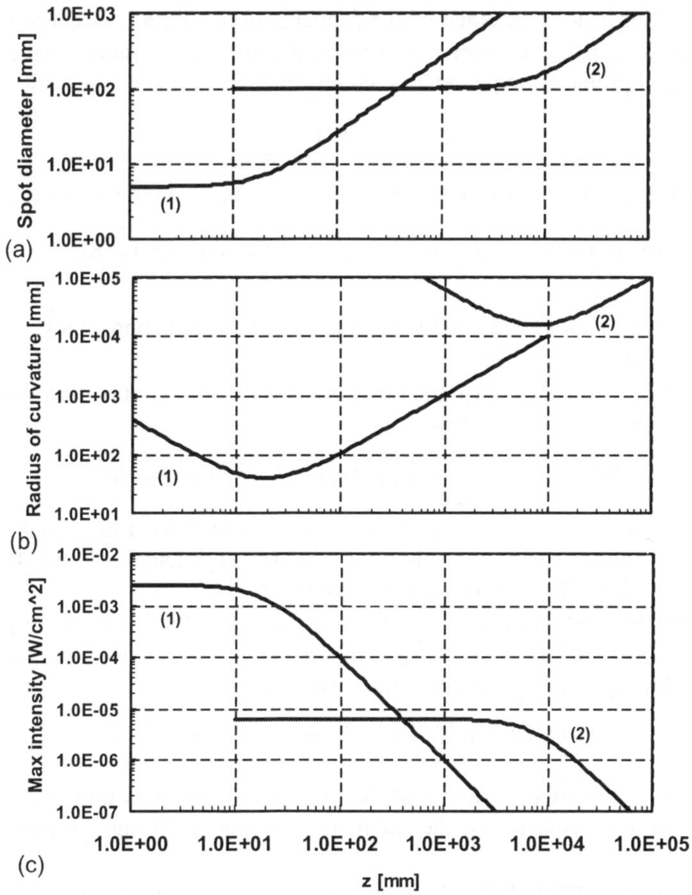

Fig. 11. Gaussian parameters: (a) spot diameter ($2\cdot\omega_0$) (b) radius of curvature, and (c) on-axis intensity for two typical mm-wave or THz situations: (1) $\lambda = 1$ mm, P = 1 mW, minimum spot size at beam waist = 2.5 mm; (2) $\lambda = 1$ mm, P = 1 mW, minimum spot size at beam waist = 50 mm.

From these expressions it is clear at $\omega(z) \geq \omega_0$, so that ω_0 is the minimum spot radius and the maximum electric field occurs in the plane of constant z where $\omega(z) = \omega_0$ which defines the "beam waist." In this plane the following simple form of the electromagnetic intensity is valid:

$$I(r) = \frac{E_0^2}{2\eta} \cdot \exp[-2r^2 / \omega_0^2]$$

where η is the intrinsic impedance of the propagation medium. Because the Gaussian integral is analytic if taken from r = 0 to ∞, the following useful relationship exists between the total propagating power and on-axis intensity at the beam waist:

$$I(r=0) = P \cdot \left(\frac{2}{\pi \cdot \omega_0^2} \right)$$

Further satisfying properties of the Gaussian beam follow from Eqns 23 and 24 in the limit of positive z $>> \frac{\pi \cdot \omega_0^2}{\lambda} \equiv z_0$. In this case, R(z) $\to$ z , $\omega^2 \to \left(\frac{\lambda \cdot z}{\pi \cdot \omega_0} \right)^2 = \left(\frac{\lambda \cdot R}{\pi \cdot \omega_0} \right)^2$, $\theta \to \pi/2$, and the complex exponential approaches $j\exp[-jkz(1+r^2/2z^2)]$. Hence, provided that the observation point is such that r < z or R, the resulting electric field behaves as

$$E(r,z) \to E_0 \frac{j\pi\omega_0^2}{\lambda R} \exp[-(\pi\omega_0 r^2 /(\lambda R)^2] \cdot \exp[-j(kR)]$$

which is the expected form of a spherical wave weighted by a beam-pattern function. The important distance parameter z_0 is called the Rayleigh length. It also happens to be the distance for which R(z) equals its minimum value.

Another satisfying property is found by tracking the r = ω or 1/e profile of the beam in the "far-field" limit z >> z_0. In this case the r = ω locus asymptotically approaches a full angle (centered about the z axis) of

$$\theta_B = 2 \cdot \tan^{-1}\left(\frac{\omega}{z} \right) \to 2 \cdot \tan^{-1} \frac{\lambda}{\pi \cdot \omega_0} \approx \frac{2 \cdot \lambda}{\pi \cdot \omega_0}$$

Redefining ω_0 as the lateral extent d of the Gaussian beam at its minimum aperture, we see once again a dependence of the far-field behavior on the ubiquitous "diffraction" ratio, λ/d.

All of these properties are exemplified in the curves shown in Fig. 11 for two representative Gaussian beams propagating in free space, one with a minimum spot diameter $2\omega_0$ = 5 mm (0.2 inch) and the other with $2\omega_0$ = 100 mm (4 inch). The wavelength of both is 1 mm and beam waists occur at z = 0. For the smaller-waist beam, the Rayleigh length z_0 is only $\approx$ 20 mm so the beam quickly diverges to a full divergence angle of about 14°. For the larger-waist beam, z_0 is about 8 m, so the beam remains highly collimated out to this distance and then begins to approach a divergence angle of just under 1°. Note that this larger beam would not be too difficult to support by a relatively simple telescope of aperture about 8 inches or more in diameter – that is, a man-portable instrument.

The behavior of the larger beam in Fig. 11 illustrates an important potential advantage of mm-wave and THz propagation over that in the lower RF bands. Namely, in applications where the remote sensor supports a Gaussian beam and the object or target is at a *short range* not much greater than the Rayleigh length, the divergence of the beam between the two can be very small. And thus the intensity will drop far slower than the $1/r^2$ spherical-wave behavior in the "far-field" of every common antenna. As will be shown later in Sec. X, this has important implications on the sensitivity and spatial resolution of practically any remote sensor.

E.3. Transformation of Gaussian Beams: A Representative System Example

A second benefit of the Gaussian beam approach is its tendency to remain Gaussian through transformation by various optical two-port components, such as lenses and mirrors. As in microwave network theory, passive optical two-ports can be represented by a number of different 2x2 matrix formulations depending on the physical formulation of the propagating electromagnetic mode. A common formulation in optics is the "ray," represented by a column or row vector [r(z), r'(z)], where r is the distance from the propagation axis and r' is the slope of the ray with respect to this axis. Optical components are represented by 2x2 ABCD matrices, and an input ray is transformed according to

$$\begin{pmatrix} r_{out} \\ r'_{out} \end{pmatrix} = \begin{pmatrix} A & B \\ C & D \end{pmatrix} \begin{pmatrix} r_{in} \\ r'_{in} \end{pmatrix}$$

Some good examples of such matrices are: (1) free space path of length L, A=1,C = 0, and D = 1; (2) thin lens of focal length f: A = 1, B = 0, C = -1/f, D = 1. The accuracy of this formulation is best for "paraxial" rays, i.e., those propagating close to the optical axis.

Remarkably, the ABCD representation also applies to Gaussian beam propagation through the definition of a complex Gaussian beam parameter

$$\frac{1}{q(z)} = \frac{1}{R(z)} - j\frac{\lambda}{n\pi\omega^2}$$

The transformation equation is then given by

$$q_{i+1} = \frac{Aq_i + B}{Cq_i + D}$$

where q_i is the beam parameter in a plane $z = z_i$. Note that the free-space ABCD matrix simply transforms as $q_{i+1} = q_i + L$.

As an illustrative example of the Gaussian beam approach, Fig. 12 shows the modeling of the radiation propagation in a system familiar to the author: the THz photomixing spectrometer. In this case, coherent mm-wave and THz radiation is generated selectively by beating two frequency-offset lasers in a small photoconductive element mounted at the driving point of a planar antenna. The planar antenna is located on a semi-insulating GaAs or InP substrate – both having very low absorption in the THz region but difficult to make antennas with because of their high dielectric constant, $\varepsilon_r \approx 13$. Therefore, the photomixer substrate is abutted to the back-side of a high-resistivity Si hyperhemisphere as discussed in Sec. III.B . The radiation coming out of the hyperhemisphere will likely be diverging, so a second focusing lens (e.g., plastic) is added at some distance to focus the radiation down to a beam waist. Since the purpose of the spectrometer is to provide radiation to a sample cell for THz spectroscopic analysis, an interesting question is if and where this beam waist will occur, and how big the minimum spot size will be.

The solution is found by first estimating the pattern coming out of the planar

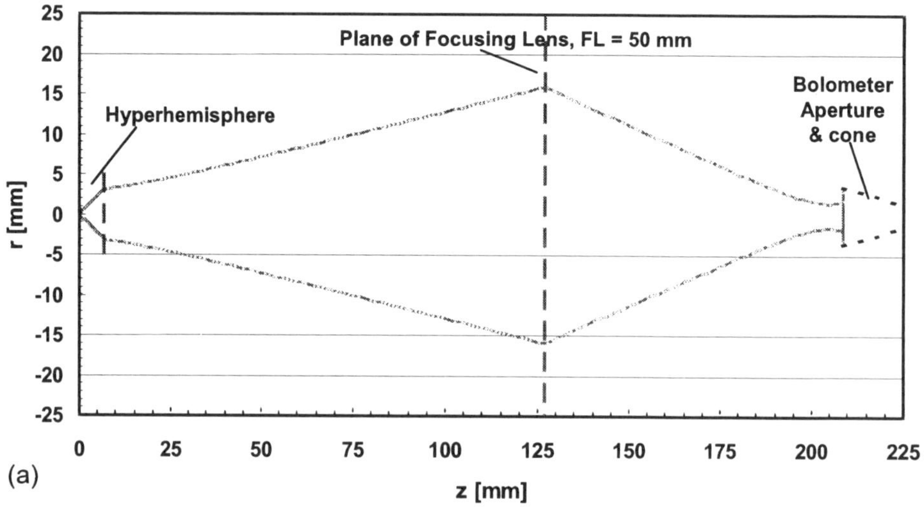

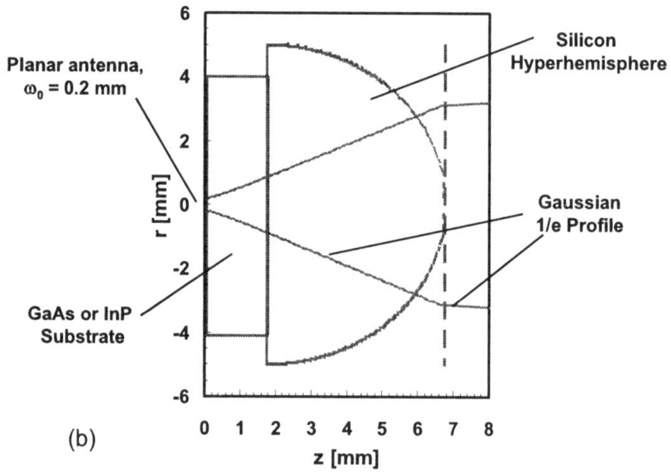

Fig. 12. (a) Results of Gaussian-beam design of THz photomixer spectrometer as a representative problem in system-level free-space radiation transformation and coupling. (b) Exploded view of photomixer region consisting of planar antenna, semi-insulating substrate and Si hyperhemisphere.

antenna as an equivalent Gaussian beam as shown in the exploded view of Fig. 12 . The two curved loci in this view represented the $r = \omega$ points. The beam propagates through the GaAs or InP substrate, into the Si hyperhemisphere, and is then transformed into free space using an ABCD matrix appropriate to a spherical-dielectric interface.[33] To avoid significant total-internal reflection, the hyperhemisphere can provide only a slight transformation of the beam, which remains diverging after passage through the Si-air interface.

The free-space Gaussian beam then propagates to the (plastic) plano-convex lens which transforms the diverging Gaussian beam to a converging beam through the application of the thin-lens ABCD matrix. By judiciously varying the hyperhemisphere-to-plano-convex separation, we achieve the beam waist shown in Fig. 12. The waist shown has a minimum spot size $\omega_0 \approx 1.5$ mm, consistent with the given frequency (300 GHz), the planar-antenna spot size (0.2 mm), the radius and set-back of the Si hyperhemisphere (5.0 and 1.76 cm, respectively), and the focal length and diameter of the plastic lens (2.0 and 2.0 inch, respectively).

F. Friis' Free-Space Transmission Equation

As discussed above, sensors in the THz region often use antennas or antenna-like structures to couple radiation to and from free space. If the antennas are unimodal, as is often the case, one can apply the concepts and terminology of RF free-space link theory, developed originally by Friis for communications systems operating at much lower frequencies. The theory assumes that the target or sample the sensor is detecting is in the "far field" of the transmit antenna, and the receive antenna is in the far field of the target or sample. In this case, even if the radiation is in the form of a Gaussian beam, the transmitted and receive waves both approach a spherical wave form, $E = E_0e^{-jkr}/r$, so that the Poynting-vector magnitude goes as $1/r^2$. And then because of the inherent reciprocity of all antennas, it makes sense to represent the receive antenna by an effective aperture A_{eff}, which is defined by

$$P_{rec} = A_{eff} \mid \bar{\vec{S}}_r(\theta_r, \phi_r) \mid \cdot \varepsilon_p$$

where P_{rec} is the power available to the antenna for delivery to a load, $\bar{\vec{S}}_r(\theta_r, \phi_r)$ is the average Poynting vector for incoming radiation along the direction $(\theta_r.\phi_r)$ in the spherical coordinates centered at the *receiving* antenna, and ε_p is the polarization coupling efficiency. Although this expression is prevalent in electromagnetics and communications texts, it is strictly valid *only when* $\bar{\vec{S}}_r(\theta_r, \phi_r)$ *is aligned with the direction of the beam-pattern maximum.* When there is mis-alignment, another factor is required which is the just the receive beam-pattern

$$P_{rec} = A_{eff} \cdot F_r(\theta_r, \phi_r) \mid \bar{\vec{S}}_r(\theta_r, \phi_r) \mid \cdot \varepsilon_p \ .$$

Now suppose that this *received* Poynting vector is generated by a second, *transmitting* antenna. We can relate the received power to the properties of the transmitting antenna by

$$\mid \bar{\vec{S}}_r(\theta_r, \phi_r) \mid \equiv \mid \bar{\vec{S}}_t(r, \theta_t, \phi_t) \mid = \tau \frac{P_{rad} D_t \cdot F_t(\theta_t, \phi_t)}{4\pi r^2} \equiv \tau \frac{P_{inc} G_t \cdot F_t(\theta_t, \phi_t)}{4\pi r^2}$$

where the subscript "t" is for transmitting, P_{rad} is the total radiated power, P_{inc} is the power used to drive the transmitting antenna (in the matched case, equal to P_{rad}), θ_t and ϕ_t are the spherical angles in the spherical coordinate system centered at the transmitting antenna, r is the distance between transmitter and receiver, and τ is the path power transmission factor. In writing this expression it is understood that F_t is taken in the direction (θ_t, ϕ_t) pointing towards the receiver, which is not necessarily the direction of the

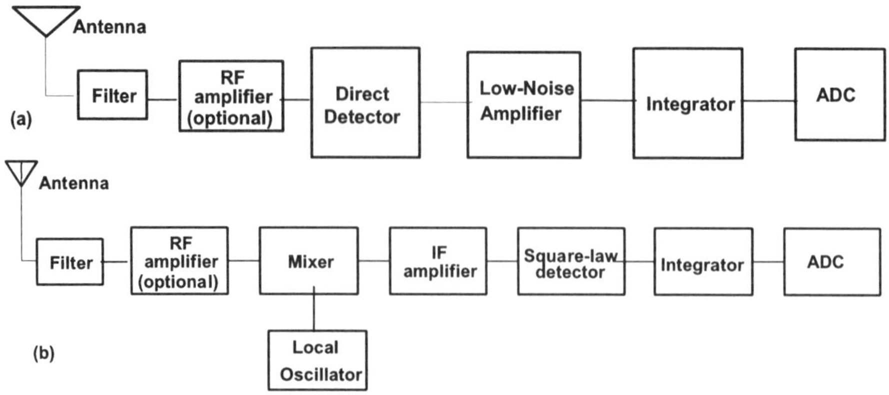

Fig. 13 (a) Direct detection receiver and (b) heterodyne (or homodyne) receiver.

maximum of F_t. Substitution into the expression for P_{rec} yields the relationship

$$P_{rec} = A_{eff} \frac{P_{in} G_t \cdot F_r(\theta_t, \phi_t) F_t(\theta_r, \phi_r)}{4\pi r^2} \tau \cdot \varepsilon_p$$

This can be simplified further in terms of the (ostensibly) known parameters of the receiving antenna using the relationships,

$$P_{out} = P_{rec} \frac{G_r}{D_r} \equiv P_{rec} \frac{G_r}{4\pi A_{eff} / \lambda^2}$$

where P_{out} is the power delivered to the load of the receiving antenna. Substitution of P_{rec} into P_{out} yields

$$P_{out} = P_{in} \left(\frac{\lambda}{4\pi r_1} \right)^2 G_r G_t \cdot F_r(\theta_r, \phi_r) F_t(\theta_t, \phi_t) \cdot \tau \cdot \varepsilon_p$$

an expression commonly known as Friis' formula, in which the *dependence on A_{eff} is cancelled out* by going from the input port to the output port. Friis' formula effectively treats the combination of antennas like a two-port network with the pattern angular dependence, polarization dependence, and path transmission included explicitly. The term $(\lambda/4\pi r)^2$ is called the free-space loss factor, which is of considerable practical and historical importance.

IV. THz Receiver Types and Performance Metrics in the Presence of Noise

A. Architectures

As discussed in the Introduction, an important defining criterion for THz sensor modality is receiver architecture. There are two types, incoherent and coherent, both shown schematically in Fig. 13. The incoherent architecture is as old as RF technology itself and the coherent came soon thereafter, dating back to the early part of the 20th century. Like many other system architectures, they persist largely by the ability of engineers to continually improve performance by perfecting the components.

A.1. Direct

The block diagram of a generic direct receiver is shown in Fig. 13(a). The incoming radiation from the target, be it thermal emission or transmitted power from the sensor itself, is collected by the receiver where it is rectified from RF (THz) to baseband by a "direct" detector. In most practical cases the baseband is defined by amplitude or frequency modulation of the incoming signal to reduce the effect of gain drifts and 1/f noise that occurs in the THz electronics. The rectified THz signal is then amplified and demodulated down to DC using synchronous detection. For AM modulation the synchronous detection is often carried out using a lock-in amplifier.

In the THz region the direct detector is almost always a power-to-voltage or power-to-current converting device. That is, it is a device that puts out a voltage or current in proportion to the incoming power. There are many examples of such devices, but the most popular are field detectors and bolometers. Field detectors, such as Schottky diodes, respond directly to the THz electric field and generate an output current or voltage through a quadratic term in their current-voltage characteristic. Bolometers are composite devices consisting of a THz absorber and a thermistor. The THz absorber is generally isolated thermally from the environment so that the absorbed THz power raises the temperature both of the absorbing layer and an attached thermistor. The thermistor is, by definition, a device that displays a large change of resistance to a small change of temperature. In some bolometers, such as the composite type, the absorber and thermistor are separate elements. In other bolometers, such as the hot electron type, they are integrated into the same device.

A key factor in all direct detectors is spectral bandwidth. As in most analyses of signal processing, we assume here that this is band limited between ν_o and $\nu_o + \Delta\nu$. This can be a real bandwidth defined by a THz bandpass filter, or it might be an approximation to a real spectrum.

A.2. Preamplified Direct

One of the most successful areas of RF electronics during the past decade has been monolithic microwave integrated circuits (MMICs). By integrating active devices, some passives, and matching circuits on the same semi-insulating substrates, it has become possible to fabricate low-noise amplifiers (LNAs) up to frequencies of 100 GHz and beyond.[34,35] For example, LNAs having a gain of 17 dB, bandwidth of 30 GHz, and a noise figure of 6 dB have been fabricated and tested around 94 GHz.[36] The advantage of an LNA is that if it has adequate gain, it can dominate the noise figure of the following square-law detector, leading to a much lower NEP than can be achieved by direct conversion using the same square-law detector. As will be shown later, the sensitivity of the preamplified direct receiver can then approach the photon-noise (quantum) limit.

A.3. Heterodyne

In the coherent system of Fig. 13(b), incoming radiation from the target, be it thermal emission or transmitted power from the sensor itself, is combined with power from a local oscillator (LO) on a THz mixer. If the signal and LO frequency are different, there will be a beat-note generated at an intermediate frequency (IF) between the two. This is called *heterodyne* conversion. It the signal and LO frequency are equal, the beat tone degenerates to dc, and the process is called *homodyne* conversion. Independent of the conversion process, all coherent detectors require a device that can generate an *efficient* conversion of the RF power to the IF band. The most popular mixers are field-type devices having a strong quadratic nonlinearity. Good examples are Schottky diodes, superconductor-insulator-superconductor (SIS) tunnel junctions,[37,38] and superconducting hot-electron bolometers.[39]

Coherent down-conversion has several unique features that distinguish it from direct detection. First, mixing a weak signal with a relatively strong LO effectively amplifies the received signal relative to the receiver noise floor, which can greatly improve the sensitivity compared to direct detection. Second, for the typically weak signals in the THz region, the mixing process is linear. That is, the signal power at the IF frequency is linearly proportional to the signal power at the input. Therefore, the receiver passband can be defined by an IF band pass filter, which is generally much lower in cost and has much higher performance than any THz filter. This feature tends to make coherent receivers the favored approach in applications requiring high spectral resolution, such as molecular spectroscopy.[40] But as will be shown below, the direct receiver tends to be preferable in wideband applications such as thermal imaging because of its superior spectral bandwidth and simplicity.

A.4. Pre-Amplified Heterodyne

An intriguing possibility in the millimeter-wave band and lower end of the THz range is a preamplifier feeding a mixer element. With recent advances in MMIC solid-state power amplifiers and the possibility of integrating them with high-frequency Schottky mixers monolithically, one can envision a receiver in which an antenna couples radiation to an LNA that, in turn, is coupled to a high-frequency mixer. The mixer then down-converts the radiation to whatever IF band makes sense, be it narrowband for spectroscopic applications, or wider band for thermal imaging applications. Intuitively, one could design the LNA with just enough gain so that the overall receiver sensitivity was not affected significantly by the mixer or following IF electronics. As will be shown later, this provides excellent overall performance if the LNA noise figure is acceptably low. And not surprisingly, it is the same architecture used at lower frequencies in communications and radar receivers alike, perhaps most commonly in the handsets of nearly every mobile telephone made today at PCS wireless frequencies.

B. Signal-to-Noise Ratio and Noise Equivalent Power

Once the THz signature is defined and the free-space link is established, it is the job of the sensor to process the signal with maximum fidelity and minimum contamination or masking by noise. From the above discussion it should be clear that the signal is the information-bearing portion of the incident THz average power, whether the power is coherent or incoherent. The noise is then all other electromagnetic mechanisms that cause the power to fluctuate about the average and, therefore, become less certain. And from Sec. II such noise is always present, even in the absence of electronic noise. So as in all other electromagnetic sensors, the analysis of sensor performance must be carried out by accounting for the noise, and it is important to define a metric for the sensitivity in the presence of the noise.

In principle, at any point in either sensor architecture of Fig. 13, it is possible to define the average power associated with the signal and the rms fluctuations in power associated with the noise. So a useful metric of the ability of the sensor to distinguish the signal from the noise is the *power* signal-to-noise ratio (SNR),

$$\frac{S}{N} = \frac{<P>}{\sqrt{<(\Delta P)^2>}} = \frac{<P>}{S_P \cdot B_{ENB}}$$

where S_P is the power spectral density and B_{ENB} is the equivalent noise bandwidth at that point in the sensor. B_{ENB} is generally dictated by sensor phenomenology, such as the resolution requirements and measurement time.

As discussed in Secs. I and IV above, there are two types of receivers used in the THz region: (1) incoherent, or direct, and (2) coherent, or heterodyne. From a signal processing standpoint, the *detector* in both types of receivers is the device that converts the signal from the RF (i.e., THz) region to baseband (usually around DC) where it can be positively identified in comparison to the residual baseband fluctuations that comprise the noise. For incoherent receivers, the *detector* is often the first device in the RF front-end, connected directly to the antenna. In coherent receivers, the *detector* is generally a device in the IF part of the circuit, after an RF front-end that uses a mixer to down-convert the RF signal from the THz region to an IF band at much lower frequency. The commonality between the incoherent and coherent detectors is that they are generally the first *nonlinear* element in the receiver chain – at least the first nonlinear element by design. This may appear contradictory to the definition of a mixer until we realize that the signal power in any coherent receiver is almost always much weaker than the local oscillator power. Under this condition, a mixer acts as a linear (in amplitude) transducer that simply translates the frequency.

As in any RF system, it is the signal-to-noise ratio *after detection* that matters most. This is what determines the bit-error-rate in communications systems and the probability of detection in radar systems, for example. This will become more clear later in the Sec. VIII.B on receiver operational characteristics. So for the present discourse, we will always seek the SNR after detection. Clearly, the signal-to-noise ratio depends on the signal strength so is not by itself a good metric for comparing detector or sensor types.

C. Noise Equivalent Power

A good figure-of-merit for overall sensor sensitivity comes from the fact that, generally speaking, only post-detection SNR values of order unity or higher are useful. This is intuitively obvious, but can also be proven mathematically by statistical decision theory. A more useful metric for sensor performance is to fix the SNR at some value and then solve for the signal power that achieves this. The fixed value of SNR universally accepted by sensor engineers and scientists is unity. The resulting metric is the noise-equivalent power spectral density, NEP, which by definition is the *input* signal power to the sensor required to achieve a SNR of unity after detection (AD). As we will show later, square-law detection of a signal buried in additive white Gaussian noise has the effect of increasing the SNR by

$$\left(\frac{S}{N}\right)_{AD} = \left(\frac{S}{N}\right)_{BD}\sqrt{\frac{\Delta\nu}{2\Delta f}}$$

where $\Delta\nu$ and Δf are the pre- and post-detection bandwidths, respectively, and $(S/N)_{BD}$ is the SNR before detection. Thus, the NEP is given by (in units of W)

$$NEP_{AD} = N_{BD}\cdot\sqrt{\frac{2\Delta f}{\Delta\nu}} \equiv NEP_{BD}\cdot\sqrt{\frac{2\Delta f}{\Delta\nu}}$$

For the purpose of comparing different sensor technologies, it is conventional to divide out the post-detection bandwidth effect (or equivalently, setting it equal to 1 Hz). This yields the normalized NEP_{AD} [in units of $W/(Hz)^{1/2}$], which we define as

$$NEP'_{AD} = N_{BD}\cdot\sqrt{\frac{1}{\Delta\nu}}$$

Generally, the pre-detection NEP scales with $\Delta\nu$, so we introduce a normalized NEP_{BD},

$$NEP'_{AD} = NEP'_{BD}\cdot\sqrt{\Delta\nu}$$

Note that these definitions are not spelled out explicitly in the most of the literature on remote sensing. But we do so here to help students or new researchers to the field since the NEP is one of the most bewildering of all system metrics.

A useful feature of the NEP is its additivity. If there are N mechanisms contributing to the noise at a given node in the receiver, if the mechanisms are uncorrelated to the signal and to each other, if they obey Gaussian statistics, then the total NEP is the uncorrelated sum

$$NEP_{TOT}^2 = NEP_1^2 + NEP_2^2 + ...NEP_N^2$$

This property applies to any node, pre- or post-detection, and will be used explicitly when we discuss the contribution from electronic noise. In reality there are cases where a noise mechanism is correlated to the signal (e.g., radiation noise) or to another noise mechanism (current and voltage noise in transistors), so one must be careful in applying this addition formula. In such cases, one can always fall back on the SNR as a useful measure of overall system performance.

D. Noise Equivalent Delta Temperature

For radiometric and thermal imaging systems, it is sometimes convenient to express the sensitivity in terms of the change of temperature of a thermal source that produces a post-detection SNR of unity. The resulting metric is the noise equivalent delta temperature, or NEΔT, which is given by[41]

$$NE\Delta T = \frac{NEP_{AD}}{dP_{inc}/dT\,|_{P_B}}$$

where P_{inc} is the incident power. In the Rayleigh-Jeans limit and assuming that the source has unity emissivity and fills the field-of-view of the sensor, we find $P_{inc} = M\, k_B T_B\, \Delta\nu$, so the NEΔT becomes

$$NE\Delta T = \frac{NEP_{AD}}{M \cdot k_B \cdot \Delta\nu}$$

E. Noise Figure and Friis Formula

For the linear elements of a sensor, including both active and passive components, a more universal metric is the noise figure, F. This is defined for all linear two-ports as

$$F = \left(\frac{(S/N)_{IN}}{(S/N)_{OUT}} \right) \quad ; 1 < F < \infty$$

In other words, the noise figure quantifies the degradation in SNR as a signal passes through a component in a linear chain. When combined with the gain of each component, the noise figure can be calculated simply from a sequence of two-ports through the expression

$$F_{TOT} = F_1 + \frac{F_2 - 1}{G_1} + \frac{F_3 - 1}{G_1 \cdot G_2} + \ldots . \frac{F_n - 1}{\prod_{i=1}^{n-1} G_i}$$

where G_i is the power gain of the ith element. Note that this gain accounts for impedance mismatch between elements. Physically, this means that components located further down a chain tend to be less important if the earlier components have high gain.

V. THz Signal and Noise Processing

Once radiation associated with the object signature is coupled into the sensor receiver, it generates the signal – a voltage or current in a network that must be processed from the antenna front-end to the data-collection or decision-making (logic) devices in the back end. Independent of the coherence of the incoming radiation, there are five functions in this process that are rather generic in THz sensors, and all electromagnetic sensors for that matter, and will be discussed in some detail here: (1) rectification, or detection; (2) integration or averaging (3) matched filtering; (4) frequency conversion; and (5) amplification; either before or after detection. Rectification is always necessary to extract the information about the object of interest, which that usually resides on the THz envelope. Therefore, it is discussed first and in some detail. The two common means of rectification in THz receivers are square-law and envelope detection. Integration is almost always carried out after the rectifier to add multiple samples of the information extracted and therefore improve the quality of the data or the confidence of the decision. Matched filtering is always carried out in the linear part of the sensor before detection. Frequency down-conversion is always necessary because logic circuits do not (at least not yet) work at THz frequencies. Amplification can be carried out either in the linear part of the receiver before detection or in the IF or baseband portion after detection.

Remarkably, all five of these functions are carried out in the two common receiver architectures discussed in the Introduction: (1) incoherent, and (2) coherent. We start with a discussion of the detection function since this is the centerpiece of most THz sensors.

A. Classical Square-Law Detection and Integration

Simply stated, a square-law detector is a device or circuit that takes an input signal and produces an output that is proportional to its square, $X_{out} = AX_{in}^2$, where X_{out} could be a current or voltage and A is a proportionality constant. The utility of such a device, just like a mixer, is in rectification, or frequency down-conversion. This is most easily seen in the special case of a coherent signal $X_{in}(t) = B\cos(\omega t + \phi)$. If put through a square-law detector, the output becomes $X_{out}(t) = AB^2 (1/2)\{1+ \cos[2(\omega t+\phi)]\}$. So if the square-law detector is followed up by a low pass filter (i.e., a time-domain integrator) with integration time $\tau >> 2\pi/\omega$, then the second term will not contribute and the output of the filter will be

$$X_{out} = AB^2/2 \equiv \Re P_S \tag{25}$$

where $\Re$ is a constant and P_S is the average absorbed input signal power. In the language of sensor theory, $\Re$ is usually called the *responsivity* and is measured either in A/W or V/W. Presumably, A is known, so that B can be determined from the X_{out} dc term. Square-law detectors are preferred in RF systems over cubic and other possible detectors for this reason. The proportionality constant A is dependent only on the detector characteristics and not on the power level, at least up to a level where saturation and higher-order effects begin to occur. A single calibration of A in the "small-signal" regime is all that is required to use the square-law detector over a wide range of input power.

We assume that the output noise from the detector is caused by an input white noise spectrum unrelated to the signal. We note that Fourier components at different frequencies correspond to different longitudinal modes, which in a random-noise (e.g., thermal) power spectrum are uncorrelated. Thus, such components will not produce an average dc component, or signal. But just as in the case of quantum noise, they will produce a fluctuation of the dc term about its average, i.e., noise. The magnitude of this noise can be determined as a sum over all possible mixing terms whose difference frequency lies within the passband of the low pass filter. If the input power spectrum S_X is flat across the pass band $\Delta\nu$, it can be shown that[42]

$$S_{X_{OUT}}(f) = A^2 \int_{\nu}^{\nu+\Delta\nu} S_{X_{IN}}(\nu+f)\cdot S_{X_{IN}}(\nu)\cdot d\nu \approx 2A^2(S_{X_{IN}})^2\cdot\Delta\nu\cdot(1-\frac{f}{\Delta\nu}) = 2\Re^2(\Delta P_N)^2\frac{1}{\Delta\nu}\cdot(1-\frac{f}{\Delta\nu}) \tag{26}$$

P_N is the noise power. The factor of two here comes from the fact that two spectral components in S_X contribute to $X(\nu')$ - one above and one below ν.

At this point, we see the merit of an integrator in the signal processing. Such a device can be analyzed as a low-pass filter of bandwidth Δf. If $\Delta f << \Delta\nu$, which is almost always the case in practice, then S_{Xout} can be considered flat over the integrator passband and we get, by linear signal processing,

$$(\Delta X_{OUT})^2 \equiv \int_0^{\infty} H(f)S_{X_{OUT}}(f)df \approx 2A^2(S_{X_{IN}})^2\cdot\Delta\nu\cdot\Delta f = 2\Re^2\cdot(\Delta P_N)^2\cdot\Delta f/\Delta\nu \tag{27}$$

And by taking the ratio of Eqn 25 to Eqn 27, we end up with the output signal-to-noise ratio

$$(SNR_{out})^2 = \frac{X_{OUT}^2}{<(\Delta X)^2>} = \frac{\Re^2(\bar{P}_S)^2}{\Re^2(\Delta P_N)^2}\frac{\Delta\nu}{2\Delta f} \equiv (SNR_{in})^2 \cdot \frac{\Delta\nu}{2\Delta f}$$

Square-law detectors also get used for *detection* when the input signal is *random noise*, as in the case of passive radiometers or noise radars. Now the input signal has a white power spectrum superimposed on the noise spectrum. In this case, the "signal" is just the portion of the power spectrum of Eqn 26 that gets converted by the detector down to baseband, assumed in this context to be dc. We know that any Fourier component of the noise spectrum is, by definition, self-correlated. The square law detector will produce a dc output from each component in accordance with the following expression

$$X_{out} = A\int_{\nu_0}^{\nu_0+\Delta\nu}\sqrt{S_{X_{in}}(\nu)}\sqrt{S_{X_{in}}(\nu)}d\nu = A\int_{\nu_0}^{\nu_0+\Delta\nu}S_{X_{in}}(\nu)d\nu$$

If the input power spectrum is flat across the passband, this results in

$$X_{out} = A\cdot S_{X_{IN}}\cdot\Delta\nu = \Re\cdot\bar{P}_N \tag{28}$$

just as for coherent signal, but now written in terms of the average noise power. Hence

$$(\Delta X_{OUT})^2 \equiv \int_0^\infty H(f)\cdot S_{X_{OUT}}(f)df \approx 2A^2(S_{X_{IN}})^2\cdot\Delta\nu\cdot B_{\text{int}} = 2\Re^2\cdot(\Delta P_N)^2\cdot\Delta f/\Delta\nu \tag{29}$$

And by taking the ratio of Eqn 28 to Eqn 29, we end up with the output signal-to-noise ratio

$$(SNR_{OUT})^2 = \frac{X_{OUT}^2}{<(\Delta X)^2>} = \frac{\Re^2(\bar{P}_N)^2}{\Re^2(\Delta P_N)^2}\frac{\Delta\nu}{2\Delta f} \equiv (SNR_{in})^2 \cdot \frac{\Delta\nu}{2\Delta f} \tag{30}$$

which is the same as for the coherent input signal.

The utility of Eqns 29 and 30 are manifest. They get applied in nearly every mm-wave and THz sensor whose *signal* is actually *random noise*. They can be understood in a different way by noting that the sampling time of the integrator is $\tau_s = 1/(2\Delta f)$, so that there are $\Delta\nu\ \tau_s$ independent samples per second. Because noise is random, the effect of adding these multiple signals becomes the sum of the powers, or the square root of the number of samples. Hence we can re-write Eqn 30 as

$$SNR_{OUT} = SNR_{in}\cdot\sqrt{\Delta\nu\cdot\tau_s} = SNR_{in}\cdot\sqrt{N_S}$$

From the standpoint of modern electronics, it is important to recognize that the operation of a square-law detection is a form of autocorrelation. This is easy to see in the case of thermal-noise, or AWGN in general. The Fourier components at different frequencies corresponding to different longitudinal modes can then mix together, so the output must be considered as a sum over all possible mixing terms whose difference frequency lies within the pass band of the low pass filter, $1/\tau$. A little thought shows that this is given by

$$s_{out} \propto \int s_{in}(t)s_{in}(t+\tau)dt$$

the autocorrelation function.[43]

B. Quantum Square-Law Detection and Integration

There exist a class of THz detectors that display quantum-mechanical response to incoming radiation. They tend to be the devices that have, for some reason, a band gap in their electronic density-of-states whose energy is of order hν . Two examples are hydrogenic-impurity semiconductors (i.e. extrinsic photoconductors), and low-T_C superconductors. Their quantum mechanical response is described by the photoelectric effect, whereby the detector absorbs photons that liberate charge-carrying particles at a rate

$$r = \frac{P}{h\nu}$$

If these charged particles are electrons or holes that are liberated in an electric field, they can contribute to a net photocurrent given by

$$i = \frac{egP}{h\nu}$$

where g is the photoelectric (more commonly called "photoconductive") gain, which may be greater or less than unity. If we think of this expression in the context of a THz photon detector, the output power must be proportional to i^2 and therefore to P^2. In other words, the photoelectric effect naturally obeys a square-law behavior. If such a "photodetector" is used as the first stage of a THz sensor, directly coupled to the incoming radiation, one will get a response to the signal power of

$$I_{dc} = \frac{eg\overline{P}}{h\nu} \tag{31}$$

But the response to the radiation fluctuations is more complicated. Following the same line of reasoning as applied in Sec. V.A above, the photodetector can mix fluctuations of different spectral components by virtue of its square-law behavior. We can duplicate the analysis of Sec. V.A simply by making the analogies X → i and ℜ → eg/hν.

$$S_I(f) = A^2 \int^{\nu+\Delta\nu} S_{X_{IN}}(\nu + f)\cdot S_{X_{IN}}(\nu)\cdot d\nu \approx 2\cdot\left(\frac{eg}{h\nu}\right)^2 (\Delta P_{in})^2 \frac{1}{\Delta\nu}\cdot(1-\frac{f}{\Delta\nu})$$

In most sensors at THz frequencies or above, the frequency f occurs in the electronic circuit that has a maximum frequency much less than Δν. In this case, the power spectrum of current fluctuations is given by

$$S_I(f) \approx 2\left(\frac{eg}{h\nu}\right)^2 \frac{(\Delta P_{in})^2}{\Delta\nu} \tag{32}$$

so that processing by a low-pass filter (integrator) yields on output mean-square current of

$$(\Delta I)^2 \approx 2\cdot\left(\frac{eg}{h\nu}\right)^2 \frac{(\Delta P_{in})^2}{\Delta\nu}\Delta f \tag{33}$$

Taking the ratio of Eqn 31 to Eqn 33 leads to the signal-to-noise ratio

$$SNR_{OUT} = \frac{I_{OUT}^2}{<(\Delta I)^2>} = \frac{(eg/h\nu)^2(\overline{P}_{in})^2}{(eg/h\nu)^2(\Delta P_{in})^2}\frac{\Delta\nu}{2\Delta f} \equiv SNR_{in}\cdot\frac{\Delta\nu}{2\Delta f}$$

the same as the classical case.

It is illuminating to examine Eqn 32 for a coherent incident signal, which using the result from Sec. III.D.2. for the mean-square input power fluctuations leads to

$$S_I(f) \approx 2\cdot\left(\frac{eg}{h\nu}\right)^2\left(\frac{\sqrt{h\nu\overline{P}_{in}}}{\sqrt{\Delta\nu}}\right)^2\cdot\Delta\nu$$

Substitution of Eqn 31 then yields

$$S_I(f) \approx 2egI_{dc}$$

This is the same form as the famous Schottky expression for electron shot noise in an electronic device, discussed later in Sec. VII.A. Therefore, it is called the *photon shot noise* expression. "Full" photon shot noise corresponds to g = 1. "Suppressed" and "enhanced" shot noise correspond to g < 1 and g > 1, respectively. This result lends credibility to our analysis of Sec. II in terms of power fluctuations.

This result has been the source of confusion to students and professionals alike because it is identical in form to the current fluctuations in electronic devices in which shot noise occurs, a good example being bipolar transistors. This has led many to believe that it is the fluctuating nature of the electrons in the photodetector that give rise to the photon shot noise. Hopefully, the present derivation makes it clear that the source of photon shot noise is fluctuations from measurement of the radiation itself and the inherent ability of photon detectors to mix down these fluctuations into an electronic circuit.

$$(\Delta i)^2 \equiv \int_0^\infty H(f)S_{X_{OUT}}(f)df \approx 2A^2(S_{X_{IN}})^2B_{\text{int}}$$

C. Classical Heterodyne and Homodyne Conversion and Detection

As discussed in Sec. IV, heterodyne and homodyne detection entail the mixing of an input signal, be it coherent or incoherent, with a local oscillator. The mixing device is usually one that, although not necessarily perfect square-law, has a large quadratic coefficient such that the current or voltage in the output port is proportional to the input current or voltage squared, $X_{out} = AX_{in}^2$, where A is a proportionality constant. This is exactly the relation we had for a rectifier. For a mixer, we change X_{in}^2 to $X_{in}X_{LO}$, and for $X_{in} = B\cos(\omega_{in} t)$ and $X_{LO} = C\cos(\omega_{LO} t)$, we find

$$X_{out}(t) = A\left[\frac{B^2}{2}(1+\cos 2\omega_{in}t) + \frac{C^2}{2}(1+\cos 2\omega_{LO}t) + BC\cos(\omega_{LO}-\omega_{in})t + BC\cos(\omega_{LO}+\omega_{in})t\right]$$

The factor $\cos(\omega_{LO}-\omega_{in})t$ is the intermediate-frequency (IF) term. In the THz region the mixer is generally operated such that $\omega_{LO} >> |\omega_{LO}-\omega_{in}|$ and $\omega_{LO}+\omega_{in} >> |\omega_{LO}-\omega_{in}|$, so that the only remaining terms are

$$X_{out}(t) = A\left[\frac{B^2}{2} + \frac{C^2}{2} + BC\cos(\omega_{LO}-\omega_{in})t\right] = \eta\Re\left[P_{in} + P_{LO} + 2\sqrt{P_{in}P_{LO}}\cos(\omega_{LO}-\omega_{in})t\right] \quad (34)$$

where P_{in} and P_{LO} are the incident input and LO powers, and η is the fraction of each that is usefully absorbed. The output power at the IF frequency for the heterodyne case is given by

$$\overline{P}_{out} = D\overline{X}_{out}^2 = D\cdot(\eta\Re)^2\cdot 2P_{in}\cdot P_{LO} \tag{35}$$

since the long-term time average of $\cos^2(\omega t)$ = ½ for any ω. The quantity D is the mixer-to-IF circuit matching factor that accounts for any mismatch between the mixer and IF load impedance. This expression clearly displays a linear input-output relationship. It is customary to lump all parameters other than P_{in} and η into one quantity called the mixer conversion gain, G_{mix}

$$\overline{P}_{out} = D\cdot(\eta\Re)^2\cdot 2P_{in}\cdot P_{LO} \equiv \eta^2\cdot G_{mix}\cdot\overline{P}_{in} \tag{36}$$

In other words, G_{mix} represents the fraction of all of the absorbed incident power in one sideband that is converted to the IF. In all mixer types, $\Re$ decreases as P_{LO} increases, and the impedance match can never be made perfect. Hence the maximum attainable single-sideband conversion gain is usually less than unity and the gain factor really represents a loss. Special types of mixers, such as parametric converters and certain quantum mixers can display a conversion gain > 1, but it is rare and difficult to achieve. Similarly, the output power for the homodyne case is

$$\overline{P}_{out}\alpha\overline{X}_{out}^2 = (\eta\Re)^2\cdot 4P_{in}\cdot P_{LO}$$
$$\overline{P}_{out} \equiv 2\cdot\eta^2 G_{mix}\cdot\overline{P}_{in}$$

which, again, is a linear input-output relationship.

The noise processing of the coherent converter is a bit more subtle. One noise process that must always be considered is the fluctuations in X_{out} caused by the quantum-mechanical fluctuations in the two *direct-detection* terms in Eqn 35, P_{in} and P_{LO}. In most if not all THz mixers, $P_{LO} >> P_{in}$, so from Sec. V we can write

$$(\Delta X_{OUT})^2 = 2\Re^2\cdot(\Delta P_{in})^2\cdot B_{IF}/\Delta\nu$$

where the IF bandwidth B_{IF} replaces the integration bandwidth for direct detection. For practical reasons it is often true that the LO power is unimodal, so that from Eqn 12 we can write $(\Delta P_{in})^2 = \sqrt{\eta\cdot h\nu_{LO}\cdot P_{LO}\cdot\Delta\nu}$, where P_{LO} is the incident LO power. Thus we find

$$(\Delta X_{OUT})^2 = 2\Re^2\cdot\eta\cdot h\nu_{LO}\cdot P_{LO}\cdot B_{IF} \tag{37}$$

or

$$(\Delta P_{OUT})^2 = 2D\cdot\Re^2\cdot\eta\cdot h\nu_{LO}\cdot P_{LO}\cdot B_{IF} \equiv G_{mix}\cdot\eta\cdot h\nu_{LO}\cdot B_{IF} \tag{38}$$

Taking the ratio of Eqns 38 and 36, we get the heterodyne radiation-noise-limited signal-to-noise ratio:

$$\left(\frac{S}{N}\right)_{IF} = \frac{\eta^2\cdot G_{mix}\cdot\overline{P}_{in}}{G_{mix}\cdot\eta\cdot h\nu_{LO}\cdot B_{IF}} = \frac{\eta\cdot\overline{P}_{in}}{h\nu_{LO}\cdot B_{IF}} \tag{39}$$

for the heterodyne case, and

$$\left(\frac{S}{N}\right)_{IF} = \frac{\eta^2 \cdot G_{mix} \cdot \overline{P}_{in}}{G_{mix} \cdot \eta \cdot h\nu_{LO} \cdot B_{IF}} = \frac{2\eta \cdot \overline{P}_{in}}{h\nu_{LO} \cdot B_{IF}}$$

for the homodyne case.

D. Quantum Coherent Heterodyne and Homodyne Conversion and Detection

As in the case of direct detection, the coherent conversion with quantum (photo) mixers follows from the classical analysis by the substitutions X → i and $\Re \rightarrow$ ge/hv. The LO oscillator and input electric fields are treated classically. So from Eqn 34 we can write

$$i_{out}(t) = \eta \frac{eg}{h\nu}\left[P_{in} + P_{LO} + 2\sqrt{P_{in}P_{LO}}\cos(\omega_{LO} - \omega_{in})t\right]$$

This leads to the heterodyne IF signal output

$$\bar{i}^2_{out} = (\eta \frac{eg}{h\nu})^2 \cdot 2P_{in} \cdot P_{LO} \tag{40}$$

and the homodyne IF signal output

$$\bar{i}^2_{out} = \left(\eta \frac{eg}{h\nu}\right)^2 \cdot 4P_{in} \cdot P_{LO} \tag{41}$$

The same substitutions into Eqn 37 lead to a mean-square IF current fluctuation arising from LO radiation fluctuations of

$$(\Delta i_{out})^2 = 2(eg/h\nu)^2 \cdot \eta \cdot h\nu_{LO} \cdot P_{LO} \cdot B_{IF} = 2egI_{LO} \cdot B_{IF} \tag{42}$$

for the heterodyne and homodyne cases.

Taking the ratio of Eqn 40 or 41 to 42 we get the "quantum-limited" SNR expressions

$$\left(\frac{S}{N}\right)_{IF} = \frac{\eta \cdot \overline{P}_{in}}{h\nu_{LO} \cdot B_{IF}}$$

for the heterodyne case, and

$$\left(\frac{S}{N}\right)_{IF} = \frac{2 \cdot \eta \cdot \overline{P}_{in}}{h\nu_{LO} \cdot B_{IF}}$$

for the homodyne case. Note that the quantum mechanical nature of the radiation enters these expressions through the photoelectric effect and the photon fluctuations – not through the mixing process. And because neither expression depends on the photoelectric gain or on the LO power, both are identical to the analogous expressions for classical coherent detection.

VI. Radiation-Noise Limits on Sensitivity

A. Radiation-Noise-Limited NEP_{BD} of THz Direct-Detection Sensors

Before radiation ever enters a receiver and gets converted to information, there are fundamental limits imposed on the sensitivity simply by the fact that every sensor must carry out a measurement. Measurements perturb the radiation field and, as such, introduce uncertainty. This uncertainty is evident through fluctuations of the radiation described in this section. These fluctuations are rather independent of the sensor architecture and depend, instead, on the nature of the radiation being detected.

A.1. Coherent Signal

In practice a THz sensor that receives a coherent signal must contend with radiation noise at least at the level of photon shot noise from the signal. For radiation in a single mode, this leads to a maximum power signal-to-noise ratio of

$$\frac{S}{N} \equiv \frac{P_{abs}}{\sqrt{(\Delta P_{abs})^2}} = \frac{\eta \cdot P_{inc}}{\sqrt{\eta \cdot h\nu \cdot P_{inc} / \Delta\nu \cdot \Delta\nu}}$$

Solving for P_{inc} with S/N = 1, we find the before-detection NEP_{BD} of

$$NEP_{BD} = \frac{h\nu}{\eta}\Delta\nu ,$$

which is the famous photon- or quantum-noise limited expression.

The NEP after classical or quantum noise-free square-law detection follows from Sec. V.A or V.B,

$$NEP_{AD} = NEP_{BD} \cdot \sqrt{\frac{2\Delta f}{\Delta\nu}} = \frac{h\nu}{\eta}\sqrt{2\Delta\nu \cdot \Delta f} \qquad (43)$$

and the specific NEP is simply

$$NEP_{AD}^{'} \equiv \frac{NEP_{AD}}{\sqrt{\Delta f}} = \frac{h\nu}{\eta}\sqrt{2\Delta\nu}$$

A.2. Thermal Signal

A common application in THz and infrared sensors alike is remote detection of a thermal signal. In this case the maximum signal-to-noise ratio occurs when the only source of noise is the radiation noise of the signal. Again, we assume that the passband is narrow enough that the Planck factor is the same for all modes, and that each mode is equally well matched to the sensor. We then find

$$\overline{P}_{abs} = [\sum_{m}^{M} \eta_m] \cdot h\nu_0 \Delta\nu \cdot f_P$$

$$< (\Delta P_{abs})^2 >= [\sum_{m}^{M} \eta_m] \cdot (h\nu_0 \Delta\nu)^2 f_P (f_P + 1)$$

$$\left(\frac{S}{N}\right)_{BD} = \frac{P_{abs}}{\sqrt{(\Delta P_{abs})^2}} = \frac{[\sum_m^M \eta_m]\cdot h\nu_0 \cdot f_P \cdot \Delta\nu}{h\nu_0 \cdot \Delta\nu \cdot \sqrt{[\sum_m^M \eta_m]\cdot f_P(f_P+1)}} \quad (44)$$

$$= \frac{\sqrt{\sum_m^M \eta_m \cdot f_P}}{\sqrt{f_P(f_P+1)}} = \sqrt{\sum_m^M \eta_m} \cdot \exp(-h\nu_0 / 2k_B T)$$

where the last step follows from the definition of f_P and remains valid at any temperature or passband center frequency.

This remarkable result is the basis for many important considerations about thermal sensing in the THz region. First, because hν is generally << $k_B T$ in terrestrial THz sensing, it implies that the signal-to-noise ratio before detection is always less than unity for a unimodal sensor, but can be greater than unity if $\Sigma\eta_m > 1$, as normally occurs in infrared detectors. Second, the modal dependence would appear, on first glance, to support an arbitrarily high signal-to-noise ratio given a suitable coupling structure. In other words, it is not clear whether or not there is a quantum-limit.

To elucidate the quantum limit for this case, we must calculate the NEP_{BD} but do so in light of the fact that both the signal and noise derive from the same incident photons. Because $P_{inc} = Mh\nu\, f_P\, \Delta\nu$, one can write

$$\left(\frac{S}{N}\right)_{BD} = \frac{[\sum_m^M \eta_m]\cdot P_{inc}}{\cdot\sqrt{[\sum_m^M \eta_m]\cdot P_{inc}(P_{inc} + M\cdot h\nu_0 \cdot \Delta\nu)}}$$

Setting the $(S/N)_{BD}$ to unity and solving for P_{inc}, it is easy to show

$$NEP_{BD} = h\nu_0 \cdot \Delta\nu \cdot \frac{M}{\sum_m^M \eta_m - 1} \equiv NEP_{QL} \qquad NET_{BD} \equiv \frac{h\nu_0 \Delta\nu}{k_B}\cdot\frac{M}{\sum_m^M \eta_m - 1}$$

where NET_{BD} is the noise equivalent temperature of a single-mode thermal source. For the unimodal case, this leads to a negative value - clearly not allowed on physical grounds and a result of the fact that S/N cannot equal 1.0. It diverges for $\Sigma\eta_m = 1$, and for $\Sigma\eta_m > 1$ it goes positive. In the special case that the sensor accepts a large number of modes, each having the same value of $\eta_m \equiv \eta$ such that $\Sigma\eta_m = M\eta >> 1$, we find NEP → $h\nu\Delta\nu/\eta$ – the same photon-noise limit as for a coherent signal.

The quantum-limited NEP after classical or quantum noise-free square-law detection follows from Eqn 43

$$NEP_{AD} = NEP_{BD} \cdot \sqrt{\frac{2\Delta f}{\Delta\nu}} = \frac{h\nu_0 M}{\sum_m^M \eta_m - 1}\sqrt{2\Delta\nu \cdot \Delta f}$$

and the specific NEP is simply

$$NEP'_{AD} = \frac{h\nu_0 M}{\sum_m^M \eta_m - 1}\sqrt{2\Delta\nu}$$

A.3. Arbitrary Signal, Thermal Background

A third important case of radiation-noise limited sensitivity occurs when the direct detector is coupled to many more spatial modes than those of the signal. Those beyond the signal modes are called "background" modes and can occur for a variety of reasons, such as imperfect coupling between the receiver antenna and the external coupling optics. If the radiation power from these modes dominates the radiation noise from the signal modes and can be associated with a brightness temperature T_B , then one can write from Eqn 44,

$$\left(\frac{S}{N}\right)_{BD} = \frac{P_{abs}}{\sqrt{(\Delta P_{abs})^2}} = \frac{\sum_m^{M_S} \eta_m \cdot P_m}{h\nu_0 \cdot \Delta\nu \cdot \sqrt{[\sum_n^{M_B} \eta_n] \cdot f_{P,B}(f_{P,B}+1)}}$$

where $f_{P,B}$ is the background Planck function, M_S (M_B) is the number of signal (background) modes, and P_m is the power per signal mode, be it coherent or thermal (e.g., if thermal $P_m = h\nu_0\, \Delta\nu\, f_P$). In general this expression is difficult to evaluate, but reduces to simplicity if all modes have the same coupling efficiency, η. Then one can write the "background-limited" expression

$$\left(\frac{S}{N}\right)_{BD} = \frac{\eta \cdot P_{inc}}{h\nu_0 \cdot \Delta\nu \cdot \sqrt{M_B \cdot \eta \cdot f_{P,B}(f_{P,B}+1)}}$$

where P_{inc} is the sum of signal power over all signal modes. This leads to the NEPs

$$NEP_{BD} = \frac{h\nu_0\Delta\nu \cdot \sqrt{M_B \cdot \eta \cdot f_P(f_P+1)} \cdot}{\eta} = \sqrt{\frac{h\nu_0 \cdot \Delta\nu \cdot P_B(f_{P,B}+1)}{\eta}} \equiv NEP_{back},$$

and

$$NEP_{AD} = \sqrt{\frac{2h\nu_0 \cdot \Delta f \cdot P_B(f_{P,B}+1)}{\eta}}$$

In the Rayleigh-Jeans limit where $f_{P,B} >> 1$, one obtains

$$NEP_{AD} \approx \sqrt{\frac{2k_BT_B \cdot \Delta f \cdot P_B}{\eta}}$$

and in the Wien limit where $f_{P,B} << 1$

$$NEP_{AD} \approx \sqrt{\frac{2h\nu_0 \cdot \Delta f \cdot P_B}{\eta}}$$

This is the familiar background-limited NEP_{AD} valid in the near-infrared or visible region,[44] which confirms the present formalism.

When the background thermal noise power is comparable to the signal noise power, both mechanisms must be considered. In calculating the NEP one can assume that the radiation noise mechanisms are uncorrelated, but must be careful to account for the correlation between the signal and its quantum noise. So it is wise to start with the SNR, which can be written in the following form if all modes have the same η,

$$\left(\frac{S}{N}\right)_{BD} = \frac{\eta \cdot P_{inc}}{h\nu_0 \cdot \Delta\nu \cdot \sqrt{M_S \cdot \eta \cdot f_{P,S}(f_{P,S}+1) + M_B \cdot \eta \cdot f_{P,B}(f_{P,B}+1)}}$$

where $f_{P,S}$ ($f_{P,B}$) signal and background Planck functions. By comparison to the above background-limited expression, this becomes

$$\left(\frac{S}{N}\right)_{BD} = \frac{\eta \cdot P_{inc}}{\sqrt{h\nu_0\Delta\nu \cdot \eta \cdot P_{inc}(f_{P,S}+1) + h\nu_0\Delta\nu \cdot \eta \cdot P_B(f_{P,B}+1)}}$$

$$= \frac{P_{inc}}{\sqrt{h\nu_0\Delta\nu \cdot P_{inc}(f_{P,S}+1)/\eta + NEP_{back}^2}}$$

where the number of background modes is implicitly contained in P_B. The Planck function $f_{P,S}$ is correlated to P_{inc} through the source temperature, so a better form is

$$\left(\frac{S}{N}\right)_{BD} = \frac{P_{inc}}{\sqrt{P_{inc}(P_{inc} + M_S h\nu_0\Delta\nu)/(M_S\eta) + NEP_{back}^2}}$$

Setting the SNR to unity and P_{inc} to the NEP, one gets the quadratic equation:

$$NEP_{BD}^2 - NEP_{BD}\frac{M_S h\nu_0\Delta\nu}{M_S\eta - 1} - \frac{M_S\eta}{M_S\eta - 1}NEP_{back}^2 = 0$$

By recognizing the coefficient of the middle term as the quantum-limited NEP_{BD} when $M_S\eta > 1$, the only physically-allowable (positive) solution is:

$$NEP_{BD} = \frac{1}{2}\left[NEP_{QL} + \sqrt{NEP_{QL}^2 + 4NEP_{back}^2 \cdot \frac{M_S\eta}{M_S\eta - 1}}\right]$$

and $NEP_{AD} = NEP_{BD}\,(2\Delta f/\Delta\nu)^{1/2}$. When $NEP_{back} >> NEP_{QL}$ and $M_S\eta >> 1$, NEP_{BD} approaches NEP_{back}. And when $NEP_{back} << NEP_{QL}$ and $M_S\eta >> 1$, it approaches NEP_{QL}. So a good approximation to the exact expression is obtained by the NEP summation formula from Sec. IV.C,

$$NEP_{AD} \approx \sqrt{NEP_{QL}^2 + NEP_{back}^2} = \sqrt{\left(\frac{h\nu_0 M_S}{M_S\eta - 1}\right)^2 2\Delta\nu \cdot \Delta f + \frac{2h\nu_0 \cdot \Delta f \cdot P_B(f_{P,B}+1)}{\eta}}$$

which is always usefully accurate if $M_S\eta >> 1$

For detection of thermal radiation, the most useful metric is NEΔT.[45] From the definition given in Sec. IV, we can compute this easily in the Rayleigh-Jeans limit using $P_{inc} \approx M_S\, k_B T_S\, \Delta\nu$ so that $dP_{inc}/dT = M_S\, k_B\, \Delta\nu$ and one can write from the approximate expression

$$NE\Delta T \approx \sqrt{\left(\frac{h\nu_0}{k_B(M_S\eta - 1)}\right)^2 \frac{2\Delta f}{\Delta\nu} + \frac{2T_B \cdot \Delta f \cdot P_B}{k_B\eta(M_S\Delta\nu)^2}} \tag{45}$$

It is interesting that there is no dependence on the source temperature.

An interesting application of Eqn 45 is to examine the wavelength dependence of NEΔT for a multimode THz thermal sensor of fixed effective aperture. To do this in a meaningful way we need to account for the dependence of M on frequency, as discussed in Sec. III.D and written in the convenient form, $M = \mathrm{Int}(\nu/\nu_R)^2 + 1$. If we assume that $M_S\eta >> 1$ we find from (45)

$$NE\Delta T \approx \frac{1}{\mathrm{Int}(\nu_0/\nu_R)^2 + 1}\sqrt{\left(\frac{h\nu_0}{k_B \cdot \eta}\right)^2 \frac{2\Delta f}{\Delta\nu} + \frac{2T_B \cdot \Delta f \cdot P_B}{k_B\eta(\Delta\nu)^2}}$$

This expression shows how important center frequency is to the sensitivity of a direct detector. From the quantum-limited NEP_{AD} expression one sees a linear dependence on

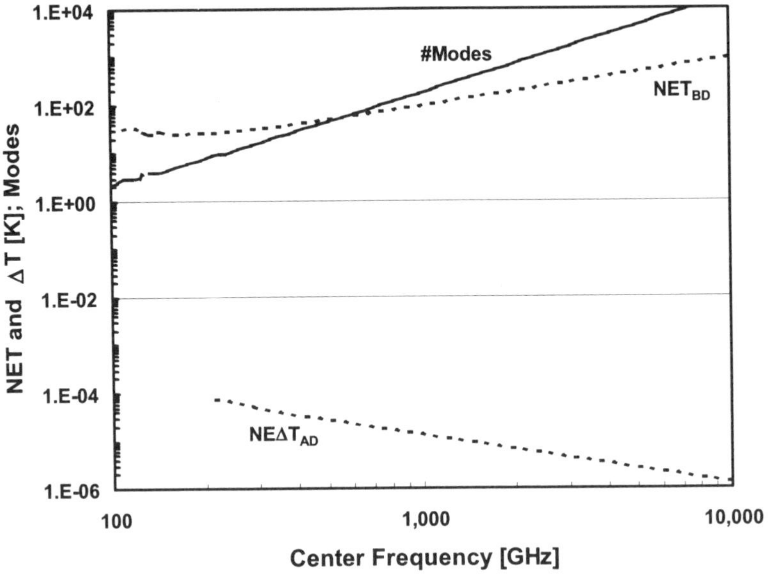

Fig. 14. Performance of a THz direct detector in the quantum limit assuming a cut-off frequency of 75 GHz and the definition $NE\Delta T_{BD} = NEP_{BD}/k_B$. For $NE\Delta T_{AD}$, an RF bandwidth of 100 GHz and a post-detection bandwidth of 30 Hz were assumed.

ν_0 and a square-root dependence on $\Delta\nu$ that appear to favor the mm-wave and THz regions over the infrared. But the NEP can be deceiving if the total signal power is also varying because of a change of mode number.

To exemplify this point, we plot in Fig. 14 the quantum-limited NEP_{BD} (expressed as a temperature NET_{BD}), the number of modes and the $NE\Delta T_{AD}$ for a direct detector having ν_R = 75 GHz. The quantum-limited curve rises linearly with center frequency as expected, but the mode density is quadratic. So the change of temperature required to match the noise decreases monotonically with frequency. This important result is sometimes misconstrued as arising from the behavior of the Planck factor instead of the mode density. And it tends to favor the infrared region in terms of ultimate sensitivity. But by the same token, the infrared direct detector often accepts so many modes that many of them couple to the background rather than the real target or the object to be imaged. This drives up the contribution from the background term in Eqn 45, leading to the common occurrence of a "background-limited infrared photodetector" or BLIP for short. In the mm-wave and THz region where the total number of modes is relatively small to begin with, the background contribution is generally easier to manage and, in good designs, make negligible.

B. Radiation-Noise-Limited NEP_{BD} of Classical and Quantum Coherent Receivers

In Secs V.C and V.D, identical expressions were derived for the IF SNR of classical and quantum coherent detection assuming that the dominant source of IF power fluctuation was radiation fluctuations in the local oscillator. Interestingly, the resulting expression, Eqn 39 depended on the input signal power but not on the LO power, so that by setting the SNR to unity, one gets for both classical and quantum coherent detection

$$NEP_{HET} = \frac{h\nu_{LO} \cdot B_{IF}}{\eta} \quad . \tag{46}$$

A similar derivation for the homodyne case would lead to

$$NEP_{HOM} = \frac{h\nu_{LO} \cdot B_{IF}}{2\eta} \tag{47}$$

These expressions are amongst the most famous in remote-sensing theory not just because of their role as a fundamental limit but because of their obvious consistency with the postulates of quantum-mechanics and classical probability theory. All sensors, even a radar operating at microwave frequencies, measure radiation by extracting photons via atomic transitions and stimulated absorption in some component of the receiver. This quantized process thus has a built-in granularity, or *uncertainty*, equal to a minimum measurable quanta, which is $h\nu/\eta$. Sensors having small η are decoupled from the environment and thus have a higher minimum quanta.

Probability theory teaches that the time rate of occurrence of an event or outcome of an experiment is just P·R, where P is the probability or "expectation value" of the event or outcome, and R is the rate of occurrence or "sampling" in the experiment. In Eqns 46 and 47, the "event" is just the stimulated absorption of a photon of energy U. From Eqn 4 the minimum expectation value is just ΔU (n=1) = $h\nu$. From sampling theory the occurrence rate is just the analog bandwidth of the measurement, B_{IF}. Now we arrive at the quantum-limited expression for the expectation-value rate, or power, associated with the measurement of radiation by any sensor at any frequency ν. Einstein is credited with the first understanding of the photoelectric effect, but apparently had little or no role in the quantum-limit of radiation measurement. Nevertheless, these expressions bear resemblance to some of his remarkable formulations – profound in concept while very simple in mathematics.

VII. Practical Limits on Receiver Sensitivity: Electronic Noise

In practice, real THz sensors are almost always limited in sensitivity by factors other than radiation noise, often by physical noise in the devices. It other cases the limit may be imposed by the atmospheric fluctuations, electromagnetic interference (EMI), power supply noise, and a myriad of other deleterious effects. Historically, it has been the environmental effects that have often made the difference between success and failure in THz deployment in the field.

A. Physical Noise of Electronics

Within every sensor system, particularly at the front end, are components that contribute noise to the detection process and therefore degrade the ultimate detectability of the signal. The majority of this noise usually comes from electronics, particularly the THz mixers or direct detectors themselves, and transistors in amplifiers that follow these devices. The majority of noise from such devices falls in two classes: (1) thermal noise, and (2) shot noise.[46] Thermal noise in semiconductors is caused by the inevitable fluctuations in voltage or current associated with the resistance in and around the active region of the device. This causes fluctuations in the voltage or current in the device by the same mechanism that causes - resistance, through Joule heating, couples energy to and from electromagnetic fields. The form of the thermal noise is very similar to that derived in Sec. II for blackbody radiation. And because the operational frequencies of electronics are generally well below the THz region, the Rayleigh-Jeans approximation is valid and the Johnson-Nyquist theorem applies. However, one must account for the fact that the device is coupled to a transmission line circuit, not to a free space mode, and the device may not be in equilibrium with the radiation as assumed by the blackbody model.

All of these issues are addressed by Nyquist's generalized theorem[47]

$$\Delta V_{rms} = [4k_B T_D \mathrm{Re}\{Z_D\}\Delta f\,]^{1/2} \qquad (48)$$

where T_D, Z_D , and Δf are the temperature, differential impedance, and bandwidth of the device. Even this generalized form has limitations since it is not straightforward to define the temperature of the device if it is well away from thermal equilibrium. Also, the differential impedance of the device is not well defined for transistors or three-terminal devices in general. This topic goes well beyond the scope of this article, so suffice it to say that Eqn 48 is useful not so much for its accuracy but because of its form. For example, the thermal noise of transistors can often be written in a way that is identical to Eqn 48 with T_D being the temperature of carriers in the active region and Z_D being the transresistance or inverse transconductance of the device.

Shot noise is a ramification of the device being well out of equilibrium. It is generally described as fluctuations in the current arriving at the collector (or drain) of a three-terminal device caused by fluctuations in the emission time of these same carriers over or through a barrier at or near the emitter (or source) of the device. The mean-square current fluctuations are given by

$$<(\Delta i)^2>= 2e\Gamma I \cdot \Delta f$$

where Γ is a numerical factor for the degree to which the random Poissonian fluctuations of emission times is modified by the transport between the emitter (or source) and collector (or drain). If $\Gamma = 1$, the transport has no effect and the terminal current has the same rms fluctuations. When $\Gamma < 1$, the transport reduces the fluctuations, usually through some form of degenerative feedback mechanism, and the shot noise is said to be suppressed. When $\Gamma > 1$, the transport increases the fluctuations, usually through some form of regenerative feedback mechanism, and the shot noise is said to be enhanced.

B. Equivalent Circuit Representations

While at first appearing to add insurmountable complexity to the theory of sensor analysis, a great simplification results from the fact that radiation noise and two forms of physical noise discussed above are, in general, statistically Gaussian. This is easy to understand in the case of thermal noise. But it remains true in the case of shot noise too provided that the number of electrons involved in the process is large. This follows from the central-limit theorem of probability theory. A very important fact is that any Gaussian noise passing through a linear component or network remains statistically Gaussian.[48] Hence, the output power spectrum S in terms of electrical variable X will be white and will satisfy the important identity

$$<(\Delta X)^2>=\int_{f_0}^{f_0+\Delta f} S_X(f)\cdot df \approx S_X(f)\cdot\Delta f$$

where Δf is the equivalent-noise bandwidth. Then one can do circuit and system analysis on noise added by that component at the output port by translating it back to the input port. In the language of linear system theory, the output and input ports are connected by the system transfer function $H_X(f)$, so the power spectrum referenced back to the input port becomes

$$S_X(in)=\frac{S_X(out)}{|H_X(f)|^2}$$

A related consequence of Gaussian noise is that it can be represented in circuit theory by an ac generator whose amplitude is the rms average of that variable and whose phase is perfectly random. This allows one to apply the tools of circuit theory.[49] Perhaps the simplest example is the ideal resistor. By the Johnson-Nyquist theorem, we know that the open-circuit voltage across a resistor fluctuates with rms value given by $(4k_BT\ \Delta f\ R)^{1/2}$. We can thus write a noise equivalent-circuit representation of this by adding in series with the resistor an ac voltage generator having amplitude $(4k_B\ T\ B\ R)^{1/2}$, a Gaussian-distributed amplitude, and a random phase. But one must remember that this Johnson Nyquist result is really based on the independent fluctuations of orthogonal photon modes, each characterized by frequency f. Hence, a more useful representation results from Fourier decomposing the total voltage generator into independent generators, each having amplitude $(4k_B\ T\ R/B)^{1/2}$, frequency f, and random phase. Then, because the generators at each f are statistically independent, we get their total voltage contribution by adding the sum of the squares. Because the noise is *white*, the total becomes

$$v_{rms}^2=\int_{f_0-B/2}^{f_0+B/2} v^2(f)df=\left(\frac{4k_BTR}{B}\right)^{1/2}\cdot B=(4k_BTR)^{1/2}$$

as expected. The power of this decomposition comes in several techniques of signal processing, such as square-law detection of Gaussian noise, as discussed in Sec. V.

C. Electrical Noise Limitations on THz Square-Law Detector

To get a complete accounting of sensor signal and noise effects for a mm-wave or THz square-law detector, we need only combine all the noise generators with the signal processing model in to one equivalent circuit. In Secs. V and VI it was assumed that the radiation was noisy but the radiation detector was noise-free. To include the noise added by and after the detector, we assume it is AWGN and simply add a variance term ΔX^2 to account for the fluctuations in the signal (voltage or current). Hence, the mean signal is given by

$$X = \Re\, P_{abs} = \Re\, [\sum_{m}^{M} \eta_m P_{inc,m}]$$

where the electrical responsivity is not mode-dependent because the detector responds the same to all absorbed power independent of how it is absorbed. The power signal-to-noise ratio after detection is given by

$$\left(\frac{S}{N}\right)_{AD} = \frac{(\Re \cdot \sum_{m}^{M} \eta_m P_{inc,m})^2}{2\Re^2 \cdot \overline{(\Delta P_{ABS})^2} \cdot (\Delta f / \Delta \nu) + \overline{(\Delta X_n)^2}}$$

A great simplification occurs when the total incident power, be it coherent or thermal, is divided equally over all modes. This is a good starting approximation for coupling thermal radiation into large coupling structures such as highly-overmoded feedhorns and integrating cavities. In this case $P_{inc,m} = P_{inc}/M$ and for AWGN in the detector and electronics one finds

$$\left(\frac{S}{N}\right)_{AD} = \frac{(\Re \cdot \frac{P_{inc}}{M} \sum_{m}^{M} \eta_m)^2}{2\Re^2 \cdot \overline{(\Delta P_{ABS})^2} \cdot (\Delta f / \Delta \nu) + \overline{(\Delta X_n)^2}} \tag{49}$$

In the common case that the detector and post-electronic noise contained in ΔX dominate the absorbed radiation noise, and assuming that none of this noise is correlated to the signal power, one can set SNR = 1 and solve for P_{inc} ($\equiv NEP_{elect}$) to find

$$NEP_{elect} = \frac{M\sqrt{\overline{(\Delta X_n)^2}}}{\Re \cdot \sum_{m}^{M} \eta_m}$$

This allows one to write the SNR and specific NEP for the general case as

$$\left(\frac{S}{N}\right)_{AD} = \frac{(P_{inc})^2}{2 \cdot \overline{(\Delta P_{ABS})^2} \cdot (M / \sum_{m}^{M} \eta_m)^2 (\Delta f / \Delta \nu) + NEP_{elect}^2}$$

As in Sec. VI.A one has to be careful in converting this to an NEP since the fluctuations in absorbed power will likely be correlated to the average signal power (e.g., radiation quantum noise). As in VI.A.C, an NEP summation formula is conveniently used as an approximation:

$$NEP_{AD} \approx \sqrt{2 \cdot \overline{(\Delta P_{ABS})^2} \cdot (M / \sum_{m}^{M} \eta_m)^2 \cdot (\Delta f / \Delta \nu) + NEP_{elect}^2}$$

When each mode has the same coupling to the detector, one has $\Sigma\eta_m = M \cdot \eta$ so that

$$NEP_{elect} = \frac{\sqrt{\overline{(\Delta X_n)^2}}}{\Re \cdot \eta}$$

and

$$NEP_{AD} = \sqrt{2 \cdot \overline{(\Delta P_{ABS})^2} \cdot \Delta f / (\eta^2 \Delta \nu) + (NEP_{elect})^2}$$

Clearly, these are the same as the "unimodal" expressions one would obtain if only one spatial mode were incident and collected by the detector. These will be used later in conjunction with passive and active direct-detection sensors.

D. Electrical Noise Limitation on Heterodyne Mixer

Characterization of mm-wave and THz mixers can be done in the similar way as for square-law detectors. To account for the noise from the mixer itself and the IF electronics following, we simply add an rms term $[\Delta P^2]^{1/2}$ to the IF power:

$$\left(\frac{S}{N}\right)_{HET} = \frac{(\eta_R)^2 \cdot G_{mix} \cdot P_{inc}}{\eta_R \cdot G_{mix} \cdot h\nu_{LO} \cdot B_{IF} + \sqrt{\overline{(\Delta P_{IF})^2}}}$$

where the radiation noise is all photon shot noise. Any mismatch between the antenna and the mixer is accounted for by η_R, and any mismatch between the mixer and the IF circuit is included in G_{mix}. By setting the S/N to unity and solving for P_{inc}, we find the specific NEP

$$NEP'_{HET} \equiv \frac{NEP_{HET}}{B_{IF}} = h\nu_{LO}/\eta_R + \sqrt{\overline{(\Delta P_{IF})^2}}\,/[(\eta_R)^2 \cdot G_{mix} \cdot B_{IF}]$$

In the special case that the mixer and post-electronic noise dominate the absorbed radiation noise, and all of this noise is AWGN, the NEP reduces to the "mixer-limited" value

$$NEP'_{mixer} = \frac{\sqrt{\overline{(\Delta P_{IF})^2}}}{(\eta_R)^2 G_{mix} \cdot B_{IF}}$$

Thus, we can write the overall heterodyne S/N ratio as

$$\left(\frac{S}{N}\right)_{HET} = \frac{(\eta_R)^2 \cdot G_{mix} \cdot P_{inc}}{\eta_R \cdot G_{mix} \cdot h\nu_{LO} \cdot B_{IF} + (\eta_R)^2 G_{mix} \cdot B_{IF} \cdot NEP'_{mixer}}$$

and

$$NEP'_{HET} = h\nu_{LO}/\eta_R + NEP'_{mixer}$$

Note that the units of this specific NEP is W/Hz, not $W/(Hz)^{1/2}$ as for the square-law detector. This is because the mixer is behaving in a linear fashion with respect to the signal power.

In the mm-wave and THz regions, the NEP'_{HET} is usually measured with two loads of the same emissivity but different temperature T_L (e.g., ambient and 77 K) and satisfying $h\nu \ll k_B T_L$. Given this fact it is useful to define a new metric called the noise-equivalent temperature, or NET defined as that load which produces a $(S/N)_{HET}$ of unity. A short derivation yields

$$NET_{HET} = \frac{1}{k_B}(h\nu_{LO}/\eta_R + NEP'_{mixer})$$

In the millimeter-wave and THz literature, this NET is more commonly known as the receiver noise temperature, T_{REC}. More specifically, in the present case this is the double-sideband noise temperature because both signal bands are assumed to be down-converted. Because in the Rayleigh-Jeans limit the noise power per mode is linear in T_L, this NET is also equal to the NEΔT that we would compute from the more common definition NEΔT = NEP/[(dP_{inc}/dT)]

For the purpose of system analysis and characterization, an even more useful metric is based on the fact that from Sec. VI.B the minimum NEP'_{HET} is just $h\nu_{LO}$ and the corresponding minimum NET is just $h\nu_{LO}/k_B$. This motivates a single parameter to represent the degradation in sensitivity caused by imperfect coupling, mixer noise, and IF electronic noise. This quantity is the "heterodyne coupling efficiency" η_{HET} , and is defined by

$$NEP'_{HET} = \frac{h\nu_{LO}}{\eta_{HET}}; \; NET_{HET} = \frac{h\nu_{LO}}{k_B \cdot \eta_{HET}}$$

VIII. Receiver Performance Limitations and Statistics

A. Optimum Signal-to-Noise Ratio Before Detection: North's Theorem

THz sensors face many of the same challenges as faced by early microwave radars trying to detect small targets or large targets over a long range. Because of the weakness of the source or the high atmospheric attenuation, the received signal in THz sensors is usually very weak and difficult to distinguish from noise. As in the radar case, there is great interest in processing the weak signal in an optimum fashion, particularly in active sensors.

In one of the most elegant theorems in sensor theory, North showed how the construct the receiver transfer function or impulse response function to maximize the signal-to-noise ratio within a given resolution element. It is similar conceptually to Shannon's theorem in the sense that both state conditions of optimization: Shannon's stating the maximum information handling capacity on a noisy channel and North's stating the maximum signal-to-noise ratio on a noisy channel. North utilized the fact that independent of the sensor type, the maximum possible RF SNR is

$$SNR_{RF} = \frac{E_S}{N_0}$$

where E_S is the RF energy per sample and N_0 is the noise spectral density in the RF part of the receiver. When phrased for a pulsed system, E_S gets replaced by E_P, the energy per pulse. But North's theorem applies to cw systems as well by recognizing that these systems sample the RF power at a rate given by the sensor instantaneous bandwidth B, so the energy per pulse gets replaced by the energy per sampling time, or P_S/B , so that $SNR_{RF} = PS/(N_0B) = S/N$.

North's theorem showed that to achieve this SNR, the linear part of the RF sensor should have the following transfer function H(f) and impulse response functions h(t) :

$$H(f) = A^*(f)$$
$$h(t) = a^*(t)$$

where A(f) is the signal power spectrum, a(t) is the signal waveform (current or voltage), and * denotes complex conjugation. The first one makes sense intuitively since it simply states that all Fourier components of the signal spectrum should get "conjugately" coupled by the receiver. It is similar to the condition for optimum power transfer between a generator of impedance Z_G and a load of impedance Z_L, which from ac circuit theory can be proven to occur when $Z_G = Z_L^*$. The second condition is not so obvious, but becomes clearer in the special case of a sinusoidal signal since then complex

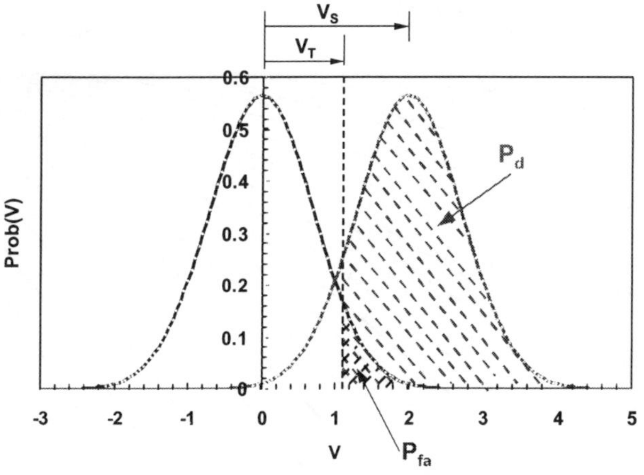

Fig. 15. Gaussian probability for the noise only and signal-plus-noise, applicable to a direct detection receiver.

conjugation is equivalent to time reversal. The combination of the two conditions is accomplished electronically with a "matched" filter – a pervasive device in radar and communications systems today at microwave frequencies and below.

B. Sensor Performance after Detection: Receiver Operating Characteristics

Given that the ultimate limits on the sensitivity any sensor depend on physical and environmental noise, the final performance can only be predicted using statistical means that properly account for each fluctuating effect. Physical noise and other fluctuations in each stage from the source of radiation to the output of the receiver detector must be ascribed a probability density function (PDF) for the relevant electrical variable at that point, and the overall PDF of the sensor in the presence of signal and noise must then be constructed.

Given the overall PDF, a common measure of the sensor performance comes from establishing a signal threshold at the output of the receiver detector.[50] The threshold approach, established early in the history of radar, is based on the observation that receivers are usually designed to "filter" the signal from the noise such that in the receiver baseband after detection the signal will have a greater amplitude than the noise. For example, if the baseband is dc, then the signal will correspond to the dc component and the noise will correspond to the residual ac power spectrum.

Given a well-define threshold, the *probability of detection Pd*, also called true positive detection, represents the probability that the sensor output PDF in the presence of the signal and all noise exceeds the threshold. The *probability of false alarm* Pfa, also called false positive detection or "false positive" for short, represents the probability that the PDF in the absence of the signal exceeds the threshold. The statistical assumption behind Pd is that a large number of such measurements are made so that Pd can be

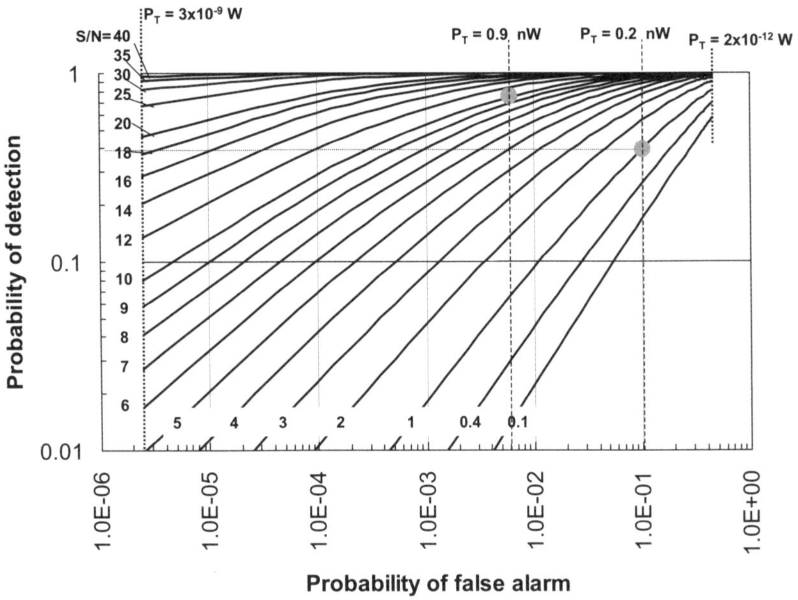

Fig. 16. Receiver-operating-characteristic diagram for direct-detection case.

thought of as either an average over time or an average over an ensemble of identical receivers processing the same signal but at all the possible noise states.

Plotted together, Pd vs Pfa form what is called the "receiver operating characteristic", or ROC for short. Having its roots in radar detection, ROC diagrams are now universal and get applied to remote and point sensors alike in all parts of the electromagnetic spectrum and to other sensor domains, such as acoustic and chemical. At present their most common application appears to be in the biomedical industry where the difference between a true and false positive from a sensor often has serious human consequences.

For electromagnetic remote sensors there are two commonly used sensor PDF functions and resulting ROC diagrams that apply to the ideal direct-detection sensor and heterodyne (or homodyne) sensor, respectively.

B.1. Direct Detection Sensor

The analysis of Sec. V showed that the output noise spectrum after direct detection could be treated in a simple way if the rf bandwidth greatly exceeded the post-detection bandwidth, which is almost always true. In this case, the output consists of a dc component superimposed on white noise whose spectral density is related to the sensor PDF. In the ideal case, all of the noise is related to "physical" fluctuations of the radiation or in the electronics so has an AWGN character. The overall PDF for the noise alone is then Gaussian, meaning that the probability of measuring a particular value of X (voltage or current) at the at any point in time is given by

$$P(X) = \frac{1}{\sqrt{2\pi \cdot \overline{(\Delta X)^2}}} \exp[-(X)^2 / 2\overline{(\Delta X)^2}]$$

where $\overline{(\Delta X)^2}$ denotes the variance. The overall PDF in the presence of signal is simply

$$P(X, X_S) = \frac{1}{\sqrt{2\pi \cdot \overline{(\Delta X)^2}}} \exp[-(X - X_S)^2 / 2\overline{(\Delta X)^2}]$$

where X_S is the average value, or signal. Note that the signal strength amplitude X_S is always positive while X goes both positive and negative – consistent with the way that square-law detectors process signals.

These two functions are plotted in Fig. 15 where a threshold X_T is also drawn. Now the definitions of Pd and Pfa are defined graphically in terms of the cross-hatched or shaded regions below the two curves and above the threshold. Mathematically they are given by

$$Pd = 1 - \frac{1}{\sqrt{2\pi \overline{(\Delta X)^2}}} \int_{X_T - X_S}^{\infty} \exp[-X^2 / 2\overline{(\Delta X)^2}] \cdot dx$$

and

$$Pd = \frac{1}{\sqrt{2\pi \overline{(\Delta X)^2}}} \int_{X_T}^{\infty} \exp[-X^2 / 2\overline{(\Delta X)^2}] \cdot dx$$

which can be written easily in terms of error functions. One can compute these integrals for various values of $(SNR)_{AD} = (X_S)^2 / \overline{(\Delta X)^2}$ and X_T. One then arrives at the ROC plot of Pd vs Pfa shown in Fig. 16. Note that this plot is *universal* in that it does not depend on the specific type of direct detector or overall performance. The only assumptions are that the noise is entirely AWGN in nature, and that the receiver architecture is based on direct detection.

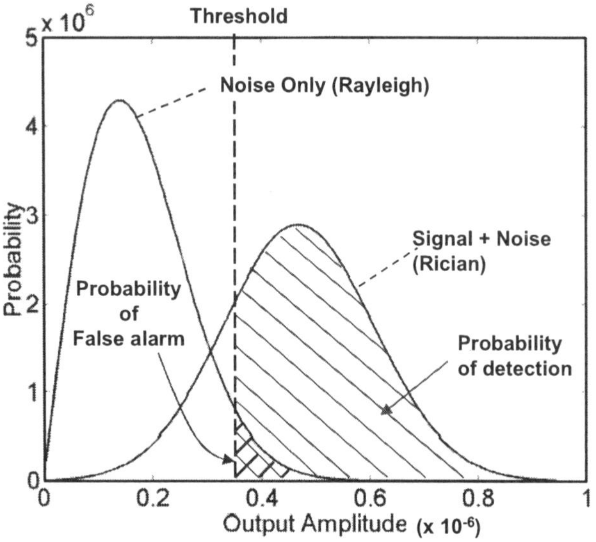

Fig. 17. Probability distribution functions for noise only and signal-plus-noise

The universal ROC curve clearly shows how the combination of SNR and X_T impacts Pd and Pfa in a trade-off fashion. It also shows that an SNR of unity is rather impractical, reaching a Pd of 0.9 only when the Pfa is greater than 0.1. This is in spite of the engineering convention of using an SNR = 1 as a criterion of good detection in sensors.

B.2. Heterodyne (or Homodyne) Sensor

The heterodyne (or homodyne) sensor is more complicated because mixing AWGN with a local oscillator will have a different result after square law detection than AWGN along. The PDF now is the combination of two Gaussian functions, one representing the fraction of the receiver noise in phase with the LO before detection, and one representing the fraction of the receiver noise in quadrature (i.e., 90° displaced) from the LO. The resulting post-detection PDF for the noise alone, decomposed into these two components, is the classic Rayleigh distribution

$$P(X) = \frac{X}{\overline{(\Delta X)^2}} \exp[-(X^2)/2\overline{(\Delta X)^2}] \cdot$$

The post-detection PDF for the signal plus noise is given by the famous Rician distribution,[51]

$$P(X, X_S) = \frac{X}{\overline{(\Delta X)^2}} \exp[-(X^2 + X_S^2)/2\overline{(\Delta X)^2}] \cdot I_o[X \cdot X_S/\overline{(\Delta X)^2}]$$

where I_0 is the modified Bessel function of order zero. These two functions are shown in Fig. 17 where the Pd and Pfa have similar graphical interpretations as given for the direct-detection case.

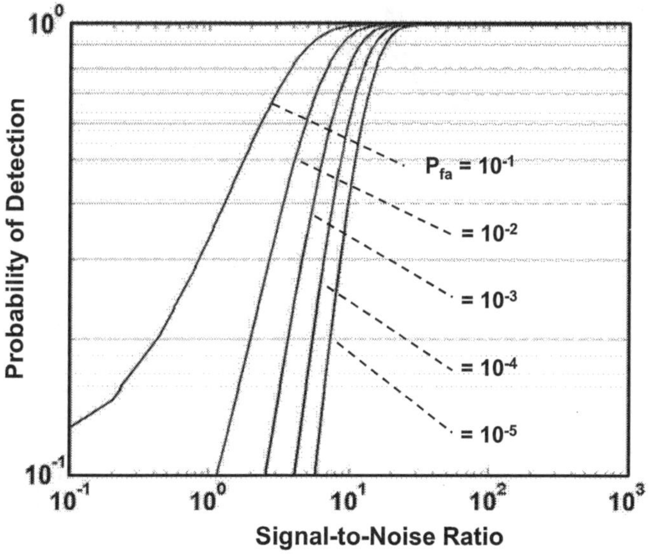

Fig. 18. Receiver operation characteristic (ROC) diagram for coherent case.

Because the Rayleigh distribution is integrable, one can easily derive an analytic form for the Pfa as

$$P_{fa} = \int_{X_T}^{\infty} P(X)dX = \exp[-(X_T^2)/2\cdot\overline{(\Delta X)^2}]$$

but the Pd is more complicated and is best done numerically,

$$P_d = \int_{X_T}^{\infty} P(X, X_S)dX$$

One form of the numerical results are plotted in Fig. 18 where Pd is plotted SNR with Pfa as a parameter. This form eliminates the explicit dependence on a threshold. The resulting curves show that, as in the case of direct detection, an SNR of unity is not very useful, achieving values of Pd > 0.1 only when the Pfa is greater than about 0.01. The key advantage heterodyne (or homodyne) over direct detection, as will be shown below, is the much greater SNR that can be achieved for a given incident signal power. This is the same reason that heterodyne receivers are still the standard today in radio communications.

IX. Overall Performance of Four Types of Passive Sensors

In a passive sensor the "signal" is generally the time-averaged thermal radiation in some spectral bandwidth that propagates between an object of interest and the sensor. The advantage is clearly that the source radiation comes for free. The downside is that the atmospheric attenuation and fading effects can quickly degrade the sensor performance as will be shown below. It is very important to construct a realistic

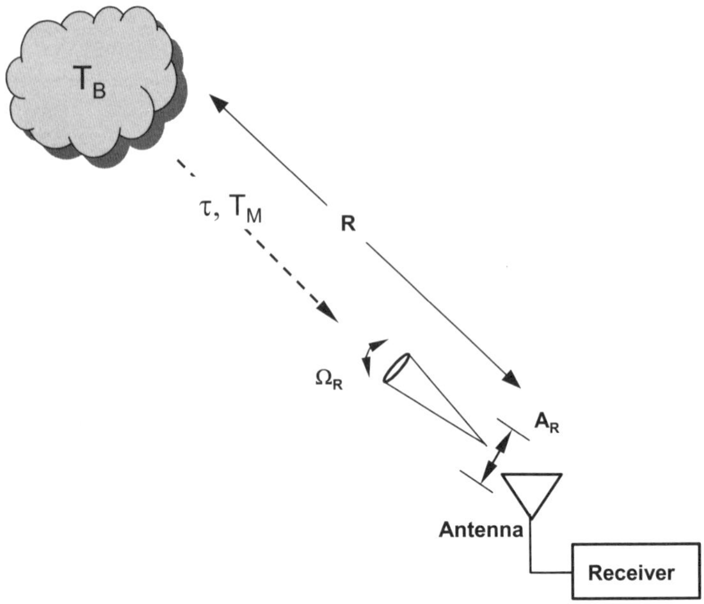

Fig. 19. Block diagram of canonical receiver with receiving-antenna effective aperture A_R, solid angle Ω_R, and separation from target R.

scenario of the object and the intervening atmosphere to properly predict the remote sensing performance.

A. Sensor Scenario

Perhaps the most common passive remote-sensing scenario in the mm-wave and THz regions is the thermal signal with a thermal background, as shown in Fig. 19. Between the target (brightness temperature T_S) and the sensor detector there is assumed to be an attenuating atmosphere whose physical temperature is T_M. In addition, telescopes or other optical components placed between the target and detector add to the attenuation, creating an overall transmission factor τ_A which can be significantly less than unity. For simplicity, the object is assumed to fill the field-of-view of the passive sensor so that each spatial mode of the receiver is subtended by the target. We assume further that the temperature of the target, the atmosphere, and the background are all sufficiently high that $h\nu << k_B T$. Hence the following expression can be used from Sec. II for the thermal power incident in each of the M spatial modes

$$P_{inc,m} = k_B \cdot \Delta\nu \cdot [\tau_A \cdot T_B + (1-\tau_A)\cdot T_M] \ .$$

where only the term dependent on T_B is actual signal. From the analysis of Sec. II.D, the fluctuations in this thermal radiation absorbed by the detector can be estimated by

$$\overline{(\Delta P_{abs})^2} = [\sum_m^M \eta_m](h\nu_0 \Delta\nu)^2 f_P(f_P+1) \approx M\cdot\eta\cdot(k_B T_{eff}\cdot\Delta\nu)^2$$

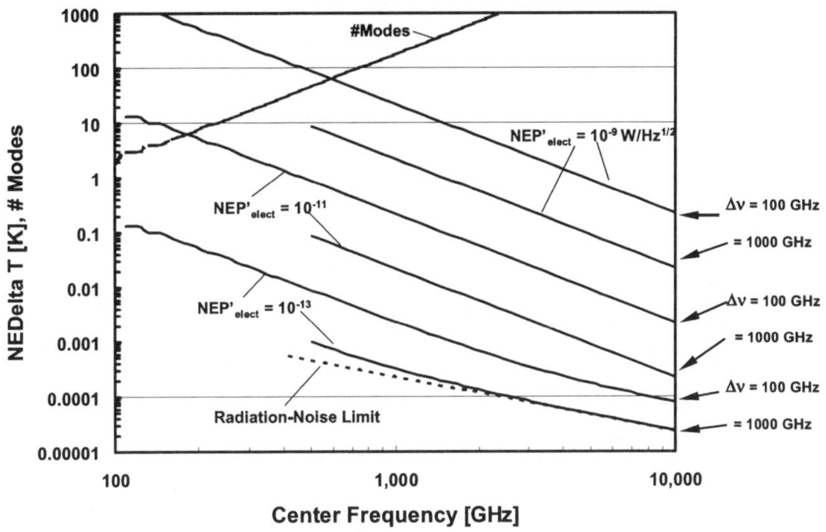

Fig. 20. Curves of NEΔT vs ν_0 parameterized by various values of the NEP_{elect} and $\Delta\nu$. Also shown is the number of spatial modes assuming ν_R= 75 GHz. Other important parameters are: Δf = 30 Hz (consistent with live-video requirements), τ_A = 1.0, and η = 0.5 (all modes), and $T_S = T_M$ = 290 K.

where the effective temperature is defined by $T_{eff} = \tau_A T_B + (1-\tau_A) T_M$. This expression will include the quantum-noise limit as a special case. The following sub-sections will evaluate four passive sensors with this thermal signal and radiation noise as the basis.

B. Direct Detection with Classical Square-Law Detector

B.1. General Analysis

In Sec. VII.C an approximate general expression was derived for the NEP of a mm-wave or THz direct detector in terms of absorbed radiation fluctuations and the electrical NEP. For the present passive scenario in the Rayleigh-Jeans limit and assuming all spatial modes have the same value of η, this expression is given by

$$NEP_{AD} \approx \sqrt{2 \cdot \overline{(\Delta P_{ABS})^2} \cdot \Delta f / (\eta^2 \Delta\nu) + (NEP_{elect})^2}$$

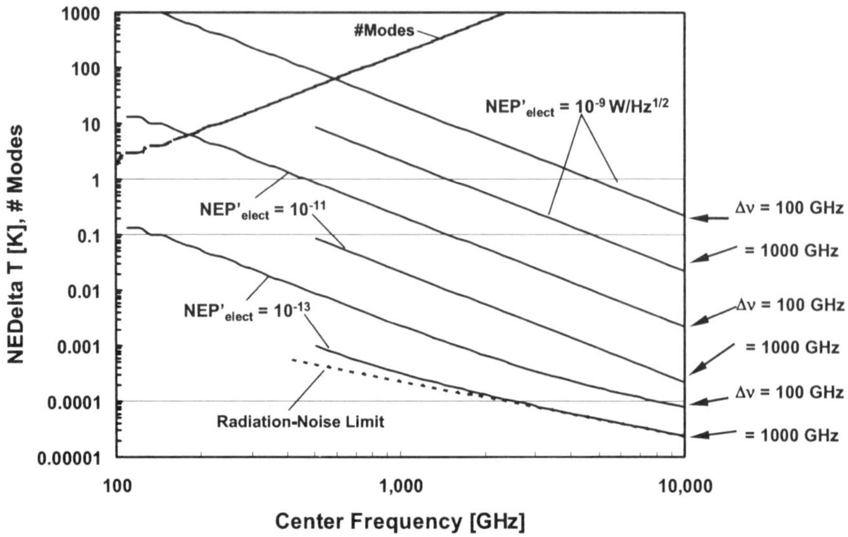

Fig. 21. Curves of NEΔT vs NEP_{elect} parameterized by various values of Δν and τ_A. Other important parameters are: ν_0 = 500 GHz, Δf = 30 Hz (consistent with live-video requirements), η = 0.5 (all modes), M = 45, and $T_S = T_M$ = 290 K.

$$= \sqrt{2 \cdot (k_B T_{eff})^2 \cdot M \cdot \Delta f \cdot \Delta \nu / \eta + (NEP_{elect})^2} \qquad (50)$$

The appropriate sensitivity metric in this case is the NEΔT defined as in Sec. IV but now in terms of just the source brightness temperature. From the expression above we have $dP_{inc}/dT_B = M \cdot k_B \cdot \tau_A \cdot \Delta\nu$, so the NEΔT is given by

$$NE\Delta T \equiv \frac{NEP_{AD}}{dP_{inc}/dT_B} = \tau_A^{-1} \sqrt{2 \cdot T_{eff}^2 \Delta f / (M \cdot \eta \cdot \Delta\nu) + [NEP_{elect} / (M \cdot k_B \cdot \Delta\nu)]^2} \qquad (51)$$

In the event that the first term in the radicand dominates, we obtain

$$NE\Delta T = \frac{T_{eff}}{\tau_A} \sqrt{2 \cdot \Delta f \cdot (M \cdot \eta \cdot \Delta\nu)^{-1}}$$

The condition on receiver sensitivity for reaching this limit is simply that the electrical NEP satisfies

$$NEP_{elect} < k_B T_{eff} \sqrt{2M \cdot \Delta f \cdot \Delta\nu / \eta}$$

Eqn 51 becomes useful in evaluating the performance of direct-detection sensors once the parameters are known. From Sec. III the number of spatial modes can be estimated using $M \approx \Omega_{FOV}/\Omega_B \approx \mathrm{Int}(\nu_0/\nu_R)^2 + 1$ where ν_R is a reference frequency, assumed here to be 75 GHz. The external coupling efficiency per mode is assumed to be

$\eta = 0.5$. The post-detection bandwidth is assumed to be 30 Hz consistent with live-video requirements. The center frequency ν, instantaneous bandwidth $\Delta\nu$, and atmospheric transmission are best left as parameters since these are the most flexible or variable parameters from an engineering or phenomenology standpoint. Fig. 20 shows the NEΔT vs ν_0 for three different NEP'$_{elect}$ values and two different $\Delta\nu$ values. Also shown is the number of modes vs ν_0 for $\nu_R = 75$ GHz. In all cases NEΔT drops monotonically with ν_0 simply because M increases quadratically – there is no other ν_0 dependence in Eqn 51. An increase of $\Delta\nu$ decreases NEΔT at any given ν_0, but a decrease in NEP$_{elect}$ is effective only down to roughly 10^{-13} W/Hz$^{1/2}$. This is because of the radiation noise arising from the 290-K source and atmosphere. The reader may note that the literature is replete with cryogenic mm-wave and THz bolometers having far lower values of NEP, consistent with operation in upper-atmospheric and astronomical sensing where the background radiation noise is far lower than in the present scenario.

Fig. 20 also reiterates the point made in Sec. VI.A.3 of how important center frequency is to sensitivity. A detector having NEP = 10^{-9} W/Hz$^{1/2}$ is relatively easy to obtain for room-temperature operation, but can provide an NEΔT only between roughly 1 and 20 K at a 1000-GHz center frequency – not considered sensitive enough for most applications. But this same detector can provide NEΔT between about 0.01 and 0.2 K if operated at a center frequency of 10 THz ($\lambda = 30$ μm). The explanation is, again, the increase in the number of spatial modes, which also explains why "uncooled" detectors can be used as the basis for sensitive cameras at roughly $\lambda = 10$ μm and shorter wavelengths. But the reader should note that the present analysis assumes all spatial modes contribute both signal and noise, which is increasingly difficult to achieve when the frequency goes well into the infrared.

A second interesting plot from Eqn 51 is NEΔT vs NEP'$_{elect}$, parametrized by $\Delta\nu$ and τ_A. The center frequency, number of modes, and mode coupling are fixed at 500 GHz, M=45, and $\eta = 0.5$, respectively. The resulting curves in Fig. 21 show, again, the asymptotic approach toward the radiation-noise-limited NEΔT for NEP'$_{elect} \leq 10^{-13}$ W/Hz$^{1/2}$. To get a useful sensitivity of, say NEΔT ~ 0.1 K with $\tau_A = 1.0$, the NEP'$_{elect}$ can be no worse than about 10^{-11} W/Hz$^{1/2}$ – a performance that is hard to achieve at room temperature if one also requires M = 45 and $\eta = 0.5$. As τ_A degrades, the required NEP'$_{elect}$ also drops until $\tau_A \sim 0.01$ (-20 dB), when the radiation noise then limits the NEΔT to values above 0.1 K. Hence for passive thermal imaging it is imperative in the THz region to choose the center frequency carefully as a trade-off between number NEP$_{elect}$, number of modes, and atmospheric transmission. From the atmospheric transmission curve of Fig. 5(a), a compelling choice of center frequency would be the highest-frequency "good window" centered around 9 cm^{-1} (270 GHz).

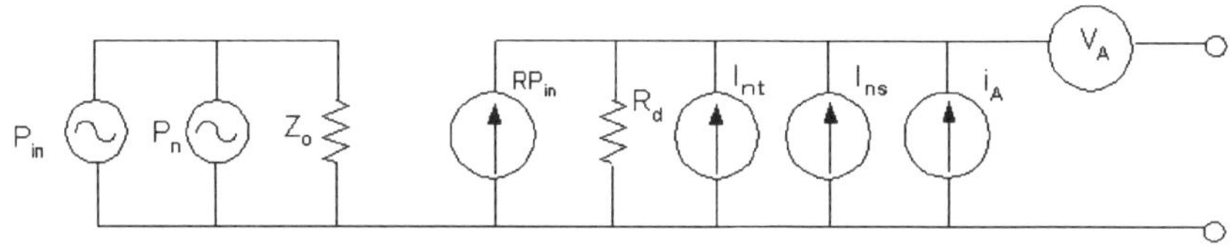

Fig. 22. Equivalent circuit of mm-wave/THz direct detector.

B.2. Detector-Specific Analysis

In many types of THz direct detectors, the more specific model of Fig. 22 is applicable which allows one to make first-principles estimates of the NEP and other metrics. The input RF impedance is represented by Z_{in}, and the rectification process is represented by power-dependent-current source of rms amplitude, i_P, which is related to the average power absorbed in Z_{in} through the responsivity $\Re$ and the coupling efficiency η_m for each incident spatial mode

$$i_P = \Re\, P_{abs} = \Re\, [\sum_m^M \eta_m P_{inc,m}]$$

The thermal and shot-noise generators represent the noise contributed by the detector itself. In a more general analysis, there would be a third generator representing cross-correlation between these two. But in many THz detectors, either the thermal or shot-noise generator dominates the other so the cross-correlation becomes insignificant. The noise contributed by all of the remaining electronics after the detector is lumped into one thermal generator, T_n. This term is generally dominated by the first amplifier that occurs in the post-detection circuit. Note that this circuit model handles the RF noise as well as the signal entering the device. Hence, given the thermal model for fluctuations of the radiation absorbed by the device and assuming, again, that $P_{inc,m} = P_{inc}/M$, we can write

$$\left(\frac{S}{N}\right)_{AD} = \frac{(\Re \cdot \frac{P_{inc}}{M} \sum_m^M \eta_m)^2}{2\Re^2 \cdot (\sum_m \eta_m) \cdot (k_B T_{eff})^2 (\Delta\nu \cdot \Delta f) + \overline{(\Delta i_n)^2} + \overline{(\Delta v_n)^2} \cdot (G_D)^2 + 4k_B T_0 \cdot \Delta f / R_A},$$

so that the NEP_{AD} can be written

$$NEP_{AD} = M \cdot \sqrt{\frac{2 \cdot (k_B T)^2 \cdot \Delta\nu \cdot \Delta f}{\sum_m^M \eta_m} + \frac{\overline{(\Delta i_n)^2} + \overline{(\Delta v_n)^2} \cdot (G_D)^2 + 4k_B \cdot T_0 \cdot \Delta f / R_A}{(\Re \cdot \sum_m^M \eta_m)^2}} \qquad (52)$$

Under the common condition that the incident noise is dominated by the electrical noise terms, we get the electrically-limited NEP:

$$NEP_{AD}^{elect} = \frac{M}{\Re \sum_{m}^{M} \eta_m} \sqrt{\overline{(\Delta i_n)^2} + \overline{(\Delta v_n)^2} \cdot (G_D)^2 + 4k_B T_0 \cdot \Delta f / R_A} \tag{53}$$

These expressions are particularly useful when the individual noise mechanisms for a direct detector are already understood.

When the noise mechanisms are not understood but the NEP of a direct detector is already known or are readily measured experimentally, the following forms are more useful and easily derived from Eqns 52 and 53 above:

$$NEP_{AD} = \sqrt{\frac{2 \cdot M^2 \cdot (k_B T)^2 \cdot \Delta v \cdot \Delta f}{\sum_{m}^{M} \eta_m} + (NEP_{AD}^{elect})^2}$$

B.3. An Illustrative Example: Schotty-Diode Direct Detector

The Shottky diode is one of the oldest and best understood of the THz detectors. Detection and mixing depend on the nonlinear rectification properties of a metal-semiconductor junction – a phenomenon that has been exploited for over half a century. The intrinsic sensitivity and frequency conversion properties of the diode are well described by classical rectifier theory.[2] One of the strengths of the Schottky diode is that it can be made to have useful sensitivity over an enormous wavelength range extending from microwaves to about 100 μm. In addition, it is capable of very large instantaneous bandwidth (limited only by the IF circuit parameters), displays excellent performance characteristics at room temperature and can be fabricated bin a variety of geometries to suit various applications. The three most prevalent geometries are the whisker-contacted honeycomb, the beam-lead diode and the planar or surface-oriented diode. Each of these will be discussed later in terms of their applications as submillimeter-wave mixers.

Independent of the particular geometry, the I-V characteristic of a Schottky diode is fit quite well by the expression[52]

$$I = I_S \left[\exp\left[\frac{eV}{\eta k T_J} \right] \right] - \exp\left[\frac{(\eta^{-1} - 1) \cdot V}{k T_J} \right] \tag{54}$$

where η is the ideality factor (≥ 1.0), T_0 is the junction temperature, and $I_S = RT_J^2 \exp(-e\phi_{bi}/k_B T_J)A$ where R is the Richardson coefficient (= 4.4 $A/cm^2/K^2$) and ϕ_{bi} is the built in potential. Typically in 100-GHz to 1 THz region the ideality factor of Schottky diodes is 1.2 to 1.3.[53] The first exponential term represents the normal thermionic emission current while the second term is mainly due to two tunneling components, field emission and thermionic field emission. The relative importance of each of these terms depends on both the temperature and the impurity concentration in the epitaxial layer. At room temperature and relatively low impurity concentration, the thermionic component usually dominates. A satisfactory fit to the I-V curve is then obtained by the more familiar expression $I = I_S \left[\exp\left\{ {eV}/{\eta kT} \right\} - 1 \right]$.

According to the seminal theory of Torrey and Whitmer,[54] the low-frequency short- circuit voltage responsivity can be estimated directly from the I-V curve by the expression

$$S \approx \frac{1}{2}\frac{d^2I/dV^2}{dI/dV} \tag{55}$$

The dependence on the second derivative generally leads to maximum sensitivity for forward bias at the "knee" of the I-V curve. The short-circuit physical noise after detection can be approximated by the three terms: (1) Schottky-diode "full" shot-noise $(\Delta i)^2 = 2eI_D\,\Delta f$, (2) the Schottky-diode thermal noise term that goes as $(\Delta i)^2 = 4k_BT_J\,G\,\Delta f$, and electronic noise all lumped into the current generator $(\Delta i)^2 = (4k_BT_0/R_A)\,\Delta f$. Note that the Schottky thermal noise is intended to account for both the series resistance and differential resistance of the device. We also assume that the radiation noise power absorbed by the detector is negligible, so that the specific NEP can be approximated as,

$$NEP'_{AD} = \frac{1}{\Re}\sqrt{2eI_D + 4k_BT_J \cdot G_D + 4k_BT_0/R_A} \tag{56}$$

It is illuminating to compute Eqns 54, 55, and 56 for a typical THz Schottky diode made of GaAs and coupled at 100 GHz to an antenna having a THz impedance of 100 Ω. We assume the diode has an area of 2 square micron and an ideality factor of 1.2. We assume the output amplifier has a noise voltage of 1 $nV/Hz^{1/2}$ and a noise current of 1.0 $pA/Hz^{½}$. The resulting I-V curves at 77 K, 200 K, and 300 K are shown in Fig. 23(a). The short-circuit responsivity of Torrey&Whitmer is shown in Fig. 23(b), including mismatch with the antenna. The corresponding curves of NEP vs bias voltage are shown in Fig. 23(c). These curves demonstrate an important reality about direct detection: it is difficult to make the detector noise dominate the electronic noise that follows it, and it is even more difficult to make the detector noise dominate the radiation noise.

The limitations introduced by the electrical noise of direct detectors and the following electronics has been a problem since the early days of the THz field. Traditionally, this problem has been approached by the use of cryogenic detectors, particularly those operating at 4.2 K (vaporization point of liquid helium) and below. There is a wealth of literature on cryogenic direct detectors, particulary a myriad of different types of bolometers. It is not the purpose of this article to reiterate , but rather to examine the alternative receiver architectures for what may operate at room temperature. This leads us to the next three architectures, all of which have the potential to significantly out-perform the direct receiver under room temperature operation.

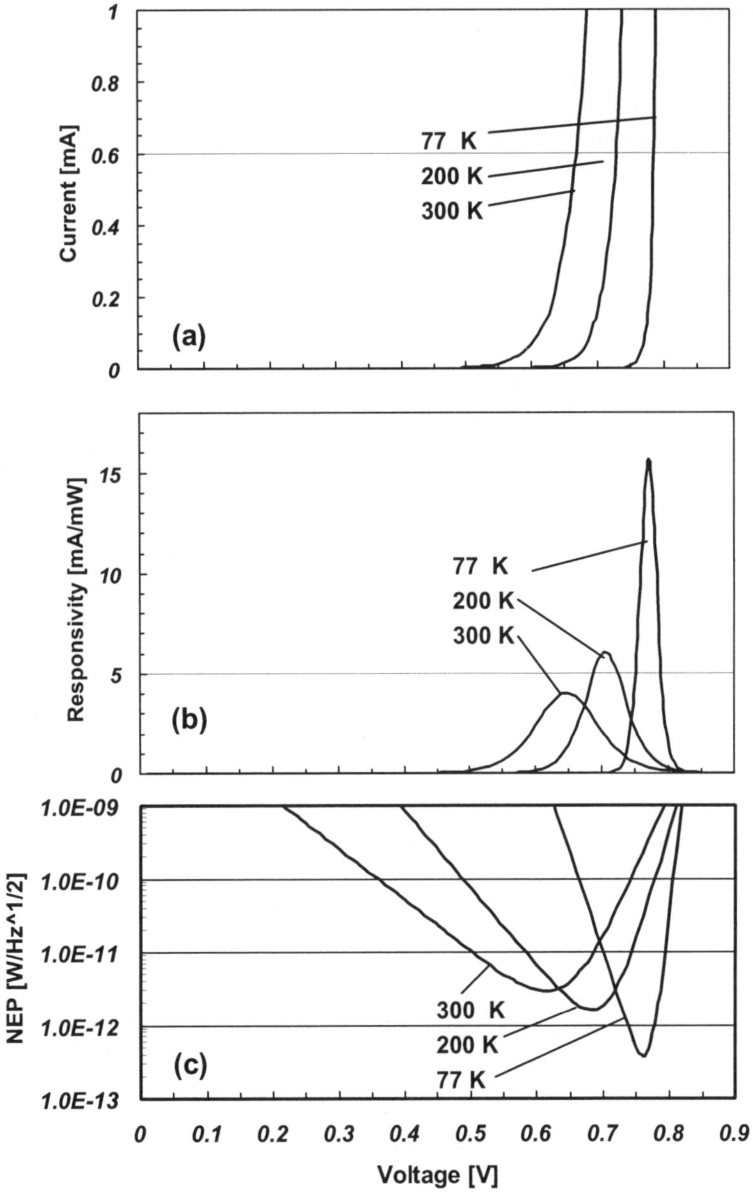

Fig. 23. Analytic results for a generic Schottky diode at 3 temperatures; (a) I-V curves, (2) electrical responsivity (including mismatch between the antenna and the diode, and (3) electrical NEP.

C. Heterodyne

In the interest of brevity, we analyze only the case of classical heterodyne detection as occurs routinely with Schottky-diode mixers, hot-electron bolometers, and other mixer types. From Sec. VII we have an expression for the SNR in the IF section immediately after down-conversion that includes the noise contribution from the mixer itself and the following IF electronics

$$\left(\frac{S}{N}\right)_{HET} = \frac{(\eta_R)^2 \cdot G_{mix} \cdot P_{inc}}{\eta_R \cdot G_{mix} \cdot h\nu_{LO} \cdot B_{IF}}$$

We multiply each term by the gain of the entire IF amplifier chain G_{IF} along with the responsivity of the square-law detector. To the LO "shot-noise" of the denominator we add the noise figure of the entire IF amplifier chain according to the Friis formula of Sec. III. The implicit assumption here is that, as for most heterodyne receivers, the IF gain is high enough that the noise contribution of the square-law detector and electronics following it are negligible. Because the IF band is generally at much lower frequency than the mm-wave or THz radiation, any losses in coupling to the square-law detector can be lumped in G_{IF}, and we obtain

$$\left(\frac{S}{N}\right)_{AD} = \frac{[\Re_D \cdot G_{IF} \cdot (\eta_R)^2 \cdot G_{mix} \cdot P_{inc}]^2}{2\Re_D^2 \cdot \{G_{IF} \cdot \eta_R \cdot G_{mix} \cdot h\nu_{LO} \cdot B_{IF} + (\eta_R)^2 \cdot G_{mix} \cdot B_{IF} \cdot G_{IF} \cdot NEP'_{mixer}\}^2 \cdot (\Delta f / B_{IF})} \quad (57)$$

$$= \frac{(\eta_R{}^2 \cdot P_{inc})^2}{2 \cdot \{\eta_R \cdot h\nu_{LO} + (\eta_R)^2 \cdot NEP'_{mixer}\}^2 \cdot (B_{IF} \cdot \Delta f)}$$

This leads to the following useful expression for the NEP in active detection

$$NEP'_{AD} = \{h\nu_{LO} / \eta_R + NEP'_{mixer}\} \cdot (2B_{IF})^{1/2} = \frac{h\nu_{LO}}{\eta_{HET}} \cdot (2B_{IF})^{1/2}$$

Assuming once again that the incident signal is thermal noise contained in the same (single) mode as the local oscillator, we have $P_{inc} = k_B T_B \, \Delta f = k_B T_B B_{IF}$, so that in passive detection

$$NE\Delta T'_{AD} \equiv \frac{NEP'_{AD}}{dP_{inc} / dT} = \frac{h\nu_{LO}}{\eta_{HET} \cdot k_B} \sqrt{\frac{2}{B_{IF}}}$$

Fig. 24 shows curves of the NEP_{HET} before detection and the $NE\Delta T_{AD}$ after detection for various values of η_{HET} and B_{IF}.

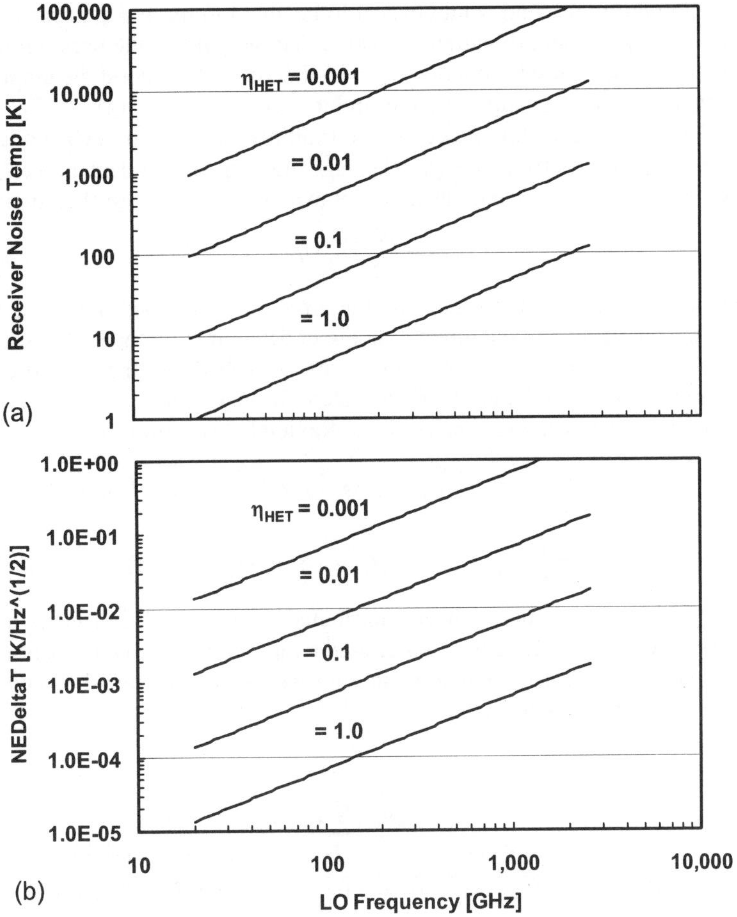

Fig. 24. Noise temperature and NEΔT vs LO frequency of double-sideband heterodyne receiver having an IF bandwidth of 10 GHz.

Several of the popular detector types can operate in both as direct detector and mixers. A good example is the Schottky diode. Based on the above analysis, the following rule-of-thumb relationship between the heterodyne and direct NEP values can be found: $NEP'_{HET} \approx (NEP'_{AD})^2/P_{LO}$. From the analysis of the generic Schottky diode in Sec. IX.B, we would have a room-temperature direct NEP'_{AD} of about $3x10^{-12}$ W/Hz$^{1/2}$ at 300 GHz. Then given a local oscillator power of 1 mW, one would predict a heterodyne NEP of $9x10^{-21}$ W/Hz, corresponding to a η_{HET} value of 0.02. These numbers are all reasonably close to the experimental reality.

D. Pre-Amplified Direct

We have seen that the electrical-noise limit in direct detection arises from the fact that the responsivity of the direct detector, at least the Schottky diode detector, is limited. So intuitively, one would expect to be able to improve the sensitivity of a direct receiver through pre-amplification, provided that the gain of the pre-amplifier is suitably high and the noise figure is suitably low. To determine how good the amplifier has to be, we apply the noise formalism to the architecture of Fig. 13(a). To get the signal absorbed by the square law detector, we multiply the P_{inc} in each mode by the input radiation coupling coefficient $\eta_{R,m}$ (i.e., the fraction of incident power in each mode absorbed by the *amplifier*) and the amplifier available gain G_m for that mode

$$S_{BD} = (\sum_{m}^{M} G_m \cdot \eta_{R,m} \cdot P_{inc,m})^2$$

Any mismatch between free space and the amplifier is lumped into $\eta_{R,m}$, and mismatch between the amplifier and the direct THz direct detector is lumped into G_m.

The mean-square power fluctuation absorbed by the amplifier depends, of course, on the type of radiation being detected. If the radiation being detected is thermal, the fluctuations can be approximated in the Rayleigh-Jeans limit by

$$\overline{(\Delta P_{abs})^2} = \left(\sum_{m}^{M} \eta_{R,m} \cdot (k_B T_B \cdot \Delta\nu)^2 \cdot G_m^2 \right) + (P_{AMP})^2$$

$$= \sum_{m}^{M} \eta_{R,m} \cdot G_m^2 \cdot (k_B T_B \cdot \Delta\nu)^2 + [G_m \cdot (F_m - 1) \cdot k_B T_{300} \cdot \Delta\nu]^2$$

where P_{amp} is the rms noise power contributed by the amplifier, F_m is amplifier noise figure for mode m, and T_B is the background temperature. We can substitute this into the mean-square power fluctuation term in Eqn 49 to get the signal-to-noise ratio at the output of the square-law detector:

$$\left(\frac{S}{N}\right)_{AD} = \frac{(\Re \cdot (\sum_{m}^{M} G_m \cdot \eta_{R,m} \cdot P_{inc,m})^2}{2\Re^2 \cdot \{\sum_{m}^{M} \eta_m \cdot G_m^2 \cdot (k_B T_B \cdot \Delta\nu)^2 + [G_m \cdot (F_m - 1) \cdot k_B T_{300} \cdot \Delta\nu]^2\} \cdot (\Delta f / \Delta\nu) + \overline{(\Delta X)^2}}$$

We also assume that the signal fluctuations in the detector and the following electronics are Gaussian, so that detector electrical NEP scales with the post-detection bandwidth Δf and we can write

$$\frac{S}{N} = \frac{[\Re \cdot (\sum_{m}^{M} G_m \cdot \eta_{R,m} \cdot P_{inc,m})]^2}{2\Re^2 \cdot \{\sum_{m}^{M} \eta_m \cdot G_m^2 \cdot (k_B T_B \cdot \Delta\nu)^2 + [G_m \cdot (F_m - 1) \cdot k_B T_{300} \cdot \Delta\nu]^2\} \cdot (\Delta f / \Delta\nu) + (\Re \cdot NEP'_{elect})^2 \cdot \Delta f}$$

At the present point in time, solid-state amplifiers only operate in the sub-THz region and, like local oscillators, are designed only for one spatial mode. This simplifies the performance analysis to the "unimodal" expression,

$$\left(\frac{S}{N}\right)_{AD} = \frac{(\Re \cdot P_{inc} \cdot \eta_R \cdot G)^2}{2\Re^2 \cdot \{ \eta_R \cdot G^2 \cdot (k_B T_B \cdot \Delta\nu)^2 + [G \cdot (F - 1) \cdot k_B T_{300} \cdot \Delta\nu]^2 \} \cdot (\Delta f / \Delta\nu) + (\Re \cdot NEP'_{elec})^2 \cdot \Delta f}$$

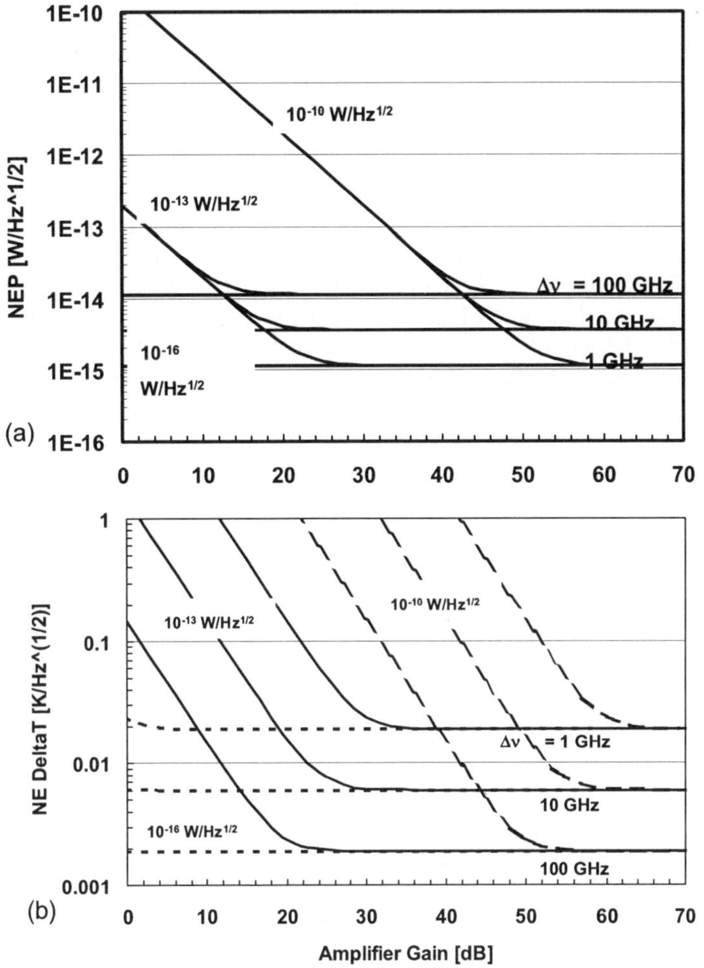

Fig. 25. (a) NEP and (b) NEΔT for pre-amplified direct detection as a function of amplifier gain and parametrized by direct detector NEP'$_{elect}$ and RF bandwidth Δν. The amplifier noise figure is fixed at 6.0 dB.

Thus the NEP is given as follows:

$$NEP_{AD} = \sqrt{2 \cdot \{\eta_R^{-1} \cdot (k_B T_B)^2 + [\eta_R^{-1} \cdot (F-1) \cdot k_B T_{300}]^2\} \cdot (\Delta\nu \cdot \Delta f) + \left(NEP_{elec}^{'} / (\eta_R \cdot G\right)^2 \cdot \Delta f}$$

or

$$NEP_{AD}^{'} = \sqrt{2 \cdot \{\eta_R^{-1} \cdot (k_B T_B)^2 + [\eta_R^{-1} \cdot (F-1) \cdot k_B T_{300}]^2\} \cdot (\Delta\nu) + \left(NEP_{elec}^{'} / (\eta_R \cdot G\right)^2}$$

The unimodal assumption and the Rayleigh-Jeans approximation also facilitate the computation of the NEΔT using the fact that the incident power per mode is $\overline{P}_{inc} \approx k_B T \cdot \Delta\nu$ and $d\overline{P}_{inc} / dT \approx k_B \cdot \Delta\nu$. We find

$$NE\Delta T \equiv \frac{P_{inc}}{dP_{inc}/dT\,|} = \sqrt{2\cdot\{\eta_R^{-1}\cdot(T_B)^2 + [\eta_R^{-1}\cdot(F-1)\cdot T_{300}]^2\}/(\Delta\nu) + \left(NEP'_{elec}/(\eta_R\cdot G\cdot k_B\Delta\nu\right)^2}$$

Fig. 25 shows plots of NEP and NEΔT for the pre-amplified direct receiver under the following conditions. Fig. 25(a) shows the NEP vs the amplifier gain parametrized by three values of RF bandwidth: 1 GHz, 10 GHz, and 100 GHz and three values of NEP for the direct detector 10^{-9} W/Hz $^{1/2}$, 10^{-11} W/Hz $^{1/2}$, 10^{-13} W/Hz $^{1/2}$, Fig. 25(a) clearly shows that the NEP curves saturate with increasing amplifier gain at levels consistent with the amplifier noise. The quantum-limited NEP under each condition is also shown according to the expression derived in Sec. VI for direct detection of a thermal signal,

E. Pre-amplified Heterodyne

An intriguing possibility in the millimeter-wave band and lower end of the THz range is a preamplifier feeding a mixer element. As in Sec. IX.D we will analyze this only for the unimodal case since that is, by far, the most effective way to design and construct amplifiers and local oscillators alike. To simplify the expression, we will also assume that the local-oscillator photon noise dominates the incoming radiation noise – an assumption that will obviously become more questionable as the amplifier gain increases

$$\left(\frac{S}{N}\right)_{AD} = \frac{[\Re_D\cdot G_{IF}\cdot\eta_R\cdot G_{mix}\cdot G_{RF}\cdot P_{inc}]^2}{2\Re_D^2 G_{IF}^2 G_{mix}^2\{\eta_R\cdot(G_{RF}\cdot h\nu_{LO}\cdot B_{IF})^2 + [G_{RF}(F-1)\cdot k_B T_{300}\cdot B_{IF}]^2 + (B_{IF}\cdot NEP'_{mixer})^2\}\cdot(\Delta f/B_{IF})}$$

$$= \frac{[\eta_R\cdot G_{RF}\cdot P_{inc}]^2}{2\cdot\{\eta_R\cdot(G_{RF}\cdot h\nu_{LO})^2 + [G_{RF}(F-1)\cdot k_B T_{300}]^2 + (\cdot NEP'_{mixer})^2\}(B_{IF}\cdot\Delta f)}$$

This leads to a performance in active sensing of

$$NEP'_{AD} = \frac{1}{\eta_R}\sqrt{\{\eta_R(h\nu_{LO})^2 + [(F-1)\cdot k_B T_{300}]^2 + (NEP'_{mixer}/G_{RF})^2\}\cdot(2\cdot B_{IF})} \qquad (58)$$

and a performance in unimodal passive sensing of

$$NE\Delta T = \frac{1}{\eta_R\cdot k_B}\sqrt{\{\eta_R(h\nu_{LO})^2 + [(F-1)\cdot k_B T_{300}]^2 + (NEP'_{mixer}/G_{RF})^2\}\cdot(2/B_{IF})}$$

Eqn 58 has the following satisfying properties. First, if the gain of the preamplifier is high enough to overcome the mixer noise, the resulting NEP has the form

$$NEP'_{AD} = \frac{1}{\eta_R}\sqrt{\{\eta_R(h\nu_{LO})^2 + [(F-1)\cdot k_B T_{300}]^2\}\cdot(2\cdot B_{IF})}$$

If the noise figure is also high enough that the amplifier noise dominates the photon shot noise

$$F > 1 + \frac{\eta_R(h\nu_{LO})}{k_B T_{300}} = 1 + \frac{\eta_R\nu_{LO}}{6.248x10^{12}}$$

and we get the "amplifier-limited" expression

$$NEP'_{AD} = \frac{(F-1)\cdot k_B T_{300}}{\eta_R}\sqrt{2\cdot B_{IF}}$$

Presently, all known solid-state room-temperature amplifiers operating in the millimeter or THz regions satisfy this condition. The plots of amplified-heterodyne receiver performance are shown in Fig. 26. For active sensing, the NEP in Fig. 26(a) clearly increases monotonically with increasing B_{IF} for any value of G or NEP_{mixer}. And each case shows a monotonically decreasing value of NEP' down to a minimum value defined by the amplifier noise limit given above. The NEP also drops inversely with receiver coupling efficiency.

In a passive sensor, the amplified-heterodyne receiver provides the performance shown in Fig. 26(b). The NEΔT now decreases with increasing B_{IF} just as in the canonical heterodyne receiver. Again, the NEΔT saturates at a minimum level corresponding to the amplifier limit. In comparison to the amplified-direct or canonical heterodyne, the amplified-heterodyne provides one advantage in terms of reaching the amplifier-limited performance with a relatively poor mixer performance. As shown in Fig. 26(a), the limit defined by a 6-dB-noise-figure amplifier can be achieved with a mixer having an η_{HET} =0.01 (NEP = 6×10^{-18} W/Hz) and an amplifier gain of just ~ 10 dB. But the receiver can still reap the benefits of a superheterodyne receiver in terms of frequency selectivity defined in the IF rather than the RF. At the present time, this seems particularly advantageous for all-room-temperature operation where LO power is either difficult to generate or difficult to distribute to the mixer. One example is imaging arrays where LNAs would be simpler to implement than local oscillators. The reason for this is that LNAs have become monolithic integrated circuits whereas local oscillators in the mm-wave and THz regions are still discrete components. Worse yet, the local oscillators often require low frequency synthesizers to multiply up harmonically and act as the reference for the fundamental oscillator at the LO frequency.

X. Issues and Performance of Active Sensors

By the definitions given in Sec. I, active sensors are those that provide their own illumination of separated targets and objects using a sub-system called the transmitter. The job of the transmitter is to provide enough coherent or incoherent radiation at the target to get a measurable reflection or transmission of this radiation at the position of the receiver. The source of the radiation has long been the bane of the mm-wave and THz regions, tending to become more difficult and weaker in power as the frequency increases above 30 GHz. Historically, the best sources for this radiation have been vacuum-tube devices, such as oscillators driving traveling wave tube amplifiers below about 100 GHz, and fundamental vacuum oscillators (e.g., backward wave oscillators) above 100 GHz. Such sources are notoriously expensive and cumbersome once the high-voltage power supply is factored in. Researchers have long sought after solid-state sources to provide the transmit power. Until recently, the power levels and limited

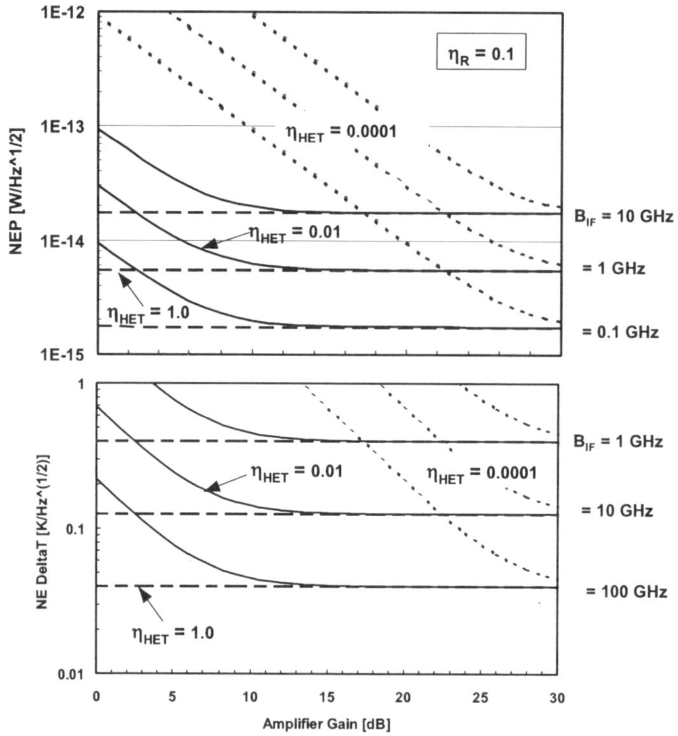

Fig. 26. (a) NEP and (b) NEΔT for pre-amplified heterodyne receiver as a function of amplifier gain and parametrized by IF bandwidth and η_{HET}. The frequency is 100 GHz, the pre-amplifier noise figure is fixed at 6.0 dB, the quantum-limited NEP is $\approx 6.6\text{x}10^{-21}$ W/Hz, and the external coupling efficiency is 0.1.

tunability have generally been sufficient to make only local oscillators, not free-space transmitters.

A. Active Sensor Scenario

By combining a transmitter (Tx) with a receiver (Rx) oriented as in Fig. 19, a coherent or quasi-coherent signal is sent out by the Tx, is reflected or transmitted by the object or target, and is then collected by the Rx where it is rectified from RF (THz) to baseband either by a direct or coherent receiver. The target might be a cloud or a partially transparent object, for example. For purpose of maximizing the received signal, the relative orientation of the Tx and Rx is very important.

If the object or target is best observed in transmission, then for good performance the Rx is best located along the direction of the transmitter radiation but on the opposite side of the object. An alternative (see Sec. X.D below) is to place a retroreflector on the opposite side of the object and co-locate the Rx next to the Tx. If the object or target is best observed in reflection because of a strong back-scatter effect, then Tx is best configured to produce a quasi-collimated beam, similar to the "large-waist" beam plotted in Fig. 11.

A common goal of all three of these Tx-Rx orientations is that the received power, as defined in the Friis analysis of Sec. III.F, should fall with separation r between the Tx and Rx by $\sim 1/r^2$ or slower. This is in contrast to the typical $1/r^4$ decay that occurs in conventional radar, where the $1/r^2$ factor from the far-field Tx radiation is multiplied by a second $1/r^2$ factor based on assumed isotropic scattering from the target or object. Although the $1/r^4$ factor is considered accurate for long-range radar, particularly at microwave frequencies and below, it is unnecessarily pessimistic for the relatively-short range applications being pursued with mm-wave and THz sensors. Two good examples – concealed-weapons detection and all-weather navigation – are addressed later in Sec. XI.

To get the receiver SNR and ROC performance, one first needs to know how much of the transmitted power is collected by the receiver. For scatter-free atmospheric propagation, the instantaneous signal power at the receiver (Rx) can be estimated analytically given reasonable assumptions about the Tx power and beam patterns. For simplicity, we make the following three assumptions: (1) the Tx radiates only the one fundamental spatial mode of its antenna, (2) the width of the Tx and Rx beams are much smaller than the lateral extent of the target at every point where they pass through, and (3) the Tx and Rx antennas are perfectly aligned such that the peaks of their patterns are anti-collinear. The Rx incident power can then be approximated by the Friis formula (III.18) in the case of perfect beam and polarization alignment:

$$P_{inc}(\nu) = \frac{P_T G_T G_R \lambda^2}{(4\pi)^2 R^2} \tau(\nu) \tag{59}$$

This can be considered as a worst-case estimate if the Tx fundamental mode is a Gaussian beam having a Rayleigh length comparable or greater than R.

In the active sensor with direct- or pre-amplified direct detection, the Rx noise power will generally be the sum of at least two terms in addition to the electrical noise of the front-end electronics: (1) radiation fluctuations in the coherent received power and (2) radiation fluctuations of the background thermal radiation. In the Rayleigh-Jeans limit these become:

$$\overline{(\Delta P_{abs})^2} = \sum_{m}^{M} \eta_m \cdot \{(k_B T_{eff} \cdot \Delta\nu)^2 + (h\nu_0 P_{inc} \Delta\nu)\}$$

In the active sensor with heterodyne- or preamplified-heterodyne down-conversion, these must be added to the local-oscillator photon fluctuations, which is done below.

It is interesting to solve for the incident power that makes the photon noise dominate the thermal noise in the above expression. Under this condition one would expect to achieve a photon shot noise limit similar to that discussed in Sec. VI. Substitution from Eqn 59 yields the condition

$$P_T > \frac{(k_B T_{eff})^2 \cdot \Delta\nu (4\pi)^2 R^2}{h\nu_0 \cdot G_T G_R \lambda^2 \cdot \tau(\nu)}$$

We estimate this for a typical terrestrial case: $T_{eff} = 300$ K, $\Delta\nu = 1$ GHz, R = 100 m, $\nu_0 =$ 300 GHz, $G_T = G_R = 1000$, $\lambda = 1$ mm, and $\tau = 0.1$. The result is $P_T > 1.4$ mW – an available power level, even from solid-state technology.

A facilitating consequence of the limited THz Tx power is that the coherent radiation term can usually be neglected in the above expression. As shown below, this greatly facilitates the sensitivity analyses and leads to the "background limited" case for active mm-wave and THz sensors.

B. Transmitter Types

B.1 Continuous Wave

The preferred source of cw radiation practically anywhere in the RF bands is that derived from solid-state electronics. The situation for solid-state sources in the mm-wave and THz region has changed dramatically in recent years with the development and continual improvement of MMIC-based solid-state power amplifiers operating around 100 GHz. The reader is referred to a review article for more details.[55] Although outside the scope of this chapter, it is important to recognize that SSPAs are a great enabler for sensors in the mm-wave and THz regions alike. Around 100 GHz they can be used as the basis for power-combining networks to produce power levels far above that possible from fundamental oscillators. Around 300 GHz and above, SSPAs can produce useful power levels (> 1 mW) and high tunability by acting as the driver for varactor multiplier chains.

Besides cost, size, and ability to integrate, another benefit of solid-state sources is modulation. Sensors, like communications systems, general provide some form of analog or digital modulation on the carrier so that the transmitted signal has, for the analog case, the form $S(t) = A(t)\cos[\omega\ t + \phi(t)]$, where A is the amplitude and ϕ is the phase. Pure amplitude modulation implies ϕ (t) is a constant and pure phase (or frequency) modulation imples that A(t) is a constant. In either case, modulation allows improvement of performance or mitigation of the atmospheric and fading effects described in Sec II.

The standard methods of detecting modulated radiation exist as readily at mm-wave and THz frequencies as they do at microwave frequencies and below. For an AM-encoded carrier, the standard demodulation technique is either envelope detection or square-law detection. It can be shown that an envelope detector is the better performer when the pre-detection signal-to-noise ratio is greater than unity, and the square-law detector is better when the pre-detection SNR is below unity.[56] Because mm-wave and THz sensors have generally operated under the latter condition, square-law detection as discussed in Sec.V should be optimal or nearly optimal in most demodulation applications.

B.2. Pulsed: T-Rays

Pulses transmitters were first developed for radar to get range information through the classic time-of-flight and range-gating techniques. This has not become an issue yet in THz systems made with solid-state sources. However, pulsed transmitters have become interesting for a different reason – if the width of a pulse becomes significantly less than 1 ps, then Fourier analysis shows that it contains significant spectral density in the THz region. Mode-locked lasers, particularly those based on dyes or Ti:sapphire crystals have been developed to the point where they routinely generate ~200 fs pulses or shorter. If such a pulse is used to excite an ultrafast- photoconductive switch (i.e., Auston switch), then the photocurrent from this switch will have THz components. And if the photocurrents are, in turn, coupled to a THz antenna, there will be significant free-space THz radiation.

Important questions are how much power a T-ray transmitter can produce and what the sensitivity of the receiver can be. The Tx power can be estimated roughly from the way photoconductive switches work, which is the application of enough peak energy from the mode-locked laser pulse to short-circuit the switch for some fraction of the laser pulse width t_p. If the switch is biased with voltage V_B and is coupled to a THz antenna with radiation resistance R_A, then the peak THz power available for radiation from the antenna will be roughly $(V_B)^2/R_A$ and the THz pulse energy will be roughly $(t_p/2)(V_B)^2/R_A$. The average THz power P_{ave} will then be $P_{ave} \approx f_{rep}(t_p/2)(V_B)^2/R_A$ where f_{rep} is the laser repetition frequency. Note that the mode-locked laser power does not enter this estimate provided that it is great enough to short the photoconductive switch at the peak point.

For example, the typical ultrafast photoconductive switch is made of a short-lifetime (~ 1 ps or less) material, such as low-temperature-grown GaAs, exposed between two low-capacitance electrodes, such as a square gap in a THz planar transmission line. To make the switch easily shorted by a mode-locked laser pulse, it can not be made too large in area, 100 square microns being typical. The safe bias voltage is then ~ 20 V. A Ti:sapphire mode-locked laser typically produces 200 fs pulses at $f_{rep} \approx 100$ MHz. These values results in a peak THz pulse power of 4 W, a THz pulse energy of 0.4 pJ, and an average THz pulse power of 40 μW. This is not too much greater than the maximum values that have been measured by several researchers in the field, which is typically in the range of 1 to 100 μW.[57] While very useful for laboratory spectroscopy and imaging, these power levels are not yet considered to be high enough for remote sensing applications. With the continuous advances being made in photonics and optoelectronics, T-rays are an intriguing prospect for future pulsed THz applications.

C. Receiver Types

C.1. Direct Conversion with Classical Square-Law Detector

To get a useful estimate of the sensitivity of an active sensor in the present scenario, we start with the generic expression for the SNR of a direct-detection receiver from Sec. VII, written for one spatial mode. To this we add the two above expressions for the incident power and the radiation fluctuations, resulting in

$$\left(\frac{S}{N}\right)_{AD} = \frac{\left[\eta \cdot \frac{P_T G_T G_R \lambda^2}{(4\pi)^2 R^2} \tau(\nu)\right]^2}{2 \cdot \eta \cdot \{(k_B T_{eff} \cdot \Delta\nu)^2 + (h\nu_0 P_{inc} \Delta\nu)\} \cdot (\Delta f / \Delta\nu) + (\eta \cdot NEP_{elect}^{'})^2 \cdot \Delta f}$$

As discussed above, the range, RF bandwidth, and antenna-gain requirements in mm-wave and THz sensors along with the limited Tx power often lead to the condition where the thermal-background radiation noise dominates the coherent incident radiation noise. This allows us to simplify the SNR to the following form

$$\left(\frac{S}{N}\right)_{AD} = \frac{\left[\eta \cdot \frac{P_T G_T G_R \lambda^2}{(4\pi)^2 R^2} \tau(\nu)\right]^2}{2 \cdot \eta \cdot (k_B T_{eff} \cdot \Delta\nu)^2 \cdot (\Delta f / \Delta\nu) + (\eta \cdot NEP_{elect}^{'})^2 \cdot \Delta f} \tag{60}$$

This expression has great utility, but for the sake of brevity only one form will be given here. This is the noise equivalent transmit power, NEP_T, analogous to the NEΔT for a passive sensor observing a thermal source. Setting the SNR to unity, we find

$$NEP_T = \frac{(4\pi R)^2}{\tau \cdot G_T G_R \lambda^2} \sqrt{2 \cdot (k_B T_{eff})^2 (\Delta f \cdot \Delta\nu) / \eta + (NEP_{elect}^{'})^2 \cdot \Delta f}$$

or in the more convenient specific form

$$NEP_T^{'} = \frac{(4\pi R)^2}{\tau \cdot G_T G_R \lambda^2} \sqrt{2 \cdot (k_B T_{eff})^2 \Delta\nu / \eta + (NEP_{elect}^{'})^2} \quad [\mathrm{W/Hz^{1/2}}]$$

Note that this expression could also be arranged in terms of the range at which the sensor SNR would be unity at each possible transmit power. This is the normal way to think about active sensors (i.e., radars) at microwave frequencies where transmit power is not so hard to come by.

In the event that the first term in the radicand dominates in either expression above, we obtain a "background-limit" analogous to that of passive sensor:

$$NEP_T = \frac{(4\pi R)^2 k_B T_{eff}}{\tau \cdot G_T G_R \lambda^2} \sqrt{2 \cdot \Delta\nu \cdot \Delta f / \eta}$$

Fig. 27 shows a plot of the NEP'_T as a function of range parametrized by the electrical NEP and the atmospheric transmission factor. It is clear that even at moderate range, the worst-case direct detector having an NEP'_{elect} of 10^{-12} can not provide useful sensitivity under any atmospheric condition unless the transmit power is >> 10 mW. So one would be advised in an active receiver with direct detection to use an NEP no worse than roughly 10^{-13} W/Hz$^{1/2}$. According to Eqn 60 this produces an SNR of 1.0 with approximately 1 mW of transmit power at a range of 1 km.

C.2. Heterodyne and Homodyne Conversion with Classical Square-Law Detector

In passive sensors the source spectrum is generally thermal radiation so is naturally very broad. Hence, mixing with a local oscillator will generate a broad IF spectrum and heterodyne conversion works very well. By contrast, with an active

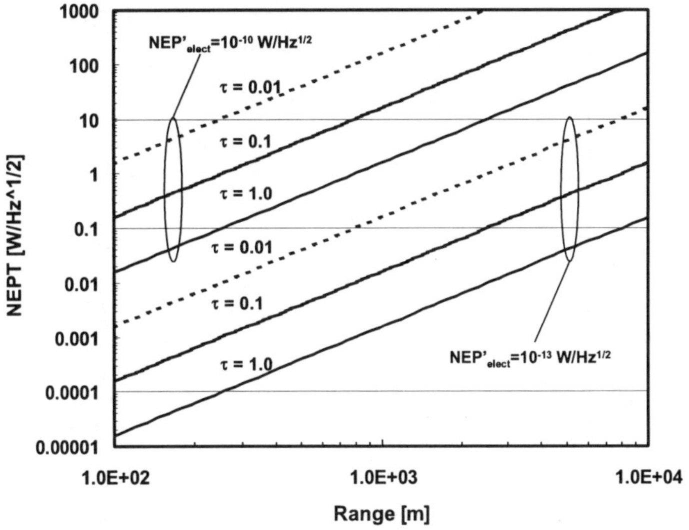

Fig. 27. Noise equivalent transmit power for the active sensor with a direct-detection receiver as a function of range separating the transmitter and receiver. The lines are parametrized by the net atmospheric transmission factor τ and the direct-detector NEP'_{elect}. The other parameters in the curve are $G_T = G_R = 100$, $\eta = 0.5$.

sensor heterodyning can occur only if the Tx and Rx local oscillator are deliberately offset in frequency, which then requires a more sophisticated receiver (e.g., in-phase and quadrature processing) to maintain good sensitivity. This is practical in the millimeter-wave region but becomes expensive and cumbersome at THz frequencies. In the latter region there is a tendency to build sensors that use the same oscillator for both the Tx and LO functions. This results in a homodyne receiver, as described briefly in Sec. IV.

To get a useful estimate of the sensitivity of an active sensor in the present scenario, we note that both heterodyne and homodyne receivers can be described by the generic Eqn 57 for the post-detection SNR in a single spatial mode with radiation noise dominated by the LO photon shot noise. To this we add the above expression for the incident power to get the surprisingly simple results

$$\left(\frac{S}{N}\right)_{AD} = \frac{\left\{H \cdot \eta_R{}^2 \cdot \frac{P_T G_T G_R \lambda^2}{(4\pi)^2 R^2} \tau(\nu)\right\}^2}{2 \cdot \{\eta_R \cdot h\nu_{LO} + (\eta_R)^2 \cdot NEP'_{mixer}\}^2 \cdot (B \cdot \Delta f)} \equiv \frac{\left\{H \cdot \frac{P_T G_T G_R \lambda^2}{(4\pi)^2 R^2} \tau(\nu)\right\}^2}{2 \cdot \{h\nu_{LO} / \eta_{HET}\}^2 \cdot (B \cdot \Delta f)} \qquad (61)$$

where B is the post-mixer electrical bandwidth (centered about zero for the homodyne case) and the factor H = 1 and 2 for heterodyne and homodyne conversion, respectively.

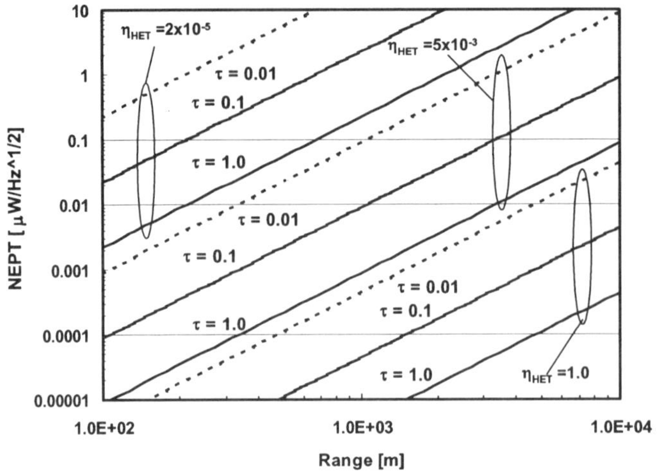

Fig. 28. Noise equivalent transmit power for the active sensor with a heterodyne receiver as a function of range separating the transmitter and receiver. The lines are parametrized by the net atmospheric transmission factor τ and the heterodyne quantum efficiency η_{HET}. The other parameters in the curve are $G_T = G_R = 100$, $\eta = 0.5$.

This then leads to the noise equivalent transmit power

$$NEP_T^{'} = \frac{(h\nu_{LO}/\eta_{HET})(4\pi R)^2\sqrt{2\cdot B}}{\tau\cdot G_T G_R \lambda^2}$$

Similar to the direct-detection case, this can be plotted vs range with τ and η_{HET} as parameters. The results are shown in Fig. 28 where the vertical axis is now shown in units of $\mu W/Hz^{1/2}$. Here we see a very large improvement in performance compared to the direct detection case. On first inspection this seems a bit incredible with even relatively poor heterodyne or homodyne mixers resulting in outstanding sensitivity. For example, a sensor in which the receiver mixer has an η_{HET} of 0.5% achieves an SNR of unity for a transmit power of 0.01 μW per $(Hz)^{1/2}$ of post-detection bandwidth over a range of 1 km and an atmospheric transmission of 0.1. This means that for a more practical transmit power of 1 mW and a post-detection bandwidth of 1 MHz – enough to accommodate AM or FM modulation to help mitigate atmospheric and gain fluctuations - the SNR would be approximately $(10^{-3})/(0.01\text{x}10^{-6})/(10^6)^{1/2} = 100$.

D. An Illustrative Example: Design of an Active Sensor to Measure Absorption Signatures

Recent measurements of the electromagnetic transmission through Bacillus subtillus, an anthrax surrogate, has revealed absorption signatures at sub-THz

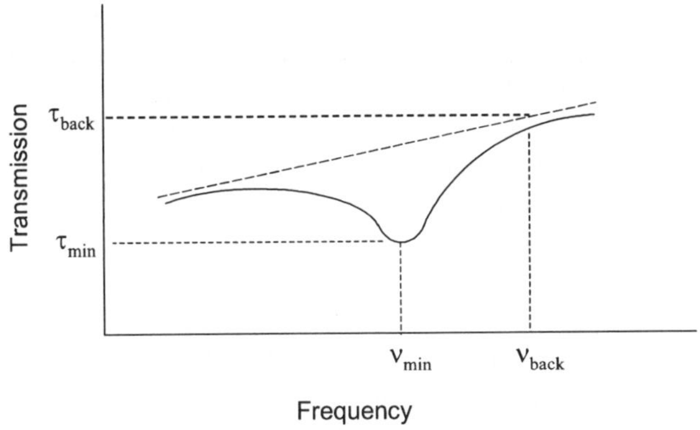

Fig. 29. Representative absorption signature in the terrestrial atmosphere.

frequencies.[58] The physical origin of these signatures is still being investigated, but from the measured absorption strength and signature one can begin to estimate their detectability in a remote sensor. This is not to imply that the resulting THz sensor will be better than competitive sensors in other (e.g., IR and visible) spectral regions. Rather, this problem serves as an interesting example of how the THz region may offer unique phenomenology in the form of unique absorption *signatures* not present in other electromagnetic regions. This is similar to the recent use of THz radiation for medical imaging. In that case, the THz provides unique contrast between tissue types that is proving useful in diagnostic imaging.

D.1. Absorption Signature Characteristics

A representative absorption signature is shown in Fig. 29 as it might appear in the power transmitted through a cloud of bioparticles. As discussed above, the background transmission in the THz region generally has a complex behavior. Although shown as a slowly increasing transmission with frequency, the background could display the opposite slope or be resonant depending on the proximity to water absorption lines. In any case the bioparticle absorption introduces a broad "dip" in the transmission spectrum that is characterized by a minimum transmission τ_{min} at frequency ν_{min} , The goal is to characterize the signature in a way that can be used in the active sensor scenario.

Table I. THz absorption signatures of *B. Subtillus*

Freq, ν_{min}	α_0 (ν_{min}) (cm^{-1})	ν_{back} , $\Delta\nu$ (GHz)	τ (ν_{back})
421.5 GHz	0.7	430.5,10.0	0.25
619.5 GHz	1.3	600.0, 19.5	$\sim 10^{-11}$
940.05 GHz	1.7	930.0,10.5	$\sim 5x10^{-4}$
1075.5 GHz	1.2	1057.5,180	$<10^{-30}$

To do this we characterize the dip by a depth $\Delta\tau$, and a half-width $\Delta\nu$. $\Delta\tau$ is the absolute difference between τ_{min} and the background transmission τ_{back} measured at a frequency ν_{back} on whichever side of ν_{min} that τ is greater (e.g., in Fig. 29 $\nu_{back} >> \nu_{min}$). Hence, $\Delta\nu \equiv |\nu_{min} - \nu_{back}|$. We assume further that the concentration of bioparticles ρ is low enough for their attenuation to be described by a coefficient α that is linearly dependent on ρ and that affects the transmission through the Lambert-Beer law $\tau(\nu) = \exp[-\alpha_0(\nu)L\rho/\rho_0]$, where L is the thickness of the bioparticle cloud and α_0 is a reference attenuation coefficient measured at a concentration ρ_0 that may be much different than the actual ρ. Then if the background is slowly varying over $\Delta\nu$, we can write

$$\Delta\tau \equiv \tau_{back} - \tau_{\min} = \exp[-\alpha_0(\nu_{back})L\rho/\rho_0] - \exp[-\alpha_0(\nu_{\min})L\rho/\rho_0]$$
$$\approx \tau_{back}\{1 - \exp[-\alpha_0(\nu_{\min})L\rho/\rho_0]\}$$

Listed in Table I are the values of ν_{min} , $\alpha_0(\nu_{min})$, ν_{back} and $\Delta\nu$ derived from laboratory transmission measurements through dry films of *B. subtilis* (an antrax surrogate) containing $\rho_0 \approx 1\text{x}10^{12}$ cm^{-3} – a density that is necessary to get an accurate measure of $\Delta\alpha$, but is much larger than expected in airborne bio-warfare agents. The results listed in Table I describe four different absorption features having center frequencies between 421 and 1075 GHz. In the THz region, the remote detection of these signatures depends critically on the atmospheric transmission $\tau(\nu_{back})$. Upon analysis of the results of PCLnWin shown in Fig. 5, the most transparent frequency by far is 421.5 GHz through which the single pass attenuation is approximately 0.25 (i.e., -6.0 dB) over a path length of 1 km at sea level and at 60% relative humidity .

D.2. Sensor Design: Direct and Homodyne Differential Absorption Radars

To take advantage of the $\Delta\tau$ we imagine an active sensor such as that shown in Fig. 30. The transmitter contains a frequency agile source that can hop between the two frequencies, keeping the transmit power constant. To make the scenario a bit more practical, we suppose the Tx and Rx are co-located and used with a retrodirective mirror that is perfectly specular. We can thus write for the difference in incident power between the two frequencies:

$$\Delta P_{inc}(\nu) = \frac{P_T G_T G_R \cdot \lambda^2}{(8\pi R)^2}\Delta\tau(\nu)$$

In practice one can readily do this frequency hopping periodically (e.g., square wave) and then synchronously demodulate in the post-detection portion of the receiver, assuming of course that Δf is greater than the modulation frequency. In this case, the post de-modulator signal-to-noise ratio will be approximately the difference in the SNRs given by Eqn 61 between the two frequency states with a reduction factor f_m for the inefficiency of simple (e.g., AM) demodulation schemes. This results in the post-detection SNR for a direct receiver of

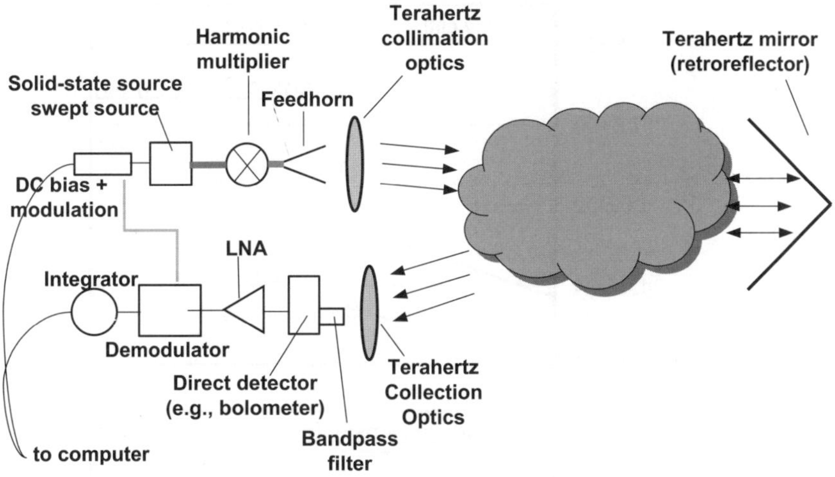

Fig. 30. Block diagram of active sensor designed for differential absorption detection of a cloud of weak bioparticles at a short standoff (~ 1 km). The receiver is based on direct detection.

$$\left(\frac{S}{N}\right)_{AD} = \frac{\left[\eta \cdot \frac{P_T G_T G_R \lambda^2}{(8\pi)^2 R^2} f_m \cdot \Delta\tau(\nu)\right]^2}{2 \cdot \eta \cdot (k_B T_{eff} \cdot \Delta\nu)^2 \cdot (\Delta f / \Delta\nu) + (\eta \cdot NEP_{elect}^{'})^2 \cdot \Delta f}$$

and an SNR for a heterodyne (or homodyne) receiver of

$$\left(\frac{S}{N}\right)_{AD} = \frac{\left\{\frac{P_T G_T G_R \lambda^2}{(8\pi)^2 R^2} f_m \cdot \Delta\tau(\nu)\right\}^2}{2 \cdot \{h\nu_{LO} / \eta_{HET}\}^2 \cdot (B \cdot \Delta f)}$$

As a practical matter, note that one can readily do this frequency hopping periodically at a much higher frequency than the rate of change of the atmospheric fading effects. Then by synchronously demodulating the signal at the receiver and taking the ratio between the two frequency states, one can cancel the effect of atmospheric fading.

Note that in a homodyne version of Fig. 30, because the transmitter and receiver are co-located at the same point in space, only a few components would have to be changed in the THz front-end. First, the direct detector would be replaced by a mixer, such as a Schottky diode. Then, a portion of the transmit power would be coupled through a THz transmission line to the mixer where it would be coupled to the mixer as an LO. Finally, an IF circuit would be added that could process both the in-phase (I) and

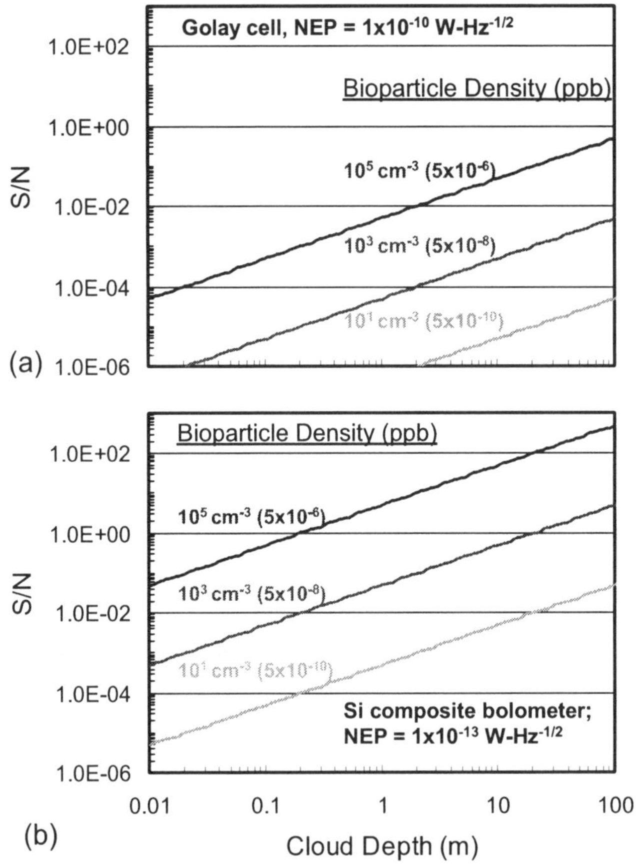

Fig. 31. SNR performance of differential-absorption radar with direct-detection receiver. In (a) the detector is a room temperature Golay cell, and in (b) it is a cryogenic (4.2-K) bolometer.

quadrature (Q) components of the mixer output. The sum of the I- and Q-detected channels would then, in principle, recover all of the signal power of the received coherent signal, independent of phase difference(s) between the received signal and the LO.

D.3. Direct and Homodyne Sensor Performance at 425 GHz

Figs. 31 (a) and (b) show the post-demodulation SNR at three different concentrations of *B. subtilis*, $\rho = 10$, 10^3, and 10^5 cm^{-3} for the direct-detection transceiver. The results are plotted vs cloud depth for a fixed Tx-to-cloud separation of 0.5 km and a transmit power of 1 mW. The other scenario parameters are listed in Table II. In Fig.

31 (a), the Rx detector is assumed to have NEP=1x10^{-10} W-$Hz^{-1/2}$ corresponding to a state-of-the-art room-temperature bolometer (e.g., Golay cell). In Fig. 31(b) the Rx detector is assumed to have NEP = 1x10^{-13} W-$Hz^{-1/2}$ corresponding roughly to a state-of-the-art cryogenic (4.2 K) bolometer (e.g., silicon composite).

The results on overall senor performance are now determined by comparing these SNR curves to the universal ROC curves of Fig. 16. For example, according to Fig. 31(b), in a cloud 20 m thick at a stand-off of 0.5 km, the SNR 1 is 1.0 for a concentration of 10^3 cm^{-3}. Assuming that the signal threshold at 0.2 nW, we find Pd and Pfa values from Fig. 16 to be approximately 0.4 and 0.1, respectively – unsatisfactory for most applications. However, the enhanced SNR at greater cloud depth or higher concentration in Fig. 31(b) would result in more reliable detection. For example, a cloud depth of 20 m and a concentration of 10^4 cm^{-3} would yield an SNR of 10 and associated Pd and Pfa values of 0.75 and 0.006 if the threshold was also increased to 0.9 nW.

For the homodyne case, we assume the down-converting device is a Schottky-diode mixer. If driven hard enough (P_{LO} ~1 mW) so that the noise-equivalent power (NEP) is limited by LO-shot noise and the coupling efficiency η is usefully large, we find $NEP_{HET} \approx h\nu/\eta_{HET} = 5.6 \times 10^{-21}$ W/Hz (h ≡ Planck's constant) for ν = 421 GHz and η_{HET} = 5%. The calculated SNR_{AD} curves for coherent detection are plotted in Fig. 32 for the same bioparticle concentrations as in Fig. 31, and the same transmit power and range. The homodyne receiver is assumed to have an IF bandwidth B of 1 MHz and the post-detection bandwidth of 1 KHz. Note that for a given cloud depth and bioparticle concentration, the coherent post-detection SNR is approximately 300 times larger than the analogous post-detection SNR of the incoherent sensor in Fig. 31(b). In other words, a homodyne receiver operating at room temperature is approximately 300 times more sensitive than a direct receiver with a cryogenic (4.2 K) detector under the same sensor scenario.

Substitution of pre-detection SNR results of Fig. 32(a) into the ROC curve of Fig. 18 leads to the following exemplary results. When the cloud depth is 20 m and the concentration is 10^3 cm^{-3}, the SNR in Fig. 32(a) is approximately 16 and we find from Fig. 18 that $P_d \approx 0.99$ for $P_{fa} = 10^{-3}$, $P_d \approx 0.95$ for $P_{fa} = 10^{-4}$, and $P_d \approx 0.90$ for $P_{fa} = 10^{-5}$, each possibility corresponding to one particular threshold level. Clearly, the homodyne

Table II. Parameters used in sensor simulation	
Background temperature (K)	290
Detector coupling efficiency	0.5
Line center frequency (GHz)	421
Background frequency (GHz)	431
Linewidth (GHz)	10
Line center abs coeff (1/cm)	0.7
Background transmission	0.25
Spectral bandwidth (GHz)	10
Integration time (s)	1.0
Transmit power (mW)	1.0
Tx aperture (cm^2)	100
Rx aperture (cm^2)	100
Tx-Rx separation (m)	1000
Cloud thickness (cm)	variable
Cloud temperature (K)	290
Molecule density (cm^{-3})	10,10^3,10^5

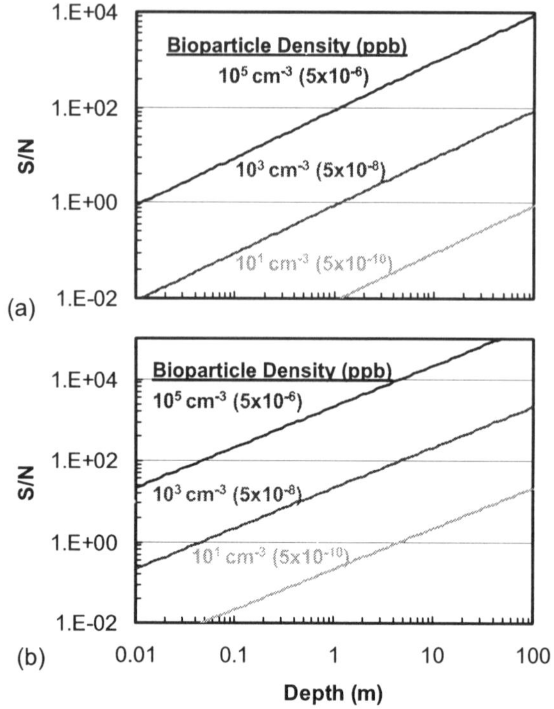

Fig. 32. (a) Pre-detection SNR of homodyne receiver in the differential-absorption radar scenario. (b) SNR after square-law detection and synchronous demodulation. The parameters are $\eta_{HET} = 0.05$, NEP = 5.6×10^{-21} W/Hz.

receiver yields a far superior performance in comparison with the direct receiver. With this superior SNR, one can envision an early-warning system in which the presence of bioparticles sets off an alarm based on the envelope of signal plus noise exceeding a threshold set in baseband.

XI. Two-Dimensional Imaging and the Quest for Popular Applications

A. *Heuristics*

A holy grail of sensor technology in the mm-wave, THz or any other band of the electromagnetic spectrum is the capability to do imaging of objects at a useful stand-off in real or near-real time. This capability is motivated, of course, by the appeal and utility of live-video and television. The appeal needs no explanation. The utility goes well beyond the entertainment or practical impact of live video and involves such issues as pattern and object recognition. The fact remains that the human visual-cognitive system

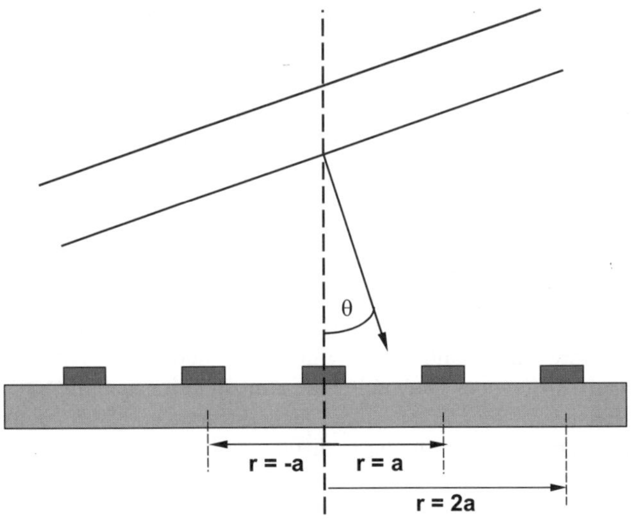

Fig. 33. Schematic diagram of optical spatial sampling

(i.e., the combination of the eyes and the brain) is still much better than computers or other information machines at recognizing objects surrounded by other objects or hidden in "clutter." And a good, if not the best, way to couple the imagery to the visual-cognitive system is a two-dimensional live image.

The live-video advantage has driven sensor technology since the beginnings of television with continual development of better cameras operating across the visible band. The early technology, such as that of the *vidicon*, was essentially an electronic version of photographic film. The technological breakthrough that revolutionized visible imagers during the past few decades has been the discretization the imaging process. The concept is simple: emulate the behavior of photographic film by placing individual detectors in a dense two-dim array in the focal plane of an optical imaging system. The detector density required to obtain image contrast comparable to that of the photographic film are well defined by a version of Nyquist's sampling theorem applied to free-space plane waves and imaging – the Whitaker-Shannon (W-S) theorem. This important result is discussed briefly below and in great detail in many books on Fourier optics and optical imaging.

The benefits of live video expanded significantly with the realization that discrete imagers could be readily extended into other parts of the electromagnetic spectrum where the phenomenology behind free space propagation are different. Perhaps the best example is the infrared region where it was realized quickly that objects that might be unapparent to a visible-band imager could be readily apparent to an infrared imager, particularly one constructed to operate in one of the well-known IR atmospheric windows from 3-to-5 micron or 8-to-12 micron wavelength. And the phenomenology behind this realization is as old as quantum physics itself – it is just blackbody, or "thermal" radiation. Terrestrial objects tend to have a physical temperature between roughly 273 and 300 K. According to the blackbody spectral power distribution of Eqn 8, the peak of emission tends to occur between approximately 8 and 10 micron. If the object in question has a temperature sufficiently lower or higher than the background and has *a sufficiently high emissivity*, it will be apparent to the camera either as a dark or bright image, respectively.

The same reasoning applies equally well to making thermal imagers in the mm-wave and THz region with the obvious difference that objects at any terrestrial temperature generally emit much less power than they do in the infrared. On the other hand, as shown above, the radiation noise is a lot lower in magnitude, particularly the photon shot noise. So the receiver sensitivity can be much lower in the mm-wave and THz region. More importantly, the mm-wave and THz regions display some phenomenological differences beyond the atmospheric and materials propagation differences discussed in Sec. II.E. One such difference relates to the difference in texture and scattering caused by the large difference in wavelength. Air-solid interfaces that appear rough and, therefore, diffusive, in the infrared can appear relatively specular in the mm-wave and THz regions A good example is a man-made wall made from a common building material, such as drywall.

B. Imaging by Spatial Sampling

Conceptually the W-S theorem states that the discrete imager can recover the same spatial information as from a continuous film if the discrete pixels sample at least twice per *spatial frequency* of the highest spatial frequency contained in the image. The definition of spatial frequency follows from the ability to decompose any spatial distribution of radiation into a Fourier series of plane waves propagating at different angles with respect to the axis of propagation. This is readily seen by the construction shown in Fig. 33. which shows an arbitrary plane wave incident on a periodic imaging array of period a. From fundamental electromagnetics, the plane wave can be represented by $E_0\exp[-j(\mathbf{k}\cdot\mathbf{r} - \omega t)]$ where **k** is the propagation vector and **r** is the position of the measurement point in some reference system whose origin is conveniently chosen as the center of the array. In this case, the vector inner product is $\mathbf{k}\cdot\mathbf{r} = (2\pi/\lambda)\sin\theta\,(m\cdot a)$, where λ is the free-space wavelength and m is an integer. In order for this phase factor to be unambiguous between neighboring elements (m = 1) of the array, it must lie in the range $-\pi$ to π since this comprises one entire branch of the Argand plane. Hence, the (un)ambiguity condition becomes $(2\pi/\lambda)\,|\sin\theta|\,a < \pi$, or

$$a < \lambda/(2\,|\sin\theta|)\ .$$

This is often written as

$$|\sin\theta|\,/\lambda < 1/2a$$

which has the same form as Nyquist's sampling theorem if we recognize 1/a as the spatial sampling rate, $\sin\theta/\lambda$ as the spatial frequency, and 1/2a as the Nyquist sampling rate.

These simple equations have several interesting properties, one of which relates to the maximum allowed spacing of the elements. For an antenna or optical front-end providing a very wide angular field-of-view, θ will vary from nearly $-\pi/2$ to $\pi/2$, the maximum $|\sin\theta|$ will approach 1, and thus the element spacing must be $\leq \lambda/2$ for unambiguous imaging. This is a familiar result from phased-array antenna theory, where it is usually stated in the reciprocal sense: i.e., the maximum spacing that array elements can have and still radiate "unambiguous" beams is $\lambda/2$. The ambiguity is manifest in the antenna performance through radiation directed into grating lobes.

Spatial sampling places just an upper limit on a, so that one is free to space the elements arbitrarily closer with no penalty on performance. This is analogous to "over-sampling" in signal processing, and can be unnecessarily expensive and complicated. So the actual element spacing is usually determined, as in visible and infrared imaging arrays, by spatial resolution requirements, which are a strong function of diffraction effects in the optical or antenna components in front of the array. Once the diffraction effects are determined, which is usually done by spatial resolution requirements of the sensor as a whole, there is no point in placing the elements in a center-to-center spacing smaller than the diffraction-limited diameter for a plane-wave incident on the sensor.

Note that strictly speaking, the above considerations apply only to a sensor having phase sensitive, or coherent, elements. In other words, each element must be connected to coherent receiver. The analogous conditions for an incoherent imaging system are a factor-of-two different and discussed in great detail with an emphasis on mm-wave and THz imaging in the excellent review article by Rutledge et al.[59]

C. *Applications and Examples of MM-Wave Imagers*

A practical advantage of mm-wave and THz imagers over visible and infrared ones is packaging. Because of the much greater maximum pixel spacing that generally applies, the former imagers can be constructed with "macroscopic" spacings of the order of 1 mm. This fact has been utilized for decades through the construction of first one- and then two-dimensional arrays of incoherent thermal detectors. Perhaps the first such array and associated camera system was engineered in the 1970s using a linear-array of pyroelectric detectors and later productized by Spiricon, Inc. The single-pixel NEP_{DIR} values were approximately 10^{-8} W/Hz $^{1/2}$. While good enough for some applications, such as THz laser beam diagnostics, the sensitivity was not adequate for most remote sensing. Later it was realized that the mm-wave and THz coupling efficiency to the pyroelectric elements was not very good because of low absorption coefficient in the typical (ferroelectric) materials. And the old trick of blackening the surface of the elements with graphite-loaded paint was not nearly as effective as in the mid-infrared region where pyroelectric detectors in "uncooled" bolometer arrays continue to be very useful.

By the late 1970s it was realized that to make widely useful mm-wave and THz imaging cameras, significant improvement would be required in the radiation coupling efficiency. Considerable research was conducted on planar antennas coupled to free space through dielectric substrates, cavities, and by other techniques. This culminated in the first room-temperature two-dim incoherent imaging arrays made from bow-tie antennas coupled to bismuth microbolometers.[60],[52] Somewhat later the first coherent imager was developed and demonstrated, this time using a linear array of Schottky diode mixers coupled to bow-tie antennas.[61] Both were developed and demonstrated primarily for plasma diagnostics – an area of widespread research, particularly after the energy crisis of the1970s.

During the mid to late1980s the strong push for plasma imaging had waned, and the primary system pull for mm-wave and THz imagers came from the radio astronomy community.[62] It was recognized that some of the cryogenic detector types that had been developed in the 1970s for high-sensitivity astronomy applications were amenable to fabrication in planar arrays. Two examples were LHe-cooled composite bolometers for direct detection,[63] and superconductor-insulator-superconductor tunnel junctions for coherent detection.[64] Although these concepts have proven to be successful, they were not quick to develop largely because of the difficulty in fabricating and packaging cryogenically cooled devices and read-out electronics.

The interest and activity in imagers increased significantly in the mid 1990s with two developments. First, monolithic microwave integrated circuits based on GaAs devices improved to the point where they could provide superior performance to discrete components in the mm-wave band. Perhaps the first example of this was the low-noise amplifier (LNA). Second, two compelling mm-wave applications where identified that offered a compelling advantage over visible and infrared imagers: (1) concealed weapons detection and (2) all-weather imaging for aircraft landing.

The sensing of weapons concealed behind human clothing is a natural one for mm-wave and THz sensors since clothing is, by design, opaque in the visible and more or less opaque in the infrared depending on the fabric. For the same reason that plastics are quite transparent to mm- and THz waves, so are most of the common fabrics. One of the first, if not the first, demonstrations of concealed weapons was made by engineers at the Millimetrix Corp[65],[66] using both active and passive mm-wave arrays. Although not highly integrated, these demonstrations showed that images of concealed guns could be of high enough quality to allow identification by a human observer.

Around this same time engineers began pursuing mm-wave imagers for all-weather imaging systems. People had long known that this was a natural application because of the size advantages of mm-wave components over lower-frequency RF ones, and the superior mm-wave propagation compared to that in the visible or infrared. The drawback of this application was always size and cost. The development of MMIC LNAs has made preamplified direct detection possible up to at least W band (75-110 GHz) in array architectures. Probably the most successful of these has been the passive camera developed and demonstrated by TRW for all-weather landing.[67,68] Below 100 GHz, it utilizes their GaAs pHEMT process to produce LNAs having $<$ 3 dB noise figure up to approximately 100 GHz. It has also been developed for 140 GHz using InP-based MMIC technology.[69]

More recently, a two-dimensional array of Nb microbolometers has been used in an active imaging array.[70] This work has shown excellent imagery at real-time acquisition rates.

Acknowledgement

The author is indebted to Dr. Dwight Woolard who commissioned this article and then practiced great patience while it was being composed. The chapter was a challenging exercise since much of the information is not well known and not covered in textbooks or review articles. And some had to be derived from scratch, leading perhaps to original expressions and arguments. The author apologizes in advance for not referencing the same derivations or expressions if they have appeared previously in the open literature. Unfortunately, there is simply not enough time to review the entire embodiment of publications related to remote sensing – the plethora from the rf side of the fence and the plethora from the photonic side. One goal of this article was to emphasize the dual nature of the THz field, drawing on concepts and techniques from both sides of the fence in a rather unpredictable way.

The author also thanks his many other DoD sponsors over the years, many of whom have supported the author in research related to this article. Among these are Dr. Jim Harvey, Dr. Dev Palmer, Dr. Edgar Martinez, Dr. James Nichter, Dr. Daniel Radack, Dr. Jim Murphy, Dr. Elias Towe, Dr. Henry Everitt, Dr. Gerald Witt, and Dr. Jane Alexander.

The author also thanks his many close colleagues in the mm-wave and THz field over the years, starting with his mentors at the Hughes Space and Communications Group (group of Mr. Frank Goodwin), his advisers and teachers at Caltech (group of Prof. T.G. Phillips), his colleagues and associates at MIT Lincoln Laboratory, his fellow program managers and government associates at DARPA, and his colleagues and students at UCLA and UCSB.

References

[1] T.G. Blaney, "Radiation Detection at submillimetre wavelengths," Journal of Physics E (Sci. Inst), vol. 11, p. 856, 1978.
[2] J. C. Wiltse, "History of millimeter and submillimeter waves," IEEE Trans. Microwave Theory Tech., vol. MTT-32, pp.118-127, Sept. 1984.
[3] P.H. Siegel, "Terahertz Technology," IEEE Transactions on Microwave Theory and Tech, vol. 50, no. 3, 2002.
[4] M. Exter and D. Grischkowsky, "Characterization of an optoelectronic terahertz beam system," IEEE Trans. Microwave Theory Tech., vol. 38, pp. 1684-1691, Nov. 1990.
[5] B. B. Hu and M. C. Nuss, "Imaging with terahertz waves," Opt. Lett., vol. 20, no. 16, pp.1716-1718, Aug. 15, 1995.
[6] D. M. Mittleman, R. H. Jacobsen, and M. C. Nuss, "T-ray imaging," IEEE J. Select. Topics Quantum Electron., vol. 2, pp.679-692, Sept. 1996
[7] D. Marcuse, Principles of Quantum Electronics (Academic Press, New York, 1980).
[8] F. T. Ulaby, R.K. Moore, and A.K. Fung, "Microwave Remote Sensing," vol. 1, sec. 5.15 (Addison-Wesley, Reading, MA), sec. 5-15
[9] R. Eisberg and R. Resnick, "Quantum Physics of Atoms, Molecules, Solids, Nuclei, and Particles, 2nd Ed. (Wiley, New York, 1985).
[10] W.B. Davenport and W.L. Root, "An Introduction to the theory of random signals and noise," (IEEE Press, New York, 1987).
[11] R. H. Kingston, "Detecton of Optical and Infrared Radiation" (Springer, 1978).
[12] PcLnWin, market by Ontar, Inc., Andover, MA
[13] J.D. Kraus, "Radio Astronomy", (McGraw-Hill, New York, 1966), Sec. 3-16.
[14] F.T. Ulaby, R.K. Moore, and A.K.Fung, "Microwave Remote Sensing," (Addison Wesley, New York, 1981), V. 1.
[15] J-A. Kong R-T. Shin, L. Tsang, "Theory of Microwave Remote Sensing," (Wiley, New York, 1985).
[16] see, www.josephson.terahertz.co.uk/QMCI/GOLAY.HTM
[17] see Infrared Laboratories, Inc., www.irlabs.com
[18] R.S. Elliott, "Antenna Theory and Design," (Prentice Hall, Englewood Cliffs, 1981).
[19] H. R. Fetterman, P. E. Tannenwald, B. J. Clifton, W. D. Fitzgerald, and N. R. Erickson, "Far IR heterodyne radiometric measurements with quasioptical Schottky diode mixers," Appl. Phys. Lett., vol. 33, no. 2, pp. 151-154, (1978).
[20] D. Rutledge, D. Neikirk, and Kasilingham, "Printed Circuit Antennas" in Infrared and Millimeter Waves, vol.10, ed. By K. J. Button (Academic, New York, 1983)., p. 1.
[21] L. Katehi and N. Alexopolous, IEEE Trans. Ant. And Propag, Vol. 31, p. 34 (1983).

[22] K.S. Yngvesson, D.H. Schaubert, T.L. Korzeniowski, E.L. Kollberg, T. Thungren, and J.F. Johansson, "Endfire tapered slot antennas on dielectric substratres," IEEE Trans. Antennas and Prop., Vo. 33, p. 1392-1400 (1985)
[23] M. Kominami, D.M. Pozar and D.H. Schaubert, "Dipole and slot elements and arrays on semi-infinite substrates," IEEE Trans. Antennas and Propagation, vol. AP-33, pp. 600-607 (1985).
[24] D. F. Filipovic, G.P. Gauthier, S. Raman, and G.M. Rebeiz, "Off-axis properties of silicon and quartz dielectric lens antennas," IEEE Trans. On Antennas and Propagation, vol. 45, pp. 760-766 (1997).
[25] For an excellent review of this topic, see "Classical Electromagnetic Radiation," J. Marion (Academic, New York, 1965), Chapter 12.
[26] R. J. Mailloux "Phased Array Antenna Handbook," (Artech House, Norwood, MA, 1994), p. 15.
[27] A. Yariv, "Quantum Electronics," (Wiley, New York, 1975), p.
[28] P. Goldsmith, "Quasi-Optical Systems: Gaussian Beam Quasi-Optical Propagation and Applications " (Wiley IEEE Press, 2001).
[29] T. G. Phillips and J. Keene, "Submillimeter Astronomy," Proc. IEEE, vol. 80, pp. 1662-1678, Nov. 1992
[30] P.D. Potter, "A new horn antenna with suppressed sidelobes and equal beamwidth," Microwave J. Vol. 6, pp. 71-78 (1963).
[31] R. Winston and W. T. Welford, "High-Collection Non-Imaging Optics," (Academic, New York, 1989).
[32] A. Yariv, "Quantum Electronics, 2nd Edition", (Wiley, New York, 1975), Chapter 6.
[33] A. Yariv, IBID.
[34] H. Wang, L. Samoska, T. Gaier, A. Peralta, H. H. Liao, Y. C. Leong, S. Weinreb, Y. C. Chen, M. Nishimoto, and R. Lai, "Power amplifier modules covering 70-113 GHz using MMICs," IEEE Trans. Microwave Theory Tech., vol. 49, pp. 9-16, Jan. 2001.
[35] S. Weinreb, T. Gaier, R. Lai, M. Barsky, Y. C. Leong, and L. Samoska, "High-gain 150-215 GHz MMIC amplifier with integral waveguide transitions," IEEE Microwave Guided Wave Lett., vol. 9, pp. 282-284, July 1999
[36] TRW reference.
[37] A. H. Dayem and R. J. Martin, "Quantum interaction of microwave radiation with tunneling between superconductors," Phys. Rev. Lett., vol. 8, pp. 246-248, Mar. 1962.
[38] G. J. Dolan, T. G. Phillips, and D. P. Woody, "Low noise 115 GHz mixing in superconductor oxide barrier tunnel junctions," Appl. Phys. Lett., vol. 34, pp.347-349, Mar. 1979.
[39] D. E. Prober, "Superconducting terahertz mixer using a transition-edge microbolomoeter," Appl. Phys. Lett., vol. 62, no. 17, pp. 2119-2121, 1993.
[40] J. W. Waters, "Submillimeter-wavelength heterodyne spectroscopy and remote sensing of the upper atmosphere," Proc. IEEE, vol. 80, pp.1679-1701, Nov. 1992.
[41] Kingston, ibid.
[42] Davenport and Root, ibid.
[43] Davenport and Root, ibid.
[44] Kingston, ibid, Sec. 2.4.
[45] J. D. Kraus, "Radio Astronomy", (McGraw Hill, New York, 1966), Chap. 7.
[46] A. van der Ziel, "Noise in Solid-State Devices and Circuits," (Wiley, New York, 1986).
[47] A. van der Ziel, ibid.
[48] W.B. Davenport and W.L. Root, ibid.
[49] C.D. Motchenbacher and F.C. Fitchen, "Low-Noise Electronic Design," (Wiley, New York, 1973).
[50] J. L. Lawson and G. E. Uhlenbeck, Threshold Signals, L. Ridenour and G. Collins, Eds. New York: McGraw-Hill, 1950.
[51] S.O. Rice "Mathematical analysis of random noise," Bell System Tech Journal, vol. 24, pp. 46-156, January 1945.

[52] M. Schneider, "Metal Semicondutor Junctions as Frequency Converters," in *Infrared and Millimeter Waves*, vol. 6, Chapter 4, p. 209 (1982).
[53] Dr. Peter Siegel, private communication.
[54] H. C. Torrey and C. A. Whitmer, Crystal Rectifiers, Ed. By L. Ridenour and G. Collins (New York, McGraw-Hill) 1950.
[55] P.H. Siegel, ibid.
[56] Davenport and Root, ibid.
[57] X-C. Zhang, "Terahertz wave imaging: horizons and hurdles," Phys. Med. and Biol., vol 47, pp. 3667 – 3677 (2002)
[58] D. Woolard, R. Kaul, R. Suenram, A. H. Walker, T. Globus, and A. Samuels, "Terahertz electronics for chemical and biological warfare agent detection," in IEEE MTT-S Int. Microwave Symp. Dig., Anaheim, CA, June 13-19, 1999, pp.925-928.
[59] D. Rutledge, D. Neikirk, and Kasilingham, "Printed Circuit Antennas" in Infrared and Millimeter Waves, vol.10, ed. By K. J. Button (Academic, New York, 1983)., p. 1
[60] D.P. Neikirk, P.P. Tong, D.B. Rutledge, H. Park. And P.E. Young, Appl. Phys. Lett. Vol. Vol. 41, pp. 329-331 (1982); D. P. Neikirk, W. W. Lam, and D. B. Rutledge, "Far-infrared microbolometer detectors," Int. J. Infrared Millimeter. Waves, vol. 5, no. 3, pp. 245-278, Mar. 1984.
[61] C.-E Zah, D. Kasilingam, J.S. Smith, D.B. Rutledge, T-C Wang, and S.E. Schwarz, "Millimeter-Wave Monolithic Schottky Diode Imaging Arrays," Int. J. Infrared and MM Waves, 1985.
[62] T. G. Phillips and J. Keene, "submillimeter astronomy," Proc. IEEE, vol. 80, pp. 1662-1678, Nov. 1992.
[63] N. S. Nishioka, P. L. Richards, and D. P. Woody, "Composite bolometers for submillimeter wavelengths," Appl. Opt., vol. 17, no. 10, pp. 1562-1567, May 1978.
[64] T.G. Phillips and D.M. Watson, "Baseline Study for a Large Deployable Reflector," private communication.
[65] G.R. Hgeuenin, "Millimeter-wave concealed weapons detection and through-the-wall imagining systems," Proc. SPIE, Vol. 2938, (1996).
[66] G.R. Huguenin, E.L. Moore, S. Bandia, and J.J. Nicholson, "A Millimeter-Wave Monolithic load switching twist reflector for compact imaging cameras," IEEE Trans. Microwave Theory and Tech, vol. 44, p. 2751 (1996).
[67] M. Shoucri, R. Davidheiser, B. Hauss, P. Lee, M. Mussetto, S. oung, and L. Yujiri, "A Passive MM-Wave Camera for Landing in Low Visibility Conditions," Proc. of the 1994 IEEE Digital Avionics Conference, p. 93.
[68] "Passive millimeter-wave camera," L. Yujiri, M. Biedenbender, M. R. Flannery, B.I. Hauss, R. Kuroda, P. Lee, H.H. Agravante, G.S. Dow, S. Fornaca, R.L. Johnson, K. Jordan, D. Low, B.H. Quon, T.K. Samec, K.E. Yokoyama, A. W. Rowe, M Shoucri, and J. Yun, SPIE Proceedings Vol. 3604, April 1997.
[69] L. Yujiri, S. Fornaca, B. Hauss, R. Kuroda, R. Lai, and M. Shoucri, "140-GHz Passive Millimeter-Wave Video Camera," SPIE Proceedings, Vol. 3703, 1999, p. 20.
[70] E.N Grossman and A.J. Miller, "Active Millimeter-Wave Imaging for Concealed Weapons Detection," Proc. SPIE

International Journal of High Speed Electronics and Systems
Vol. 13, No. 4 (2003) 1099–1128

World Scientific
www.worldscientific.com

TERAHERTZ EMISSION USING QUANTUM DOTS AND MICROCAVITIES

GLENN S. SOLOMON, ZHIGANG XIE, and MUKUL AGRAWAL
Solid-State Photonics Laboratory, Stanford University
Stanford, CA 94305-4085, USA

Semiconductor solid-state lasers based on conduction-valence band recombination are now commonplace for low-power red emission, and are available commercially at nearly continuous wavelengths throughout the near-UV and near-IR communications bands. Furthermore, new lasers and broadband spontaneous emission sources are available through a wide wavelength range, including 3-8μm based on both conduction-valence band and intersubband transitions, and up to 70μm using cascaded intersubband transitions. Competing processes make the design of these semiconductor lasers extremely difficult when extended to the very long, 300μm wavelength regime corresponding to low terahertz frequencies. We discuss material and device-design considerations for extending semiconductor lasers to this regime. We suggest a new set of device structures based on a semiconductor quantum dot (QD) gain medium, where the lasing occurs through discrete conduction states. In one implementation, two QDs are coupled to make a coupled-asymmetric quantum dot (CAD) laser. In another implementation, an ensemble of non-coupled QDs is selectively placed in a high quality cavity, called a microdisk, which is resonant with an intersublevel QD transition. We demonstrate the initial fabrication of these structures.

Keywords: quantum dots, self assembly, intersubband, lasing, terahertz

1. Introduction

The terahertz (THz) frequency window has recently been of scientific, commercial, and military interest for a variety of reasons. While optical sources are available a frequencies greater than 10 THz, and electronic oscillator circuits are available at frequencies in the 500 GHz range, the THz band, for the purposes of this chapter between 0.8-3 THz, remains elusive. The region is conceptually interesting because of the overlap of sources and technologies from the optical and electronic domain. From a solid-state physics perspective, there is a growing interest in THz spectroscopy for research in fundamental band structure [1], excitonic physics [2] and multi-particle states such as Bose-Einstein Condensation and quantum liquids [3]. In chemistry and biology there is interest in low THz sources (1 THz) for molecular detection [4]. The frequency regime is also of technological interest such as THz-band communications. New military sensing capabilities, including radar and spectroscopic sensing, will be developed in the THz frequency range. New spectroscopic sensing can be used effectively to enhance security. Finally, sources emitting in this region can used in medical imaging and biological sensing.

One type of source will not meet all these needs, but needs fall into two or three categories. First, a coherent source of sufficient power is necessary. In some instances an electronic circuit with a THz-resonant local oscillator is desired, while in other circumstances an optically based source, such as a laser is more applicable. In both cases power in the milliwatt (mW) range is more than likely necessary. Finally, a broadband source will be useful for spectroscopy. While spectroscopic needs can be met with a series of well-designed lasers, in some situations where lower power is acceptable, a broadband source will be more effective.

Semiconductor-based solid-state lasers and light-emitting diodes (LEDs) are ubiquitous. These sources typically operate near the visible spectral range, at wavelengths between 400 nm out to a few microns, although there are sources available at shorter and longer wavelengths. What limits the emission wavelength and what can be done to bring the emission wavelength out to the 100 μm to 300 μm THz regime? There are two general categories that conspire to reduce the emission efficiency in both the UV and far IR that are based on the solid-state nature of the lasing material: competing parasitic processes and poor material properties. Completing parasitic processes, such as free-carrier absorption and Auger recombination in long wavelength sources, and highly electrically resistive layers in short wavelength sources are generally intrinsic, while poor materials properties are technology related. The most mature optoelectronic material system, based on GaAs, turns out to have significantly advantageous materials properties, in part because of years of development, but also because of available lattice matched alloys with good refractive index contrast and bandgap differences, and relatively easily produced substrates.

Take as an introductory example the case of UV lasing based on, for example GaN. Since the transition energy is large (UV), the bandgap or heterostructure band offsets also must be large. Typically, the optically active transitions occur in quantum wells (QWs), which have more favorable density of states, thus, the QW barrier materials must have even larger bandgaps. The crystal growth processes are very difficult in these wide bandgap semiconductors because good substrates are not available and there is typically a significant lattice mismatch between alloys, making heterostructures with large confinement difficult to fabricate. In many material systems it is difficult to find stable regions in the crystal growth parameter space where materials can be successfully deposited, either do to phase segregation or excessively high vapor pressure at ideal growth temperatures. Furthermore, the index contrast between nearly lattice-matched materials is small, making waveguiding difficult. Finally, because the bandgaps are large, the intrinsic carrier concentration is low and high activated doping levels are difficult to achieve, resulting in more highly resistive structures and poor electrical contacts.

In contrast, for near-IR based devices, GaAs and InP are available. The excellent crystal growth technologies developed around GaAs and InP has been fortuitous for sources emitted in the visible to near IR. Despite being some of the first compound semiconductors investigated, the excellent crystal growth properties are rather unique. While Si is unique in the formation of a stable, high quality oxide, GaAs and to some extent InP are unique in optical heterostructure devices. In the GaAs-AlAs system there is a favorable band offset which is over 50% of the GaAs bandgap, and yet the materials are nearly latticed matched with good index contrast.

For the narrow energy transitions used in long-wavelength lasers, either inter- or intraband transitions can be used. The same types of material issues are present here as in the wide bandgap materials. Lattice mismatch is often a problem, and index contrast is once again an issue. Now the materials' refractive indices become smaller with increasing operating wavelength. Furthermore, although it is not difficult to achieve the proper doping, excess carriers contribute to parasitic free-carrier absorption at long wavelengths. Besides free-carrier absorption, other processes become favorable at longer wavelengths. In the 30 meV (41.3μm) range optical phonons are typically active in multi-atom basis crystals, and multi-particle processes, such as Auger recombination also become more significant.

In the near to mid-IR, where free-carrier absorption is still a small parasitic there is a competition between interband and intraband lasers. The general problems with interband lasers are the competing process of Auger recombination, and the crystal growth issues such as available substrates, lattice matching and alloy uniformity. As the interband energy reduces, Auger processes, becomes more favorable. The crystal growth problems are typical of those mentioned above with respect to both UV and IR devices. Such is the case for typical IR materials such as HgCdTe, where specially alloyed substrates of the $Cd_{0.96}Zn_{0.04}Te$ are necessary to minimize the lattice mismatch [5]. Furthermore, controlling the alloy uniformity has been difficult. However, lasers are made and will continue to improve, although the pace has been slow compared to GaAs and InP. Operating at 2.6 μm, edge emitting devices with threshold current densities of 420 A/cm^2 have been achieved on approximately 800 μm long structures under pulsed conditions at 77K [5]. Parallel to these efforts has been the intraband based structures using either GaAs- or InP-based material systems. These efforts leverage the successes of these materials system at shorter wavelengths along with those of quantum well (QW) physics. These are truly artificially structured crystals, where the transition energies and rates are determined by designing barrier heights and QW widths to tune individual QW wavefunctions. For lasers, the result of these efforts has been the cascade lasers [6]. These devices are unipolar, typically utilizing the conductor band, and can in principle operate through the electromagnetic spectrum except for the previously raised issues of competing parasitic processes and index contrast.

A terahertz optical emission source is proposed here that will offer both broadband spontaneous emission and lasing for a variety of needs in the THz frequency regime. The source is based on the emission from a quantum dot (QD) active region to reduce non-radiative loss and which can be coupled to a high quality optical cavity. The THz source can be optically pumped with a low power, compact laser source, or electrical injection can be implemented.

In one implementation two layers of QDs are quantum mechanically coupled by tuning the size/shape of the QDs in each layer. We call this structure a Coupled Asymmetric Dot (CAD) laser. It directly addresses the problem of the removal of carriers from the ground state of the optical transition. In a different device, the emphasis is on creating a high quality waveguide to either focus broadband emission or to allow feedback and gain for lasing, and uses a microdisk cavity. In the microdisk, cavity modes are present at the circumference of the disk containing the QDs. These circumferential modes, called whispering gallery modes, exist due to total internal reflection from the disk sidewalls. Because the device emits in-plane, thick vertical layers of material are not necessary, easing fabrication. QD emission that is coupled to these optical modes can be guided and extracted from the device.

Other sources have recently become available in the THz regime. A quantum cascade laser source has been made in the THz regime [7]. Other sources approaching, but not achieving THz emission frequencies have been demonstrated in the electronic domain [8]; however, clear demonstration of THz emission has not, to our knowledge, been demonstrated in the electronic domain, and it is unlikely these sources will provide the coherent emission power of a laser.

2. The Terahertz Emission Regime

Here, we discuss three particular issues that are important in long-wavelength optoelectronic sources. These are (1) the long-wavelength active region; (2) losses particular to the ultra long wavelength regime; and (3) materials and fabrication issues associated with device geometry of long-wavelength sources.

2.1 THz emission sources

Two possible THz emission/absorption active regions are available: *Interband, i.e.,* direct bandgap transitions and *intraband* (which we prefer to call *intersubband*), *i.e.,* transitions in quantum-confined structures between confined conduction or valence states. These transitions are shown in Fig. 1. For direct bandgap, interband transitions in the THz regime a near zero bandgap semiconductor is required. For intersubband transitions more conventional wider bandgap materials can be used, but require precise, often completed structures. The wavefunctions of the crystal can be written as Bloch functions, $\Psi = u(k,r)e^{ik\cdot r}$, where $u(k,r)$ is the basis function, with periodicity on the order of the crystal lattice and $e^{ik\cdot r}$, is the envelope function. For an optical transition in the crystal, under the assumptions of a weak electromagnetic field and where the photon momentum is small compared to the crystal momentum, the Hamiltonian is the momentum operator, $i\hbar\nabla$. This derivative operator necessitates that for non-zero transitions the initial and final wavefunctions must have a non-zero inner product. The

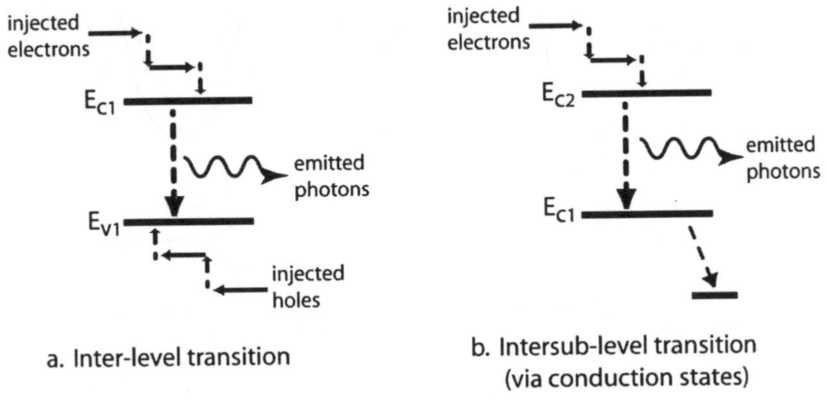

Fig. 1. Energy band diagram of interband transition (a), and intersubband transition (b) under carrier injection. Electrons and holes must be injected in the interband transition, and the ground state is the annihilation of carriers. Intersubband radiative transitions are unipolar, but the lower state of the transition is occupied by carriers which must be removed through another transition.

interband transition is a transition between two bands with orthogonal basis states and overlapping envelope states, such as the conduction band and valence band transition, shown in Fig. 1a. If the transition is intersubband the basis states overlap and the envelope states are orthogonal.

In Fig. 2 schematics of the band structure of the direct bandgap transition in bulk semiconductor (Fig. 2a) and the intersubband transition in a quantum well (QW) structure are shown (Fig. 2b), where the curly lines represent photon emission. In Fig. 2a the density of states in conduction and valance bands are continuous, while in Fig. 2b a QW structure is shown having discrete jumps in the density of states due to subband formation in conduction and valence bands. The direct bandgap transition can be used in either bulk material, shown, or with a QW. Indeed, in most applications a QW will be used. Again, the intersubband transition *requires* quantum confinement since the subbands form from the confinement.

While direct bandgap structures are in concept simple and the output power often higher than intersubband structures, there are increased intrinsic losses in these structures. The materials development of these very narrow bandgap semiconductors is in its infancy. If the development of mid IR interband sources is any guide, THz interband sources will have a significant development period.

Intersubband transitions overcome a large portion of the Auger and inter-valance band absorption losses (both discussed below) associated with long wavelength direct bandgap

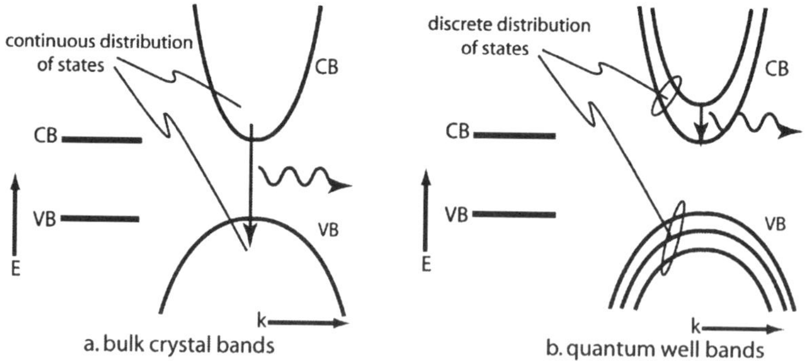

Fig. 2. Band structure of a direct bandgap semiconductor (a) and of a quantum well direct bandgap semiconductor (b). In (a) the direct transition is band to band, from the conduction band (CB) to the valance band (VB) with the emission of a photon (curly line). In (b) the direct transition is from intersubbands in the conduction band. While band to band emission (CB-VB in (a)) can be enhanced using a QW, quantum confinement is necessary in the intersubband transition of (b).

structures. QWs can be made with high precision using well-established materials, typically GaAs or InP, and the transition energies can be tuned through the well-developed molecular-beam epitaxy (MBE) growth process. The MBE process has local (100 μm) resolution on the order of a monolayer (2.5Å) in the crystal growth direction. However, the transition strengths from intersubband band transitions are typically lower than direct bandgap transitions, reducing the output power. Even though the dipole strength is typically larger than in direct bandgap transitions, because the spatial extent of the envelope states, the overall overlap integral between the states is often less.

There are also addition design constraints in intersubband structures. In many cases, for example symmetric QWs, normal emission or absorption is forbidden. Also, the optical transitions involve carriers from a single band; the transitions are thus unipolar. In unipolar, intersublevel devices carriers transition from the upper radiative state to the lower radiative state. Conduction band electrons do not recombine with valence band holes as in interband transitions, and thus a population of lower state carriers remains. If the decay of the lower state population is not sufficient the quasi-equilibrium population will reduce the density of available lower states, clamping the radiative decay process. Thus, the intersublevel QD (or QW) emission source must be designed in a way that reduces the ground state quasi-equilibrium occupation. An addition decay path is shown from the lower emission state in the intersubband diagram in Fig. 1b.

As of June 2002, the cascade laser does provide emission in the THz regime [6]. The laser operates at 3 THz with an output power of 1 mW at 40K. THz emission has not been demonstrated from a direct bandgap transition in a narrow bandgap semiconductor. Indeed, the cascade laser has been a major achievement, the result of over a decade of focused research. The structure is complicated and can only be made at long wavelengths in a small number of laboratories worldwide.

2.2 Free carrier, phonon, Auger and intervalence band absorption losses

One of the most difficult problems in designing long wavelength sources is the increase in absorption and parasitic recombination related losses. These losses increase nearly exponentially with increasing wavelength as schematically illustrated in Fig. 3. Free-carrier absorption is not in-principle limited to a single band, but for the purposes of our discussion we will limit it to a single band. This is especially relevant because of our focus on low-energy, THz absorption. Since the momentum of each carrier is uniquely defined within a band except for spin degeneracy, and the photon momentum is

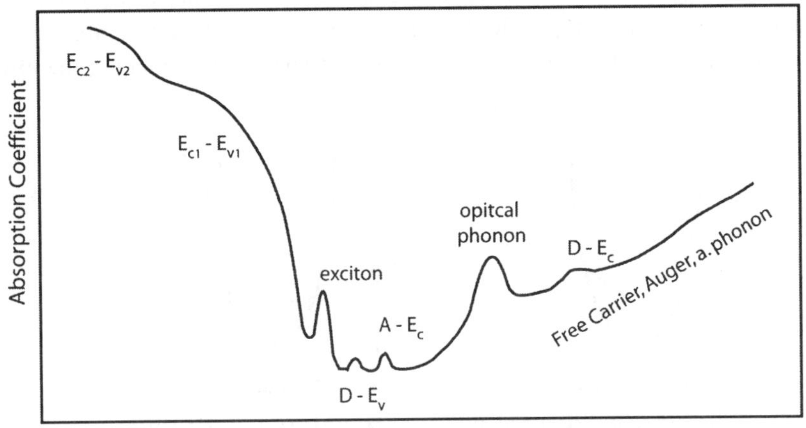

Fig. 3. Schematic of absorption as a function of wavelength indicating the increasing effect of free-carrier absorption at long wavelengths. E_c-E_v is the conduction-to-valence band transition (two bands are shown), while *D*, *A* are donor and acceptor levels. (Not to scale.)

negligibly small compare to the crystal momentum, the intraband free carrier absorption necessities an additional momentum scattering interaction. Additional scattering could include lattice scattering (phonon) or impurity scattering. Also, necessary is a partially filled band. In the semi classical picture, the electric field of the incident radiation accelerates the free carriers in a band, which are eventually decelerated by scattering. If the scattering is from the lattice, the energy is converted to heat. Thus, from this semi classical picture the free carrier absorption is related to the high-frequency conductivity [9,10]

$$\sigma = \sum_i q^2 \frac{n_i}{m_i^*}\tau_i^* = \sum_i q^2 \frac{n_i}{m_i^*}\left\{\left\langle \frac{\tau_m}{1+\omega^2\tau^2}\right\rangle - i\omega\left\langle \frac{\tau_m^2}{1+\omega^2\tau^2}\right\rangle\right\}, \tag{1}$$

where q is the charge of the carrier in the ith band, n is the carrier concentration, m^* is the effective mass in the ith band, τ^* is the effective relaxation time, defined as a function of the incident frequency, ω and the relaxation time, τ_m. For free-carrier absorption where a photon is absorbed by an electron in the lowest conduction band or hole in the highest valence band the absorption becomes

$$\alpha_c = \left(\frac{\lambda^2 q^3}{4\pi^2 c^3 n_r \varepsilon_0}\right)\left[\left(\frac{n}{m_c^2 \mu_c}\right) + \left(\frac{p}{m_h^2 \mu_h}\right)\right], \tag{2}$$

where λ is the wavelength,, n_r is the refractive index, ε_0 is the vacuum dielectric constant, n (p) is the electron (hole) concentration in the conduction (valance) band, μ is the conductivity carrier mobility ($\mu_i = q\tau_{m,i}/m_i$). A quantum mechanical treatment [11] confirms the semi classical picture when $\hbar\omega < kT$ (or E_F). The process is depicted schematically in Fig. 4 below. Free carrier absorption becomes increasingly important at long wavelengths, increasing with the square of the wavelength. Since n (p) is proportional to $m^{3/2}$, the bracketed terms are proportional to $m^{1/2}$, making the effect of

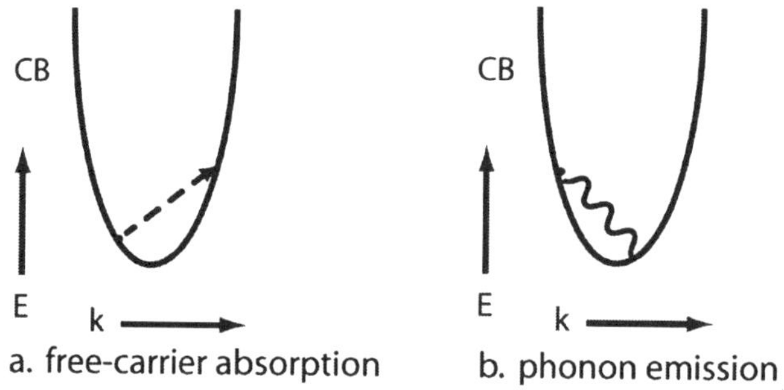

Fig. 4. Free-carrier absorption within a single bulk band is shown in (a), where the transition is k-conserving. In (b) a phonon emission is shown, reducing the energy of the electron. This transition is not necessarily k conserving.

holes larger for similar τ. In narrow bandgap structures, where the conduction band electron mass is typically smaller than in wider bandgap materials, the free carrier absorption due to conduction electrons is slightly reduced, but the λ^2 is the dominant feature as seen in Fig. 3. In interband structures made of wider bandgap materials, free-carrier absorption is reduced because of increased restrictions on allowed momentum in the confinement direction.

Phonon absorption or emission losses involve an electron or hole, and an absorbed or emitted phonon, and do not require momentum conservation. Phonon emission with the relaxation of an electron is shown in Fig. 1,2b. While the emission of a phonon can occur at any temperature, phonon absorption requires the presence of a phonon population, which is given by Bose statistics. The optical phonon (LO) has a peak energy in its dispersion ~ 30-50 meV at $\boldsymbol{k}=0$. The acoustic and optical phonon branches are continuous except for a small energy gap near the Brillouin zone edge. For GaAs the **k**=0, LO phonon energy is 32 meV, or just into the THz regime. Thus, when band structure of the system is designed around this energy to produce direct bandgap THz radiation, the energies are also close to the LO phonon, k=0 resonance, dramatically reducing THz photon emission.

While the free-carrier and phonon absorption processes are not momentum conservation, the Auger and intervalence band absorption process involve are at least three particle processes to conserve momentum. Auger recombination can take many forms in semiconductor materials. An example is an electron initially in the conduction band relaxing to a valance band hole state, and another electron in the conduction band moving to a higher conduction band state (Fig. 5a). Another Auger process is the recombination of an electron in the conduction band with a hole in the valance band and the relaxation of a hole deeper in the conduction band (Fig. 5b). In these processes, the energy of the

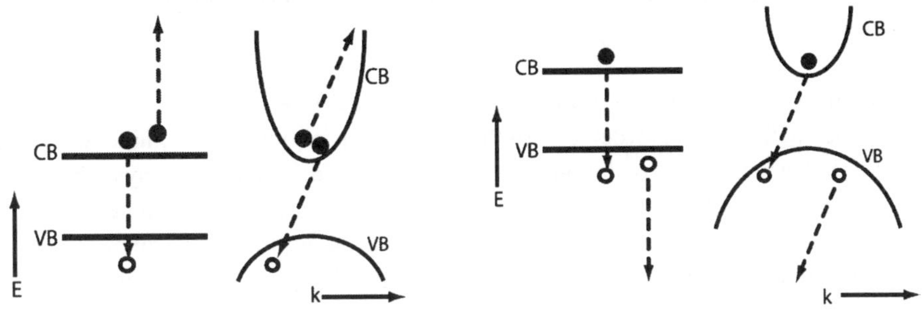

Fig. 5. Energy and energy-momentum diagrams showing two common Auger processes in narrow bandgap semiconductors. In (a), an electron in the conduction band (CB) relaxes to the valance band (VB) and the energy and momentum is converted to exciting an electron in the CB. In (b) the energy from the electron relaxation process excites a hole deeper into the VB. Since these processes involve three particles, the transition rates are proportional to n^2p in (a), and np^2 in (b).

two transitions is equal and the total momentum is conserved. Note that here the relaxation from the conduction band is not k conserving and thus a photon could not be emitted without another particle interaction to conserve momentum.

Since these Auger processes involve three particles, the transition rates are proportional to the concentrations of all the carriers involved, *i.e.*, n^2p in (Fig. 5a), and np^2 in (Fig. 5b), and can be represented in the form [12]:

$$R_{Auger} = c(T)n^2p, \text{ where } c(T) = c_0\left(\frac{E_a}{kT}\right), \text{ with } E_a = \frac{m_c E_g}{m_c + m_h} \tag{3}$$

where the Auger process is represented in Fig. 5a. The constant c_0 is a function of the dielectric constant, and m_c and m_v. The value of E_a depends of the details of the multiparticle process and there is some discrepancy [9,12] on the form of E_a. From the above equation as the bandgap narrows the Auger process increases exponentially. The process is further enhanced by the generally smaller electron mass in narrow bandgap materials.

Finally, intervalance band absorption (IVBA) processes are important when there are valance band energy differences that are less than the wavelength of emitted light. The valance band is composed of heavy and light hole bands, which are degenerate in cubic bulk semiconductors, and the spin split-off band. These processes are similar to free carrier absorption, and correspond to transitions between subbands in the valance band.

We would like to make two important points associated with the losses discussed. First, losses that are relatively minor, but still can impact device performance in the near IR such as Auger recombination, can dominate fundamental device processes in long-wavelength, THz devices. Second, QW confinement can usually reduce these effects.

2.3 Materials, fabrication and geometry issues in THz devices

With wavelengths on the order of 30-300 μm, active regions, waveguide and cavity designs, and general device layout must be considered. In many ways, these long wavelengths offer relaxed processing accuracy; however, when vertical components are a necessary feature in the device design the crystal growth process becomes expensive, time consuming and variable. For example, vertical cavity surface emitting lasers (VCSELs) in the THz regime are not realistically possible. With half-wavelength cavities and mirror pairs numbering over 30, the overall vertical size is on the order of 150 μm and would take over 100 hours to make by molecular-beam epitaxy (MBE). However, good cavities are needed for these devices to overcome the large losses

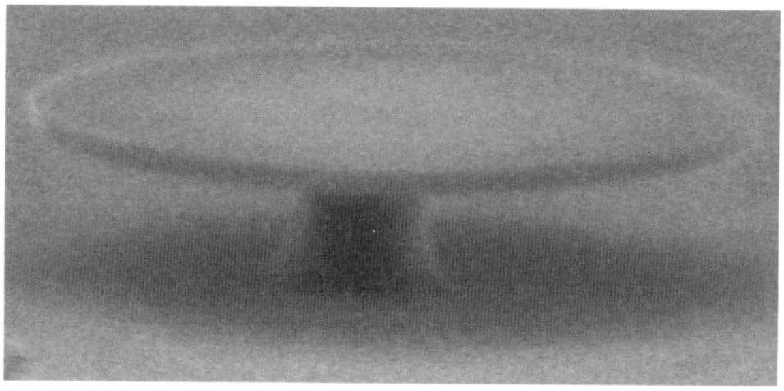

Fig. 6. GaAs/AlGaAs microdisk cavity make by the authors. The cavity mode is on the perimeter of the disk [13].

previously discussed. Thus, a high-quality in-plane cavity is necessary. For this purpose we use a microdisk cavity. This cavity possesses the very high quality factors necessary for THz emission, yet the in-plane geometry of the microdisk reduces materials costs and time, and simplifies fabrication. A scanning-electron microscope (SEM) image of such a cavity is shown in Fig. 6. For THz emission the in-plane cavity diameter will be approximately 100 μm and the height will be 5 μm (for 6 THz). The large in-plane diameter simplifies fabrication while the height is very reasonable for any crystal growth process.

3. QD and Cavities as Terahertz Sources

We have made a case for the use of intersubband transitions for THz emission, instead of the direct bandgap transitions in narrow bandgap materials. While the choice of either interband or intersubband structures is not obvious for mid-IR applications, the choice is much more obvious in terahertz applications. While several mid-IR direct bandgap materials are under development for many years for interband devices, there is no such terahertz material. If the development of interband materials begins in the terahertz range, it will require substantial development lead time. In direct narrow bandgap materials the material control is poor and the intrinsic losses large. For intersubband transitions standard materials can be used and the epitaxial control that has been developed over the last two decades can be leveraged to produce well-engineered structures. Furthermore, the losses, so problematic in long-wavelength devices are reduced in the structures.

What is gained by further reducing the system dimensionality from the 1-dimensional confinement of QWs to the 3-dimensional confinement of QDs? First, the QD structure does provide reduced losses in the categories discussed in Section 2.2. While LO phonon absorption is inhibited at frequencies below 8 THz, free carrier absorption and Auger recombination are prominent, as well as acoustic phonon absorption at longer wavelengths. These losses are reduced in intersubband structures because of the reduced dimensionality in momentum space; however, these losses are not eliminated as there remains many possible available transitions. However, in QDs, with their atomic-like transitions, there is no crystal momentum dispersion in the band structure, putting sever restrictions on the losses discussed in Section 2.2. This can be seen in Fig. 7. An additional advantage of QD THz emitters is the relaxed non-normal emission requirements associated with QWs.

In certain promising geometries QWs simply will not function optimally. QWs and bulk materials are not particularly compatible with many nanostructures because of the nonradiative recombination that occurs at free surfaces. Bulk-crystal band structure is based on an effectively infinite crystal with no surface effects. Free surfaces are discontinuities in bulk materials, the causing breaking of the crystal symmetry and modified local bonding (surface reconstruction), and thus create new 'surface states.' Nearly all surfaces have surface states, and a large number of these states exist in the energy gap of band structures, creating band bending and recombination sites for carriers. Because of their 3-dimensional confinement, QDs can be uniquely isolated from free surfaces with minimal effort while still remaining spatially close to these surfaces. The

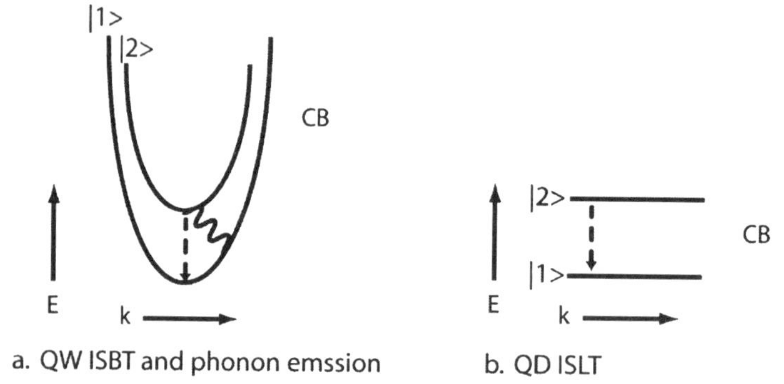

Fig. 7. Intersubband recombination processes and the band structure of a QW and QD are compared. (a) Radiative and nonradiative (phonon) relaxation processes in intersubband transitions (ISBT) in a QW. Because there is a distribution of energy differences between the two subbands, there is likely a match to the energy. (b) Because there is no momentum dispersion, the intersublevel transition (ISLT) has a fixed energy and is likely not to correspond to the phonon energy.

microdisk cavity is one such nanostructure (Fig. 6). The cavity mode is very close to the microdisk edge and nonradiative recombination can become a parasitic in the critical active region. Towards the center of the disk, away from the cavity region there is no surface recombination, but any QW recombination in this region is poorly coupled to the cavity mode. With QDs in the cavity mode, carriers diffuse from the pumping region to the QDs at the microdisk edge where they recombine without nonradiative surface recombination.

Like QWs, optical emission can occur in QDs through either interband or intersubband transitions. Since there is no *band* is the QD electronic structure we often use *level*, such as an intersublevel transition, but for convenience also often retain the band usage. However, just like interband structures formed from direct bandgap materials, interlevel, conduction to valence state QD transitions in the THz regime requires tuning the QD confined conduction and valence state differences to below 20 meV (5 THz). Any substantial quantum confinement will be on the order of 10-20 meV, implying the QD material bandgap must be near zero. While semimetals with bandgaps of 10 meV are possible, as mentioned above the materials growth technologies will be substantial. Thus, our structures will be based on intersublevel QD transitions.

As in QWs, designing a QD, or ensemble of QDs with an intersublevel transition is in principle not difficult. However, in unipolar, intersublevel devices carriers transition from the upper radiative state to the lower radiative state. Since the structure is unipolar, conduction band electrons do not recombine with valence band holes, and thus a population of lower state carriers remains. If the decay of the lower state population is not sufficient the quasi-equilibrium population will reduce the density of available lower states, clamping the radiative decay process. Thus, the intersublevel QD (or QW) emission source must be designed in a way that reduces the ground state quasi-equilibrium occupation.

While QDs active regions can provide reduced losses and can be integrated into high performance cavities such as microdisk, current fabrication suffers from large inhomogeneous size distributions. Using crystal growth techniques such as MBE, QWs can be fabricated with precision because the critical confinement dimension is in the growth direction, where MBE has local, monolayer thickness control. However, for QDs two additional confinement dimensions are needed. First generation QD structures were fabricated from QW processed into QDs using lithography. However, these structures suffered from surface recombination due to defects associated with directly etching the active device region. An alternative approach was developed in the mid 1990s, and uses lattice mismatch strain to form heterostucture islands of quantum-size dimensions in a larger bandgap host. When a lattice mismatched material such as InAs is deposited on host crystal such as GaAs (lattice mismatch~7%), the lattice mismatch strain is absorbed by the host as biaxial compression of the epitaxy (InAs). Since the InAs lattice constant is larger than the GaAs host, when a small amount of material is deposited it is compressed in-plane (biaxial compression), expanding in the growth direction according to Poisson's ratio, but remaining essentially planar. However, as more lattice mismatched material is deposited, the accumulated strain energy increases to a level where a different strain accommodation is energetically more favorable: the lattice mismatched epitaxy creates islands. These islands provide surface curvature to laterally relax the mismatched material. The crystal surface has excess bond energy compared to the interior of the crystal since the bonding environment of the bulk has been disturbed by the surface. Without the excess strain energy from the lattice mismatched epitaxy, the surface area would be minimized, minimizing the total crystal energy. However, as the total strain energy increases it is energetically more favorable to create more surface area to reduce the strain energy. The islands formed remain once covered by the host (although their shape may change) and if the proper materials are chosen, can be used as QDs. A layer of QDs has a large inhomogeneous size distribution because the local strain environment driving the process is not unique and is difficult (or impossible) to control. For example, InAs-based QDs are formed in a GaAs host with in-plane dimensions of 20-30 nm and vertical dimensions of 4-5 nm. Photoluminescence at 4K on isolated single QDs indicates linewidths on the order of 100 μeV, while the spectral linewidth of the full ensemble is typically 25 meV near the 1.3eV ensemble-average spectral peak emission.

Such a large QD ensemble linewidths is detrimental for three reasons. First, the emission efficiency is directly related to the density of optical emitters, and since losses are large in the THz, it is particularly important in these applications that the emitter density at the designed frequency is large. The designed spectral linewidth is rarely large. Second, to overcome losses and facilitate gain a high-quality optical cavity is used, which has a narrow resonance. Thus, to optimize coupling between the emission source and the cavity the emission source should have a spectral linewidth on the order the spectral linewidth of the cavity. While microdisk cavities, for example, do not have resonances of 100 μeV they can be on the order 1000 μeV. The final problem with the size variation in strain-induced QDs is the inability to precisely tune the coupling between QDs in different layers. It has been shown [14] that layers of QDs separated by thin layers of the host crystal can be coupled. It is possible to observe this coupling for an individual coupled QD pair [15], but it is currently not possible to engineer this coupling for the entire ensemble because the QD size control is not precise. This last problem is particularly detrimental when designing QD intersublevel structures, since designing short lower radiative state lifetimes is important and is often accomplished in QW systems by coupling the lower state to another structure. Such is the case in the coupled-asymmetric quantum dot emission device described below. To realize this device, new QD fabrication techniques will be necessary.

4. The Coupled Asymmetric Quantum Dot (CAD) Laser

Our vision of an intersublevel QD laser employs a coupled asymmetric QD (CAD) structure. The band structure is shown in Figure 8. In this structure a large, QD will be deposited on top of a smaller dot by our vertical aligned QD column technique. The size of each QD will be controlled by adjusting the alloy composition. We have successfully used this technique to align up to 10 QDs of the same approximate alloy composition in vertical columns [16]. Three QD conduction energy levels will be used from these QDs to form the core portion of the THz laser. Lasing will occur between the excited state and ground state in the conduction levels of the smaller dot, and we denote this transition as |3>-|2>. Holes must be absent from this device in the first excited valence state or conduction-valence level radiative recombination from excited electron and hole states becomes a competing process. For fast relaxation out of the electron ground state (|2>) a second, larger QD will be used. Its electron energy level will be tuned an LO phonon away from the |2> level and will provide extremely fast relaxation for the ground state of the smaller QD. This transition is denoted as |2> - |1>. While it may appear that the electron-hole recombination will adequately empty the lasing ground state, we do not believe the spontaneous emission decay, denoted by |2>-|1'> will effectively drain this level. This direct optical recombination process is slow, on the order of 1 ns, and even though electrons will be relaxing through the ground states of two QDs, not one, it will be too slow to effectively drain carriers. It is therefore necessary to couple the electron into a different reservoir than the vacuum electromagnetic field through radiation. We do this by adding a QW to the end of the asymmetric dot structure (Fig. 8). The lasing levels are shown in Fig. 9.

While the crystal growth of such a structure cannot currently be accomplished, the structure is conceptually realistic in a GaAs host with InAs QDs. A simulation based on

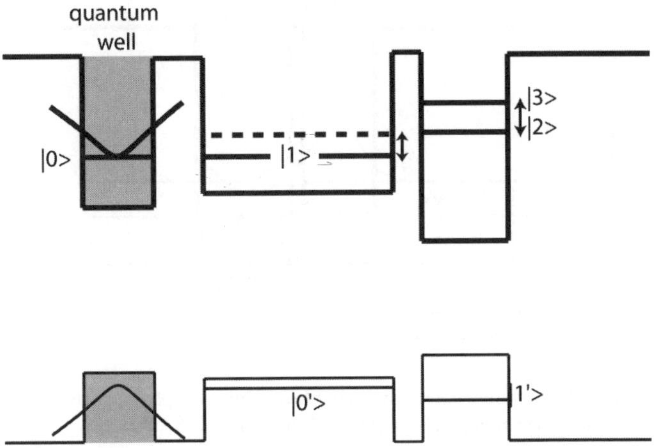

Fig. 8. Schematic diagram of the band structure of the Asymmetric Coupled QD laser.

this design has been made; the wavefunctions for an InAs coupled QDs structure is shown in Fig. 10. The device emits at approximately 7 THz.

Because of the coupling between the two QD, such a structure can be viewed as a QD *molecule* coupled to a reservoir (the QW). In such a structure a layer of such molecules are used. To increase the output, several layers of these QD molecules can be employed. As in the cascade quantum well laser, the quantum dot molecules can be vertically aligned and biased so that the larger dot aligns with a new smaller dot. The band structure of such a device structure is shown in Figure 11. As in the QW cascade laser,

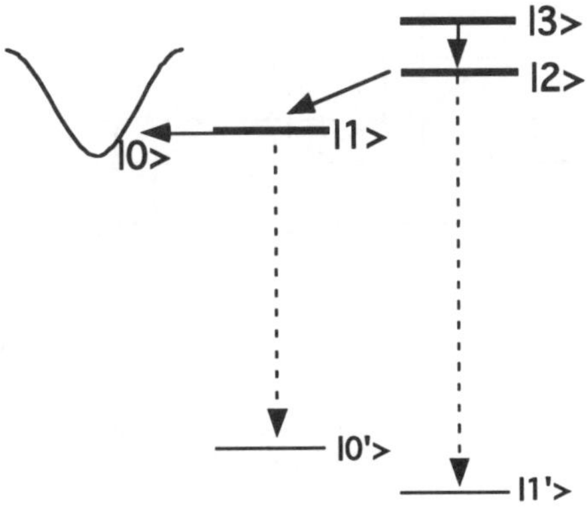

Fig. 9. Radiative and nonradiative transitions involved in the lasing process are shown.

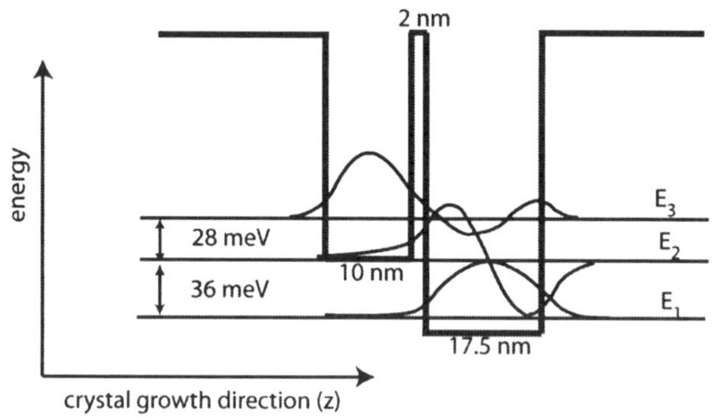

Fig. 10. Simulation of the wavefunctions and energy levels for a coupled QD structure. Radiative emission is from the |3>-|2> transition (courtesy of Cun-Zheng Ning, NASA Ames)

the output of the ground state pumps the upper state of the next QD molecule and the QW reservoir is not necessary.

In this cascade structure we must not only align the large and small QDs in each molecule, but also align the molecules into columns. Such a structure will be a sever test for emerging QD crystal growth techniques.

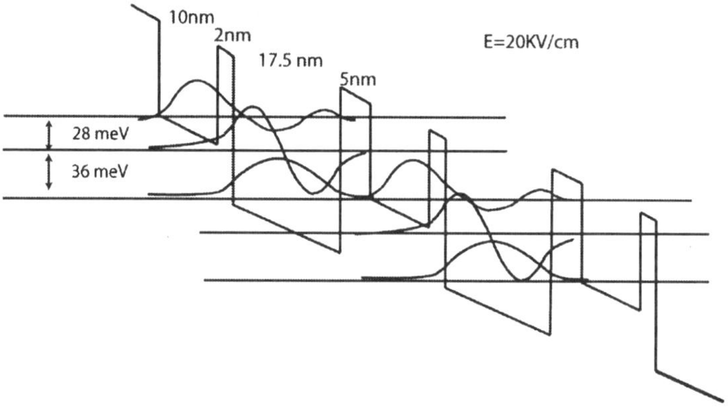

Fig. 11. The wavefunctions of a cascade of coupled QDs is shown, where the ground state of the higher potential molecule cascades to the upper state of the next molecule. (courtesy of Cun-Zheng Ning, NASA Ames)

5. Quantum Dot Microcavity Terahertz Source

A different approach from the coupled QD molecule (CAD) emitter source is a device structure designed around optical cavity coupling of the QD emitters. Unlike the CAD structure, these QDs are not quantum mechanically coupled. Here the cavity is principally used to provide optical feedback to QD radiative transition. However, the cavity additionally reduces the spontaneous emission lifetime of the radiative transition, increasing the spontaneous emission rate and reducing the lasing threshold. While the focus in the CAD structure is reducing the quasi-equilibrium ground state carrier density, here we focus on enhancing feedback. While the microdisk cavity is tuned to the radiative transition resonance, in a next-generation structure a second cavity can be formed vertically to enhance the ground state radiative decay rate and hence reduce the ground state quasi-equilibrium carrier density. Thus, the function of the primary cavity is to enhance feedback for stimulated emission in the intersublevel transition as well as to increase the spontaneous emission decay at this transition to increase the optical emission efficiency, while the second vertical cavity, tuned to the conduction-valence state transition, enhances spontaneous emission decay from the lasing ground state. Despite its apparent increased complexity such an addition is not difficult.

The QD microcavity THz emission source has two components: (1) An ensemble of QDs where the direct transition between the first excited state and the lowest lying state in the conduction band is used as the THz emission source; and (2) a passive cavity, the microdisk, which is used to couple light from the QD emitter source, provide enhanced spontaneous emission and feedback.

The terahertz microdisk has a diameter on the order of 300μm, yet the peak of the cavity mode is approximately 25μm from the disk edge. If QDs are uniformly distributed across the microdisk most of the QDs will not spatially couple to the cavity mode. In fact, measurement and theory by coworkers at Northwestern University show [17] that one explanation for the less than expected spontaneous emission enhancement is due to collected emission from QDs not spatially coupled to the cavity mode. Furthermore, the spatially uncoupled QDs towards the disk center trap carriers impeding carrier injection and diffusion into the cavity region. To reduce these effects we position QDs near the microdisk edge using selective area crystal growth.

We will break the discussion of this device into four parts: a description of the cavity effect and spontaneous emission; the mode distribution resulting from the microdisk cavity; placement of QDs selectively in the cavity mode; and rate equations describing the lasing process.

5.1 Enhancement of spontaneously emission in microcavity

Spontaneous emission from a source such as a QW exciton, or atom is often thought of as an inherent property of the emitter, but is in fact a result of the interaction between the emitter dipole and the vacuum electromagnetic fields. Therefore, the radiation emitted from the source can be altered by suitably modifying the surrounding vacuum fields with a cavity. [18, 19] In 1946, Purcell first proposed the concept with respect to controlling the spontaneous emission (SE) rate of a quasi-monochromatic dipole by using a cavity to tailor the number of electromagnetic modes to which it is coupled. Later cavity quantum-electrodynamics (cavity-QED) has provided a firm theoretical and experimental basis for this idea. The ability to enhance the SE rate is so called Purcell Effect. Thus, besides the feedback to enhance gain, we can use our cavity to decrease the SE lifetime to reduce the lasing threshold.

First consider a general electric dipole transition given by the Fermi golden rule [20]

$$\frac{1}{\tau}=\frac{4\pi}{\hbar}\rho(\omega)\left\langle\left|\vec{d}\cdot\vec{\varepsilon}(\vec{r})\right|^2\right\rangle, \tag{4}$$

where τ is the spontaneous emission lifetime, $\vec{\varepsilon}(\vec{r})$ is the vacuum electric-field amplitude at the location $\vec{r}$ of the emitter and $\vec{d}$ is the electric dipole. $\rho(\omega)$ is the density of vacuum states at the emitter's angular frequency ω. However if the emitter is put into a cavity, the final density states $\rho(\omega)$ will be greatly modified and instead of being a constant value it is given by a normalized Lorentzian in the cavity case. So at resonance frequency ω_0, the density of modes could be very high depending the quality factor of the cavity.

Purcell originally considered this enhancement of spontaneously emission at the "weak coupling regime". That is, a localized dipole (wavelength λ_e, linewidth $\Delta\lambda_e$) placed on resonance with a single cavity mode (wavelength λ_c, linewidth $\Delta\lambda_c$, quality factor Q=$\lambda_c/\Delta\lambda_c$), and since $\Delta\lambda_e \ll \Delta\lambda_c$, the escape time of SE photons out of the cavity is much shorter than the radiative lifetime so re-absorption is negligible. In the weak-coupling regime, the emitter interacts with a quasicontinuum of modes and the SE rate can be expressed using Fermi Golden rule. A simple derivation shows that the SE rate in the cavity mode, referenced to the total SE rate in a homogeneous medium, is given by the Purcell factor $F_p = 3Q\lambda_c^3/4\pi^2 n^3 V$, where n is the refractive index of the medium and V is the effective mode volume (V is given by the spatial integral of the vacuum field intensity for the cavity mode, divided by its maximum value). To obtain the maximum enhancement for a given by the Purcell factor, the emitter must be on exact resonance, located at the antinode of the vacuum field, with its dipole parallel to the vacuum electric field. Thus, the Purcell factor is the maximum enhancement from a cavity. But for a practical microcavity, the spontaneous emission enhancement can be expressed as

$$\frac{\gamma}{\gamma_0}=\frac{3Q(\lambda_c/n)^3}{4\pi^2 V}\frac{\Delta\lambda_c{}^2}{4(\lambda-\lambda_c)^2+\Delta\lambda_c{}^2}\frac{|E(\vec{r})|^2}{|E_m|^2}2\eta^2=F_p g(\lambda)h(\vec{r})2\eta^2 \tag{5}$$

Where γ_0 is the spontaneous emission rate of dipole without the cavity; $E(\vec{r})$ is the electric field amplitude of the cavity mode at the position of the dipole emitter; $E_m = (h\nu / 2\varepsilon_0 n^2 V)^{1/2}$ is the maximum value of the electric field amplitude. The first term F_p is the Purcell factor. The second term $g(\lambda)$ and the third term $h(r)$ describe the spectral and spatial matching between the dipole emitter and the cavity mode. The factor of 2 comes from the two-fold degeneracy of the cavity mode. η describes the orientation match between the emitter dipole and the polarization of the cavity mode. We have shown the effect of the cavity on the spontaneous emission lifetime [21] and the that this effect is governed predominately by the spectral relationship to of the cavity and QD emitter in vertical, micropost cavities with cavity quality factors on the order of 1000 [22].

5.2 Calculation of THz optical mode in a microdisk

For ultralow laser thresholds, our microcavity design incorporates both high reflectivity and matching between a small gain volume and a single optical mode. The disk thickness is less than $\lambda / 2n_D$ (to guarantee single mode in the vertical direction), but larger than $\lambda / 4n_D$ (to guarantee one mode), and for most cases, it is selected to be $\lambda / 3n_D$. The index of refraction, n_D is calculated in the terahertz frequency range in the Appendix. Since we are interested in a wavelength of ~100 μm, the thickness $d \approx 10\mu m$. Normally, the modes allowed in this kind of cavity are the combination of guided modes and non-guided modes, and the solution is very complicated. But in the case of R~λ, there are only several TE and TM modes allowed in the cavity. For example, we choose R=150 μm.

5.2.1 Allowed modes in the cavity:

The TE modes have lower threshold because of higher n_{eff}, so emission from TE modes will in general only be observed. We can divide the space into three regions: the guided region (in the disk), the evanescent region (outside disk, but close to the disk edge) and the free-propagation region (far from the edge). Inside the disk, the field of TE_{ml} can be expressed as a Bessel function, J_m [23]:

$$\begin{cases} H_z = H_0 n_{eff} J_m(k_0 n_{eff}\rho)\cos(k_0\sqrt{n_D^2 - n_{eff}^2}\, z)e^{im\theta} \\ H_\rho = -H_0\sqrt{n_D^2 - n_{eff}^2}\, J'_m(k_0 n_{eff}\rho)\sin(k_0\sqrt{n_D^2 - n_{eff}^2}\, z)e^{im\theta} \\ H_\theta = -iH_0 m\sqrt{n_D^2 - n_{eff}^2}\,\dfrac{J_m(k_0 n_{eff}\rho)}{k_0 n_{eff}\rho}\sin(k_0\sqrt{n_D^2 - n_{eff}^2}\, z)e^{im\theta} \end{cases} \tag{6}$$

$$\begin{cases} E_z = 0 \\ E_\rho = -\mu_0 c H_0 m \dfrac{J_m(k_0 n_{eff} \rho)}{k_0 n_{eff} \rho} \cos(k_0 \sqrt{n_D^2 - n_{eff}^2} z) e^{im\theta} \\ E_\theta = -i\mu_0 c H_0 J_m'(k_0 n_{eff} \rho) \cos(k_0 \sqrt{n_D^2 - n_{eff}^2} z) e^{im\theta} \end{cases} \tag{7}$$

$$m \approx 2\pi R n_{eff} / \lambda = 18$$

Here different mode index l corresponds to different effective index of refraction, n_{eff}. E and H are the electric and magnetic fields in polar coordinates, z, ρ, Θ, μ_0 is the permeability of free space, c the speed of light, k_0 the wave vector, and R is the disk radius. To obtain the allowed modes in the cavity, we solve the eigenfunction:

$$\sqrt{n_D^2 - n_{eff}^2} \tan\left[\frac{k_0 d}{2} \sqrt{n_D^2 - n_{eff}^2}\right] = \sqrt{n_{eff}^2 - 1} \tag{8}$$

with the boundary condition:

$$J_m(k_0 n_{eff} R) = 0 \tag{9}$$

We can now calculate the effective index and allowed wavelengths in the cavity. We express the quality factor Q (due to radiation loss only) as:

$$Q = b \cdot \exp(2mJ),\ \mathrm{J} = \tanh^{-1}(S) - S,\ \text{and}\ \mathrm{S} = \sqrt{1 - 1/\mathrm{n}_{\mathrm{eff}}^2} \tag{10}$$

where $b \sim 1/6.5$ in this formula is from an estimation of tunneling rates using the WKB approximation. We observe that: (1) a higher n_{eff} will lead to a higher Q and a lower threshold. This is the reason we only see TE modes instead of TM modes; (2) higher index, m means the field intensity will be closer to the edge, where the incidence angle is larger and so the radiation loss is smaller; (3) the FWHM of the edge-emission angle is estimated to be $2/\sqrt{m}$. So for higher order of TE_{ml}, we obtain a narrower edge-emission angle.

From the above analysis, we see that only the TE_{m1} or TE_{m2} mode is allowed in the cavity. Now, we can calculate λ, n_{eff} and Q for TE_{m1} (m=*17, 18, 19, 20, 21*) and TE_{m2} (m=*13, 14, 15, 16, 17*). The results are shown in Table 1. These absolute values determined by the calculations are only of relative meaning, because the cavity quality factor, Q will depend on the quality of the regrown cavity as well as the calculation above. We see that the wavelengths of $TE_{m+4,1}$ and $TE_{m,2}$ modes are very close to each other.

	$TE_{17,1}$	$TE_{18,1}$	$TE_{19,1}$	$TE_{20,1}$	$TE_{21,1}$
λ (μm)	108.76	105.12	101.75	98.60	95.67
n_{eff}	2.5587	2.5940	2.6274	2.6592	2.6893
$Q(\times 10^9)$	1.3	7.7	47.2	296.6	1891.4
	$TE_{13,2}$	$TE_{14,2}$	$TE_{15,2}$	$TE_{16,2}$	$TE_{17,2}$
λ (μm)	109.52	105.58	101.95	98.60	95.49
n_{eff}	2.5514	2.5895	2.6254	2.6592	2.6911
$Q(\times 10^9)$	0.0055	0.0308	0.1764	1.0354	6.2040

Table 1. Calculated wavelength, effective index and quality factor for the first and second order TE modes in a microdisk cavity resonant in the 100 μm wavelength range.

An example of the microdisk cavity spatial mode distribution is shown in Fig. 12 for a microdisk resonant at approximately 100 μm. The spatial position of the cavity mode is approximately 25 μm from the microdisk edge.

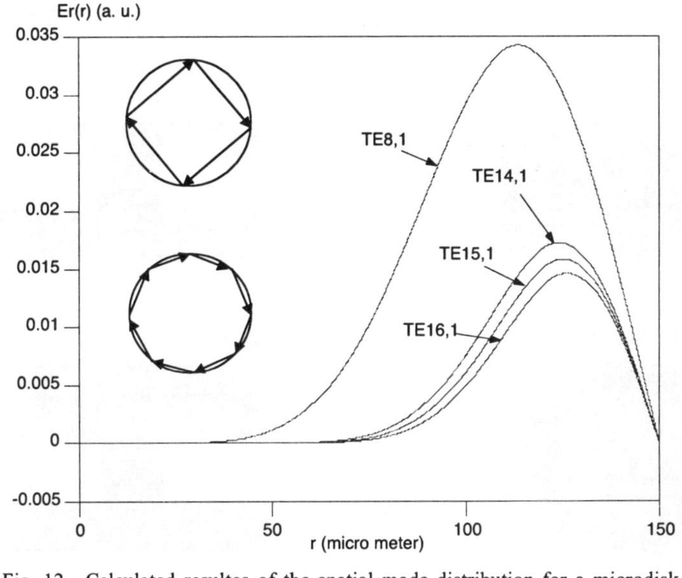

Fig. 12. Calculated resultes of the spatial mode distribution for a microdisk cavity resonant at a wavelength of approximately 100μm.

5.3 QD placement in the microdisk

To align QDs near the microdisk perimeter, we chose to design a fabrication and crystal growth process where the QDs are selectively located along this perimeter. The process is achieved through an understanding of the strain-induced QD formation process. Since the QD formation relieves lattice mismatch stress by creating lateral free surface and the microdisk edge has a natural free surface, the microdisk edge is an optimum place for QD nucleation. In the normal QD formation process, when the strained material is deposited a thin region can be accommodated by planar biaxial compression, then after some thickness as been established the lowest energy state for the crystal is one with QDs. If there is a free surface to relax the strain material, the extra energy associated with creating lateral free surface is reduced: the extra lateral surface is already present. Thus, QDs will form near the microdisk perimeter first. Thus allows us to terminated the crystal growth after the QDs have formed on the microdisk perimeter but before they form on the interior region.

The fabrication steps are shown in Fig. 13. First, the epitaxial structure containing the undercut region (AlGaAs) and 1/2 of the cavity region (GaAs) is made by MBE (Step 1 Fig. 13). In Step 2 lithography and wet-chemical etching is used to define the microdisk. Here, a non-selective etch is used to etch down to the GaAs substrate, then a selective etch is used to etch the AlGaAs and undercut the GaAs. In Step 3, Fig. 13 the processed

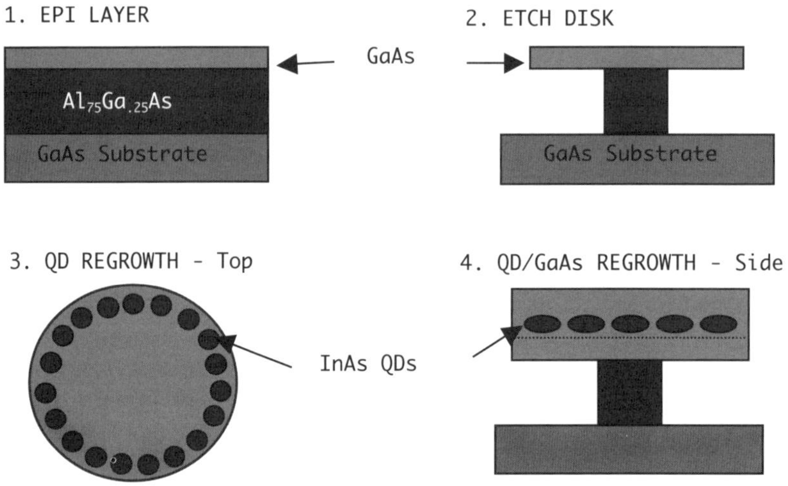

Fig. 13. Schematic of regrowth processing steps to form a microdisk cavity. (1) MBE is used to form the AlGaAs etch layer and approximately half the GaAs cavity. In (2) an isotropic etch etches down to the AlGaAs layer, after which a selective etch undercuts the AlGaAs to form the microdisk post. In (3), the regrowth is made after a final cleaning etch; QDs are aligned along the edge only, and (4) the GaAs top cavity is added.

sample is re-inserted into the MBE machine and the InAs QD layer is made, and then covered with GaAs (Step 4).

The full process has already been demonstrated and is shown in Fig. 14 on the next page. In the figure atomic-force microscope (AFM) images show InAs QDs aligned at the edge of the microdisk. An AFM image of the disk is shown in Fig. 14b, where the light region at the disk edge is a near continuous, single chain of QDs. An enlargement of the image is shown in Fig. 14a. The QDs are approximately 40 nm in diameter. In Fig. 14c an

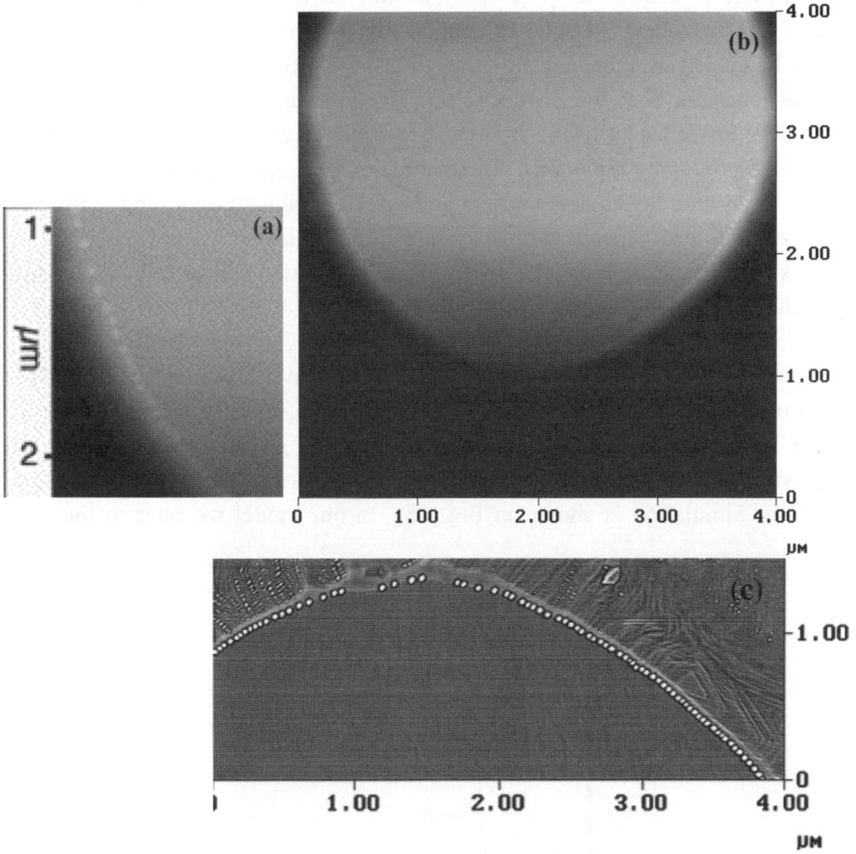

Fig. 14. AFM images of 4μm diameter GaAs micro-post, showing InAs QDs at the post edge and no QDs in the interior. (a,c) are magnified view of the post edge highlighting the QDs. Notice the very good edge alignment of QDs. Scale is in microns.

AFM image of a micropost is shown. Here, there is no undercut. The image is clearer because the AFM is more stable without the less ridged undercut layer. Here, the microdisk is only 4 μm and the GaAs top-half of the cavity was not deposited in order to study the QD alignment. To fully investigate the structure we have designed the device to operate in near IR range with 900 nm wavelength operation. To design for this wavelength regime smaller disks are more useful. We are currently investigating the QD alignment process using the regrowth process on a variety of microdisk sizes.

5.4 Tuning optical transitions for lasing

The cavity is present for two purposes: (i) to capture photons, provide feedback and increase gain; and (ii) to enhance the spontaneous emission (SE) decay rate, γ, increasing the light output from the QD emitters. Point (i) is necessary in all semiconductor-based lasers. Point (ii) is helpful in all lasers but can be used in a subtler way in the intersublevel QD microcavity laser. In general, by coupling to electromagnetic cavity modes γ can be enhanced, reducing the onset to lasing. However, in designing the THz laser we must consider both the SE decay rate from the upper lasing level and the SE decay rate of the lower lasing level. In a band-to-band, electron-hole laser electrons and holes recombine producing a photons. Intersubband lasers are unipolar devices, excited state carriers decay into lower state carriers and do not annihilate each other. Therefore, the lasing threshold also depends the decay rate from the lasing ground state. In some cases such as in the cascade laser, the ground state population can be reduced by strong coupling to the LO phonon resonance.

At least initially, a choice must be made, to make the cavity resonant with the lasing wavelength or the lower state transition wavelength. To investigate options we have modeled the lasing threshold as a function of excited and ground state SE decay rates. We do this with a simple set of rate equations that include loss. The QD structure for the purposes of the simulation is shown in Fig. 15. In our model we refer to the lowest

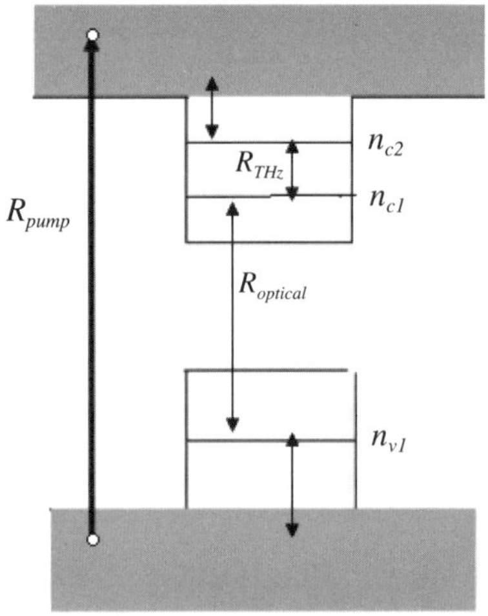

Fig. 15. Schematic of the QD used to establish a set of rate equations.

bound energy state of electron as n_{c1} and the next higher level is called n_{c2}.. The only confined level of holes is n_{v1}. The electrons are pumped from very deep in valance band to very high in conduction band with a pumping rate of R_{pump} carriers per unit time per unit volume. The electrons and holes scatter down the bands and subsequently are captured in the confined electron and holes states, n_{c2} , n_{v1} respectively. Let us say that the corresponding capture rates are $R_{e,capture}$ and $R_{h,capture}$ carriers per unit time per unit volume respectively. Hence we can write a simple set of rate equations for describing the time evolution of carrier density in the three confined levels of QDs:

$$\frac{dn_{c1}}{dt} = R_{THz} + R_{non-radiative} - R_{Optical}$$

$$\frac{dn_{c2}}{dt} = R_{e,capture} - R_{THz} - R_{non-radiative} \tag{11}$$

$$\frac{dn_{v1}}{dt} = R_{Optical} - R_{h,capture},$$

where, R_{THz} is the net rate of the intersublevel THz transition – the rate at which electrons in the higher energy level n_{c2} drop to an empty lower energy level n_{c1}, emitting a THz photon (either spontaneously or by stimulation), minus the rate at which electrons in the lower energy levels n_{c2} absorb a THz photon and jumps to an empty higher energy level n_{c2}. Adding all three components, R_{THz} can be written as:

$$R_{THz} = \frac{(n_{c2})(n_{QD} - n_{c1})}{\tau_{c2,c1}} + \frac{(n_{p,THz})(n_{c2} - n_{c1})}{\tau_{c2,c1}} \tag{12}$$

Similarly, $R_{Optical}$ is the net rate of confined conduction band-to-valence state transitions that absorb or emit a photon. This is the more typical interband transition. It is equal to the rate at which electrons in the lower energy level of the conduction band (n_{c1}) drop to an empty energy level in the valence band (n_{v1}), emitting an optical photon (either spontaneously or by stimulation), minus the rate at which electrons in the valance band energy level n_{v1} absorbs an optical photon and jumps to an empty conduction band level n_{c1}. Adding all three components $R_{Optical}$ can be written as:

$$R_{Optical} = \frac{(n_{c1})(n_{QD} - n_{v1})}{\tau_{c2,c1}} + \frac{(n_{p,Optical})(n_{c1} - n_{v1})}{\tau_{c1,v1}} \tag{13}$$

$n_{p,THz}$, and $n_{p,Optical}$ are the photon densities in the THz mode and optical mode of the cavity. This cavity can be singly resonant at the THz frequency, or doubly resonant to enhance the optical transition rate. $\tau_{c2,c1}$ is the spontaneous emission lifetime. $R_{non-radiative}$ is the rate of non radiative intersublevel transitions.

Since the rate of change of the carrier densities in all three confined energy levels strongly depends on the densities of photons in both the cavity modes, it becomes necessary to know how the photon densities themselves evolve with time. The time evolution of photon densities in the two cavity modes can be described by:

$$\frac{dn_{p,THz}}{dt} = -\frac{\omega_{THz}}{Q_{THz}} n_{p,THz} + R_{THz} \tag{14}$$

$$\frac{dn_{p,Optical}}{dt} = -\frac{\omega_{Optical}}{Q_{Optical}} n_{p,Optical} + R_{Optical} \tag{15}$$

where, ω_{THz} and $\omega_{Optical}$ are the resonant frequencies of THz and optical modes. Q_{THz} and $Q_{Optical}$ are the corresponding Q-factors. These coupled rate equations for carriers and photons in two modes can easily be simulated. The simulation is used to predict the required threshold pumping rate.

For the purposes of the simulation, we set the loss rate higher than the SE rate for any states to model the high losses in the THz range. A set of L-I curves is shown in Fig. 16. The horizontal axis is the pumping rate of electron-hole pairs into the bulk crystal, and the non-radiative loss lifetime is set at 40ps, the upper state, THz SE lifetime is 1000ps

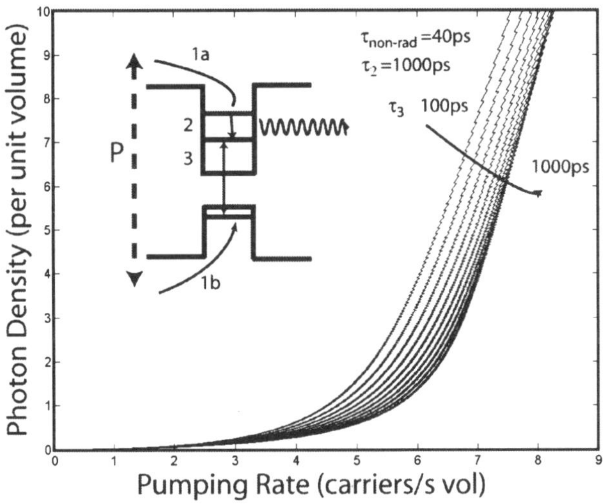

Fig. 16. Results of rate equations modeling of QD intersublevel laser. The lasing level is 2, and the non-radiative lifetime is set at 1/25 of the SE lifetime of the lasing state. The ground state lifetime is varied and clamps the threshold when approaching the SE lifetime of the lasing state.

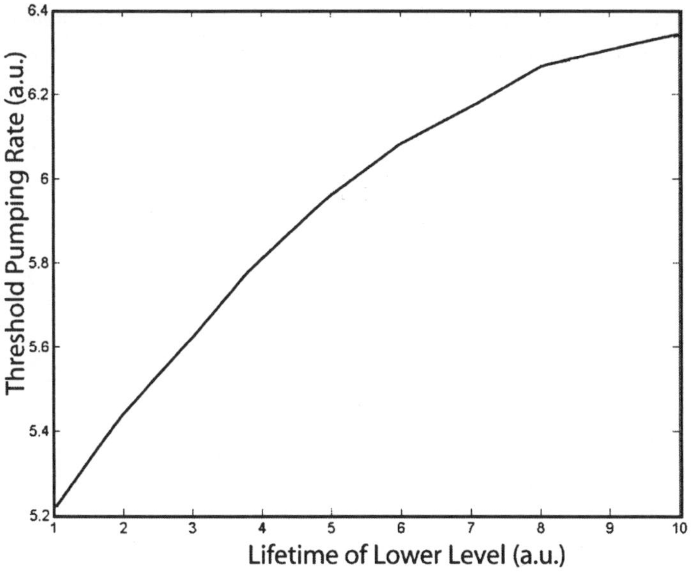

Fig. 17. The modeled pump rate necessary to achieve lasing is plotted against the lifetime of the lower level quantum dot state. As the lifetime of this state increases, so does the pump rate necessary for lasing.

and the lower, optical SE lifetime varies between 100 and 1000 ps. As the lower state lifetime increases the lasing threshold becomes clamped because the lower state cannot remove carriers as fast as they enter the upper state. This is highlighted in Fig. 17, where the pumping rate required for lasing (the threshold rate) is plotted against the lifetime of the lower state. If the upper state SE lifetime is sufficiently small, the microdisk must be designed to be resonant with lower state wavelength. Conversely, if the upper state SE lifetime is long compared to the lower state SE lifetime, the cavity must be designed to be resonant with the upper state.

We have made single QD SE lifetime measurements of the ground state emission. An example is shown in Fig. 18. The QD is optically pumped with energy above the GaAs bandgap. The quick rise time indicates carriers relax to the ground state remarkably fast. In Fig. 18 the rise time of the ground state is approximately 40 ps. The SE lifetime of the ground state however is approximately 500 ps. Thus, our initial conclusion is that the ground state SE lifetime may limit the threshold if adjustments to our cavity design are not made. It is possible that the cavity could be resonant at both the intersublevel transition and the ground state band-to-band transition however, while the microdisk will have more resonances in the band-to-band region it is unlike that these resonances will align with both transitions of the QD. It is possible to incorporate a vertical cavity structure into the microdisk cavity, where the vertical cavity is composed of distributive-Bragg reflector (DBR) mirror pairs tuned to the ground state transition (approximately wavelength, 1 μm). This should not affect the microdisk cavity, which remains resonant in the THz range. A schematic of this proposed structure is shown in Fig. 19.

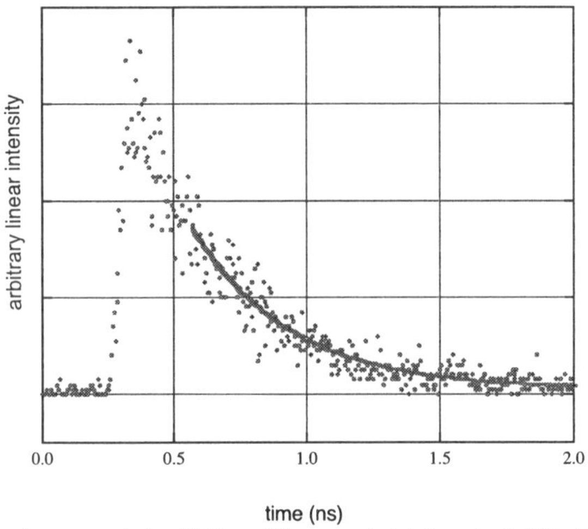

Fig. 18. Spontaneous emission lifetime of the ground state in a single QD at 4K. The fast rise time indicates fast relaxation into the ground. The ground state lifetime is approximately 500 ps.

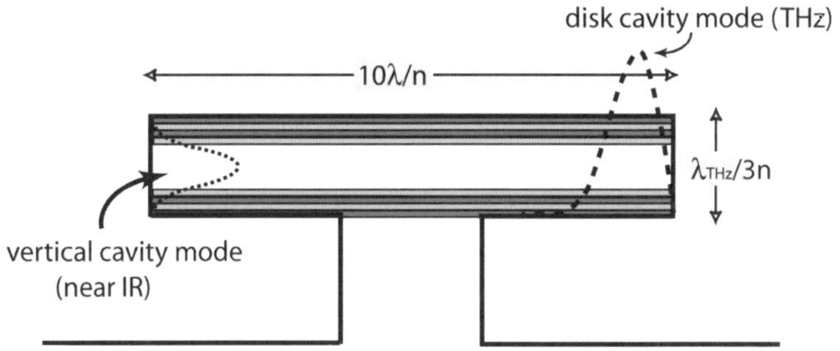

Fig. 19. Microdisk cavity with vertical, planar distributive-Bragg reflector (DBR) mirror pairs. The microdisk cavity is resonant in the terahertz frequency range, and couples to the intersublevel QD transition. The vertical cavity is resonant in the near-IR and couples to the conduction-valence transition. The vertical cavity reduces the spontaneous emission lifetime, and removes carriers from the lasing ground state.

6. Conclusions

Terahertz emission using optical-based sources is an emerging field. New techniques will certainly be required to overcome a particularly difficult combination of materials issues and nonradiative losses. We have proposed two rather unique terahertz sources based on semiconductor QDs. QD active regions offer reduced losses because of their three-dimensional confinement. In addition, QDs can be inserted into a variety of novel structures. One QD-based THz device is the CAD laser, and is similar to the quantum cascade laser. Realizing this device will require a new crystal growth method in order to properly tune the coupled QD electronic states. A second QD THz device is the QD microdisk, where QDs are positioned in the cavity mode of a microdisk. Here the coupling of the cavity with THz QD resonance provides optical feedback and enhances the spontaneous emission decay process, providing more efficient spontaneous emission and reducing the lasing threshold. While further development of this device is necessary many of the individual components of this device have been designed and fabricated.

Acknowledgments

This work was initiated from a fruitful collaboration with Hui Cao of the Physics department at Northwestern University using QDs in microcavities and QD and cavity research conducted with Y. Yamamoto of Stanford University; both efforts are in the near IR wavelength range. We would like to thank D. Woolard for helping us initiate this program in the terahertz domain and for his continued support. GSS has enjoyed many technical discussions with J. Harris at Stanford University on a broad range of topics including those present in this chapter for many years. His enthusiastic support is appreciated by all the authors. We would like to thank D. J. Miller for editing the manuscript. This work is supported by the Army Research Office.

References

[1] J. Kono, S.T. Lee, M.S. Salib, G.S. Herold, A. Petrou, and B.D. McCombe, Phys. Rev. B **52**, R8654 (1995).

[2] J. Cerne, J. Kono, M.S. Sherwin, M. Sundaram, A.C. Gossard, and G. E. W. Bauer, Phys. Rev. Lett. **77**, 1131 (1996).

[3] Y. Naveh and B. Laikhtman, Phys. Rev. Lett. **77**, 900 (1996).

[4] D. Woolard, *et. al.*, Phys. Rev. E 65, 1903-1914 (2002).

[5] M. Zandian, J. M. Arias, R. Zucca, R. V. Gil, and S. H. Shin, Appl. Phys. Lett. **59,** 102 (1991); R. Zucca, M. Zandian, J. M. Arias and R. Gil, SPIE Proc., **1634**, 161 (1992); J. M. Arias, M. Zandian, R. Zucca, and J. Singh, Semicond. Sci. Technol., **8**, S 255 (1993).

[6] J. Faist, F. Capasso, D. L. Sirtori, A. L. Hutchinson, and A. Y. Cho, Science **264**, 553 (1994).
[7] M. Rochat, *et al.,* Appl. Phys. Lett. 81 1383 (2002).
[8] E. R. Brown, *et al.,* Appl. Phys. Lett. **58** 2291 (1991).
[9] R. A. Smith, *Semiconductors, Second Edition,* Cambridge University Press, Cambridge (1978).
[10] H. Y. Fan, Rep. Progr. Phys. **19**, 107 (1956).
[11] W. P. Dumke, Phys. Rev. **124**, 1813 (1961).
[12] A. R. Adams, E. P. O'Reilly and M. Silver, *Strained Layer Quantum Well Lasers,* in *Semiconductor Lasers, Vol. I*, edited by E. Kapon, Academic Press, San Deigo, CA (1999).
[13] H. Cao, J. Y. Xu, W. H. Xiang, Y. Ma, S. –H. Chang, S. T. Ho and G. S. Solomon, Appl. Phys. Lett. **76** 3519 (2000).
[14] G. S. Solomon, J. A. Trezza, A. F. Marshall, and J. S. Harris, Jr., Phys. Rev. Lett. **76**, 952, 1996.
[15] M. Bayer, P. Hawrylak, K. Hinzer, S. Fafard, M. Korkusinski, R. Wasilewski, O. Stern, A. Forchel, Science **291**; 451 (2001).
[16] G. S. Solomon, S. Komarov , J. S. Harris, Jr. and Y. Yamamoto, J. Cryst. Growth, **175/176**, 707 (1996).
[17] W. Fang, J. Y. Xu, A. Yamilov, H. Cao, Y. Ma, S. T. Ho and G. S. Solomon, Opt. Lett. **27**, 948 (2002).
[18] E. M. Purcell, Phys. Rev. 69, 681 (1946).
[19] Drexhage, K. H. *Progress in Optics* (ed. by Wolfe, E., Vol. XII, 165-232 North-Holland, Amsterdam, 1974).
[20] J. M. Gerard, Phys. Rev. Lett. **81**, 1110 (1998).
[21] G. S. Solomon, M. Pelton, and Y. Yamamoto, Phys. Rev. Lett. **86**, 3903 (2001).
[22] G. S. Solomon, M. Pelton, and Y. Yamamoto, Phys. Rev. Lett., accepted for publication (2003).
[23] Ruo Peng Wang, *et al.*, J. .Appl. Phys. **81**, 391 (1997).

Appendix: Calculation of the Refractive Index

From E.D. Palik, Handbook of Optical Constants of Solids, Academic, New York, 1985, the refractive index of GaAs as a function of frequency is:

$$n^2 = 1 + \frac{A}{\pi}\ln\frac{E_1^2-(\hbar\omega)^2}{E_0^2-(\hbar\omega)^2} + \frac{G_1}{E_1^2-(\hbar\omega)^2} + \frac{G_2}{E_2^2-(\hbar\omega)^2} + \frac{G_3}{E_3^2-(\hbar\omega)^2}$$

where

$E_0 = 1.428eV$, $E_1 = 3.0eV$, $G_1 = 39.194eV^2$, $E_2 = 5.1eV$, $G_2 = 136.08eV^2$,

$E_3 = 0.0333eV$, $G_3 = 0.00218eV^2$,

$A = 0.5858$

So the calculated n=3.6257 for a wavelength of 100 μm.

International Journal of High Speed Electronics and Systems
Vol. 13, No. 4 (2003) 1129–1148

TERAHERTZ TRANSPORT IN SEMICONDUCTOR QUANTUM STRUCTURES

S.J. ALLEN and J.S. SCOTT

Center for Terahertz Science and Technology
University of California, Santa Barbara 93106

Photon assisted transport, dynamic localization and absolute negative conductance appear in the terahertz photoconductivity in semiconductor quantum structures and are close analogs of quasi-particle transport in microwave irradiated superconducting junctions. By embedding superlattice devices in quasi-optical arrays and integrating them into terahertz cavities, the dynamical conductance of electrically biased superlattices can be measured. Models including the complications of electric field domains can account for the results in a semi quantitative manner. Uniform electrically biased superlattices appear to be potentially important as a terahertz gain medium.

Keywords: *terahertz, photon assisted transport, resonant tunneling, dynamic localization*

1. Introduction

The terahertz part of the electromagnetic spectrum represents a transition region for solid-state devices from electronics to photonics. At the one end of the spectrum, below 100 – 200 GHz, solid-state devices are best described as transport devices. While the material parameters like effective masses, band gaps, and g-factors require detailed quantum

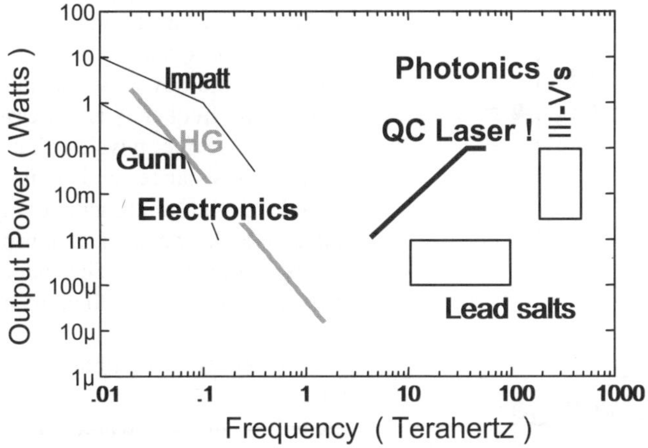

Fig. 1. Commercial solid-state electronics bracket the terahertz part of the electromagnetic spectrum. The low frequency end is marked by the high frequency limits of electronics including harmonic generators (HG); the high frequency end is marked by the low frequency limits of quantum transition devices.

mechanical, band structure models, the device physics is essentially controlled and modeled by electrical transport in the diffusive, drift or ballistic limits using combinations of Boltzmann transport equations and Poisson's equation to describe space charge distributions and electric fields. The essential device physics is quasi-classical. At the other end of the spectrum, above ~ 10 terahertz, solid state devices are clearly quantum transition devices, involving transitions between quantum mechanical levels in a semiconductor or band gap engineered semiconductor heterostructure.

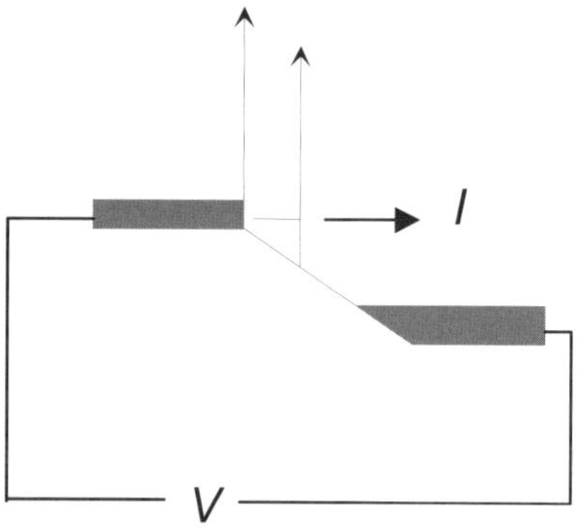

Fig. 2 Schematic diagram of the conduction band profile in a double barrier resonant tunneling diode under bias.

This is most clearly demonstrated by examining the operating ranges of state of the art commercial high frequency microwave sources like Gunn diodes or IMPATT diodes and low frequency laser diodes like lead salt lasers or the most recently developed quantum cascade laser.[1,2,3] There appears a decade wide frequency gap in solid-state technology that marks the transition between transport devices and quantum transition devices. (See Fig. 1)

It appears that the device physics that will underlie a successful technology that will bridge the apparent gap in terahertz technology will combine both transport physics and quantum transitions, *quantum transport devices*. Indeed, the resonant tunneling diode is a quantum transport device that exhibits negative resistance and can be made to oscillate at frequencies as high as 700 GHz.[4] (However, the power available becomes quite small as the frequency is increased.) But here, while the device physics that produces the negative resistance and gain is intimately related to the alignment of quantum well states, the device does not produce gain by quantum transitions. The device remains a transport device rather than quantum transition device.

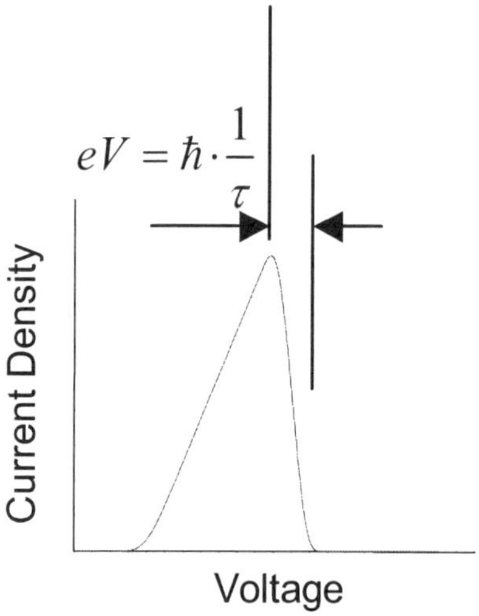

Fig. 3 Schematic current voltage characteristic if the resonant tunneling diode depicted in Fig. 2.

A necessary condition for the device to have evolved into a "quantum transition device" is $\omega\tau > 1$, where ω is the operating frequency and τ is a measure of the relaxation time or other line broadening mechanisms.[5] For the resonant tunneling diode the relevant broadening is probably the voltage width of the negative resistance region in the current-voltage (I-V) characteristic. At room temperature carrier relaxation times are variable, but if we assume .1 psecs (equivalent to a GaAs mobility of 2500 cm^2/volt·sec) then $\omega\tau = 1$ at a frequency of ~ 1.5 THz. Quantum transport devices will be quantum transition devices in the terahertz part of the electromagnetic spectrum.

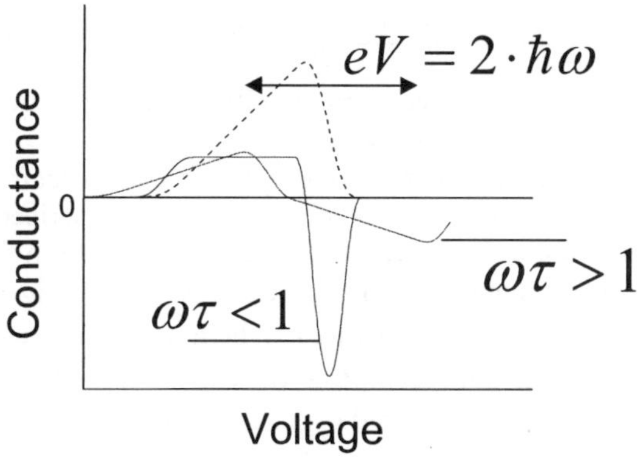

Fig. 4 The differential conductance at high and low frequencies.

In the following we will discuss quantum transport devices in the high frequency limit, $\omega\tau > 1$, using the device physics that was developed for superconducting electronics. We will review the various experiments that have exposed, in semiconductor quantum structures, the rich phenomena that had previously been the exclusive domain of superconducting electronics.

2. High Frequency Quantum Transport

Experiments on microwave induced changes in the I-V characteristics of superconducting weak links and tunnel junctions[6] stimulated the theory of quantum transport in high frequency fields developed by Tien and Gordon[7]. Here we apply it to resonant tunneling in semiconductor quantum structures.

A schematic diagram of the conduction band profile of a resonant tunneling diode is shown in Fig. 2 and the I-V characteristic in Fig. 3.[8] Ideally, current flows only when the quantum well state is aligned with filled states, or Fermi sea, in the emitter. Under electrical bias, as the quantum well state falls below the bottom of the conduction band the current falls to zero in a span of voltage that is related to the width of the quantum well state and various leverage arms in the device. This picture is ideal and real devices have extrinsic and intrinsic sources of leakage, which can carry current even when the device is biased beyond resonant tunneling.

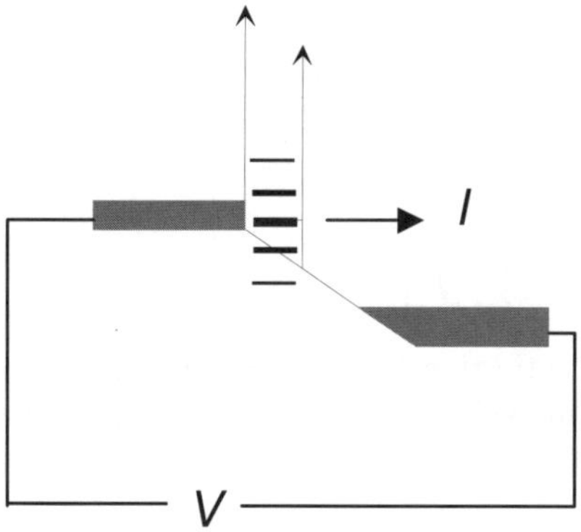

Fig. 5 In the presence of a high frequency field the quantum well state that provides the channel for resonant tunneling splits into a manifold of states each separated from its neighbor by $\hbar\omega$.

The differential conductance of the device is shown in Fig. 4. At low frequencies it is simply the $\frac{dI}{dV}$ of the static I-V shown in Fig. 3. At high frequencies, defined as $\omega\tau > 1$, the current-voltage characteristic and differential conductance is expected to be more interesting.

Following the original work of Tien and Gordon [7], we assume that the transport in the presence of a high frequency field can be described by resonant tunneling through a manifold of quantum well states each separated from the state of the quiescent structure by an integer number of photon energies, $n\hbar\omega$, Fig. 5. We assume that the resonant tunneling structure experiences a high frequency electric field given by $E\cos(\omega t)$ causing the quantum well state to oscillate up and down with respect to the emitter by $eaE\cos(\omega t)$. The manifold of states conserves state density; the weight of each is given by the squared Bessel function, $J_n^2\left(\frac{eaE}{\hbar\omega}\right)$ and $\sum_{n=-\infty}^{n=\infty} J_n^2\left(eaE/\hbar\omega\right) = 1$.

If the high frequency field is weak we need only consider the lowest order Bessel functions. Following Tucker [9] we find that the differential conductance is written as

$$G = \frac{I_{DC}\left(V + \hbar\omega/e\right) - I_{DC}\left(V - \hbar\omega/e\right)}{2\hbar\omega/e} \tag{1}$$

The differential $\frac{dI}{dV}$ is replaced by a finite difference. This is shown in Fig. 4. The most striking feature is the extension of the negative resistance or gain region to high voltage, where there is no feature in the static or quiescent I-V characteristic. But this can be simply understood as gain provided by a downward transition from filled emitter states to the empty quantum well states. The system, in principle, exhibits population inversion, as shown in Fig. 6.

But, it is difficult to measure the differential conductance or high frequency impedance/admittance of nano-structured devices. Rather, photon assisted quantum transport in semiconductor devices has been exposed by measuring changes in the DC I-V characteristic in the presence of strong high frequency fields. This is much the same approach used in the early work on superconducting junctions referred to above.[6]

The current voltage relation then becomes

$$I(V,E) = \sum_{n=-\infty}^{\infty} J_n^2\left(\frac{eaE}{\hbar\omega}\right) I\left(V - n\hbar\omega\right) \tag{2}$$

a is an effective length from the emitter to quantum well that "levers" the oscillatory motion of the quantum well.

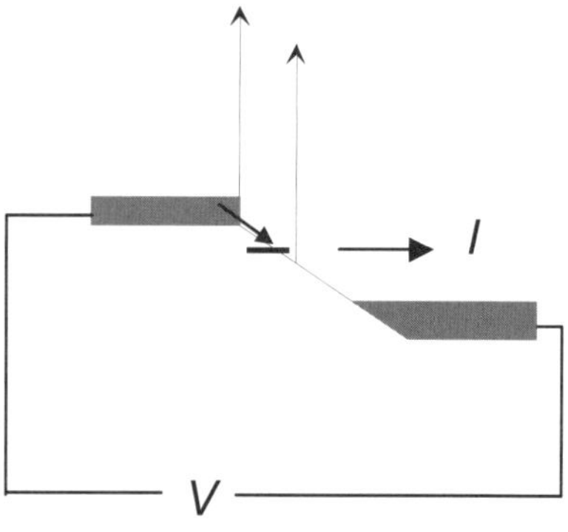

Fig. 6 The resonant tunneling diode is a gain medium when biased beyond current cut-off.

3. Photon Assisted Transport

One of the most striking examples of photon-assisted transport in semiconductor quantum structures appears in triple barrier, double quantum well structures.[10] Fig. 7 shows the appearance of the new channels

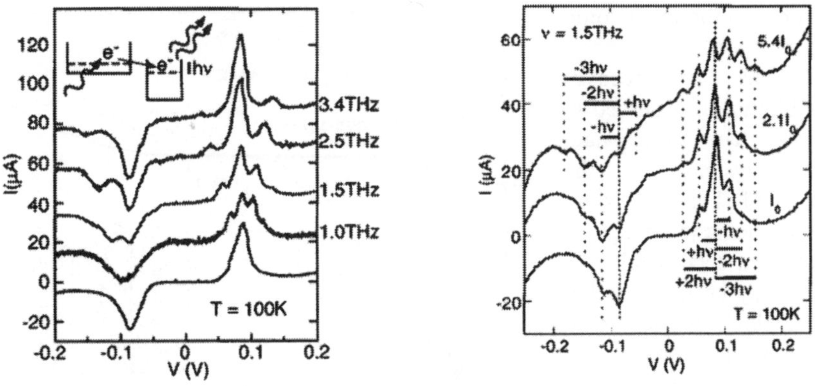

(a) (b)

Fig. 7 Photon assisted transport in a triple barrier / double quantum well resonant tunneling diode. (a) Dependence on terahertz frequency. (b) Appearance of new channels at high terahertz power.

for resonant tunneling with irradiation with terahertz radiation. In Fig. 7a, the displacement of the new features from the static peak increases with increasing frequency, as expected. In Fig. 7b, high order channels or multi-photon processes appear as the terahertz power or field strength is raised.[11]

Two conditions must be met in order to make these observations. First, the system needs to be driven at terahertz frequencies; this point we have made earlier. Second, to transfer appreciable state density into the photon side bands, the argument of the Bessel function needs to be roughly *n*, the order of the Bessel function. That is to say, $eaE/\hbar\omega \geq 1$.

Since $\hbar\omega \sim 2\ meV$ and $a \sim 10\ nm$, field strengths of the order of $E \sim 2000\ volts/cm$ are required. If this field were in free space it would correspond to ~ 100 watts/mm^2. Tunable sources at terahertz frequencies delivering 100's of watts are found only at Free Electron Laser facilities such as the Center for Terahertz Science and Technology, Santa Barbara.[12, 13]

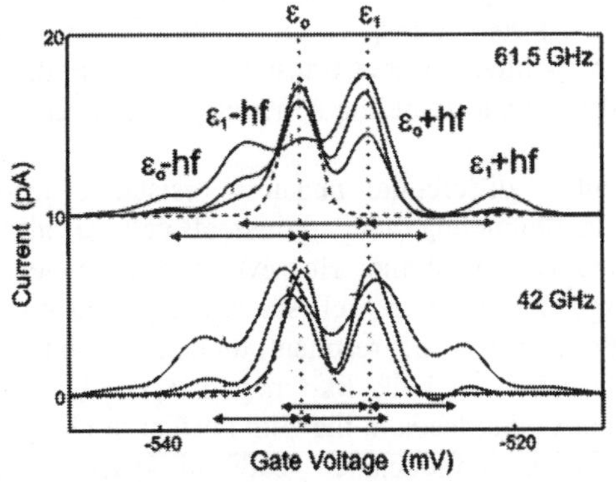

Fig. 8 Photon assisted transport through a quantum dot. Two channels are present marked ε_0 and ε_1. New channels appear when excited by microwave radiation. After Oosterkamp et al.[15].

On the other hand the terahertz regime is optimum. We can clearly dramatically improve the $\omega\tau > 1$ condition by increasing the frequency, but the high frequency field and required power will increase accordingly, as the frequency or frequency squared, respectively. The terahertz regime is the part of the spectrum where these features will emerge in semiconductor quantum structures and at field strengths that do not threaten to destroy the device. It is also the regime that would most benefit from new device concepts based on terahertz photon assisted transport.

Attempts to recover features like those observed in the triple barrier resonant tunneling system in the simpler double barrier system, described schematically above, have not been successful. A key issue may be the fact that in a double barrier/single quantum well system the emitter does not have well defined quantum levels, as is the case for the triple barrier system. Qing Hu and co-workers have addressed this issue.[14] They find in simulations that photon assisted features emerge in systems with extended states only if the radiation field itself is localized near the junction. However, features are recovered in lateral defined quantum point contacts. Some of the most striking photon assisted transport in laterally defined semiconductor quantum structures was demonstrated by Oosterkamp et al.[15] and shown in Fig. 8. These experiments were carried out at 200 mK and come to life at microwave frequencies due to the long relaxation times in the high mobility 2-dimensional electron gas.

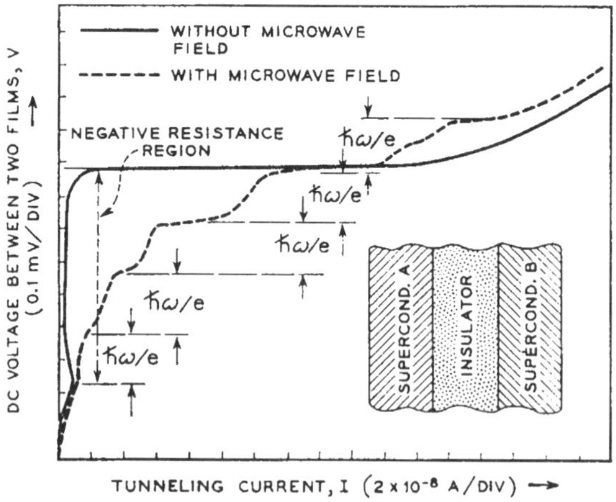

Fig. 9 Photon assisted quasi-particle tunneling in superconductor-insulator-superconductor junctions, Dayem and Martin[6].

For completeness we also display the original work of Dayem and Martin[6] on superconducting junctions. This is shown in Fig. 9.

4. Dynamic Localization and Absolute Negative Conductance

The fact that triple barrier/double quantum well systems display strong photon assisted features invites a discussion of photon-assisted transport in multi-quantum well superlattices. Indeed, the first evidence of photon assisted transport in semiconductor quantum structures appeared in sequential resonant tunneling superlattices and reported by Guimares, Keay et al.[16,17,18]

Transport through superlattices exhibits differential negative resistance whose origin is the same as that found in the resonant tunneling diode. For small voltages all the quantum well states are aligned and the conduction is ohmic. However, when the electric field in the superlattice is sufficiently large the quantum well states are pulled out of resonance and the current falls. However, unlike the resonant tunneling diode the superlattice is best described by a distributed or bulk like negative conductance, analogous to the negative conductance in GaAs that leads to the Gunn effect in GaAs. It is well known that bulk negative differential conductance leads to space charge instabilities [19,20], which in turn produce electric field inhomogeneities – electric field domains. As is the case in bulk GaAs, these domains can be static or dynamic and can lead to a rich variety of non-linear dynamics that have been the subject of much research. [21]

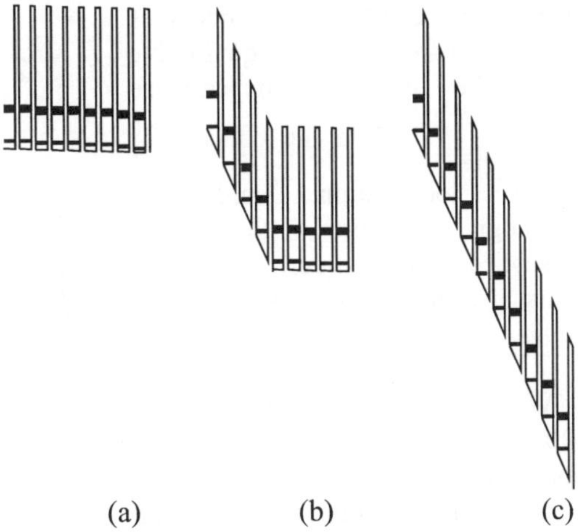

Fig. 10. Electric field distributions. (a) Low field. (b) Intermediate field with domains. (c) High field.

A typical I-V characteristic for a superlattice is shown in Fig. 11. This ten period superlattice exhibits ohmic behavior at low fields, domain formation at intermediate fields and resonant tunneling into excited states at high field. The different regimes are shown schematically in Fig. 10. In the intermediate field regime the current remains approximately constant while the increasing voltage across the sample is determined by the relative size of the high and low field domains. In fact, in Fig. 11, the ten discontinuities in the I-V characteristic mark the jump like motion of the domain wall separating high and low field. In the presence of intense terahertz radiation Keay, Galán et al.[17] discovered that the photon side bands could be sufficiently strong that domains could be stabilized by transport through a photon assisted channel.

But perhaps the most interesting behavior occurs near zero bias in strong terahertz electric fields. If the terahertz field is strong enough the current or more properly the conductance can be completely suppressed. This corresponds to terahertz electric field strengths such that the argument of the zeroth order Bessel function $eaE/\hbar\omega$ is equal to its first zero. This behavior is shown in Fig. 12.

As the terahertz field strength is increased the conductance drops as does the saturated current. But at the critical field, defined above, the conductance is actually negative. The current is pumped back into the voltage source.

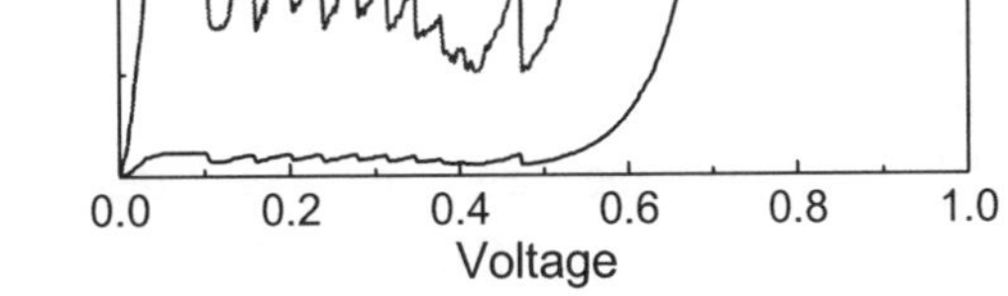

Fig. 11 I-V characteristic for a ten period superlattice. Upper trace is 10 x lower trace.

It is important to distinguish this phenomenon from rectification or photovoltaic effects. In these cases, some asymmetric feature in the I-V characteristic determines the photon induced current flow and its direction. It could flow in a direction that could oppose or aid the current flow from the voltage source. Here the current always flows into the voltage source, the current reverses itself as the applied voltage is reversed. This phenomenon is called *dynamic localization*. As the terahertz field is increased further the conductance becomes positive again following in a qualitative manner the expected dependence $J_0^2\left(eaE/\hbar\omega\right)$. While the concept of dynamic localization and its terminology implies that the conductance should be zero, we see in fact that it actually becomes *absolutely negative*.

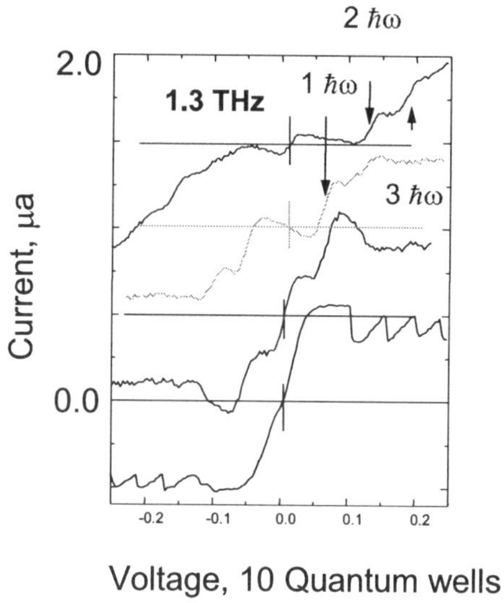

Fig. 12 I-V characteristic as a function terahertz field strength (1.3 THz).

When the system is dynamically localized, resonant tunneling into various photon side bands can still carry the current. The zero photon side band, or zero photon channel, is quenched. The electron can only tunnel or jump with emission or absorption of a photon. For small voltages, voltages less than that required to produce resonance with the first photon sideband, the electron absorbs a photon and climbs the Stark ladder producing absolute negative conductance. When the voltage drop is comparable or greater than the photon energy current reverses and flows "down hill" in the direction dictated by the voltage source. The cross over from negative to positive should occur when the voltage drop is equal to the terahertz photon energy. Fig. 13 shows the crossover from negative to positive current flow increasing with terahertz frequency.

In the presence of a terahertz field that produces absolute negative conductance the diode should be absolutely unstable. In other words, with no voltage applied, an open circuit, the negative conductance should cause the diode to charge itself, developing charge on the terminals and a voltage across the device until the current returns to zero, the frequency dependent cross over points shown in Fig. 13. Under these conditions the device will charge itself to a voltage proportional to the terahertz frequency. It would become a monolithic frequency to voltage converter.[22] The spontaneous charging does not happen here, presumably, because the experiments were carried out under constant voltage bias, which pins the device voltage.

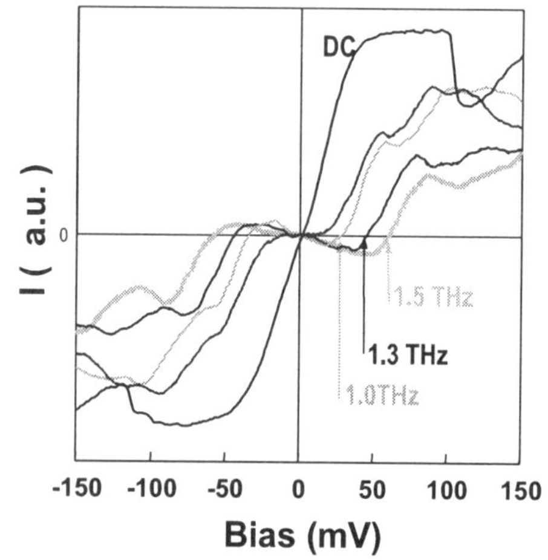

Fig. 13 The cross over from absolute negative current to positive is proportional to frequency.

The appearance of absolute negative conductance at a determined terahertz frequency and field amplitude has some important implications. In fact Dunlap et al. anticipated the above.[23] In their theoretical paper they envisioned a high mobility 2-dimendional electron gas subjected to an intense microwave field. There was no superlattice per se; rather the model considered the intrinsic periodicity of the GaAs. The required high frequency field strength is inversely proportional to the periodicity and comparing the 10 nm superlattice to the .27 nm lattice constant we see that this phenomenon could be recovered with field strengths ~ 40 times larger, scaling the high frequency incident power up by more than 10^3. In the engineered superlattice we have sacrificed scattering time and mobility but introduced a larger period that brings the fields to reasonable values. But access to the UCSB free-electron lasers is crucial here.

We should not lose sight of the fact that the superlattice system that produced the absolute negative conductance in Fig. 13 was definitely in the sequential limit. Transport occurs via motion, or jumping, up and down a Stark ladder with states localized in a given quantum well. The model developed by Dunlap et al was in the coherent band limit which for us would be a coherent miniband limit, originally proposed by Esaki and Tsu.[24] But a miniband superlattice that supports coherent transport will break up into a Stark ladder in the limit that Bloch oscillation is relevant. This Stark ladder in the miniband case will differ from the Stark ladder in the sequential resonant tunneling superlattice only by the fact that in the latter the states are localized to a quantum well whereas the Stark ladder in the Bloch oscillator limit may extend over several quantum wells. The qualitative comparison of the results in Fig. 13 with early discussion by Dunlap is still valid.

Further to this point, recent Monte Carlo simulations by Willenberg et al [25] have shown that the dynamical response or frequency dependent conductivity of a coherent miniband superlattice in a large DC electric field can be described either by Stark ladder states (a quantum mechanical approach) or by solving a Boltzmann transport model (quasi classical approach). While the latter should only be valid when the Stark splittings are much less than the miniband width, the solutions are in quantitative agreement even when the Stark ladder exceeds the miniband width. This is an important result, especially as we attempt to use a uniform Stark Ladder as a gain medium.

A superlattice device that is inserted in an effective open circuit, such as that envisioned in the frequency to voltage converter opens new degrees of freedom that have important consequences. Charge build up at the terminals is produced by the displacement of the electron plasma in the superlattice. A correct description of the open circuit dynamics at terahertz frequencies needs to include the restoring force for the plasma displacement and plasma oscillations play an important role in determining the overall dynamics. Campbell and colleagues[26] have explored the non-linear dynamics of an open circuited superlattice and find dissipative chaos for certain ranges of terahertz field strength and frequency. Most recently the model including plasma oscillations has been used to find stable self-bias points, readdressing the problem that Dunlap discussed, and the results are shown in Fig. 14.[27]

For frequencies approaching and exceeding the plasma frequency the semiconductor superlattice diode will self-bias at the appropriate values. This is not surprising; above the plasma frequency the plasmon dynamics play an ever-decreasing role. On the other hand, at low frequencies, where the dynamics of the space charge that is responsible for the self-bias is controlled by the plasma frequency the results are not clean. Presumably this regime is related to the early discussion by Alekseev et al. [26] on the dissipative chaos.

It must be also noted that as interesting as these results are, the models indicate that these effects will emerge in systems with little damping, typically $\omega\tau \geq 10$. To date the appearance of self-bias voltages in the presence of strong terahertz drive related to the drive frequency have not been observed experimentally. It should be apparent however that the superlattice systems that produced Fig. 13 are a good place to explore these effects.

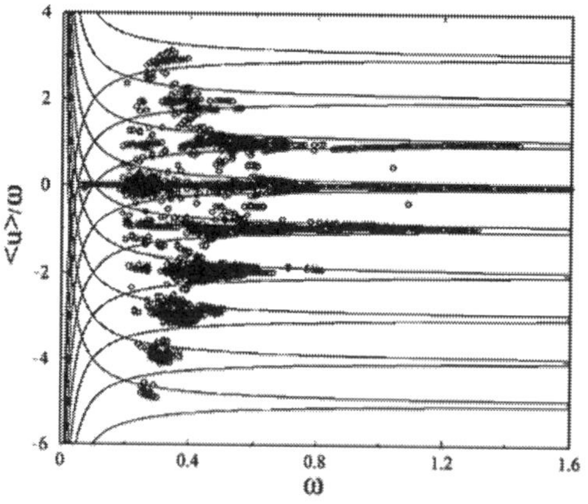

Fig. 14 Self bias voltage in units of $\hbar\omega/e$ vs applied frequency normalized by the plasma frequency. From Alekseev et al. [27]

5. Coherent Miniband Superlattices

In principle, transport in a superlattice should not depend in a qualitative way on the barrier height and width, and quantum well thickness. To be sure, for large barriers the "miniband" will become narrow, the miniband effective mass large and conductivity depressed. But in real systems if the tunneling rate is too slow, the miniband too narrow, the electron will lose its phase memory before it tunnels back or returns. These are conditions for sequential resonant tunneling; there is no coherent "miniband transport". While much of the theoretical discussion in the previous section concerned miniband transport in strong fields, the experiments concerned superlattices in the sequential resonant tunneling limit. Here we wish to explore terahertz photon assisted transport in superlattices that are in the miniband limit.

At first blush ones intuition might indicate that little can happen. We argued earlier that the drive frequency should exceed any quantum level broadening in order to capture photon side bands and photon assisted transport. If the drive frequency is less than the miniband width, we might assume that since the intrinsic width of these states exceeds the photon energy no sharp features will emerge. On the other hand, we also pointed out that in the presence of an electric field the miniband system will develop a Stark ladder, albeit with states with a spatial extension over several quantum wells, with widths set by various scattering mechanisms and not the tunneling rate or miniband width. The miniband

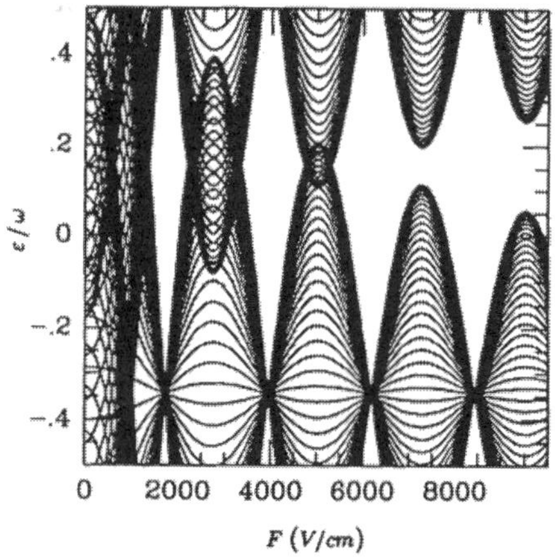

Fig. 15 The envelope of the lowest miniband in an extended Floquet picture vs terahertz field strength. Miniband width = 4 meV, photon energy = 1 meV. [28]

width only determines the spatial extent of the wavefunctions. Experimentally we expect the same qualitative behavior in miniband superlattices even when the photon energy is much less than the minibandwidth.

Holthaus[28] first modeled miniband collapse; results for a miniband width of 4 meV and photon energy of 1meV are shown in Fig. 15. As the terahertz field strength is increased the band of allowed states evolves in a complicated way but eventually collapses at the zeros of the zero order Bessel function, $J_0\left(eaE/\hbar\omega\right)$ just as it did for the sequential resonant tunneling case.

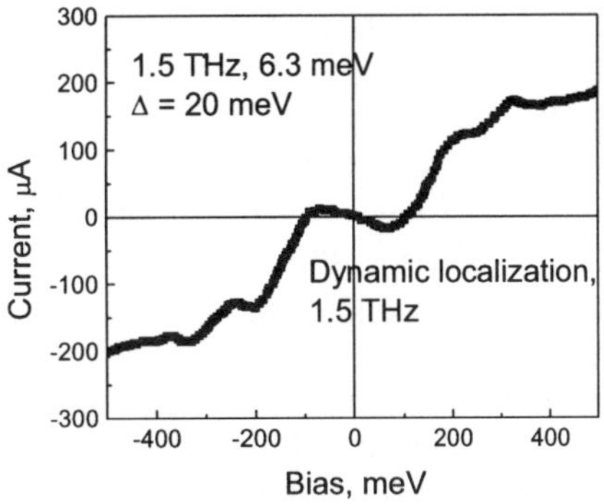

Fig. 16 Dynamic localization or miniband collapse for $\hbar\omega < \Delta$, the miniband width.

Experimentally, dynamic localization has been recovered in the miniband limit and it appears similar, in a qualitative way, to the phenomenon in the sequential tunneling limit. [29] Fig. 16 shows absolute negative conductance in a miniband superlattice with a bandwidth of 20 meV and a photon energy of 6.3 meV, approximately 1/3 of the minibandwidth. Also apparent, with the zero photon transport totally quenched, are features corresponding to one photon assisted transport. The photon assisted transport in the miniband limit resonates with Stark ladder transitions but can also be described as resonance with Bloch oscillations much like the early results of Unterrainer. [30]

6. Superlattice Devices as a Potential THz Gain Medium

A simple superlattice under electrical bias is expected to support gain without inversion (Fig. 17). If the number of quantum wells is sufficiently large we can ignore the contact or emitter and focus on the periodic quantum well structure. Further, if the level broadening caused by momentum and energy relaxation is less than the Stark splitting then we expect that the differential conductance of the electrically biased structure will exhibit gain for frequencies below the Stark splitting and loss for frequencies above.

Analytic expressions for the differential conductivity can be obtained following Ktitorov et al. [31], and we have for the differential conductivity

$$\sigma(\omega) = \sigma_0 \frac{1-\omega_B^2\tau_p\tau_e - i\omega\tau_e}{\left(\omega_B^2-\omega^2\right)\tau_p\tau_e + 1 - i\omega\left(\tau_p+\tau_e\right)}, \tag{3}$$

where τ_p and τ_e are the momentum and energy relation times respectively. σo is the conductivity in the absence of an applied electric field. As we remarked earlier, this result is in principle valid only for Stark splitting much less than the miniband width but Willenberg's Monte Carlo simulations [25] show that this result is correct even in the limit that the Stark splitting is much bigger than the miniband width.

The dynamical response is marked by negative conductance (gain) from DC to a frequency corresponding to the Stark splitting provided that $\omega_B \tau_p > 1$, Fig. 18. In this respect it behaves like the Gunn effect with the important distinction that the gain region is not limited by the energy relaxation time but rather the Bloch frequency. Equally important, the gain is resonant at frequencies just below the Bloch frequency. The superlattice under electrical bias provides a gain medium that can be tuned with the applied electric field.

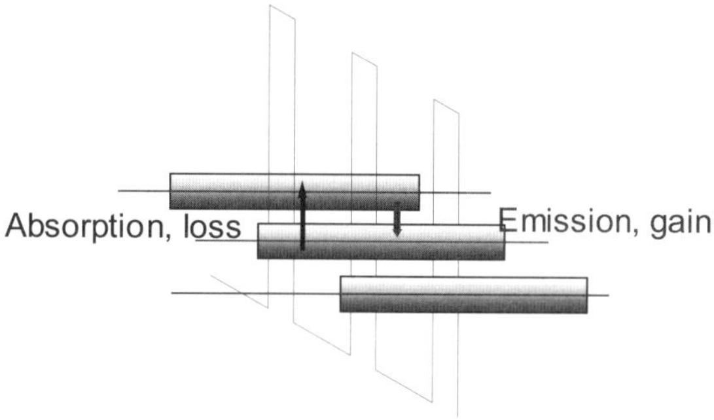

Fig. 17 An electrically biased superlattice can present gain at frequencies below and loss at frequencies above the Stark splitting.

One can develop some intuition by referring to Fig. 17. We first note that each rung of the Stark ladder should have equal numbers of electrons and there is no inversion. To zeroth order one would expect no interaction with the electromagnetic field. Indeed, it would appear that the sum rule or integrated real conductance in Fig. 18 would be zero. On the other hand, if we have line broadening as shown in Fig. 17 and thermal distribution of carriers in each state, as the figure implies, transitions at frequencies below the Stark splitting will occur between states that are inverted; there are more carriers in the upper than the lower. Transitions at frequencies above the Stark ladder splitting will occur between states that are not inverted. This leads to gain below the Stark ladder splitting and loss above as shown in Fig. 18.

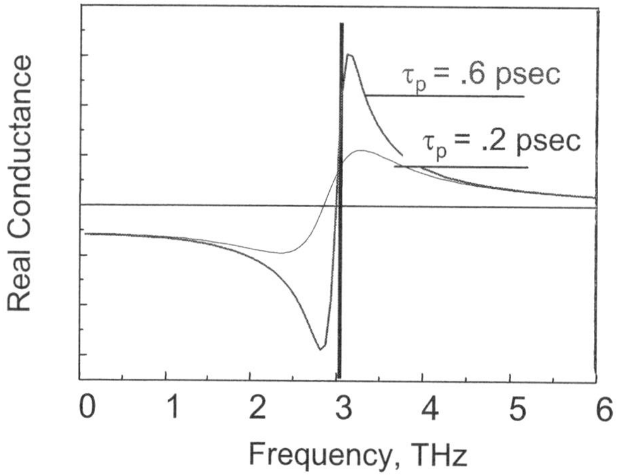

Fig. 18. The differential conductance for an electrically biased superlattice with Stark splitting of 3 THz and energy relaxation time of 1 psec.

It is worth noting that if $\omega_B \tau_p < 1$, then the gain region is limited to frequencies below the energy relaxation rate, $1/\tau_e$, and the behavior is indistinguishable from the Gunn effect.

Terahertz photon assisted transport is recovered by measuring the DC transport in the presence of an intense terahertz field. Experimentally, given access to a tunable free electron laser with relatively high power, this is relatively straightforward. One must simply measure changes in

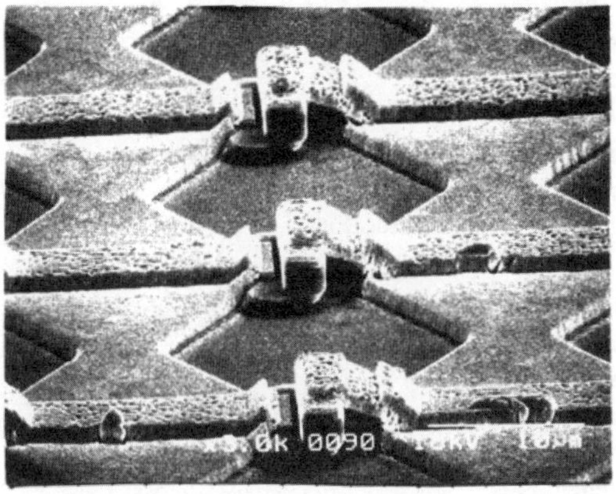

Fig. 19 Scanning electron micrograph of elements of a quasi-optical array of superlattices connected "head to foot".

DC current or the terahertz photoconductive response as a function of electrical bias, terahertz frequency or other relevant parameter. The measurement of the terahertz dynamical conductance or admittance, as shown in Fig. 18, is more difficult. This stems from the fact that the device is necessarily much smaller than a wavelength and a suitable terahertz "circuit" is required.

2-Dimensional quasi-optical arrays are an effective way to couple long wavelength radiation into micron size devices.[32,33] (Fig. 19, Fig. 20) If the wavelength of the radiation is greater than the periodicity of the array then the array presents a two-dimensional sheet admittance defined by the conductance of the array elements and all of the parasitic elements.

To enhance the sensitivity of the measurements we have placed the quasi-optical array in a terahertz cavity. The cavity-quasi-optical-array-superlattice system has the potential to be the basis of a terahertz oscillator if the electrically biased superlattice presents sufficient gain below the Stark ladder splitting. Here we wish to use it to measure the terahertz dynamical impedance, potential gain and loss, as a function of electrical bias.

The system is shown in Fig. 21. The cavity is defined by a highly reflecting inductive mesh on the back of the substrate and a spherical reflector that can be moved relative to the substrate. The active element, the superlattice loaded quasi-optical array, is located on the front of the substrate. In the experiments described below the quasi-optical array is actually a capacitive grid with a 20 μm period (Fig. 20).

Fig. 20 Scanning electron micrograph of a section of a 2 mm x 2 mm quasi-optical array loaded with GaAs /AlGaAs superlattices mesas.

Measurements of the dynamical conductance proceed by first measuring the transmission through the substrate with quasi-optical array and inductive grid, but without the external spherical reflector. The terahertz radiation from the UCSB free-electron lasers is tuned through the transmission Fabry-Perot resonance. At this frequency the external spherical reflector is inserted and the overall transmission maximized by tuning its separation from the substrate. Applying a voltage to the entire quasi-optical array then modulated the transmission through the tuned cavity system. The modulated fractional change in transmission can be projected on models that included the dynamical terahertz conductance of the superlattice mesas, the inherent parasitic elements of the quasi-optical array, and the elements of the terahertz cavity.

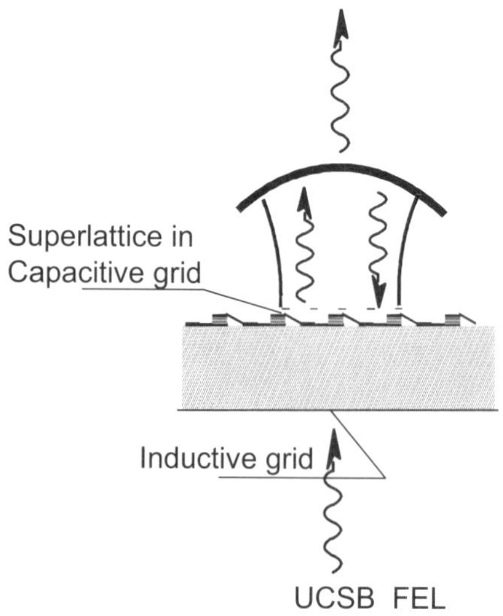

Fig. 21 A terahertz cavity is formed by an inductive grid on the backside of the substrate and a cylindrical metallic reflector. The superlattice is embedded in a quasi-optical array formed by a capacitive grid.

Using the UCSB Free-electron lasers measurements can be carried at out at frequencies from 330 GHz to 2.55 THz. We discuss only the extremes here. All experiments are carried out at room temperature.

The resonant transmission through the substrate without the external spherical reflector is shown in Fig. 22. The model reproduces the measured transmission quite well. The position of the resonance is a very sensitive function of the actual substrate thickness and effective real dielectric constant.

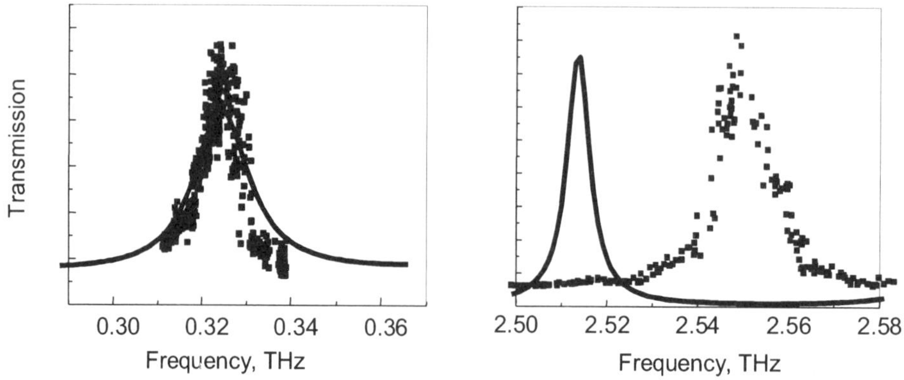

Fig. 22 Resonant transmission through the quasi-optical array / substrate / inductive grid Fabry-Perot around 300 GHz and around 2.5 THz. The solid lines are calculated with the model transmission. The failure of the model to find the correct transmission maximum for both frequencies with the same parameters is due to small errors in the effective thickness of the substrate or small changes in the real dielectric constant. The error in peak position is ~1% and of no significant consequence.

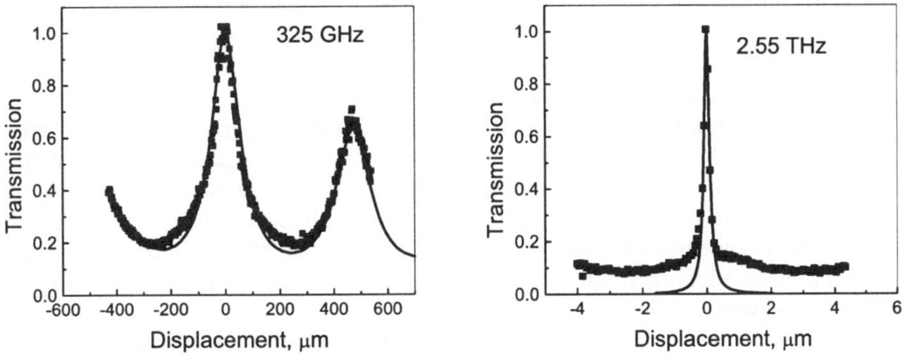

Fig. 23 Transmission through terahertz cavity as a function of displacement of spherical reflector. The solid lines are modeled by assuming an effective reflectivity for the spherical reflector.

The failure to exactly lie on top of the measured resonance is of no consequence. The width is more important and the model reproduces the measured values to within a factor of two.

With the free-electron laser tuned to the substrate resonance the external spherical mirror is then moved and the transmission through the cavity system measured (Fig. 23. Here we model the transmission/reflection of the spherical reflector (caused by diffraction around the periphery) by a reflection coefficient.

With the cavity and free-electron laser so tuned, the quasi-optical array is biased with an applied voltage and the change in transmission recorded. The results are shown in Fig. 25. At low frequencies, large, ~100%, changes in transmission are recorded while at the highest frequency the modulated transmission is only ~ 2%.

The current voltage characteristic of the quasi-optical array is displayed in Fig. 24 Since there are 100 elements in the array, the voltage drop per element is is 0.01 times that displayed on the voltage axis. Like wise, since the period of the array is 20 μm the current through an element of the array, mesa cross sectional area of 2x20 μm^2, is 0.01 times the current axis values. At large electric fields the current is expected to drop[24] and display the negative conductance evidenced at zero frequency in Fig. 18. The I-V characteristic is predicted to follow

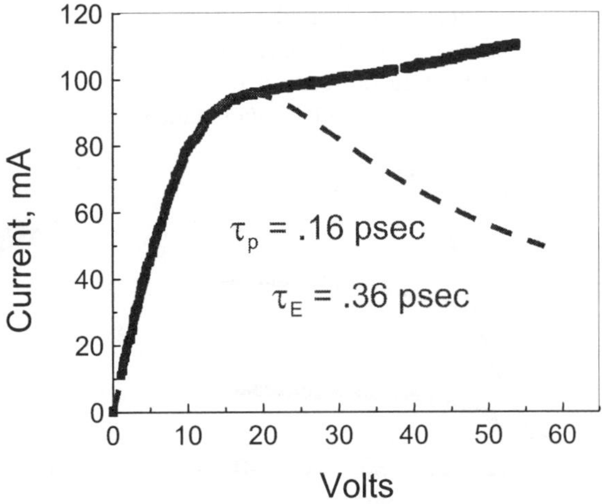

Fig. 24 Measured I-V characteristic, solid line. Modeled I-V characteristic assuming uniform electric fields, dashed line.

$$J = \frac{\sigma_{00} E}{1 + \omega_B^2 \tau_p \tau_e}. \tag{4}$$

The experimental I-V characteristic does not.

As was the case for the sequential resonant tunneling superlattice, these materials exhibit bulk differential negative resistance and are unstable against the formation of domains.[19,20] For a voltage controlled multi valued I-V these domains are electric field domains. As shown in Fig. 10, the high field domain is thought to be resonant tunneling from the ground state of a quantum well into the excited state of the neighboring well, the low field domain supports miniband transport at the point where the Stark ladder states just begin to separate.

The dashed curve in Fig. 24 is fit to the above expression for the nonlinear current vs voltage. τ_p can be determined by measuring the transverse magneto-resistance with an applied magnetic field along the current direction of the quasi-optical array (perpendicular to the current flow in the superlattice itself). τ_e is then extracted from the voltage at which the current saturates. At this point the current should remain roughly constant as the current is limited by the maximum current that can flow through the miniband and the additional voltage is dropped across the high voltage domain defined by ground to excited state tunneling. The experimental current is not precisely constant but is nearly so. We assume that the low voltage domain will shrink but present a terahertz conductance given by the model just as the superlattice develops zero conductance. The high voltage domain is assumed to be a terahertz short circuit.

We model the results for the modulated transmission as follows. In the first, we assume that the applied voltage produces a uniform field (dashed lines in Fig. 25.). In the second we assume that the superlattice breaks into high and low field domains at the point that the DC dynamical conductance is zero. The high field domain is assumed to present a terahertz short circuit.

Including the domain formation produces rough overall agreement reproducing the magnitude of the changes in transmission and the break in modulated transmission at the point where domains are formed.

It is instructive to examine the predicted conductivity versus frequency for the superlattice, assuming uniform electric field (Fig. 26). As the applied voltage exceeds that required for negative DC conductance, we expect relatively little change. At this

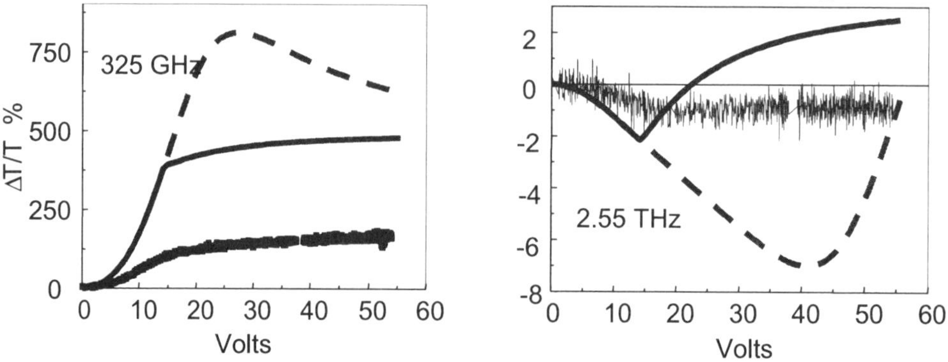

Fig. 25. Modulated cavity transmission at 325 GHz and 2.55THz. The dashed lines are modeled by assuming no electric field formation (uniform fields). The solid lines assume domain formation with dynamics as discussed in the text.

point the superlattice will break into two domains; one with an electric field at the critical point for development of negative resistance, the other with a field that will support positive conductance, presumably due to alignment of the staircase structure shown in Fig. 10. Below this critical field or voltage large changes in the low frequency transmission can be induced, but at high frequencies very small changes are expected, as shown experimentally in Fig. 25.

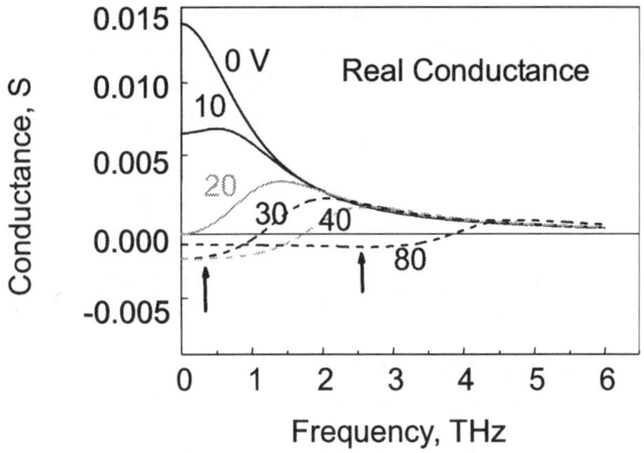

Fig. 26. Predicted changes in the conductance or admittance of the superlattice loaded quasi-optical array. The two arrows indicate the frequencies where data is displayed in Fig. 23 and Fig. 25.

There remains the fact that the fractional change in transmission at low frequencies differs by more than a factor of 2 from that predicted by the model. We believe that this discrepancy may be accounted for by including finite contact resistance between the metallization in the quasi-optical array and the superlattice itself.

It seems clear that a road to terahertz gain at room temperature is stabilization of a uniform field in the superlattice; suppression of the negative resistance at low frequencies while maintaining the high frequency gain.

7. Conclusions

At terahertz frequencies quantum structures exhibit the rich phenomena that has heretofore been the exclusive province of superconducting electronics. Photon assisted transport, dynamic localization and absolute negative differential conductance appear in multi quantum well superlattices as long as $\omega\tau > 1$. Terahertz quantum detectors and monolithic terahertz frequency to voltage conversion are potential applications.

The most important potential application is as a terahertz source based on Bloch oscillation in a uniformly biased superlattice. The multi-quantum well superlattice is a logical extension of a resonant tunneling diode but provides intrinsic power combing, an essential feature of any solid-state terahertz device that will deliver useful power.

We propose to use this voltage tunable gain in an oscillator by integrating superlattice mesa defined diodes into a quasi-optical array. This provides further power combining. The quasi-optical array is then inserted into a terahertz cavity defined by an inductive grid on the back of the substrate and mechanically tuned, millimeter size spherical reflector.

Using the tunable terahertz radiation from the UCSB Free-electron lasers we measure the transmission through the terahertz cavity, loaded with the active quasi-optical array, as voltage is applied to the superlattice. Large changes in transmission are measured at low frequencies while substantially smaller changes are recorded above

~2 THz. At present we are able to satisfactorily model the terahertz dynamical response of the complete system if we include electric field domain formation. A critical feature and a major obstacle to success is the formation of electrical field domains just as the superlattice system begins to develop useful gain. Suppressing domain formation is essential to its future application.

References

1. C. Gmachl, F. Capasso, D. L, Sivco and A. Y. Cho, "Recent progress in quantum cascade lasers and applications", Rep. Prog. Phys. **64**, 1533 (2001).

2. M. Rochat, L. Ajili, H. Willenberg, J. Faist, H. Beere, G. Davies, E. Linfield and D. Ritchie, "Low threshold terahertz quantum cascade lasers", submitted to Appl. Phys. Lett.

3. "Terahertz Sources and Systems", Proceedings of the NATO Advanced Research Workshop, Château de Bonas, June 2000, edited by R.E. Miles P. Harrison and D. Lippens, Kluwer, Netherlands (2001).

4. E. R. Brown, J. R. Söderström, C. D. Parker, L. J. Mahoney, K. M. Molvar, T.C. McGill, "Oscillations up to 712 GHz in InAs/AISb resonant-tunneling diodes", Appl. Phys. Lett. **58**, 2291 (1991).

5. For the resonant tunneling diode the relevant broadening is probably the voltage width of the negative resistance region in the current-voltage (I-V) characteristic.

6. A. H. Dayem and R. J. Martin, "Quantum Interaction of Microwave Radiation with Tunneling Between Superconductors", Phys. Rev. Lett. **8**, 246 (1962).

7. P. K. Tien and J. P. Gordon, "Multiphoton Process Observed in the Interaction of Microwave Fields with the Tunneling between Superconductor Films", Phys. Rev. **129**, 647 (1963)

8. T. C. L. G. Sollner, W. D. Goodhue, P. E. Tannenwald, C. D. Parker, and D. D. Peck, "Resonant tunneling through quantum wells at frequencies up to 2.5 THz", Appl. Phys. Lett., **43**, 588 (1983).

9. J.R. Tucker, "Quantum limited detection in tunnel junction mixers", IEEE J. Quant. Electr., **QE-15,** 1234-1258 (1979).

10. H. Drexler, J. S. Scott, S. J. Allen, K. L. Campman, and A. C. Gossard, "Photon-assisted tunneling in a resonant tunneling diode: Stimulated emission and absorption in the THz range", Appl. Phys. Lett. **67**, 2816 (1995).

11 Note that a fraction of the applied voltage appears as a voltage drop between the two quantum wells. As a result the voltage of the new channel is controlled by the terahertz frequency but appears at a voltage that is displaced from the static peak by several times $n\hbar\omega$.

12. G. Ramian, "The new UCSB free-electron lasers", Nucl. Instr. & Meth. in Phys. Research, **A318**, 225 (1992).

13. S.J. Allen, K. Craig,, C.L. Felix, P. Guimaraes, J.N. Heyman, J.P. Kaminski, B.J. Keay, B.J., A.G. Markelz, , G. Ramian, , J.S Scott, ., M.S Sherwin, ., K.L. Campman, , P.F. Hopkins, , A.C. Gossard, , D. Chow, ,M. Lui, ., T.Y. Liu, **"**Probing tetrahertz dynamics in semiconductor nanostructures with the UCSB free-electron lasers", J. of Lumin., **60-61**, 250 (1994).

14. Qing Hu, S Verghese, R A Wyss, Th. Schäpers, J. del Alamo, S. Feng, K. Yakubo, M. J. Rooks, M. R. Melloch and A Förster, "High-frequency (f~THz) studies of quantum-effect devices", Semicond. Sci. Technol. 11, 1888 (1996).

15. T.H. Oosterkamp, L.P. Kouwenhoven, A.E.A. Koolen, N.C. van der Vaart and C.J.P.M. Harmans, "Photon sidebands of the ground state and the first excited state of a quantum dot", Phys. Rev. Lett., 78 1536 (1997).

16. P.S.S. Guimaraes, B.J. Keay, J.P. Kaminiski, S.J. Allen, Jr., P.F. Hopkins, A.C. Gossard, L.T. Florez, J.P. Harbison, "Photon-mediated sequential resonant tunneling in intense terahertz electric fields", Phys. Rev. Lett. **70**, 3792 (1993).

17. B.J. Keay, S.J. Allen, Jr., J. Galan, J.P. Kaminki, K.L. Campman, A.C. Gossard, U. Bhattacharya, M.J.W. Rodwell, "Photon-assisted electric field domains and multiphoton-assisted tunneling in semiconductor superlattices", Phys. Rev. Lett. **75**, 4098 (1995).

18. B. J. Keay, S. Zeuner, and S. J. Allen, Jr., K. D. Maranowski, A. C. Gossard, U. Bhattacharya and M. J. W. Rodwell, "Dynamic Localization, Absolute Negative Conductance, and Stimulated, Multiphoton Emission in Sequential Resonant Tunneling Semiconductor Superlattices", Phys. Rev. Lett. 75, 4102–4105 (1995).

19. W. Shockley, "Negative resistance arising from transit time in semiconductor diodes", Bell Syst. Tech. J. **33**, 799 (1954). 19.

20. H. Kroemer, "Generalized Proof of Shockley's Positive Conductance Theorem", Proc. IEEE 58, 1844 (1970).

21. *Semiconductor superlattices : growth and electronic properties*, ed. by H.T. Grahn, World Scientific, Singapore ,1995.

22. It is apparent that the device would not be very accurate. The cross over point is slightly below the expected value due to the finite width of the quantum well states, which is reflected in the width and finite conductance of the photon assisted resonances.

23. D.H. Dunlap, V. Kovanis, R.V. Duncan, J. Simmons, "Frequency-to-voltage converter based on Bloch oscillations in a capacitively coupled GaAs-Ga/sub x/Al/sub 1-x/As quantum well", Phys. Rev. **B48**, 7975 (1993).

24. L. Esaki, and R. Tsu, "Superlattice and negative differential conductivity in semiconductors", *IBM J. Res. Dev.* **40**, 61 (1970).

25. H. Willenberg, private communication.

26. K.N. Alekseev, G.P. Berman, D.K Campbell, E.H. Cannon and M.C Cargo, "Dissipative chaos in semiconductor superlattices", Phys. Rev. **B54**, 10625 (1996).

27. K.N. Alekseev, E.H. Cannon, F.V. Kusmartsev and D.K. Campbell, "Fractional and unquantized dc voltage generation in THz-driven semiconductor superlattices", Europhys. Lett. **56**, 842 (2001).

28 M. Holthaus, "Collapse of Minibands in Far-Infrared Irradiated Superlattices", Phys. Rev. Lett., **69**, 351 (1992).

29. S. Zeuner, unpublished.

30. K. Unterrainer, ;B.J. Keay, M.C. Wanke, S.J. Allen, D. Leonard, G. Medeiros-Ribeiro, U. Bhattacharya, M.J.W. Rodwell, **"**Inverse Bloch oscillator: strong terahertz-photocurrent resonances at the Bloch frequency"**, .** Phys Rev. **76**, 2973 (1996).

31. S.A. Ktitorov, G.S. Simin, V.Y. and Sindalovski, "Bragg reflections and the high-frequency conductivity of an electronic solid-state plasma", *Fizika Tverdogo Tela* **13**, 2230-2233 (1971), Soviet physics – Solid State **13**, 1872-1874 (1972).

32. D.P. Neikirk, D.B. Rutledge, M.S. Muha, H. Park, and Yu Chang-Xuan, "Far-infrared imaging antenna arrays", *Appl. Phys. Lett.* **40**, 203 **(**1982) .

33. S.J. Allen, Jr.; C.L. Allyn, H.M. Cox, F. DeRosa, **"**Dispersion of the saturated current in GaAs from DC to 1200 GHz", G.E. Mahoney, Appl. Phys. Lett., **42**, 96 (1983).

International Journal of High Speed Electronics and Systems
Vol. 13, No. 4 (2003) 1149–1253

Advanced Theory of Instability in Tunneling Nanostructures

D.L. Woolard
US Army Research Office, P.O. Box 12211
Research Triangle Park, North Carolina 27709, USA
Dwight.Woolard@us.army.mil

H.L. Cui
Department of Physics and Engineering Physics, Stevens Institute of Technology
Hoboken, New Jersey 07030, USA
hcui@stevens-tech.edu

B.L. Gelmont
Department of Electrical Engineering, University of Virginia
Charlotteseville, Virginia 22903, USA
gb7k@virginia.edu

F.A. Buot
Naval Research Laboratory
Washington, DC 20375, USA
buot@estd.nrl.navy.mil

P. Zhao
Department of Electrical and Computer Engineering, North Carolina State University
Durham, North Carolina 27695, USA
pzhao@eos.ncsu.edu

This work is concerned with the quantum structure of resonant tunneling diodes, which exhibits intrinsic instability that can be exploited for the development of high-speed, high-frequency devices. The article examines in detail the physics underlying the nonliner instability, in both a one-band model and a multiple-band model. The theoretical basis of the description of electronic processes in such structures are described in some detail in terms of nonequilibrium Green's functions. Also presented here is a semi-phenomenological model of the resonant tunneling diode based on nonlinear circuit theory. Recent works and progresses in this and related areas are summarized here as well.

Keywords: Terahertz Radiation; High-speed electronics; Resonant Tunneling Devices.

1. Introduction

The physics and modeling of electron transport in semiconductor materials and devices has long been an important subject of study. Indeed, this is an important area of research both from a point of view of fundamental science and from the impact it has on device development and applications. Electron transport processes determine the basic principles of operation for many semiconductor-based devices. More importantly, they dictate the ultimate performance of almost all devices that are important to the area of very high-speed electronics. Hence, advanced theoretical techniques for the study of high-speed and high-frequency physical processes have always played an important role. For example, the previous advances in solid-state devices that utilize electron transport control and modulation mechanisms have had a revolutionary impact to electronic technology at microwave frequencies. Specifically, the invention of the solid-state transistor by Shockley and his associates and the subsequent development of such devices as bipolar junction transistors (BJT's), field-effect transistors (FET's) and metal-oxide-semiconductor field-effect transistors (MOSFET's) have forever altered high-frequency electronics. Of course, this is a direct result of semiconductor-based device superiority in functionality, reliability, miniaturization, and cost-effect integration as compared to tube-based devices for lower-power applications. Indeed, semiconductor-based transistors can collectively provide high-efficiency and low-noise operation in the lower portion of the millimeter-wave regime (i.e., millimeter-wave frequency band 30 - 300 GHz). However, conventional three-terminal semiconductor devices possess natural limitations due to their fundamental structural implementation that prohibit extension of the operation frequencies into the submillimeter-wave domain. (i.e., submillimeter-wave frequency band 300 - 3000 GHz). Hence, two-terminal semiconductor devices have emerged as the key technology for the generation, amplification and detection of electrical signals at millimeter-wave and submillimeter-wave frequencies.[1] However, the overall performance of even state-of-the-art two-terminal technologies suffers as they are extended for operation high into the so-called terahertz frequency regime (i.e., 300 GHz to 10 THz).

In fact, the last research frontier in high-frequency electronics now lies in the terahertz (or submillimeter-wave) regime between microwaves and the infrared (i.e., 0.3 - 10.0 THz). While the terahertz (THz) frequency regime offers many technical advantages (e.g., wider bandwidth, improved spatial resolution, compactness), the solid-state electronics capability within the THz frequency regime remains extremely limited from a basic signal source and systems perspective (i.e., μ watts output power levels). Historically, this limited development results from the confluence of two fundamental factors. First, extremely challenging engineering problems exist in this region where wavelength is on the order of component size. Second, the practical and scientific applications of this shorter-wavelength microwave region have been restricted in the past to a few specialized fields (e.g., molecular spectroscopy). However as we enter the new millennium, important applications of THz technology are rapidly emerging that are extremely relevant to national defense. For example, the strong absorption of electromagnetic energy above 300 GHz by atmospheric molecules is very prohibitive to communications. On the other hand, this same fundamental interaction mechanism suggests THz electronics as a potential tool for the identification and interrogation of chemical and biological

(CB) agents. Furthermore, the recent proliferation of CB agents as instruments of warfare and terrorism has led the Department of Defense to rank the development of early-warning systems for biological, and then chemical, as the highest priorities. At the same time, recent research has identified physical mechanisms within the THz regime that suggest advantages for the sensing of both chemical and biological agents.

Therefore, there are strong new motivations for developing THz-frequency systems and achieving this goal will require dramatic improvements to the solid-state source technology. While significant improvements in very high frequency signal source performance can be expected from the continued advancement of conventional technological approaches (e.g., diode multipliers and photomixers) it is wise to look to for novel approaches to this problem. One potential avenue for realizing this goal is through innovations provided by nano-scale device concepts and nanotechnology. In general, there are great expectations within the electronics community for improvements to functionality, speed and integration as semiconductor device dimensions approach the nano-scale. Indeed, the progression towards nanoelectronics and molecular electronics seems to suggest entirely new device paradigms (e.g., spin electronics) which are introduced by fundamental quantum mechanical effects. However, to fully realize the potential innovations suggested by nanoelectronics it will be necessary to overcome a number of basic theoretical challenges. For example, the ultra-small dimensions of nanoelectronics naturally introduce a large number of potentially performance limiting/enhancing effects (e.g., ballistic transport, nonequilibrium processes, dissipative phonon scattering, coulomb blockade, active molecular interconnects, etc.) and a broad control and utilization of these mechanisms will require an enhanced level of physical insight.

If semiconductor-based nanotechnology is to offer novel methods for the realization of THz-frequency sources then it will most probably involve the utilization of fundamental instability mechanisms at the mesoscopic scale (i.e., between the microscopic and macroscopic). Of course, an instability process that arises exclusively from some ultra-small physical interaction can be expected to yield very fast response times. Furthermore, if this hypothetical instability mechanism could be engineered in such a fashion that it was only weakly influenced by macroscopic effects then extremely high frequency oscillatory behavior would be the result. As heterostructures are the primary building blocks of semiconductor-based nanoelectronics, it is natural to first consider time-dependent electron-tunneling processes within one-dimensional nanostructures as a catalyst for very high frequency autonomous oscillations. As nano-scale structures exhibit extremely fast response times (e.g., picoseconds in resonant tunneling diodes) and allow for the selective control of electronic coupling and charge transport processes, they provide a potential approach for the inducement of intrinsic instabilities. Hence, tunneling nanostructures may provide an avenue for engineering a mesoscopic-type instability mechanism that can induce self-oscillations.

The accurate study of instability processes in nanostructures present new and formidable theoretical challenges. While all electron transport involves the macroscopic characterization of a time-dependent charging processes, it is often possible to incorporate the underlying quantum mechanical effects through a quasi-classical approach. Here, near equilibrium approximations and statistical averaging on time-stationary quantum mechanical states are usually employed. However, a complete and rigorous study of instability processes in nanostructures will require a detailed consideration and time-dependent quantum mechanical effects and nonequilibrium phenomenon (e.g., multi-band transport). In the sections that follow, the subject of instability processes in nanostructures will be considered. Specifically, advanced theoretical approaches will be developed and applied towards the study of one-dimensional tunneling devices that possess the potential for exhibiting self-oscillations. Section II will first consider intrinsic oscillation processes within double-barrier quantum-well structures. Section III presents a duality theory that suggests self-oscillations may be induced within staggered-bandgap heterostructures. Section IV presents a multi-band physics-based model that can be accurately applied towards tunneling studies in staggered-bandgap heterostructures. Finally, Section V will present a Green's function formalism for rigorously incorporating quantum dissipation into highly nonequilibrium and time-dependent transport processes within single- and multi-band based devices.

2. Intrinsic Oscillations in Tunneling Structures

2.1. *Introduction*

The search for compact solid-state based, high-frequency power sources have been an important research subject for many years.[2,3,4,5] Since the end of 1980's, resonant tunneling diodes (RTD) have been treated as possible high frequency power sources.[6,7,8] However, as it is well known, the traditional implementation of a RTD has not been successful as a power source at terahertz (THz) frequency.[9,10,11,12] Indeed, the output power of a RTD is on the order of μ watts at operation frequencies near 1 THz.[11] This failing is contributed by the extrinsic design manner of the oscillator that utilizes external circuit elements to induce the oscillation. This failing of the "traditional" RTD-based oscillator is tied directly to the physical principles associated with its implementations. In fact, the f^{-2} law points out it is impossible to get higher output power at tearahertz frequencies for a single device utilized in an extrinsic design manner.[5] In contrast to the extrinsic design of RTD oscillators, the intrinsic design of RTD oscillators makes use of the microscope instability of RTDs directly.[9,13] This type of an approach will avoid the drawbacks associated with the extrinsic implementation of RTD's. It is believed that if the dynamics surrounding the intrinsic oscillation can be understand and controlled, RTD sources based on the self-oscillation process should yield milliwatt levels of power in the THz regime.[9] However, the exact origin of the intrinsic high-frequency current oscillation has not yet been fully established. The transport dynamics in RTD's is governed by the quantum mechanical tunneling process that occurs through a quantum-well that is formed by a double-barrier heterostructure. The lack of knowledge related to the origin of the intrinsic instabilities in double-barrier quantum-well structures (DBQWSs) directly hampers realizing an optimal design (device and circuit) of a RTD-based oscillator.[10] Thus, it is extremely important to understand the creation mechanism of the intrinsic instability in DBQWSs.

Historically, Ricco and Azbel suggested in their qualitative arguments, that intrinsic oscillation exists in a double barrier structure for the case of one-dimensional transport.[14] Their theory attributed the instability to a process that cycled in and out of resonance. Specifically, when the energy of the incoming electrons matched the resonance energy, the tunneling current then charged the potential well and lifted its bottom, thus driving the system away from resonance. The ensuing current decrease (i.e., associated with the off-resonance) then reduced the charge in the well, bringing the system back to resonance, and a new cycle of oscillation commences. According to such a theory, there should been current oscillation at the resonance bias. However, numerical simulation results contradict this simple theory.[13,15,16]

In another theory, it was theorized that the nonlinear feedback, caused by stored charges in the quantum well, was responsible for the creation of the current oscillation.[17] However, their phenomenological theory can not explain why the nonlinear feedback caused by stored charge in the quantum-well at bias voltages lower than those associated with resonance does not lead to current oscillation.[18] In subsequent studies of RTDs, Jensen and Buot observed intrinsic oscillations in their numerical simulations of DBQWSs.[15] However, this initial work did not provide underlying explanations on the oscillation mechanism. Recently, Woolard et. al., suggested that the current oscillation might be caused by the charge fluctuation near the emitter barrier of RTD.[19] However, the cause of the charge oscillation and how the charge oscillation affects the electronic resonant tunneling were not made clear. Hence, the origin of intrinsic oscillation has eluded revelation for more than a decade. Furthermore, very high frequency electron dynamics in tunneling structures is of fundamental importance to nanoelectronics. Experimental investigations of similar time-dependent processes are also receiving more attention.[20] However, to date, there has not been a completely conclusive demonstration of intrinsic oscillations in RTD's. Hence, the development of an accurate fundamental theory that provides insight into the catalyst of the intrinsic oscillation is a key first step for the successful design of an RTD-based oscillator.

In earlier work, a new theory was presented that provided a basic idea for the origin of the intrinsic oscillation in a DBQWSs (hereafter refereed to as paper I).[18] This theory revealed that the current oscillation, hysteresis and plateau-like structure in I-V curve are closely related to the quantum mechanical wave/particle duality nature of the electrons. In addition, these effects were shown to be a direct consequences of the development and evolution of a dynamical emitter quantum well (EQW), and the ensuing coupling of the quasi-discrete energy levels that are shared between the EQW and the main quantum-well (MQW) formed by the DBQWS. Through this new understanding of the dynamical behavior of the RTD, it was possible to qualitatively predict the existence of an oscillation. However, while this initial description was able to self-consistently explain all the physical phenomena related to the intrinsic oscillation it could not provide quantitative design rules. This paper will extend the earlier theory through the application of basic quantum mechanical model. According to the fundamental theory of quantum mechanics, there are two equivalent methods for determining the energy levels of a coupled system if the system can be viewed as the combination of several sub-systems. The first method treats the subsystems separately initially and then models the interaction between the subsystems to get the combined energy-level structure of the entire system. The second deals with the system as a whole and obtains the energy-level structure by direction solution of Schrödinger's equation. The advantage of the later is that it can give an exact description of the system's energy subbands without the development of models for the sub-systems. This paper will develop multi-subband model for the describing the electron dynamics in DBQWSs. The multi-subband based theory will provide a relationship between the oscillation frequency and the energy-level structure of the system. A method for calculating the energy levels for an open quantum system is also presented. This subband model will be combined with time-dependent Wigner-Poisson simulation results to provide; (1) a quantitative explanation for the origin of the intrinsic oscillations in RTD's, and (2) a detailed design methodology for a future implementation and optimization of DBQWS-based THz oscillators.

This Section is organized as follows. In Subsection 2.2, the fundamental theory for the origin of the intrinsic instability in DBQWSs is developed. Subsection 2.3 will present numerical simulation techniques for generating the current oscillation behavior in RTD's (i.e., through the Wigner-Poisson Model) and a numerical algorithm for generating the multi-subband structure in DBQWSs (i.e., through the Schrödinger's equation model). Subsection 2.4 will use results from both numerical simulations to analysis the instability behavior and to establish underlying mechanisms influencing the intrinsic oscillations. In Subsection 2.5, general conclusions and design rules will be given.

2.2. *General Instability Theory*

Previously,[18] a qualitative explanation was given for the creation of intrinsic oscillations in a Double-Barrier Quantum-Well Structure (DBQWS). In this prior work, the oscillations were recognized to arise primarily from two interrelated processes. The first step being the creation of an emitter quantum-well (EQW) - i.e., in front of the first heterostructure barrier - which dynamically occurs just as the device is being biased into the negative differential resistance (NDR) regime. The intrinsic oscillations are then induced as a secondary result of the coupling between the EQW and the main quantum-well (MQW) that is defined by the DBQWS. Here, the creation, time-dependent fluctuation and subsequent disappearance of the EQW are key processes that determine the formation of the I-V characteristics and the intrinsic high frequency current oscillation. For example, Figs. 3 $\sim$ 5 shows the results of a simulation study on an emitter-engineered DBQWS and illustrate the dynamical behavior of the potential and current density profiles. As first revealed in another paper,[21] the intrinsic current-density oscillations result from very small quantum-based fluctuations in potential and are not driven by charge exchange between EQW and MQW. Here the potential variations within the EQW and MQW are completely in phase and this indicates that the oscillation is purely of quantum mechanical origin. Details on this phenomenon given later in the paper.

This analysis and physical interpretation of the intrinsic oscillation phenomena is directly supported by detailed numerical simulation,[18] however, it is important and insightful to offer a more fundamental and mathematically rigorous explanation for the physical argument. While it is clear that the defined energy coupling between EQW and MQW is crucial to the process, it is important to note that the previous statements are somewhat qualitative in nature. Specifically, while the EQW is the fundamental catalyst, there is a dynamical coupling between the spatially separated quantum wells that influences and ultimately determines the time-dependent energy levels of the entire system. Indeed, the time-dependent tunneling transport that occurs within the DBQWS is strictly a multi-subband physical process. In this paper, a simplified multi-band analysis will be used to derive a quantitative description of the intrinsic current oscillation behavior.

It is very important to note that the electron transport under study is occurring within a time-dependent quantum system with dissipation and that this system is subject to open boundary conditions. In this type of situation, strictly speaking, one must consider quasi-discrete electron transport where a density of available tunneling states exist across a continuous energy space. A rigorous analysis would require that a fully time-dependent quantum mechanical description be applied (e.g., Schrödinger's equation) to derive the peaks of the time-dependent tunneling probability which could then be used to predict the most probably energy states of the electron dynamics. In other words, the energy levels that we seeks to identify can not be rigorously derived as energy eigenvalues as this is not a proper eigenvalue problem - i.e., it is an open and dissipative system and one that is subject to instability. However, the quasi-discrete energy level structure can be estimated through an approximate analysis based upon the time-independent Schrödinger's equation. The justification of this approach can be derived as follows. As already stated, the transport problem under study will contain time-dependent potential energy profiles under some conditions (i.e., when intrinsic oscillations are present) and this profile may be written generally as $U(z,t) = U_0(z) + \Delta U(z,t)$ where the last term contains all the time dependency. As just noted, in this situation most of the electron transport will occur through a set of quasi-bound resonant energy levels - i.e., defined by the peaks in the transmission function. The wavefunction associated with each of these subbands may be modeled as

$$\Psi_k(z,t) = e^{\frac{-i}{\hbar}F_k(t)}\psi_k(z,t), \tag{1}$$

where $F_k(t) = \int_0^t dt' E_k(t')$ and $E_k(t)$ has been defined as a the real part of time-dependent quasi-bound energy state and $\psi_k(z,t)$ is the wavefunction amplitude. Note that the time-dependency in ψ_k from Eq (1) is introduced through dissipation effects (i.e., imaginary energy-state effects). The quasi-bound system is now described by

$$\hat{H}(z,t)\Psi_k(z,t) = i\hbar\frac{\partial\Psi_k(z,t)}{\partial t} = \frac{\partial F_k(t)}{\partial t}\Psi_k(z,t) + i\hbar\frac{\Psi_k(z,t)}{\psi_k(z,t)}\frac{\partial\psi_k(t)}{\partial t}. \tag{2}$$

In the limit of very small time-dependent potential variations, i.e., $\Delta U(z,t) \to 0$, we must have $\partial\psi_k(z,t)/\partial t \to 0$ which leads immediately to

$$\hat{H}(z,t)\Psi_k(z,t) = E_k(t)\Psi_k(z,t) \tag{3}$$

since $\frac{\partial F_k(t)}{\partial t} = E_k(t)$. Specifically, when the potential variations in time are sufficiently weak then the model coefficients (i.e., $\psi_k(z,t)$) are slowly varying. This allows one to approximate the quasi-bound energy levels using the time-independent Schrödinger's equation and to approximate the time-dependent wavefunction using Eq. (1). In this study, electron transport through a single-band system (i.e., conduction band) will be considered and the goal is to derive expressions for electron current and density in terms of the subband structure for a DBQWS. In this multiple energy-state semiconductor system, the wave-function for the electrons can be written generally as

$$\psi(z,t) = \sum_{k=1}^{N}\Psi_k(z,t). \tag{4}$$

Here, the previously derived model equation, $\Psi_k(z,t) = \psi_k(z,t)e^{\frac{-i}{\hbar}F_k(t)}$, will be utilized where $F_k(t) = \int_0^t dt' E_k(t')$ and $E_k(t)$ is the real-component of the energy associated with the kth energy level. Here, the wavefunction $\psi_k(z,t)$ is assumed independent of the $E_k(t)$. In fact, $\psi_k(z,t)$ can be viewed as the kth subband energy-level coefficient for the total wavefunction that incorporates effects of dissipative or energy gains induced either by scattering (e.g., electron-phonon) or the applied bias source, respectively. For these studies, time variations in $\psi_k(z,t)$ due to either internal dissipation or external energy gains are not on the order of those under consideration (i.e., 10^{12} Hertz). Specifically, dissipation effects will not lead to oscillatory behavior that persists and the externally applied biases are time-independent. The electron current and carrier densities can be easily derived from the subband wavefunctions as,

$$J = -\frac{i\hbar}{2m}\{\psi^*(x,t)\nabla\psi(x,t) - (\nabla\psi^*(x,t))\psi(x,t)\} \equiv <\psi|\hat{\mathrm{j}}|\psi> \tag{5}$$

$$\rho = |\psi|^2 \tag{6}$$

where $\hat{\mathrm{j}} = \frac{i\hbar}{m^*}(\nabla^+ - \nabla)$ is the current density operator. The last two equations explicitly demonstrate the importance of device energy-level structure in determining the time-dependent, or oscillatory, behavior, of electron transport through the system. Substituting Eq. (4) into Eqs. (5) and (4), yields

$$\begin{aligned} J(t) &= \, < \psi|\hat{\mathrm{j}}|\psi > \\ &= \sum_k < \psi_k(t)|\hat{\mathrm{j}}|\psi_k(t) > \\ &\quad + 2Re \sum_{k,l(l<k)} < \psi_k(t)|\hat{\mathrm{j}}|\psi_l(t) > e^{-i\frac{F_l(t)-F_k(t)}{\hbar}} \end{aligned} \tag{7}$$

and

$$\rho(t) = |\psi|^2 = \sum_k |\psi_k(t)|^2 + 2Re \sum_{k,l(l<k)} \psi_k^*(t)\psi_l(t)e^{-i\frac{F_l(t)-F_k(t)}{\hbar}} \tag{8}$$

The oscillation terms in the above equation will usually be smeared out by the cancellation effect induced by variations in phase (e.g., unequal subband structures leading to conditions such that $E_{l_1} - E_{k_1} \neq E_{l_2} - E_{k_2}$). When these conditions apply the transport can be described simply as an individual summation over single subbands. Hence, there is no coupling between the bands and no intrinsic oscillations. This is analogous to the more typical transport problem where interband coupling can be ignored and only the first terms in Eqs. (7) and (8) are necessary. In fact, it is this type of situation where the concept of band transport is most often applied and most useful. However, an accurate and useful characterization of intrinsic oscillation behavior must include effects resulting from multi-subband coupling. The time-dependent nature of the electron transport that may arise due to interband coupling can be characterized for various levels of oscillation strength. The relative strengths of the interband oscillations may be conveniently classified into three categories that will now be considered individually.

Case 1. Maximum Subband Coherence.

If we assume that conditions sufficient for intrinsic oscillations exist, the strongest oscillations (i.e., largest time-dependent variations in current density) will occur when all the energy-dependent phase factors in Eqs. (7) and (8) are equivalent and when all contribute to the instability. This condition of maximum coherence is directly defined by the resulting subband structure and is given by the relation

$$\Delta E(t) = \Delta E_{lk}(t) = |E_l(t) - E_k(t)| = const \qquad \text{and} \qquad l \in \{l_i\}, k \in \{k_j\} \tag{9}$$

where the sets $\{l_i\}$ and $\{k_j\}$ are of equal number and assume all possible values from the number sequences $1, 2, \cdots, n$ such with $(l_i < k_i)$. It is important to note that the phase factor defined in Eq. (9) must possess a natural time dependency. This is true because any condition of oscillation in current density will be accompanied by a corresponding oscillation in the band structure of the device. In turn, this will lead to time-dependent perturbations in the subbband structure and the general expression for the phase factor in Eq. (9). In this case, the current density and carrier density can be written as

$$\begin{aligned} J(t) &= <\psi|\hat{\mathrm{j}}|\psi> \\ &= \sum_k <\psi_k(t)|\hat{\mathrm{j}}|\psi_k(t)> \\ &+2Re\left\{ e^{-i\frac{\Delta F(t)}{\hbar}} \sum_{k,l(l<k)} <\psi_k(t)|\hat{\mathrm{j}}|\psi_l(t)> \right\} \end{aligned} \tag{10}$$

and

$$\rho(t) = |\psi|^2 = \sum_k |\psi_k(t)|^2 + 2Re\left\{ e^{-i\frac{\Delta F(t)}{\hbar}} \sum_{k,l(l<k)} \psi_k^*(t)\psi_l(t) \right\} \tag{11}$$

where $\Delta F(t) = \int_0^t dt' \Delta E(t')$. The superposition of all the in-phase contributions (i.e., second terms on RHS of Eqs. (10) and (11)) defines the maximum amplitudes for oscillations in current density and electron density. Recall, that the inner product terms are not capable of directly contributing to the *intrinsic* oscillation phenomenon as discussed earlier above. Note that while this condition of complete regularity in the subband structure leads to the strongest oscillations, it is also the most stringent and difficult to realize in practice.

Case 2. Partial Subband Coherence.

The next level of oscillation condition is characterized by the condition where a finite and countable number of energy-dependent phase factors in Eqs. (7) and (8) are equivalent and where each of these contribute to the instability. This condition of partial subband coherence is defined by

$$\Delta E(t) = \Delta E_{lk}(t) = |E_l - E_k| = const \quad \text{and} \quad l \in \{l_i\}, k \in \{k_j\} \tag{12}$$

where the sets $\{l_i\}$ and $\{k_j\}$ of equal number, with $(l_i < k_i)$, assume some of the values from the number sequences $1, 2, \cdots, n$. In this case, the current density and the carrier density can be expressed as

$$\begin{aligned} J(t) &= <\psi|\hat{\mathrm{j}}|\psi> \\ &= \sum_k <\psi_k(t)|\,\hat{\mathrm{j}}|\psi_k(t)> \\ &+2Re\left\{ e^{-i\frac{\Delta F(t)}{\hbar}} \sum_{k,l(l<k),l=l_i,k=k_j} <\psi_k(t)|\,\hat{\mathrm{j}}|\psi_l(t)> \right\} \\ &+ \sum_{k,l(l<k,l\neq l_i,k\neq k_j)} <\psi_k(t)|\hat{\mathrm{j}}|\psi_l(t)> e^{-i\frac{F_l-F_k}{\hbar}} \end{aligned} \tag{13}$$

and

$$\rho(t) = |\psi|^2 = \sum_k |\psi_k(t)|^2 + 2Re\left\{e^{-i\frac{\Delta F(t)}{\hbar}} \sum_{k,l(l<k),l=l_i,k=k_j} \psi_k^*(t)\psi_l(t)\right\}$$
$$+ \sum_{k,l(l<k,l\neq l_i,k\neq k_j)} \psi_k^*(t)\psi_l(t)e^{-i\frac{F_l-F_k}{\hbar}}. \quad (14)$$

Under these conditions, the in-phase contributions from only the second terms on the RHS of Eqs. (13) and (14) will contribute to the intrinsic oscillations as the last terms lack sufficient coherence.

Case. 3 Minimum Subband Coherence.

The last and weakest form of intrinsic oscillations is characterized by the condition where only a single set of subbands contributes to the instability. The energy difference of the subbands is

$$\Delta E(t) = |E_l - E_k| \quad (15)$$

where l and k can assume only one set of values from the energy level index $1, 2, \cdots, n$ and $l < k$. In this case, the current density and carrier density can be written as

$$J(t) = <\psi|\hat{\mathrm{j}}|\psi> = \sum_k <\psi_k(t)|\,\hat{\mathrm{j}}|\psi_k(t)> + 2Re\left\{e^{-i\frac{\Delta F(t)}{\hbar}} <\psi_{k_j}(t)|\hat{\mathrm{j}}|\psi_{l_i}(t)>\right\}$$
$$+ \sum_{k,l(l<k,l\neq l_i,k\neq k_j)} <\psi_k(t)|\hat{\mathrm{j}}|\psi_l(t)> e^{-i\frac{F_l-F_k}{\hbar}} \quad (16)$$

and

$$\rho(t) = |\psi|^2 = \sum_k |\psi_k(t)|^2 + 2Re\left\{e^{-i\frac{\Delta F(t)}{\hbar}t}\psi_{k_j}^*(t)\psi_{l_i}(t)\right\}$$
$$+ \sum_{k,l(l<k,l\neq l_i,k\neq k_j)} \psi_k^*(t)\psi_l(t)e^{-i\frac{F_l-F_k}{\hbar}t} \quad (17)$$

As in the prior case, the in-phase contributions from only the second terms on the RHS of Eqs. (16) and (17) will contribute to the intrinsic oscillations. For each of the three previous subband coherence cases, the current density and the carrier density can be expressed most generally as

$$J(t) = <\psi|\hat{\mathrm{j}}|\psi> = \sum_k <\psi_k(t)|\hat{\mathrm{j}}|\psi_k(t)> + 2Re\left\{e^{-i\frac{\Delta F(t)}{\hbar}}P(t)\right\} \quad (18)$$

and

$$\rho(t) = |\psi|^2 = \sum_k <\psi_k(t)|\psi_k(t)> + 2Re\left\{e^{-i\frac{\Delta F(t)}{\hbar}}G(t)\right\} \quad (19)$$

where the functions P and G are slowly-varying functions of time. Here, the incoherent subband terms in Eqs. (18) and (19) have been excluded. Also while the first terms have been retained, it should be noted that they only contribute to short term transients and to the final static components of current density and electron density.[13] P incorporates the effects from the subband coupling and actually represents the inter-subband current. Similarly, function G accounts for the contribution to the charge density from the coupling between the subbands. The expressions defined in Eqs. (18) and (19) reveal that intrinsic high frequency oscillations can arise in any quantum system from the wavefunction coupling between multi-subbands. The relative strength of that instability being specifically achieved once the subbands structure satisfies one of the criterion given in Eqs. (9), (12), or (9). Further analysis of the equations can provide a clear physical picture regarding the creation of the oscillation. The main key being the energy-dependent phase factors. It should be noted that prior simulations have shown that the self-consistent potential varies in a periodical form.[13] Thus, it is reasonable to express the energy difference as

$$\Delta E(t) = \Delta E_0 + f(\omega, t) \tag{20}$$

where $f(\omega, t)$ is a periodical function and ω is the oscillation frequency. Note that a complete analysis must also include the phase variations in P and G and these will be discussed later in this paper. ΔE_0 is defined as the average energy difference between two energy levels for a system subject to intrinsic oscillations, or the energy difference at the balance point. Obviously, the phase differences in Eqs. (18) and (9) between time t_1 and t_2 can be written as

$$\Delta\phi(t_1, t_2) = \frac{1}{\hbar}(\Delta F(t_2) - \Delta F(t_1)) = \frac{\Delta E_0}{\hbar}(t_2 - t_1) + \frac{1}{\hbar}\int_{t_1}^{t_2} dt' f(\omega, t') \tag{21}$$

Recognizing that the phase variation in one period is 2π, the oscillation frequency of electron current, due to the subband structure is given by

$$\frac{1}{T} = \frac{\Delta E_0}{h} \tag{22}$$

where T is the period of the intrinsic oscillation. The previous derivation establishes a physics-based description for the creation of the intrinsic oscillation. An accurate physical model for this instability process will be able to describe the time-dependent variations in electron density and potential energy. Consider for example an arbitrary oscillation process and assume that the density of electrons at a particular real-space point reaches its maximum value at t_0. The corresponding potential energy at this same space point will also assume its maximum value at time t_0 since we are considering the Poisson-based interaction potential. Assuming an oscillation condition exists, this variation in both electron density and electron potential energy will cycle periodically as the phase varies over 2π. The model equations in (15) and (16) for electron current-density and electron density directly exhibit this type of behavior through the energy-dependent phase factors. In turn, this model would impose variations in the potential energy through the application of Poisson's equation. Most importantly, the feedback influence of potential energy variations on the energy-dependent phase-factor (i.e., defined in Eqs. (20) - (22)) has been incorporated into the analysis.

This quantum-based model allows one to interrogate the intrinsic oscillation process to determine the underlying physical mechanisms responsible for the instability. Specifically, if detailed simulations are utilized to derive values for the subband structure and the appropriate ΔE_0 under the condition of intrinsic oscillation then insight into the fundamental catalysts can be obtained. Furthermore, as will be shown later in this paper, this information can be used to predict methods for enhancing the oscillation strength in quantum-well systems. The next two subsections of this section will present simulation tools and studies that allow for a complete analysis of the intrinsic oscillations in DBQWSs. In particular, Wigner-Possion Equations simulations and Schrödinger-based simulations will be used together to derive current density oscillations, the subband phase factor, and the subband wavefunction amplitudes. This information will be combined with the previous model to reveal the fundamental origins of intrinsic oscillations and will be used to predict structural modifications that lead to enhanced instabilities in arbitrary quantum-well systems.

2.3. *Numerical Techniques*

2.3.1. *Wigner-Poisson Model*

The Wigner function formulation of quantum mechanics was selected for this study because of its many useful characteristics for the simulation of quantum-effect electronic devices, including the natural ability to handle dissipate and open-boundary systems. The Wigner function equation was first employed in quantum device simulation by Frensley.[22] Later, Kluksdahl et al incorporated Poisson's equation (PE) and applied the model to the study of RTD with self-consistent potentials.[23] The Wigner function equation (WFE) can be derived in several ways.[24] Since the Wigner function may be defined by nonequilibrium Green's functions, the WFE may be derived from the equation of motion of the nonequilibrium Green's function.[25,26] With the lowest order approximation to scattering being considered only, we have

$$\frac{\partial f}{\partial t} = -\frac{hk}{2\pi m^*}\frac{\partial f}{\partial x}$$
$$-\frac{1}{h}\int dk' f(x,k')\int dy[U(x+y)-U(x-y)]sin[2y(k-k')] + \left.\frac{\partial f}{\partial t}\right|_{coll}, \quad (23)$$

where h is the Planck's constant, m^* is the electron effective mass and U is the conduction-band-edge. The appropriate treatment of scattering in semiconductors is very important for getting accurate transport results. Recent research has shown that the computation burden associated with a detailed consideration of electron-phonon scattering is very formidable. Typical computer CPU times required for the calculation of one point in the I-V curve are the order of 30 hours on a 100 CPU-Cray T3E machine.[27] This huge amount of computation times would severely impend a study such as the one presented here if the electron scattering was modeled from first principles. Thus, the relaxation time approximation to scattering has been employed in this paper. In terms of the relaxation time approximation to scattering, the collision terms in the above equation may be written as [25]

$$\left.\frac{\partial f}{\partial t}\right|_{coll} = \frac{1}{\tau}\left[\frac{f_0(x,k)}{\int dk f_0(x,k)}\int dk f(x,k) - f(x,k)\right], \tag{24}$$

where τ is the relaxation time and f_0 is the equilibrium Wigner function. Details regarding the physical model used for deriving τ is given in Appendix A. The device under study is subject to open boundaries at the emitter and collector, hence, the boundary conditions in the Wigner function is

$$f_{x=0,k>0} = \frac{4\pi m^* k_b T}{h^2} ln\left\{1 + exp\left[-\frac{1}{k_B T}(\frac{h^2k^2}{8\pi^2 m^*} - \mu_0)\right]\right\}, \tag{25}$$

$$f_{x=L,k<0} = \frac{4\pi m^* k_b T}{h^2} ln\left\{1 + exp\left[-\frac{1}{k_B T}(\frac{h^2k^2}{8\pi^2 m^*} - \mu_L)\right]\right\}. \tag{26}$$

The second equation used in the RTD model is the Poisson equation (PE)

$$\frac{d^2}{dx^2}u(x) = \frac{q^2}{\epsilon}[N_d(x) - n(x)], \tag{27}$$

where ϵ is the dielectric permittivity, $u(x)$ is the electrostatic potential, q is the electronic charge, $N_d(x)$ is the concentration of ionized dopants, and $n(x)$ is the density of electrons, given by

$$n(x) = \int_{-\infty}^{\infty} \frac{dk}{2\pi} f(x,k). \tag{28}$$

The current density may be written as

$$j(x) = \int_{-\infty}^{+\infty} \frac{dk}{2\pi}\frac{\hbar k}{m^*} f(x,k). \tag{29}$$

To solve the WFE-PE equations, we must discretize these two equations over the simulation domain. For the one-dimensional transport case considered here, the discretization of PE's is trivial. Therefore, only the discretization of the WFE will be discussed here. Details of this procedure are well described by Jensen and Buot.[15] Hence, only a summary of the method is given here. Assuming the simulation box is between $x = 0$ and $x = L$, the domain may be discretized as follows.

$$f(x,k) = f(x_i, k_j) = f_{ij} \tag{30}$$

$$x_i = (i-1)L/(N_x - 1) = (i-1)\delta x, \qquad \delta x = L/(N_x - 1), \tag{31}$$

$$k_j = (2j - N - 1)\delta k/2, \qquad \delta k = \pi/N\delta x. \tag{32}$$

where N_x and N are the number of x and k points on a grid in phase space. The time-dependent Wigner function equation can be written as

$$\frac{\partial f}{\partial t} = \frac{\mathcal{L}}{i\hbar} f \tag{33}$$

where

$$\mathcal{L} = i(T + V + S). \tag{34}$$

In the above equation, T, V, and S are the drift, potential, and scattering terms, respectively. Using a second-order upwind difference scheme to discrete the position derivative, they can be expressed as

$$\mathbf{T} \cdot f(x, K) = -[\frac{\hbar^2 \delta k}{2m^* \delta x}](2j - N - 1)\Delta^{\pm} f(i, j), \tag{35}$$

$$\Delta^{\pm} f(i, j) = \pm\frac{1}{2}[-3f(i) + 4f(i \pm 1) - f(i \pm 2)], \tag{36}$$

$$\mathbf{V} \cdot f = \sum_{j'=1}^{N} V(i, j - j')f(i, j), \tag{37}$$

$$V(i, j) = \frac{2}{N} \sum_{i'=1}^{N/2} sin[\frac{2\pi}{N} i'j][U(i + i') - U(i - i')], \tag{38}$$

and

$$\mathbf{S} \cdot f = \frac{\hbar}{\tau} \left\{ f(i, j) - \frac{\delta k f_0(i, j)}{2\pi\rho(i)} \sum_{j'=1}^{N} f(i, j') \right\}. \tag{39}$$

The discretized density of electrons and current density may be written, respectively, as

$$n(i) = \frac{\delta k}{2\pi} \sum_{j=1}^{N} f(i, j), \tag{40}$$

and

$$J(i + \frac{1}{2}) = \frac{\hbar \delta k}{8\pi^2 m^*} \sum_{j=1}^{N} k_j \begin{cases} 3f(i + 1, j) - f(i + 2, j), & j \leq \frac{1}{2}N \\ 3f(i, j) - f(i - 1, j), & j > \frac{1}{2}N \end{cases} \tag{41}$$

The formal solution of the Eq. (30) is

$$f(t + \Delta t) = e^{-\frac{i\mathcal{L}}{\hbar}t} f(t) = \frac{1 - \frac{i\mathcal{L}}{2\hbar}t}{1 + \frac{i\mathcal{L}}{2\hbar}t} \tag{42}$$

This equation may be written as

$$[-r + \mathcal{L}][f(t + \Delta t) + f(t)] = -2rf(t) \tag{43}$$

where $r = \frac{2\hbar}{\Delta t}$. In descreting the above equation, the boundary condition information has been incorporated into the drift term. The boundary condition does not change with time. Thus, we have

$$[-r + \tilde{\mathcal{L}}][f(t + \Delta t) + f(t)] = -2rf(t) + 2BC \tag{44}$$

where $\tilde{\mathcal{L}}$ is the same operator as defined by Eq. (30) except that the boundary conditions on Wigner function, i.e., now denoted as BC, have been factored out.

In the discretization of Eq. (24) and (27), the dielectric function of the material and the effective mass of electrons is taken to be constant throughout the structure. In these studies, the conduction band profile is first approximated by a square well potential and $n(x)$ is derived through the use of Eqs. (23) and (28). Subsequently, this value of electron density is substituted into the Poisson equation and then the new conduction band profile $U(z) = u(z) + \Delta_c(z)$ is calculated, where $\Delta_c(z)$ is the offset of the band edge. Using this new conduction band profile at next time step, the Wigner function equation is solved again. This iteration continues until steady-state or a per-assigned time value is achieved.

In order to ensure the convergence of the numerical simulation results, we have employed a very small time step in our simulation. The time step is $1fs$. Furthermore, we have tested several structures with different momentum and position-space descretizations and different simulation boxes to ensure that accurate simulation results were obtained. Details on these results are presented elsewhere.[18] It should also be noted that the numerical technique adapted in this paper is a well-established approach and has been widely used. The accuracy of this approach has also been verified before within similar simulations. [15,18,23,28]

2.3.2. *Multi-subband Model*

As will be shown, the results obtained from the Wigner-Poisson transport simulations (see Subsection 2.4) yield intrinsic current-density oscillations and a time-dependent potential energy profile. This implies that the energy level structure must change with the time variation of the double-barrier tunneling structure. Furthermore, if one energy subband exists to conduct the current before the creation of the emitter quantum well (EQW), there will be at least two energy levels available for passing the current after the creation of the EQW. According to our basic theory, the key to understanding the intrinsic oscillation process is contained within the dynamics of the subband structure after the EQW is created. Hence, direct information regarding the energy structure must be generated to reveal the underlying mechanisms.

Arguments were already given earlier in Subsection 2.2 justifying the use of a quasi-bound subband description of the electron transport for cases of time- dependent potentials with small variations in amplitude. This allows one to approximate the position of the quasi-bound energy levels using the time- independent Schrödinger's equation even if the potential energy profile of the DBQWS is time-dependent. This approach will now be applied to determine the energy subband structure for an open transport system. These non-resonant scattering states will be determined using a numerical approach similar to that used by Frensley and Liu et al.[29,30] However, the specific approach utilized here introduces strictly open boundary conditions to the problem. In discrete form, the Schrödinger's equation can be written as

$$-\frac{\hbar^2}{2m^*\Delta^2}\psi_{i-1} + \left(\frac{\hbar^2}{m^*\Delta^2} + V(x_i) - E\right)\psi_i - \frac{\hbar^2}{2m^*\Delta^2}\psi_{i+1} = 0 \tag{45}$$

where Δ is the spatial mesh spacing which is assumed constant eveywhere. Note that this formula ignores changes in effective mass across hetero-interfaces. Outside of the device boundaries, the incident, reflected and transmitted wavefunctions can be modeled as

$$\psi_{in} = e^{-ik_0x} + re^{ik_0x} \qquad \text{for} \qquad x \leq 0 \tag{46}$$

$$\psi_{out} = te^{ik_nx} \qquad \text{for} \qquad x \geq l \tag{47}$$

At the boundaries, one may enforce the conditions

$$\psi = \psi_{in} \qquad \text{and} \qquad \frac{1}{m^*}\frac{\partial\psi}{\partial x} = \frac{1}{m^*}\frac{\partial\psi_{in}}{\partial x} \qquad \text{for} \qquad x \leq 0 \tag{48}$$

and

$$\psi = \psi_{out} \qquad \text{and} \qquad \frac{1}{m^*}\frac{\partial\psi}{\partial x} = \frac{1}{m^*}\frac{\partial\psi_{out}}{\partial x} \qquad \text{for} \qquad x > l \tag{49}$$

The discrete equations at the boundaries may be defined using Eqs. (46)- (49) as

$$(ik_0\Delta - 1)\psi_1 + \psi_2 = 2ik_0\Delta r, \tag{50}$$

$$-\psi_{n-1} + (1 - ik_n\Delta)\psi_n = 0. \tag{51}$$

Defining that

$$s_j = \frac{\hbar^2}{2m^*\Delta^2}, \tag{52}$$

and

$$d_j = -\frac{\hbar^2}{m^*\Delta^2} + V(x_j), \tag{53}$$

and leads to the discrete equations on the interior points of

$$-s_j\psi_{j-1} + (d_j - E)\psi_j - s_{j+1}\psi_{j+1} = 0. \tag{54}$$

The previous discrete system of equations may be used to determine the subbands upon application of the appropriate boundary conditions. The equations that were defined earlier for transport outside the boundaries and that were applied to derive the discrete boundary relations have implicitly defined an open system - i.e., as there is an incident flux. Also, the system is derived for arbitrary levels of reflection at the input boundary (i.e., $x = 0$ point). Under these general conditions it is not directly possible to define the allowable energy levels for unknown values of reflection. Fortunately, as will be discussed in detail later, the subbands of interest in this study admit either very small or zero levels of reflection. Specifically, this study considers the DBQWS under conditions of large dc biases. Hence, the highest energy levels are approximately resonant and suffer very little reflection. Furthermore, the lower energy states are bound from the left by the emitter barrier and must have

$r = 0$ to admit physically plausible solutions (i.e., decaying to the left). Hence, for this study the reflection coefficient, r is approximated to be zero. This allows the discrete system to be written in the matrix form $G\psi = 0$ as follows

$$(ik_0\Delta - 1)\psi_1 + \psi_2 = 0 \tag{55}$$

$$-s_j\psi_{j-1} + (d_j - E)\psi_j - s_{j+1}\psi_{j+1} = 0 \qquad \text{for} \quad j \geq 2, \quad j \leq n-1 \tag{56}$$

$$-\psi_{n-1} + (1 - ik_n\Delta)\psi_n = 0 \tag{57}$$

The previous E-dependent system can now be used to determine the allowable energy states of the system. The sufficient and necessary condition that the above equations have solutions is that the coefficient determinant of the equations is equal to zero. The coefficient determinant can be recursively expanded as

$$f_1 = ik_0\Delta - 1 \tag{58}$$

$$f_2 = (d_2 - E) + s_2 \tag{59}$$

$$f_j = (d_j - E)f_{j-1} = s_3^2 f_{j-2} \qquad \text{for} \qquad 3 \leq j \leq n-1 \tag{60}$$

$$f_n = (1 - ik_n\Delta)f_{n-1} - s_n f_{n-2} \tag{61}$$

By setting the recursive determinant to zero, $f_n = 0$, the equation for determining the energy levels of the open system is obtained. Substituting the calculated energy levels into Eqs. (55) to (57), allows for the calculation of the non-normalized wave-functions. The introduction of the zero-reflection condition on the input boundary does introduce at least a small error into the subband estimation. However, two further points are noteworthy. First, the zero-reflection condition does not directly affect the accuracy of the lower energy-states (i.e., those with energy less than the conduction band energy at $x = 0$) calculation. Indeed, they are only perturbed through the estimates on the total charge in the emitter region that in turn affects the depth of the EQW. Second, the zero-reflection condition should be much more accurate than invoking a closed system condition (i.e., incident wave equal to zero) that should significantly shift the energy levels upward.

2.4. *Simulation Results and Instability Analysis*

Jensen and Buot first numerically observed the intrinsic oscillations.[15] Thereafter, Biegel and Plummer obtained very similar results upon analysis of the same double-barrier structure.[16] This identical double-barrier RTD was also previously investigated within this study and high-frequency intrinsic oscillations were also observed. These results were reported earlier in publication (i.e., Zhao et al, 2001) hereafter referred to, as Paper I. It is important to note that these prior investigations utilized uniform doping within the emitter region. Since as will be shown here, the emitter plays a crucial role in the origin of the intrinsic oscillations, these earlier works did not adequately probe the underlying physical mechanisms. In fact, very recent studies have shown that emitter-engineered structures can be used to enhance the amplitude of the intrinsic oscillation in Resonant tunneling diodes (RTD's).[21] In this device structure, an alternating doping profile was utilized to modify the conduction band at the electron injection point and to create a transport condition more conducive to intrinsic oscillations. Our simulation results show that the emitter-engineered region favors of the creation of the oscillations in the following aspects. First, the modified emitter induces an artificial barrier in front of the GaAs/AlGaAs heterojunction barrier thereby directly facilitating the formation of a shallow emitter quantum well (EQW) in front of the first barrier. As discussed in paper I, the creation of an EQW is the necessary condition for the creation of the intrinsic current oscillation. Hence, a shallow EQW exists in advance and contributes to the quantum-well inducing interference-depletion mechanism, which is a definite advantage. Second, this duel-barrier structure (i.e., now formed from the built-in barrier at the emitter contact and the first RTD barrier) within the emitter region prohibits the propagation of reflected waves back into the open-boundary reservoir. This influence tends to increase the amplitude of the electron wave in the emitter region and favors the further enhancement of the interference of the electron waves in the emitter thereby leading to a deeper EQW.

The analysis of the intrinsic instability phenomenon presented in this paper will utilize the emitter-engineered structure from reference.[21] The modified GaAs/AlGaAs double-barrier quantum-well system (DBQWS) has the following symmetric structure: total emitter length of 270 Åwith a 30 Åspacer, 30 Åbarriers, 50 Åquantum well, and total collector length of 270 Åwith a 30 Åspacer. The emitter-engineered system has 270 Ålong emitter region of the form (from left to right): 30 Ådoped 10^{18} cm^{-3}, 50 Ådoped 10^{18} cm^{-3}, 160 Ådoped 10^{18} cm^{-3}. The quantum well region is undoped and the collector is symmetrically doped with the emitter. The heterostructure barrier heights are 0.3 eV - corresponding to the GaAs / $Al_{0.3}Ga_{0.7}As$ interfaces. The effective mass of electron is assumed to be a constant and equals to $0.0667m_0$; and the device temperature is 77 K. It should be noted that a relatively low temperature must be employed to allow for electron interference within the emitter and the subsequent formation of the EQW. Bulk GaAs parameters are used to calculate the relaxation time and the chemical potential, and the compensation ration for scattering calculations is 0.3. The chemical potential is determined by $\int_0^\infty \sqrt{\epsilon} f(\epsilon) d\epsilon = \frac{2}{3}\mu(T=0)^{3/2}$, where $f(\epsilon)$ is the Fermi distribution function. The numerical simulator utilizes discretizations for momentum and position space of 98 and 82 points, respectively.

2.4.1. *Simulation Results*

Let us first consider simulation results generated from the Wigner-Poisson model that was presented in Subsection 2.3.1. Fig. 1 shows the average I-V characteristics for the emitter-engineered DBQWS. The I-V characteristic exhibits the well-known Z-like feature, which have been observed often in experimental measurement. It should be noted that instability in the current densities are present within the bias

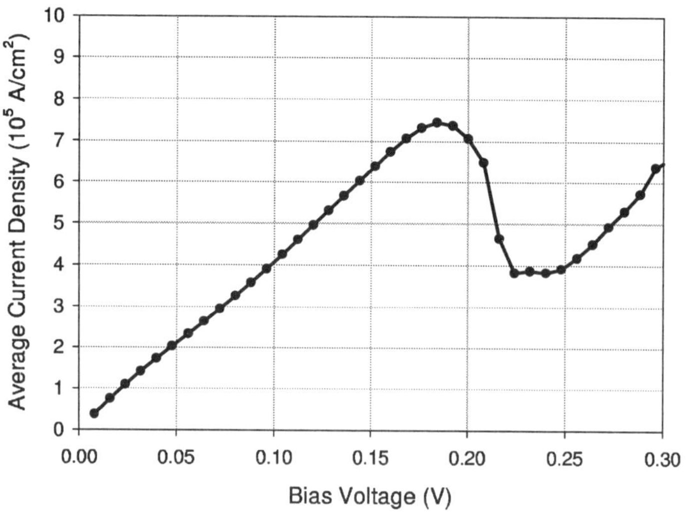

Fig. 1. The current-voltage (I-V) characteristics of the double barrier quantum well structure (DBQWS) considered in this study.

region 0.224 V to 0.240 V. Fig. 2 plots the current density as a function of time for several values of applied bias. These simulation results exhibit the following important features. First, there is bias voltage window (BVW), defined from 0.224 V to 0.240 V, in which the current demonstrates intrinsic oscillations. Second, in the vicinity of the bias voltage point 0.224 V, the current oscillations are stable (i.e., oscillatory and nondecaying). In our simulation, we have chosen the total simulation time to be 1500 ps. Completely stable oscillations were only observed within a small neighborhood around the bias point 0.224 V. Fig. 3 to 5 show time-dependent self-consistent potential and electron density at the bias voltages 0.200V, 0.224V, and 0.280V, respectively. Collectively, these figures give the potential profiles and electron density distributions in the BVW and outside the BVW. From these figures it is apparent that the current oscillation is concurrent with oscillations of potential and electron density in the whole region of the device. The oscillations have the following noteworthy features. Before the bias reaches the BVW region, the potential and electron density oscillate irregularly for a short time before settling into a stable state. This oscillatory features, which span across a significant region of bias, offers strong evidence against Ricco and Azbel's theory on the origin of the current oscillation in double barrier systems. As these oscillations are do not correspond to a bias point just before the onset of negative differential resistance. In the BVW, all the nonperiodic transients of the potential and the electron density cease after a

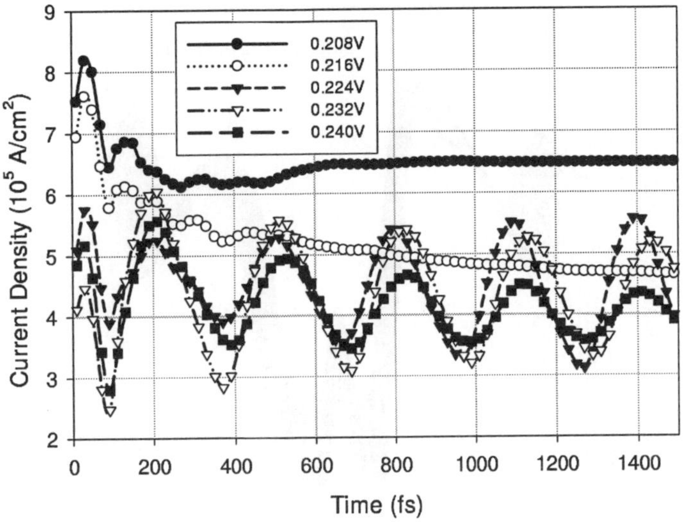

Fig. 2. Current density as a function of time for the DBQWS over the bias voltage window of 0.208V ~ 0.240 V. The current oscillation period at 0.224 V is 300 fs.

hundred femtoseconds. Furthermore, all oscillations are significantly damped outside the BVW. The time dependent results for potential profile and electron density can be used to reveal some of the underlying causes for the intrinsic oscillations in DBQWS's.

The simulation results show that upon entering the BVW an EQW is observed to form in front of the first barrier structure. According to our qualitative and time-dependent quantum-energy-level-coupling model, previously given in Paper I, the fundamental origin of the intrinsic oscillation can be understood on the following basis. After the bias voltage passes the resonant point, the sudden increase in the electron reflection coefficient associated with the DBQWS leads to a dramatic increase in the amplitude of the reflected electron wave. The interference between the injected and the reflected electron waves causes a large spatial depletion of electron density in the emitter region.[31] The depletion of electrons induces a drop in the potential and forms an EQW. Also, the depth of the EQW increases with the increase of the bias voltage and the energy level of the EQW separates from the three-dimensional states in the emitter region. Furthermore, the interaction between the energy levels (including the conduction band edge in the emitter) in the EQW and that in the main quantum well (MQW) will greatly influence the transport of electrons through the DBQWS. Several factors jointly influence the

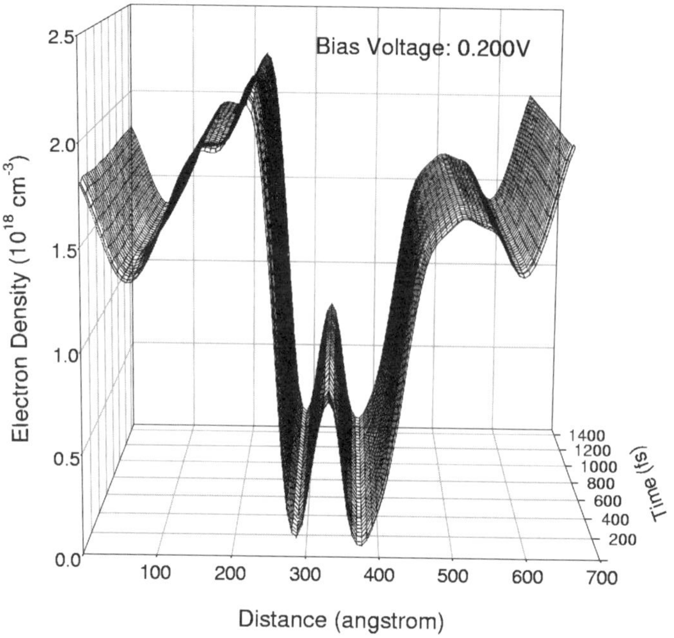

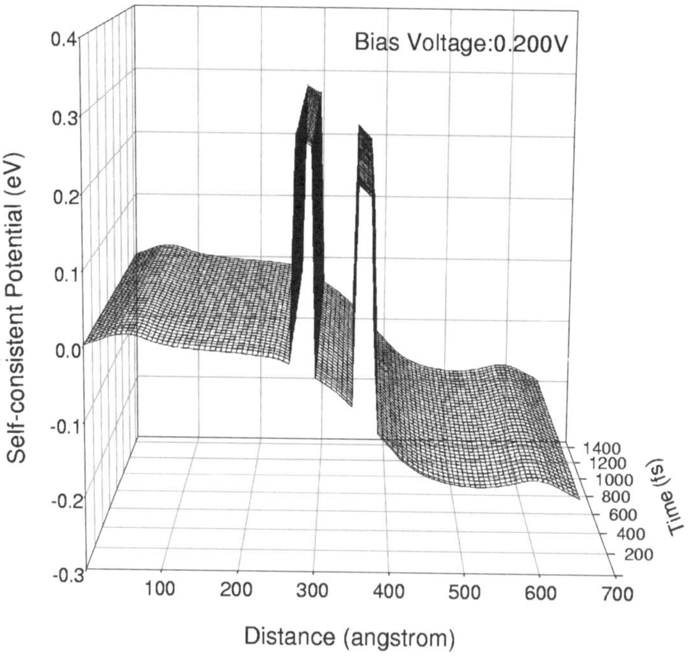

Fig. 3. The time evolution of (a) the electron density distribution, and (b) the self-consistent potential energy, at a bias voltage of 0.200 V.

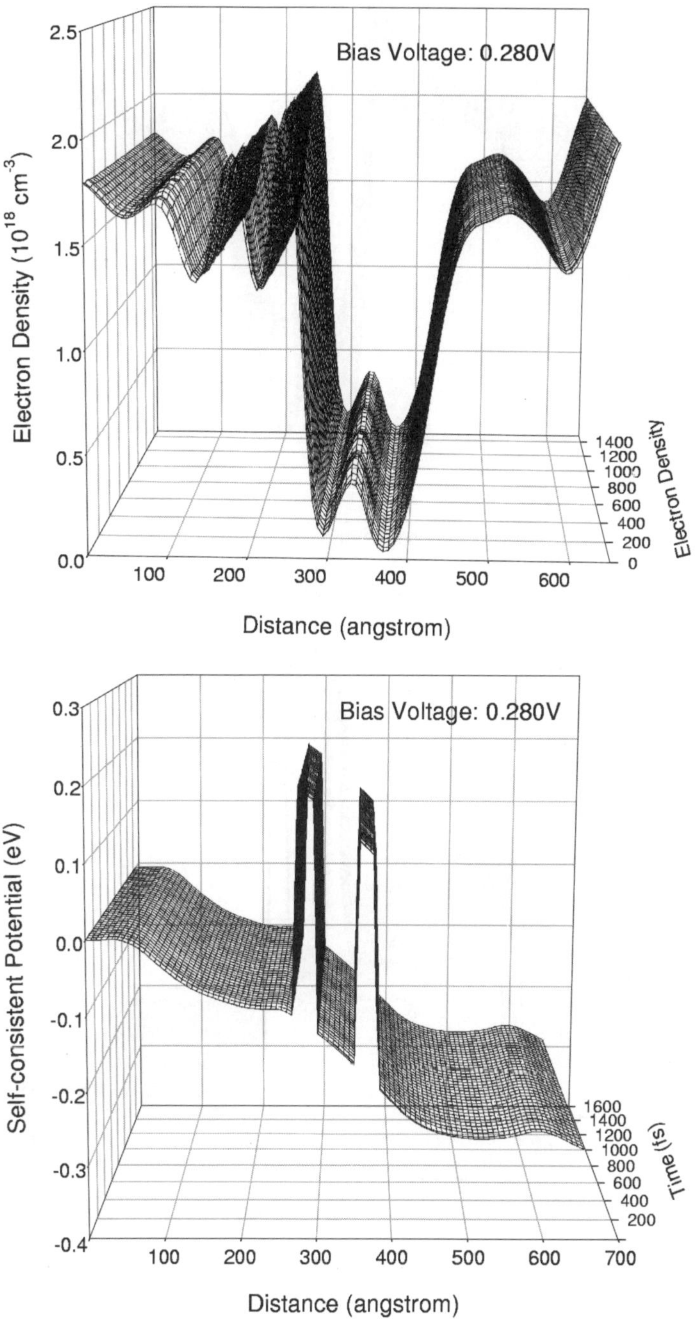

Fig. 4. The time evolution of (a) the electron density distribution, and (b) the self-consistent potential energy, at a bias voltage of 0.280 V.

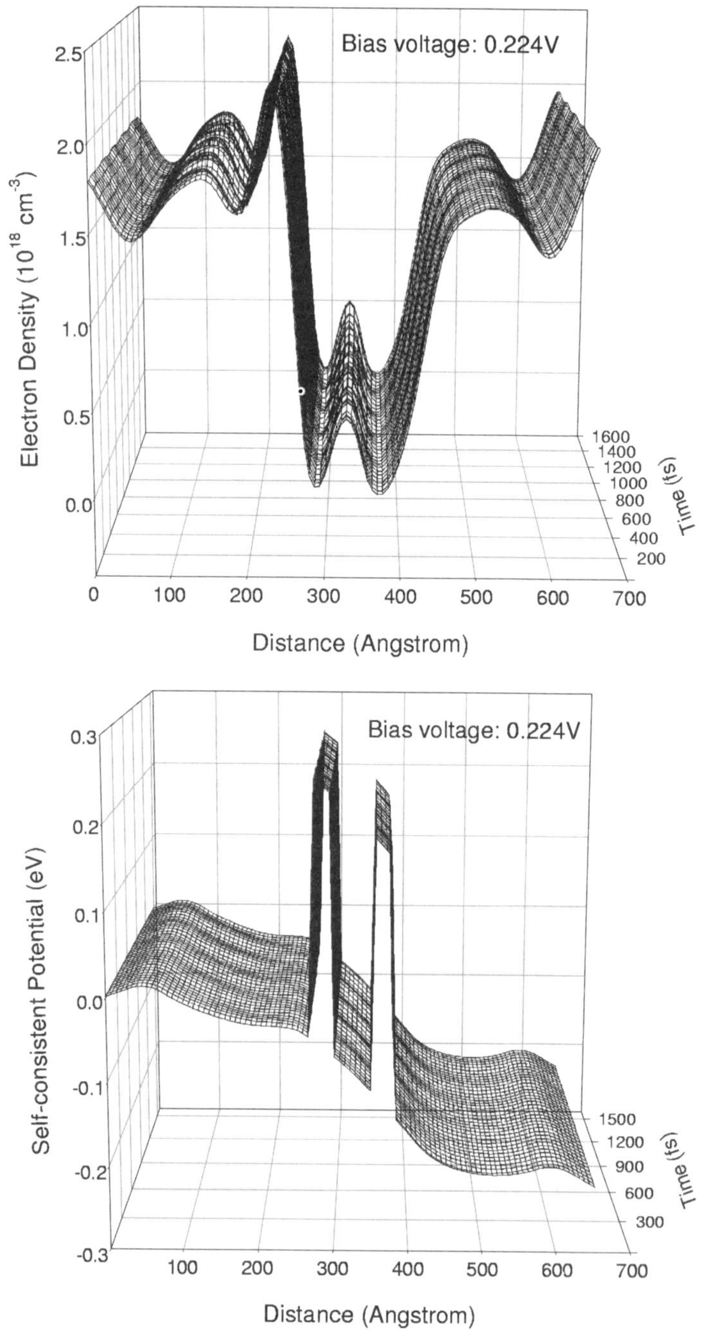

Fig. 5. The time evolution of (a) the electron density distribution, and (b) the self-consistent potential energy, at a bias voltage of 0.224 V.

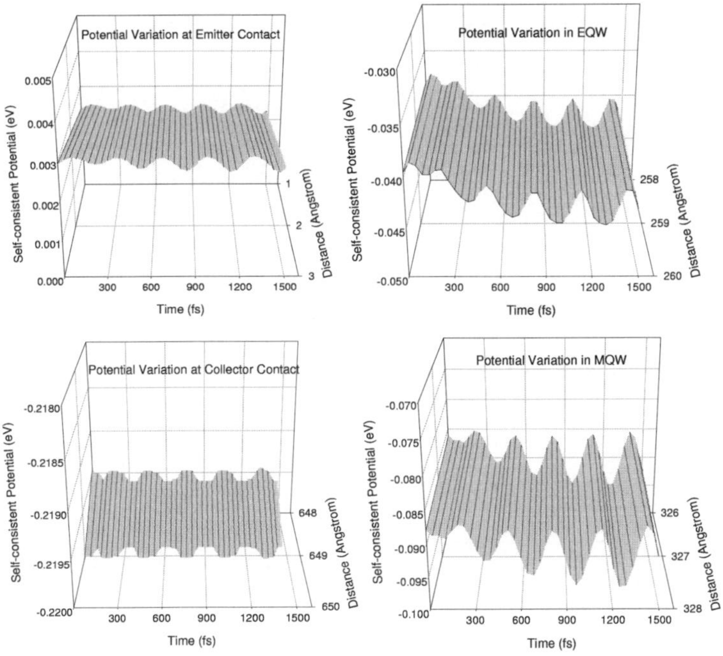

Fig. 6. The time evolution of the potential energy profile at various locations within the DBQWS.

tunneling process. The coupling between the conduction band edge in the EQW and the lowest energy level in the MQW plays a key role and tends to lift the energy level in the MQW, while at the same time it depresses the conduction band edge in the EQW. On the other hand, the applied bias has exactly the opposite effect on energy level in the MQW. The interplay of these two opposite forces determines the existence of the EQW and thereby the major features of the current-time and current-voltage characteristics. These prior arguments, based upon coupling mechanisms between EQW and MQW, do provide a qualitative explanation for the origin of the instability. However, a more detailed and in depth investigation is required to establish a robust and quantitative description of the oscillation physics.

Here, it is very important to note that the electron potential oscillations are completely in phase within the coupled well system. For example, Figs. 6 and 7 show simulation results for several regions of the device and illustrate the instability dynamics. As shown in Fig. 6, the potential energy variations within the EQW and MQW have nearly identical phase. On the other hand, Fig. 7 shows that the electron density within the EQW and the MQW are almost completely out of phase. Hence, the small fluctuations in potential energy are not induced as a direct result of cyclic charging and discharging of the coupled well system. Hence, the instability in the DBQW tunneling structure is arising directly out of quantum-

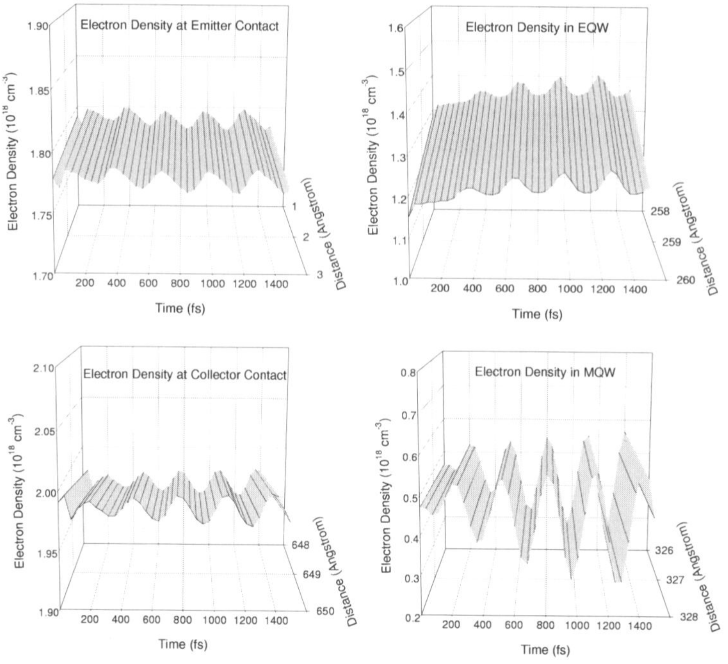

Fig. 7. The time evolution of the electron density profile at various locations within the DBQWS.

coupling between EQW and MQW. Of course, the in-phase fluctuations in the potential barrier structure leads to changes in the tunneling characteristics which induces charging (or discharging) of the MQW at the expense (or benefit) of the EQW. This process also leads to a time-dependent current density. Therefore, this is a *purely quantum mechanical oscillation process* and a complete description of the dynamics will require an investigation into the energy subband structure as derived earlier in Subsection 2.2. Furthermore, the use of the general relations developed in Subsection 2.2 will require that time-dependent subband structure be generated from numerical calculations. A numerical method for calculating the energy level structure of an open system has already been developed in Subsection 2.3.2 for this exact purpose.

The Schrödinger-equation based model of Subsection 2.3.2 was applied to derive the quasi-bound subband structure of the DBQWS. Here, the time-dependent potential-energy profiles that were generated by the Wigner-Poisson Model were utilized. These time-dependent tunneling structure profiles can be used to generate time-dependent subband behavior that is approximately consistent (i.e., except for energy-state broadening) with the observed current density oscillations. Recall that open system conditions were approximated by assuming that the reflection from the source boundary was small (i.e., $r = 0$). As will be shown, this is an accept-

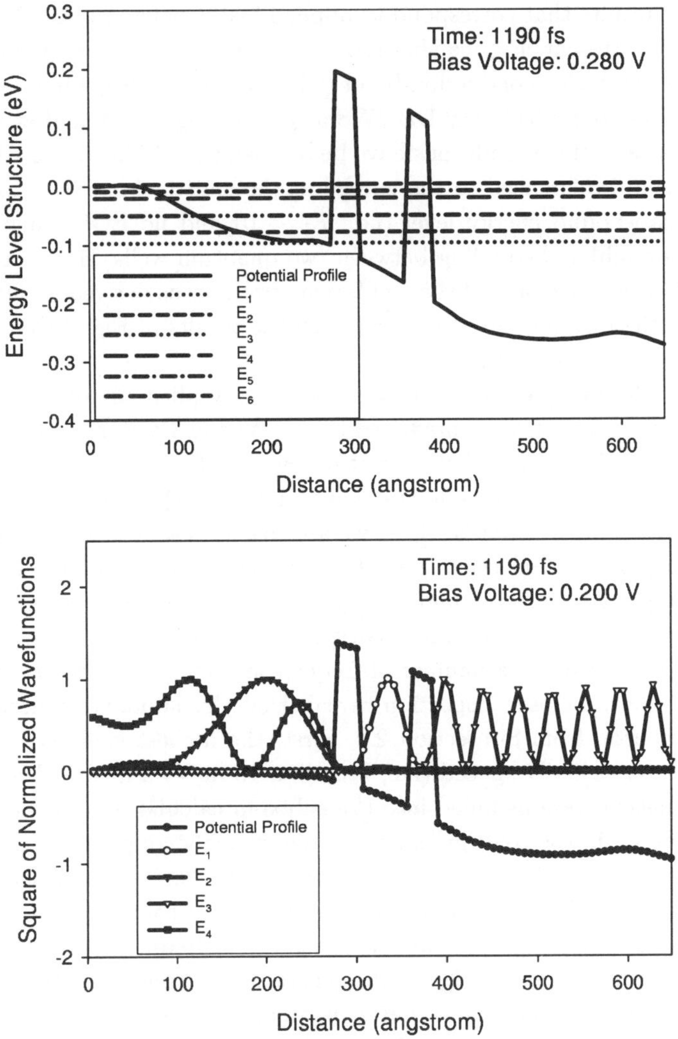

Fig. 8. (a) The energy subband structure referenced to the potential energy profile, (b) the square of normalized electron wavefunctions referenced to the potential energy profile, both at a bias voltage of 0.200 V and at time 1190 fs.

able approximation for our purposes because the most important subbands to the oscillation process are bound from the source-side when biases corresponding to the BVW are applied. Hence, the reflected-component of the emitter wavefunction must be zero for these subbands at the source boundary otherwise the solution would be nonphysical.

Consider results that correspond to applied biases below the BVW where oscillations are not present. Fig. 8a shows the time-independent energy-level structure reference to the device conduction-band (CB) profile and Fig. 6b gives the square of the wavefunctions within the DBQWS at bias voltage 0.2 V and at a simulation time where steady state conditions have be reached (i.e., 1190 fs). Here the subband energy levels are referenced to the energy at the emitter boundary. Since there is only one quasi-bound energy level (i.e., $E_1 < 0$) and no EQW then there is no possibility for subband coupling between two quantum wells. Hence, this subband structure does not meet any of the oscillation criterions as derived in Subsection 2.2. In addition, these results are consistent with the results of Fig. 2 that show no oscillations at bias voltage of 0.2 V.

Consider the next results that correspond to applied biases above the BVW where oscillations are also not predicted by the Wigner-Poisson model. Fig. 9 shows the time-independent energy-level structure (again referenced to the device CB profile) and the square of the wavefunctions within the DBQWS at bias voltage 0.280V and simulation time of 1190 fs. Here, we see that an EQW has formed due to the presence of significant reflection from the first barrier. In this particular case, there are multiple pairs of quasi-bound energy levels that have the potential to couple the EQW to the MQW. Hence, in theory, these levels have the potential to induce an intrinsic oscillation at a number of frequencies defined by the energy-level separation. However, a closer inspection reveals that this subband structure does not meet the criterion from Subsection 2.2. First, the highest subbands (i.e., E_4, E_5, and E_6) are very closely spaced (i.e., less than 10 meV) and would not be resolvable if dissipation were included into the subband calculation. Furthermore, these highest levels are in the continuum and do not strongly couple the individual wells. Second, while each of the remaining subband pairs (here, the strongest interband contribution would be from those that are nearest to each other in energy) have the potential for inducing at instablity they do so at frequencies that are relatively close (i.e., $\sim 50\%$ difference). Hence, the generation of a net oscillation of significant amplitude will require that the various components are not of random phase. Direct calculations of the coefficients (i.e., F and G) for the subband oscillation models derived in Subsection 2.2 indicate that the contributions from the quasi-bound energy levels (i.e., E_1, E_2, E_3 and E_4) have random relative phases. Hence, the subband instability criterion is not met and the model predicts no oscillations in agreement with the Wigner-Possion simulations.

Finally, consider the structure subject to a bias that induces oscillations. Fig. 10 shows the time-dependent energy-level structure derived from the subband model for a bias voltage of 0.224 V. Fig. 10b gives the energy-level structure (again referenced to the device CB profile) and Fig. 10c gives the square of the wavefunctions of the DBQWS. Results in Fig. 10b and Fig. 10c are for a bias voltage of 0.224 V and are at simulation time 800 fs. From Fig. 10b one can see that there are only two energy levels that strongly couple the EQW to the MQW (i.e., as E_3 and E_4 both have energies above the emitter injection energy and therefore lie in the continuum).

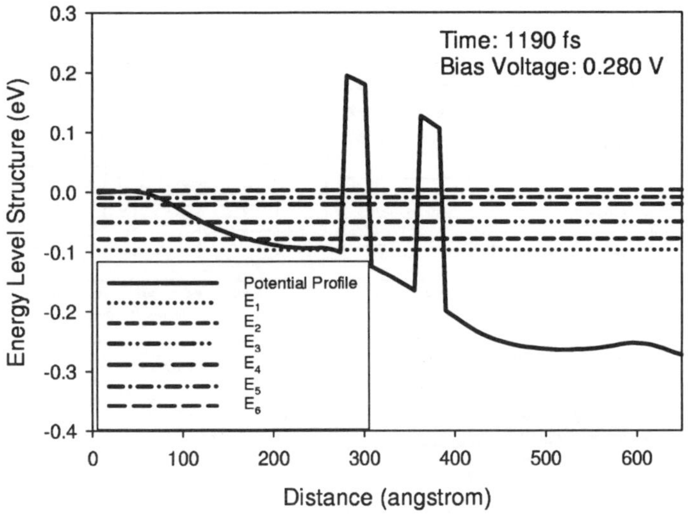

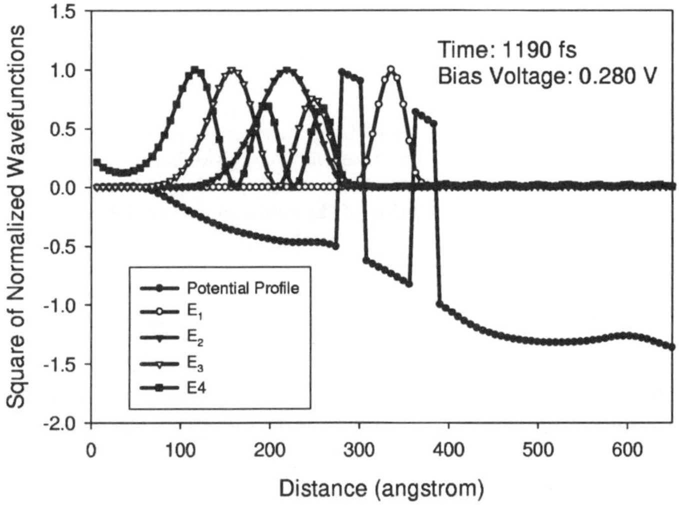

Fig. 9. (a) The energy subband structure referenced to the potential energy profile, (b) the square of normalized electron wavefunctions referenced to the potential energy profile, both at a bias voltage of 0.280 V and at time 1190 fs.

Thus, the main contribution to the intrinsic oscillation is from the lowest subband pair. Hence, this case corresponds to that of minimum subband coherence discussed in Subsection 2.2. In this case, the wavefunctions that contribute to oscillation can be written as

$$\psi(z,t) = \psi_1(z,t)e^{-i\frac{F_1}{\hbar}} + \psi_2(z,t)e^{-i\frac{F_2}{\hbar}} \tag{62}$$

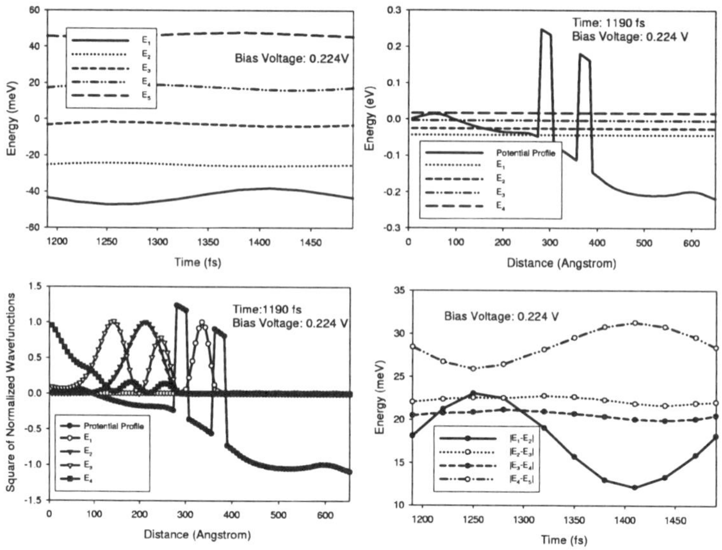

Fig. 10. (a) The time evolution of the energy subband structure at a bias voltage of 0.224 V. (b) The energy subband structure referenced to the potential energy profile, (c) the square of normalized electron wavefunctions referenced to the potential energy profile, both at a bias voltage of 0.224 V and at time 1190 fs. (d) The time evolution of the subband energy differences at a bias voltage of 0.224 V.

where $F_i(t) = \int_0^t dt' E_i(t')$ is the energy of the ith energy level. The relationship between the oscillation frequency and the energy level structure can be derived from Fig. 10a - where there is only a significant time-dependency in E_1. The general form for the relationship for a subband pair (e.g., $E_2 - E_1$) was given earlier in Eq. (20). Note that the cycle of variation in time for E_1 is identical to the current-density period of oscillation (see Fig. 2). The cycle of variation for the subband pair, $\Delta E = E_2 - E_1$, is given in Fig. 10d. The subband energy difference varies periodically and can be written as

$$\Delta E(t) = \Delta E_0 + E_m sin(\omega_E t + \alpha) \tag{63}$$

where E_m is the amplitude of the energy difference oscillation, ω_E is the oscillation frequency and the phase is given by

$$\alpha(T_E) = sin^{-1}\left[\frac{\Delta E(t_0) - \Delta E_0}{E_m}\right] - \frac{2\pi}{T_E}t_0 \tag{64}$$

where $T_E = \frac{2\pi}{\omega}$ is the period of oscillation of the subband pair. At this point, consider again the expression for the time-dependent electron density given in Eq. (19). The phase of the electron density can be defined as $\phi(t) = \frac{\Delta F(t)}{\hbar} + \beta(t)$ where $\beta(t)$ has been introduced to account Eq. (14) for phase-change due to the product of the subband wavefunctions given in Eq. (14). The difference in phase for the electron density, at any two arbitrary times, t_1 and t_2, may now be written as

$$\Delta\phi(t_1,t_2) = \frac{1}{\hbar}\left[\int_0^{t_2} dt'\Delta E(t') - \int_0^{t_1} dt'\Delta E(t')\right] = \frac{1}{\hbar}\int_{t_1}^{t_2} dt'\Delta E(t')$$

$$= \frac{1}{\hbar}\left[\Delta E_0(t_2 - t_1) + \int_{t_1}^{t_2} dt' E_m sin(\omega_E t' + \alpha)\right] + \beta(t_2) - \beta(t_1) \qquad (65)$$

where we have used Eq. (63). The period of oscillation of the electron density can now be defined by $T_\rho = t_b - t_a$ such that $\Delta\phi(t_a, t_b) = 2\pi$. It is now important to note that the period of oscillation for the current density (and equivalently the electron density) at any spatial point is exactly equal to the period of oscillation for the subband energy difference. Since $T_\rho = T_E$ one has immediately that $\int_{t_a}^{t_b} dt' sin(\omega_E t' + \alpha) = \int_0^{T_E} dt' sin(\omega_E t' + \alpha) = 0$. Therefore, evaluation of Eq (65) over the time period $t_b - t_a$ yields the relation

$$T_\rho = T_E = \frac{h}{\Delta E_0} \qquad (66)$$

for defining the period (and frequency) of the intrinsic oscillation. Fig. 10d can be used to define the average energy difference of $\Delta E_0 = 18.1 meV$. Equation (66) now predicts an intrinsic oscillation frequency of

$$\nu = \frac{1}{T_\rho} = 4.35 THz \qquad (67)$$

The oscillation frequency for current density obtained from the Wigner-Poission model was found to be 3.33 THz (i.e., see Fig. 2). Hence, the simple energy-subband description predicts an oscillation frequency within 30 percent of the that obtained using the more complex (and physically complete) simulation model. This small error is very understandable in light of some of the assumptions utilized in deriving the subband energy structure (e.g., no reflection at the emitter boundary and no dissipation). Indeed, the assumption of no reflection at the input boundary of the device will slightly overestimate the number of electrons in the EQW and lead to an overestimation the energy difference between the energy levels in the DBQWS. As the subband model given in Eq. (66) is very sensitive to changes in the energy difference (i.e., a $\Delta E = 4.135 meV$ means a 1 THz change in the frequency) this estimation for intrinsic oscillation frequency is very good. Furthermore, these results verify that the energy-subband structure is the key underlying mechanism for inducing the instability. Lastly, the previously developed model provides for basic insight into the phenomenon and offers guidance for defining new structures that will admit enhanced oscillation characteristics. These observations will be summarized in the next subsection.

2.5. *Conclusions and Design Criterion Summary*

In summary, a theory describing the origin of the intrinsic oscillations in a double-barrier quantum-well system (DBQWS) has been presented. Furthermore, a derivation based upon an approximate subband model for the quantum system has been utilized to derive a simple relationship between the oscillation frequency and the energy level structure. The theory shows that the quantum mechanical coupling between the emitter quantum-well (EWQ) and the main quantum-well (MQW) is the root cause of the instability and self-oscillations of the electron density and electron current. Furthermore, the intrinsic oscillations arise directly as a result of the coupling between the energy subbands. Here, the main drivers of the process are the in-phase fluctuations of potential energy within the EQW and MQW that arise out of the subband coupling. Therefore, the instability is a purely quantum mechanical phenomenon with the frequency of oscillation determined by the average energy-difference of the quasi-discrete subband pairs that contribute (e.g., for minimum subband coherence $\nu = \frac{\Delta E_0}{h}$). These studies establish the fundamental principles for the intrinsic oscillation mechanisms. Most importantly, these studies also provide explicit guidance for defining new structures that will admit enhanced oscillation characteristics at operation frequencies within the terahertz regime.

Any practical implementation of a quantum-mechanical based intrinsic oscillator device towards the implementation of a very high frequency source will require an analysis of the basic device with the embedding circuit. However, these fundamental studies provide important information regarding the design of the resonant tunneling structures that have the capacity for admitting the necessary instability properties. This paper has presented a Wigner-Poisson simulation tools for identifying the occurrence of intrinsic oscillations and has developed simplified subband model for interrogating the energy state structure. A summary of the basic design criterion established by these studies include:

(1) For certain applied biases the DBQWS develops an EQW that couples to the MQW defined by the double-barrier heterostructures of the resonant tunneling diode. The subband coupling then induces quantum-based fluctuations in the potential energy profile that lead to intrinsic oscillations in electron density and electron current. As these oscillations are critically dependent on the coupling of the quasi-discrete energy levels, the intrinsic oscillations will only occur at sufficiently low temperatures (e.g., 77 K) that allow for the formation of a distinct subband structure (minimum energy broadening effects). Hence, depending on the operation temperature selected, an adequate level of resolvability within the subband structure must be established to enable the instability mechanism.

(2) The demonstration of the subband-coupling as the underlying catalyst for the intrinsic instability immediately provides guidance for alternative heterostructures systems that should provide superior oscillation performance. For example, in the DBQWS under study the subband coupling develops between the MQW, which always exists, and the EQW that forms only at certain biases due to interference effects arising out of quantum reflections from the first heterostructure barrier. An alternative approach for realizing intrinsic oscillations is to utilize a double-well system constructed from a triple-barrier heterostructure system. This approach will provide more latitude in biasing the device in that oscillations should

be produced over a wider range of applied bias. There is also the possibility that multiple double-well system may be combined to realize larger oscillation amplitudes. Also, as the fundamental driver of the instability is the subband coupling, it is certainly possible to envision single well systems that yield coupled subbands (e.g., parabolic wells) with the potential for producing intrinsic oscillations. Finally, as was directly demonstrated by the studies presented here, it is possible to utilize well engineering to modify the shape of the quantum wells and to enhance the amplitude of the observed current density oscillation.

(3) As the subband-coupling has been shown to produce the instability, it should be possible to design structures that yield enhanced oscillation amplitudes through engineering of the energy-level separations and the associated phase of the quantum mechanical scattering-state functions. Specifically, the theory has demonstrated that the oscillation is a product of individual subband-pair coupling. In particular, results from this study demonstrated a case of minimum subband-coherence where only one pair of energy states contributed to the oscillation. Hence, if tunneling structures were designed such that partial or maximum subband-coherence was achieved then the amplitude of the current oscillation should be enhanced. This would specifically entail the design of structures that resulted in equal energy state spacing and coherent superposition in the corresponding wavefunction inner products. A general theoretical description of this procedure is given in Subsection 2.2.

3. Duality Theory in Solid-State THz Generation

3.1. *Introduction*

The inherent *pairing* of the relevant dynamical variables between analytical theories of self-oscillation in resonant tunneling structures (RTS's) with conventional (type I) and with staggered (type II) alignments of the heterostructure energy-band gaps is discussed. Two entirely different physical models for type I and type II RTS's were introduced and described by a unifying set of coupled nonlinear-rate equations. These coupled equations were solved for the limit cycle solution. Applying the result to type I RTS, the limit cycle predicts a rising average current whereas the non-oscillatory solution predicts a falling current as a function of bias in the current "plateau", after the resonant-current peak. The limit cycle also predicts a decreasing amplitude of current oscillation as a function of bias in the current plateau. The behavior of the fundamental frequency as a function of bias in the plateau agrees with experiments and simulations of AlGaAs/GaAs type I RTS. Applying the result to type II RTS, the limit-cycle oscillation of the barrier-well polarization and trapped charge in the barrier induce THz oscillation in the resonant tunneling current across the device, before the resonant-current peak. The time-averaged results agree with the measured current-voltage (I-V) characteristic of AlGaSb/InAs/AlGaSb type II RTS. In particular, the measured smaller current offset at forward bias compared to that of reverse bias in the I-V hysteresis loop is predicted by the physical model and limit cycle analysis. Thus, the duality theory of type I and type II RTS is hereby established. Based on this type of analysis, the equivalent-circuit model of RTS previously given by the author needs further refinement in a form of an additional small capacitor across the nonlinear circuit element, to account for the oscillating stored charge in the quantum well which is out of phase with the oscillating stored charge at the emitter, with the whole parallel branch in series with the emitter-quantization-induced inductance connecting across the emitter and drain. This has recently been used by Poltoratsky and Rychkov in their equivalent-circuit model of solid-state THz generation.

The development of heterostructure material fabrication technology established the heterojunction as the basic building block of advanced high-speed and high-frequency semiconductor devices for electronic, microwave, and optoelectronic applications. A heterojunction is an interface within the semiconductor material across which the chemical composition changes, in contrast with p-n junction where only the dopants change. Junction between GaAs and AlGaAs, junction between AlGaSb and InAs, junction between Si and GeSi are some of the examples. A host of different band-edge alignments[32] have generated different Zener effect or interband-tunneling transport devices.[33]

What is not widely recognized is that double-barrier resonant tunneling structures have the potential to exhibit autonomous oscillation at much higher frequencies (THz range) than those of impact ionization avalanche transit time (IMPATT) diodes and Gunn effect microwave sources. These tunneling devices have nanometric sizes, whereas IMPATT and Gunn effect devices are quite large in comparison. As THz sources, there are clear advantages in terms of simplicity, compactness, and monolithic integration with power combiners, matching guided-wave structures and antennas using integrated-circuit fabrication technology. For type I RTS, self-oscillation behavior was first demonstrated by Jensen & Buot,[15] and later by Biegel & Plummer,[16] through quantum transport numerical simulations. Recently, Buot et al.,[34] employing numerical, graphical, and analytical methods, investigated in more detail the dependence of the occurrence, amplitude, and frequency of oscillation on various device parameters. A strong self-oscillatory behavior of type I RTS occurs, under appropriate values of the device parameters, when the device is operating in the negative-differential-resistance (NDR) region just after the resonant current peak.

3.1.1. *Type I RTS*

Numerical quantum transport simulations and experiments of most type I RTS's revealed the characteristic peak-to-plateau-to-valley behavior of the I-V characteristics.[10,15,16,35,36,37] Double hysteresis I-V behavior was also seen with forward and backward voltage sweep.[35,37] Moreover, time-dependent simulation results of different groups revealed intrinsic high-frequency current oscillations in the plateau region of the current-voltage (I-V) characteristics.[15,16]

The current peak during forward bias is the result of the passage of the quantum well (QW) discrete energy level into the forbidden energy region of the emitter. This forbidden region does not necessarily correspond to the energy gap between the conduction and valence bands of the emitter. This forbidden region maybe created above the emitter conduction band edge by virtue of the quantization of the supply electronic states by the confining emitter quantum well (EQW). This is illustrated in the conduction energy-band edge (EBE) diagram of Fig. 11(a), which shows a triangular EQW. The passage of the QW energy level into the forbidden region of the emitter creates a sudden drop of the current across the device, producing a characteristic sharp current peak. As the electrons are build up in the emitter, the interference of the reflected electrons and the incoming electrons effectively broadens the EQW by virtue of the selfconsistency of charge and potential. The broadening is due to the redistribution of the electrons with some regions becoming positive (deficit of electrons) and some regions negative (excess of electrons) in the emitter.

This EQW renormalization broadens the EQW and the lowers the quantized energy levels in the emitter towards the conduction band edge. With this EQW broadening, the alignment of the QW discrete energy level with occupied states in the emitter is consequently restored yielding high transmission coefficients and larger currents. This is depicted in Fig. 11(b). This feedback is basically a catalytic process since the quantum well charge, through the selfconsistent potential, helps in restoring the QW energy level alignment with the occupied states in the emitter. The resulting emitter discharge and the selfconsistency of charge and potential restore the emitter potential profile which produces 2-D quantization, i.e., back to the situation depicted by Fig. 11(a). The process therefore oscillates between that of Figs. 11(a) and 11(b), and the average is responsible for the plateau-like behavior above the valley-current minimum.

The driving source of this oscillatory condition is the build up and redistribution of charge resulting in significant ripple effect due to the interference of the reflected and incoming electrons. We identify this driving source in terms of the total charge buildup,Q, at the emitter in time duration, RC, where R is the access resistance and C is the RTS capacitance. Therefore, Q/RC measures the maximum buildup rate of supply electrons at the emitter in the absence of tunneling to the QW. In the presence of tunneling, one needs to solve the proper coupled rate equations discussed in Sec IIA.

We expect the maximum values of Q/RC and oscillation amplitude just after the current peak, in the plateau region as depicted in Figs. 11. This is because there is a considerable broadening or renormalization of the EQW in going from Figs. 11(a) to 11(b), i.e., in bringing the allowed EQW allowed states in line with the QW energy level. On the other hand, well within the plateau region, there is only a further broadening of the EQW and hence the amplitude of oscillation will become smaller as the drain bias is further increased. This is indicated in Figs. 12.

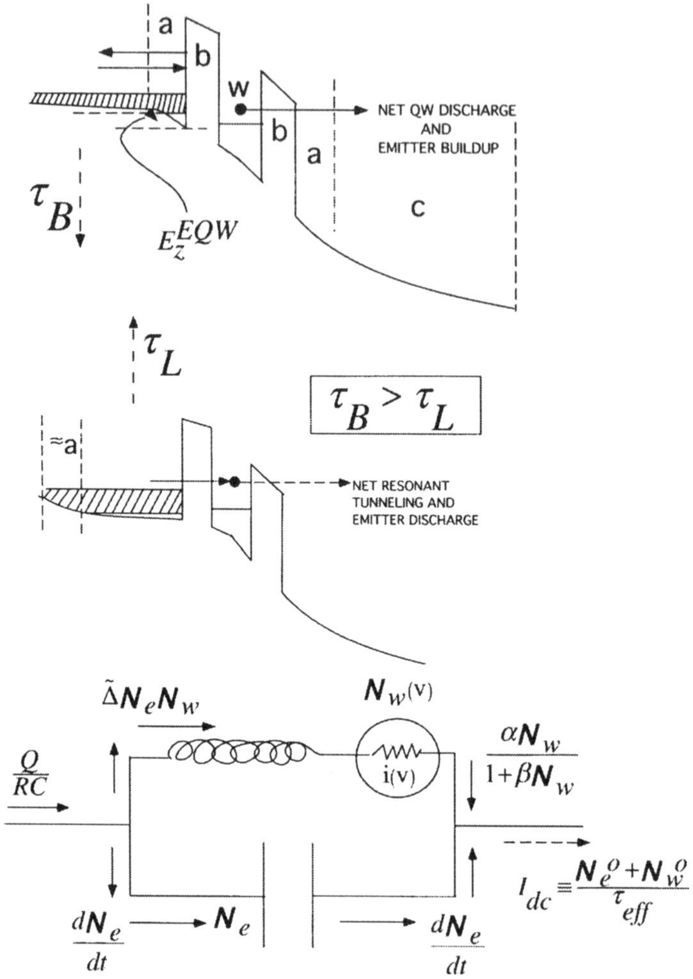

Fig. 11. (a) Conduction band edge diagram showing that emitter 2-D quantization results in a premature alignment ofQW energy level with the forbidden region of the emitter. The build-up time τ_B of the reflected charge leads to the condition in (b). (b) Realignment of QW energy level with occupied states in the emitter causes rapid depletion of built-up charge in time τ_L after which condition (a) is restored.(c) Equivalent-circuit model, where $i(v)\,[N_w(v)]$ is the current (stored charge) of an ideal RTS for a voltage drop v across the negative resistor.

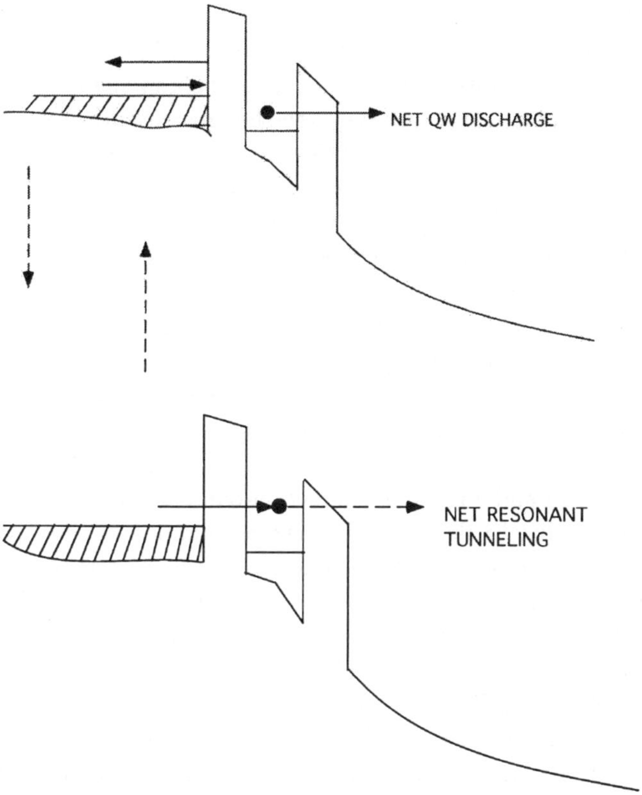

Fig. 12. EBE diagram showing that smaller amplitude of oscillation occurs within the plateau region as the QW energy level continues to shift downward with increase in applied bias. Hence Q/RC also decreases with increase in applied voltage.

3.1.2. *Type II RTS*

Another interesting but entirely different physical mechanism inducing self-oscillation occurs in type II RTS. In contrast to that in type I RTS, this oscillation occurs before the resonant current peak. The modulation of the position of the discrete energy level in the quantum well relative to the occupied levels in the emitter is driven by the dynamics of trapped holes in the barrier. The trapped holes serve as control charge (similar to the base charge of a bipolar transistor) for modulating the tunneling electron current from the emitter to the drain. The self-oscillation of the type II RTS is brought about by the oscillatory build-up and decay of the polarization pairing between the conduction-band electrons in the quantum well and the trapped holes in the barrier, as was first shown by Buot[38,39] in agreement with the experiments.[40]

We are therefore interested in Coulomb-correlated pairing between the conduction electron in the quantum well and trapped hole in the barrier, with two-body correlated motion confined transverse to the heterostructure growth direction. Since the pairing is not between the created electron and hole, we refer to this polarization pair as a '*duon*' to distinguish from *excitons* and electron-hole Cooper pairs.

The bottom-of-conduction and top-of-valence EBE diagrams of type II RTS are depicted in Fig. 13. The staggered band-edge alignment shown in the figure can be realized for example by using InAs/AlSb heterojunctions, Fig. 13(a), or InAs/AlGaSb heterojunctions, Fig. 13(b). Due to its inability to confine holes in the barrier, the structure of Fig. 13a yields a type I RTS I-V characteristic as measured experimentally.[40] Stronger hole confinement can be obtained by using InAs/AlGaSb heterojunction, Fig. 13(b), which yields a new I-V bistability before the current peak.[40] Unless otherwise specified in what follows, quantum well refers to the conduction band edge and conduction-band electrons.

When the localized valence-band electrons confined in the AlGaSb barrier see the available states in the drain region under bias, these electrons tunnel to the drain leaving behind quantized holes. The Zener transition is initiated when a matching of the discrete level, ε_n, of the right barrier with the unoccupied conduction-band states in the drain first occurs, at $(k_z^D)^2 \geq (k_F^D)^2$ in Fig. 14, with the ' $>$ ' sign holding for indirect gap Zener tunneling.[84] The Zener tunneled electrons, deposited at the spacer layer, are acted on by the field of the depletion region and quickly recombine at the drain contact. In effect this process creates a *polarization* between the barrier and the quantum-well, which establishes a high-field domain in this region. The result is a consequent redistribution of the voltage drop across the device. The time-dependent dynamics of the hole charging and discharging that follows is dictated by the self-consistency of the potential. At higher bias, this is described by Fig. 15. As hole charging occurs, Fig. 15(1), the polarization between the barrier and the quantum-well induced by the trapped hole charge [Fig. 15(2)] creates a high-field domain tending to lower the residual potential drop between the contact and the right barrier. Owing to self-consistency of the potential, further polarization leads to the 'switching' of the intravalence-band tunneling of the trapped holes from the barrier towards the quantum well region [Fig. 15(3)]. When the situation shown in Fig. 15(3) is reached, other possible mechanisms for hole discharging may also occur, namely, thermal activation of the valence electrons in the continuum to recombine with localized holes, or loss of any bound hole states in the barrier.

The *bound hole leakage* can be approximated by tunneling through a triangular potential barrier, which is likely to have a smaller barrier height than that of electron Zener tunneling (viewed as a tunneling through a potential barrier). Any or all of the hole-leakage processes mentioned above will also immediately restore the high field between the barrier edge and the right contact. The situation shown in Fig. 15(1) is thus revisited, after which the process repeats. Oscillations of the

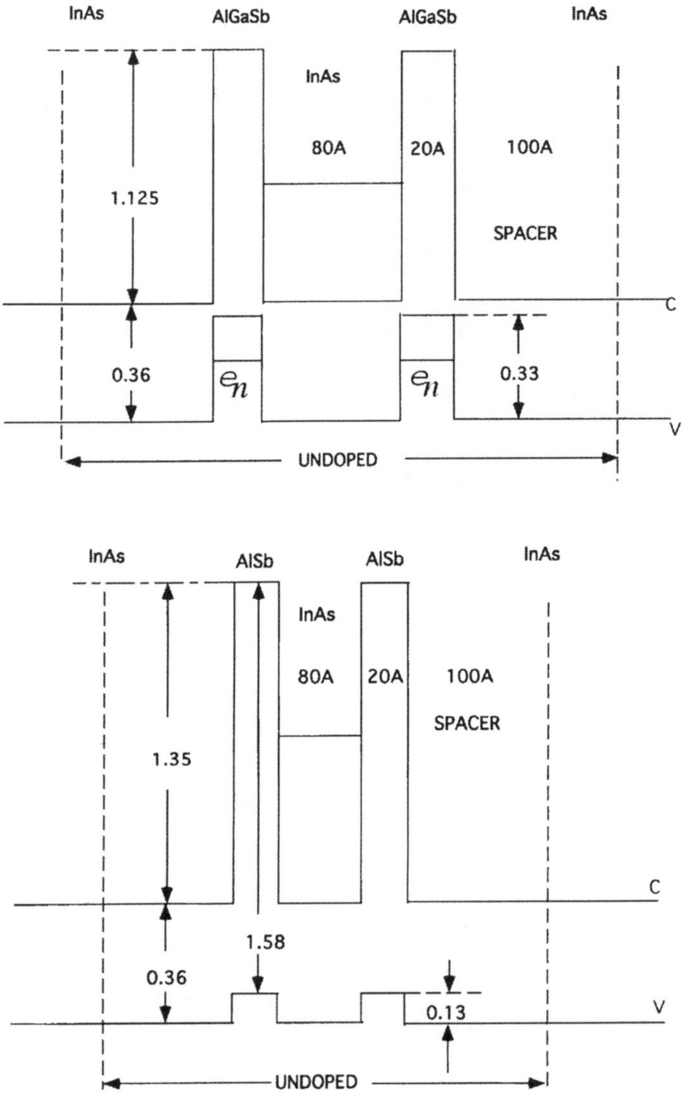

Fig. 13. (a) EBE alignment of RTD using InAs/AlSb heterojunction. (b) Energy band edge alignment of RTD using InAs/AlGaSb heterojunction. Band-edge offsets are indicated in electron volts. ε_n and ε_w are the discrete energy levels in the barrier and quantum well, respectively.

hole charging of the AlGaSb barrier can occur in the THz range, by virtue of the nanometric dimensional features of the device. The oscillatory process limits the average amount of hole charge that can be trapped in the barrier as a function of drain bias. The direct interband recombination process in the barrier is not considered since it can not compete with the conduction-band electron tunneling

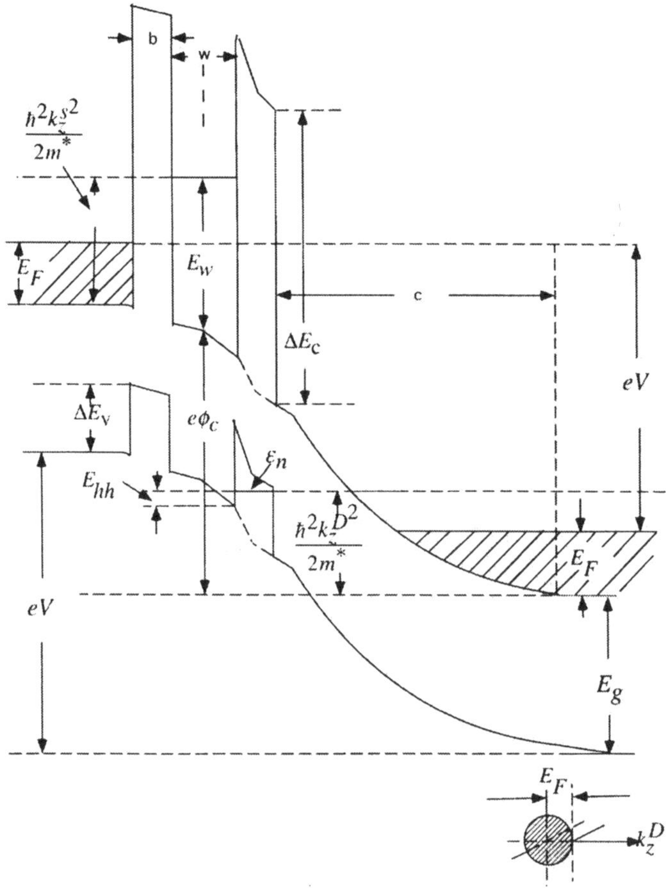

Fig. 14. Average EBE profiles with applied drain potential eV. E_F is the Fermi energy. The lower right-hand corner indicates the unoccupied transverse and longitudinal momentum states outside the Fermi sphere defined by E_F of the drain.

process. There is not enough time for direct interband recombination to take place, since the velocity of the conduction electrons at the barrier is quite large owing to the small probability of being inside the barrier region.[25,15]

3.2. *Nonlinear Quantum Dynamics*

The discharging and charging processes in both RTS discussed above are governed by the two characteristic times. These are the charge build-up time, τ_B, and the charge-leakage time,τ_L . In general, if $\tau_B > \tau_L$, then oscillatory behavior will occur, the charging process will always be lagging behind the discharging process and oscillations will result. This criterion also holds true in single-electron devices.[41]

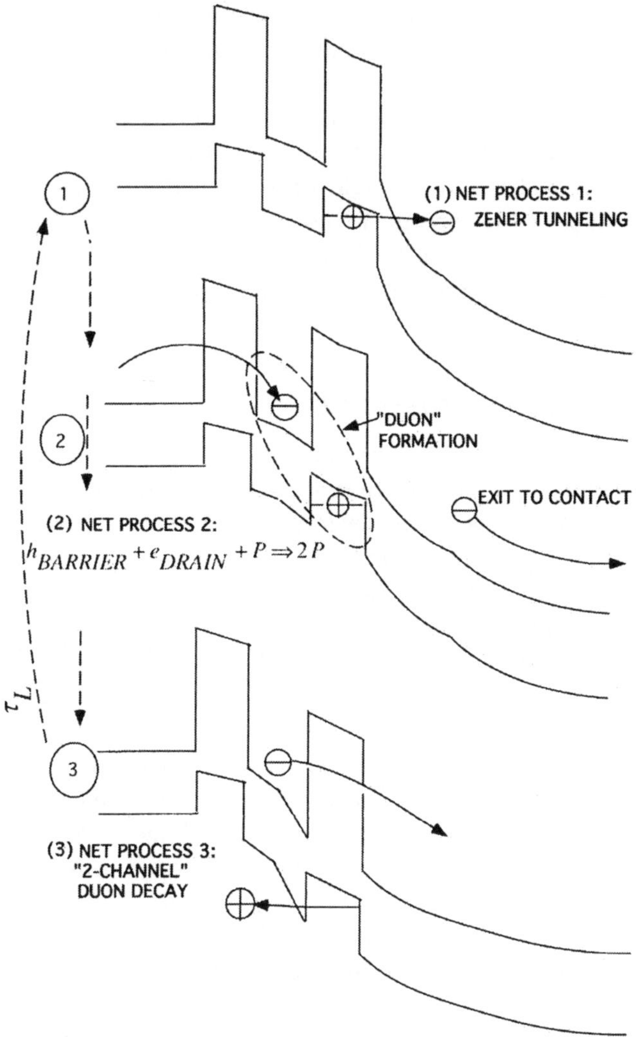

Fig. 15. EBE diagram showing the mechanism of oscillation of trapped hole charge in the barrier: (1) e-h generation by electron Zener tunneling, (2) duon generation is through an autocatalytic process similar to the stimulated production of photons in lasers, (3) there are three possible mechanisms for hole discharging, mentioned in the text, any one or all will quickly restore the situation depicted in (1).

Several nonlinear models were attempted in the literature to explain the oscillatory behavior of conventional RTS. A nonlinear model for RTS by Abe[42] only focuses on the selfconsistent electron charge in the QW and its effect on the tunneling probability, i.e., focusing only on the catalytic feedback process due to selfconsistent QW potential mentioned in the **Introduction**. The dynamical coupling with the emitter (reservoir) was completely ignored. Highly mathematical treatments of a similar model were given by Jona-Lasino et al.[43] and by Presilla & Sjöstrand.[44] A Schrödinger equation with nonlinearities concentrated in the double-barrier region was proposed. The major assumption is that the emitter region is left undisturbed, which is unrealistic in the light of all numerical results. No comparison with salient features of the experimental results of Goldman et al.[37] and Sollner[6] were attempted. The recent paper of Buot et al.[34] is the first to explain all the different salient features between the experimental results of Goldman et al.[37] and of Sollner.[6]

3.2.1. *Oscillatory Dynamics of Type I RTS*

Let N_e be the number of supply electrons at the emitter that are generated by the EQW broadening and participate in tunneling to the QW, and let N_w be the number of corresponding electrons generated in the quantum well. In the NDR, the frequency, ω_{eq}, of tunneling from the emitter to the quantum well is given by $\omega_{eq} = \widetilde{\Delta} N_e N_w$, where $\widetilde{\Delta}$ represents the tunneling probability factor which takes into account the dependence of tunneling coefficient on the barrier height and width, taking into consideration the longitudinal quantization of the supply electrons in the emitter. $\widetilde{\Delta}$ therefore could be dependent on the driving source Q/RC, which affects the existence of limit cycle solution for very small values of Q/RC. The explicit dependence of ω_{eq} on N_w is explained as follows. In the NDR, the frequency of tunneling from the emitter to the quantum well is enhanced by the presence of N_w . This is because the realignment of the allowed energy levels in the emitter and the QW discrete energy level is aided by the presence of N_w by virtue of the positive feedback due to self-consistency of the potential. This represents a catalytic feedback process. This can be symbolically defined as $N_e + N_d + N_w \rightarrow 2N_w$, describing a catalytic transport process of three interacting components, where N_d represents the electrons reaching the drain.

We can now write the 'effective' generation rate of supply electrons at the emitter as,

$$\frac{\partial}{\partial t} N_e = \frac{Q}{RC} - \frac{N_e}{\tau_{eq}}, \tag{68}$$

where $\tau_{eq} = 2\pi/\omega_{eq}$. As mentioned in the **Introduction**, Q/RC is the driving source term, Q is the electron buildup at the emitter in time constant RC to produce enough broadening of the EQW, R is the series resistance and C is the double barrier capacitance. Similarly, the effective generation rate of electrons in the quantum well is,

$$\frac{\partial}{\partial t} N_w = \frac{N_e}{\tau_{eq}} - \frac{N_w}{\tau_c}. \tag{69}$$

Equations (68) and (69) describe the situation depicted in Fig. 11(a), namely, at

$N_e = 0$ the generation rate of N_e is at its maximum while the generation rate of N_w is negative, i.e., N_w is actually decaying. The equivalent-circuit model closely describing Eqs.(68) and (69) is shown in Fig. 11(c). This is the same equivalent-circuit model for RTS with inductive delay introduced by Buot and Jensen[36] to explain the presence of intrinsic high-frequency oscillations in their numerical results.[15] This was discussed in more detail by Buot and Rajagopal.[45] The coupled rate equations above are similar to the ones used to describe an interband-tunnel high-frequency oscillator introduced by Buot[38] and discussed in more detail by Buot and Krowne.[46] The maximum generation rate of N_e, given by Q/RC, is the driving source of a 'dual' theory for type I RTS. This is expected to depend on the confining electric field at the emitter for a given bias, when the QW energy level passes into the forbidden region of the emitter. In the last equation, τ_c is the characteristic time for the decay of N_w due to tunneling to the drain or collector.

The last term of Eq. (69) can be given a more fundamental consideration by formulating the process describing the decay of N_w due to tunneling from the quantum well to the collector or drain. Let N be the total number of matching states in the drain for the electrons in the quantum well to transition to. Let N_x be the number of matching states already occupied and no longer available by virtue of Pauli exclusion principle. The production rate of N_x is proportional to the product of the available number of matching states and N_w. Let λ be this proportionality constant. And let γ the decay rate of N_x by virtue of electron drift in the depletion region followed by absorption at the metal contact. Then we can write the rate equation for N_x as,

$$\frac{\partial}{\partial t}N_x = \lambda(N - N_x)N_w - \gamma N_x, \tag{70}$$

where the first term is also the decay rate of N_w. The process described by the last term of Eq. (70) is the fastest process in the problem, N_x is therefore expected to relax much faster than N_w and N_e. Thus by adiabatic elimination of fast variables, we can let $\frac{\partial}{\partial t}N_x \Longrightarrow 0$. Then we obtain,

$$N - N_x = \frac{N}{1 + \frac{\lambda N_w}{\gamma}}. \tag{71}$$

Upon substituting the expression of Eq. (71) in the first term of Eq. (70), we obtain the decay rate of N_w given by $\lambda N N_w/(1 + \lambda N_w/\gamma)$. Thus, we can express the decay rate of the quantum-well electrons as,

$$\frac{N_w}{\tau_c} = \frac{\alpha N_w}{1 + \beta N_w}, \tag{72}$$

where $\alpha/\beta = \gamma N$, and $1/\beta$ is proportional to the sum of available states in the collector. Equation (72) is a typical decay rate for dynamical systems involving a fast relaxing 'sink', typified by the Michaelis-Menten decay law in chemical kinetics. The parameter $\alpha = \lambda N$ is the decay rate constant and α/β is the value of the saturated decay rate of N_w. Therefore, we can also write the generation rate for N_w as,

$$\frac{\partial}{\partial t} N_w = \widetilde{\Delta} N_e^2 N_w - \frac{\alpha N_w}{1 + \beta N_w}, \tag{73}$$

where the first term is also the explicit expression for N_e/τ_{eq} in Eq. (69). Comparing with Eq. (69), we obtained the following relation:$\tau_c = (1 + \beta N_w)/\alpha$. As seen in Eq.(77) below, the physical situation corresponds to $\alpha/\beta > Q/RC$. This means that the maximum discharging rate of the quantum well is larger than the build-up rate of supply electrons at the emitter. Indeed, we can estimate that $Q/\tau_B \approx Q/RC$ and $Q/\tau_L = \alpha/\beta$, where τ_B is the length of time to bring the device from the state of Fig. 11a to that of Fig. 11b and τ_L is the corresponding length of time for bringing back from state of Fig. 11b to Fig. 11a. Therefore, the physical requirement that $\alpha/\beta > Q/RC$ implies that we may arrange for R to be large enough for a given capacitance C such that $\tau_B > \tau_L$. This is the situation we are considering in our analysis of the oscillation. Note that oscillatory behavior and hysteresis very much depends on these two characteristic times as discussed by Buot and Rajagopal.[45,47]

$\tau_B > \tau_L$ implies that at steady state one would only see the average values of the built-up charge in the emitter and broadened EQW since the process of charge build up takes longer compared to the time duration for discharging the emitter (which is roughly the leakage time of the QW charge to the collector). Moreover, the average QW charge will be a slowly varying function of the bias, i.e., resembling the current plateau. This behavior of the quantum-well charge mimicking the current plateau was indeed found by Jensen and Buot in their quantum transport numerical simulation of type I RTS.[15] The stationary solution to the coupled rate equations, (68) and (73) is given by,

$$\frac{Q}{RC} = \widetilde{\Delta}\ N_e^2\ N_w = \frac{\alpha N_w}{1 + \beta N_w}, \tag{74}$$

and the average oscillatory current is approximately given by,

$$I_{dc} = \frac{N_e^0 + N_w^0}{\tau_{eff}}, \tag{75}$$

where τ_{eff} is the effective relaxation time of the sum of $N_e^0 + N_w^0$, as indicated in Fig. 11d. Here,

$$N_e^0 = \left(\frac{\alpha - \beta\frac{Q}{RC}}{\widetilde{\Delta}}\right)^{\frac{1}{2}}, \tag{76}$$

$$N_w^0 = \frac{\frac{Q}{RC}}{\alpha - \beta\frac{Q}{RC}}. \tag{77}$$

Note that N_e^0 decreases with increase of $\frac{Q}{RC}$, whereas N_w^0 increases with $\frac{Q}{RC}$.

3.2.2. *Oscillatory Dynamics of Type II RTS*

Let G be the maximum rate of e-h generation by Zener tunneling from the barrier valence band to the drain conduction band for a given bias. This constant rate G is a function only of the applied voltage, or more appropriately of the biasing field at the depletion layer of the drain region (refer to Fig. 14). It is therefore our measure of the applied bias at the drain contact for a given device structure. It is clear that the maximum rate of generation of e-h by Zener tunneling occurs in the absence of any hole charge trapped in the barrier, and hence in the absence of *duons* which we denote by P. The reason for this is that the highest field in the depletion layer for a given bias occurs in the absence of *duon* s, as indicated in Fig. 15(1). In the presence of *duon* formation, one has to solve the proper coupled rate equations. We noticed here that G corresponds to Q/RC of type I RTS but with very important difference, namely, G increases with bias whereas Q/RC decreases with bias. This difference basically expresses the duality theory of the self-oscillation between these two types of RTS, as we shall demonstrate in what follows.

The effective generation rate for electron-hole pairs (e-h) by Zener tunneling will of course decrease with polarization since the region between the quantum well and the barrier becomes a growing high-field region at the expense of the voltage drop in the depletion region. We expect this rate of decrease to be proportional to the rate of production of *duons*. On the other hand, the generation rate of the *duons* autocatalytically depends on the concentration of existing *duons*. The generation of *duons* involves three interacting components, namely, (a) trapped holes in the barrier, (b) generated conduction electrons, and (c) existing *duons* to stimulate the net transfer of conduction electron from the emitter to the quantum well to be paired with the trapped hole in the barrier to form more *duons*. One can view this autocatalytic "pairing" of conduction electron in the quantum well and trapped hole charge in the barrier as a stimulated transformation of the e-h pair generated by Zener tunneling into a *duon*, with the electron created by Zener tunneling recombining at the drain contact, leaving only the "polarization pair" (*duon*) as depicted in Fig. 15(2). This can be symbolically defined as $h_{barrier} + e_{drain} + P \rightarrow 2P$. Our model amounts to a phenomenological use of *semiconductor Bloch equations* (coupled transport equations for electrons, holes, and polarization densities) coupled with the *Poisson equation*, which requires very large-scale high-performance computer resources to simulate.

The *duon* generation rate with three interacting components can thus be expressed as,

$$\textit{Stimulated duon } (P) \textit{ generation rate} = \widetilde{\Delta}\, N_B^2\, P, \tag{78}$$

where N_B is the number of 'unpaired' holes which is equal to the number of *exiting* conduction electrons (created by Zener tunneling), $\widetilde{\Delta}$ is the rate parameter (per number of electrons and per number of holes produced by Zener tunneling as indicated in 15(2)). We can now write the *effective* generation rate of unpaired trapped holes in the barrier as,

$$\frac{\partial}{\partial t} N_B = G - \widetilde{\Delta}\, N_B^2\, P. \tag{79}$$

Note that the total concentration of trapped holes or control charge in the barrier, Q_B, at any time is given by,

$$Q_B = N_B + P, \tag{80}$$

where P is given as the *duon* concentration.

To obtain the rate equation for P, we need to formulate the process in Fig. 15(3) describing the decay of the high-field domain between the quantum well and barrier. This decay rate for P is expected to saturate to a constant rate for very large P . Let N be the total number of matching states for the holes in the barrier to transition to. Let N_P be the number of matching states in the valence band no longer available by virtue of holes already transitioning to these states. The production rate of N_P is proportional to the product of N_P and P. Let λ be this proportionality constant. And let γ be the decay rate of N_P by virtue of recombination of holes and screening electrons in the valence band. Then we can write the rate equation for N_P as,

$$\frac{\partial}{\partial t} N_P = \lambda(N - N_P) - \gamma N_P, \tag{81}$$

where the first term is also the decay rate of P. The process described by the last term of Eq. (81), i.e., recombination of holes and screening electrons in the valence band, is the fastest process in the problem. N_P is therefore expected to relax much faster than P and N_B. Thus by adiabatic elimination of fast variables, we can let $\frac{\partial}{\partial t} N_P \to 0$. Then we obtain,

$$N - N_P = \frac{N}{1 + \lambda P/\gamma}. \tag{82}$$

Upon substituting the expression of Eq. (82) in the first term of Eq.(81), we obtain the decay rate of P given by $\frac{\lambda N}{1+\lambda P/\gamma}$. Likewise, the decay rate via tunneling of conduction electrons from the quantum well to the drain is limited, through self-consistency, by the limit set on the trapped hole decay process. Thus, we can express the decay rate of the *duon* concentration as,

$$\textit{Duon (P) decay rate} = \frac{\alpha P}{1 + \beta P}, \tag{83}$$

where $\alpha/\beta = \gamma N$, $1/\beta$ is proportional to the sum of available states in the valence band of the quantum well region and the states participating in thermal recombination, otherwise it represents the actual number of hole states in the barrier in the case of the loss of bound hole state. Equation 83 is similar to the Michaelis-Menten decay law in chemical kinetics. The parameter $\alpha = \lambda N$ is the decay rate constant and is the value of the saturated decay rate of *duons*. Therefore, we can also write the generation rate for as,

$$\frac{\partial}{\partial t} P = \widetilde{\Delta}\, N_B^2\, P - \frac{\alpha P}{1 + \beta P}. \tag{84}$$

As seen in Eq. (86) below, the physical situation corresponds to $\alpha/\beta > G$. To prevent this 2-dimensional dynamical nonlinear system with a stimulated intermediate process from becoming unbounded, the maximum e-h generation rate for a given bias must be less than the maximum decay rate of *duons*. Therefore, in the formation of the high-field domain, the maximum discharging rate is larger than the maximum build-up rate. Indeed, we can estimate that $1/\tau_B \approx G$ and $1/\tau_L \approx \alpha/\beta$. Therefore $\alpha/\beta > G$ implies that $\tau_B > \tau_L$. Equations (79) and (84) describe a 2-parameter $(\widetilde{\Delta}, G)$ and 2-dimensional (N_B, P) dynamical system. The stationary solution to the coupled rate equations, Eqs. (79) and (84) is given by,

$$G = \widetilde{\Delta}\ N_B^2\ P = \frac{\alpha P}{1 + \beta P}, \tag{85}$$

and the total trapped hole concentration, Q_B, is thus given by,

$$Q_B^0 = N_B^0 + P^0 = \left(\frac{\alpha - \beta G}{\widetilde{\Delta}}\right)^{\frac{1}{2}} + \frac{G}{\alpha - \beta G}. \tag{86}$$

Note that N_B^0, which decreases with increase in G, corresponds to N_e, and P^0, which increases with G, corresponds to N_w. We should point out that neither Eqs. (68) and (69) for type I RTS nor Eqs. (79) and (84) for type II RTS account for the background current across the device which is expected to have an Ohmic behavior as a function of the bias voltage. For type II RTS where the coupled rate equations apply before the current peak, it is estimated that this background current is very small; it is also small for type I RTS under oscillating condition and operating just after the current peak.

3.3. *Stability analysis*

For the following stability and nonlinear analyses, it is convenient to unify the fundamental rate equations of type I and type II RTS's and write them in dimensionless form as,

$$\frac{\partial}{\partial t}\Pi = \Delta Q^2 \Pi - \frac{\Pi}{1 + \Pi}, \tag{87}$$

$$\frac{\partial}{\partial t}Q = G - \Delta Q^2 \Pi, \tag{88}$$

where the transformation of variables which expresses the duality of type I and type II RTS's is indicated as follows:

Dimensionless variable	Type I RTS	Type II RTS
Π	βN_w	βP
$\mathcal{Q}$	βN_e	βN_B
Δ	$\left(\frac{\widetilde{\Delta}\ \beta^2}{\alpha}\right)$	$\left(\frac{\widetilde{\Delta}\ \beta^2}{\alpha}\right)$
$\mathcal{G}$	$\frac{Q/RC}{(\alpha/\beta)}$	$\frac{G}{(\alpha/\beta)}$
τ	αt	αt

(89)

Note that $\widetilde{\Delta}, \alpha$, and β are different parameters between type I and Type II RTS's. The stationary solution to the coupled rate equations, Eqs. (87) and (88) is given by,

$$\mathcal{G} = \Delta \mathcal{Q}^2 \Pi = \frac{\Pi}{1+\Pi}. \tag{90}$$

In terms of these dimensionless variables, the stationary values of Π and $\mathcal{Q}$ are given by,

$$\Pi^0 = \frac{\mathcal{G}}{1-\mathcal{G}}, \tag{91}$$

$$\mathcal{Q}^0 = \left(\frac{1-\mathcal{G}}{\Delta}\right)^{\frac{1}{2}}. \tag{92}$$

The physical situation now corresponds to, $0 < \mathcal{G} < 1.0$, where for $\mathcal{G} > 1.0$, the dynamical system under study cannot be sustained or becomes unbounded in the presence of a catalytic process .

The question whether there is a nonstationary solution to our fundamental rate equations can first be answered by examining the stability of the stationary point in space. This is done by examining the neighborhood of the stationary point. Let us denote the coordinates of this neighborhood by,

$$\Pi = \Pi^0 + p, \tag{93}$$

$$\mathcal{Q} = \mathcal{Q}^0 + q. \tag{94}$$

The solution for the trajectories in space about the equilibrium point is given by

$$\begin{pmatrix} p \\ q \end{pmatrix} = A_1 e^{\lambda_1 \tau} \begin{pmatrix} V_1^p \\ V_1^q \end{pmatrix} + A_2 e^{\lambda_2 \tau} \begin{pmatrix} V_2^p \\ V_2^q \end{pmatrix} \tag{95}$$

where $\lambda_1$and λ_2 are the eigenvalues of the matrix defined by substituting Eqs. (93)-(94) in Eqs. (87)-(88) and retaining only linear terms in p and q [refer to Eq. (97) below, without the nonlinear terms]. The corresponding eigenvectors are V_1 and V_2, respectively. The eigenvalues are given by,

$$\lambda_{1,2} = \frac{Tr\,(M)}{2} + \frac{1}{2}\sqrt{Tr\,(M)^2 - 4\det(M)}. \tag{96}$$

The character of the stationary point can thus be determined with the help of the invariants of the matrix (M), namely, $Tr\,(M)\,, det\,(M)$, and $D\,(M) = [Tr\,(M)]^2 - 4\det(M)$. The stationary point can not be a saddle point for physical reason since $\det(M) == 2\mathcal{G}\,(1-\mathcal{G})^{\frac{3}{2}}\sqrt{\Delta} > 0$. The physical processes depicted in 11(a)-11(b) & 15 also suggest that the stationary point can only be any one of the following cases: stable focus $(Tr\,(M) < 0)$, center $(Tr\,(M) = 0)$, or unstable focus $(Tr\,(M) > 0)$. On physical grounds, we expect the limit cycle solution for uniqueness and structural stability. For the unstable focus we have to demonstrate that a limit cycle exists. The region in parameter space where the structurally stable limit cycle is possible lies in the area under the bifurcation curve (locus of $Tr\,(M) = 0$) in this space.

The trace of (M) is given by $Tr\,(M) = \mathcal{G}\left[(1-\mathcal{G}) - \{\Delta/(1-\mathcal{G})\}^{\frac{1}{2}}\right]$. Thus, $Tr\,(M) > 0$ implies $(1-\mathcal{G})^3 > 4\Delta$. In the next section, we will employ a nonlinear perturbation technique using the method of multiple time scales with values of the parameter around $Tr\,(M) = 0$. As we shall show in the following nonlinear analysis the limit cycle indeed occurs at $Tr\,(M) > 0$. The amplitude and frequency of oscillation is expected to depend on the actual values of the two parameters $\mathcal{G}$ and Δ in this region.

3.4. *Limit Cycle Solution*

Retaining nonlinear terms for p and q measured from the stationary point, the rate equations from Eqs. (87) and (88), become a matrix equation,

$$\frac{\partial}{\partial\tau}\begin{pmatrix} p \\ q \end{pmatrix} = \begin{pmatrix} \mathcal{G}\,(1-\mathcal{G}) & 2\mathcal{G}\,[\Delta/(1-\mathcal{G})]^{\frac{1}{2}} \\ -(1-\mathcal{G}) & -2\mathcal{G}\,[\Delta/(1-\mathcal{G})]^{\frac{1}{2}} \end{pmatrix}\begin{pmatrix} p \\ q \end{pmatrix} + \begin{pmatrix} N^p \\ N^q \end{pmatrix}, \tag{97}$$

where the nonlinear components of the last term are given by,

$$N^p = (1-\mathcal{G})^3 p^2 + 2\,[\Delta\,(1-\mathcal{G})]^{\frac{1}{2}}\,pq + \frac{\Delta\mathcal{G}}{(1-\mathcal{G})}q^2 + \Delta pq^2$$
$$+\sum_{n=3}^{\infty}(-1)^n\,(1-\mathcal{G})^{n+1}\,p^n, \tag{98}$$

$$N^q = -2\,[\Delta\,(1-\mathcal{G})]^{\frac{1}{2}}\,pq - \frac{\Delta\mathcal{G}}{(1-\mathcal{G})}q^2 - \Delta pq^2. \tag{99}$$

The perturbation technique employed in what follows essentially transforms the above nonlinear equation into a hierarchy of solvable and simpler equations, obtained by equating coefficients of powers of the smallness parameter, ε. Near $Tr\,(M) = 0$, we use as our smallness parameter the departure of Δ from Δ_c, where $4\Delta_c = (1-\mathcal{G})^3$. Thus, let the smallness parameter be $\varepsilon = \sqrt{\{\Delta - \Delta_c\}/\mathcal{D}}$, where $\mathcal{D}$ is determined from the expansion of Δ in powers of ε. $\mathcal{D} \approx \Delta_2$ in the analysis that follows. $\mathcal{G}$ is assumed constant at fixed bias, i.e., a function only of the external bias. We make the following expansion,

$$\Delta = \sum_{j=0}^{\infty}\varepsilon^j\Delta_j, \textit{ where } \Delta_0 = \Delta_c. \tag{100}$$

We also expand the matrix in powers of ε through direct Taylor expansion in powers of $\{\Delta - \Delta_c\}$ as,

$$(M) = (M_c) + \varepsilon\Delta_1\left(\left.\frac{\partial\,(M)}{\partial\Delta}\right|_{\Delta=\Delta_c}\right) + \frac{1}{2}\varepsilon^2\left[\Delta_2\left(\left.\frac{\partial\,(M)}{\partial\Delta}\right|_{\Delta=\Delta_c}\right)\right]$$
$$+\Delta_1^2\left(\left.\frac{\partial^2\,(M)}{\partial\Delta^2}\right|_{\Delta=\Delta_c}\right) + O\left(\varepsilon^2\right). \tag{101}$$

Using $4\Delta_c = (1-\mathcal{G})^3$, we obtain the following expressions,

$$(M_c) = \mathcal{G}(1-\mathcal{G})\begin{pmatrix} 1 & 1 \\ -\mathcal{G}^{-1} & -1 \end{pmatrix}, \tag{102}$$

$$(M_1) = \left(\frac{\partial (M)}{\partial \Delta}\bigg|_{\Delta=\Delta_c} \right) = 2\mathcal{G}(1-\mathcal{G})^{-2}\begin{pmatrix} 0 & 1 \\ 0 & -1 \end{pmatrix}, \tag{103}$$

$$(M_2) = \left(\frac{1}{2}\frac{\partial^2 (M)}{\partial \Delta^2}\bigg|_{\Delta=\Delta_c} \right) = -2\mathcal{G}(1-\mathcal{G})^{-5}\begin{pmatrix} 0 & 1 \\ 0 & -1 \end{pmatrix}. \tag{104}$$

We let the solution depends on time τ in a combination $\tau_0 = \tau$ and $\tau_1 = (\Delta - \Delta_c)\,\tau$. Thus, instead of determining the solution in terms of τ we seek the solution as a function of τ_0, τ_1, and ε. This method of doing the nonlinear perturbation analysis is well-known and is often referred to as the method of multiple time scales (Nayfeh 1981). This has the virtue that it separates the dependence of the solution into the fast and slow time scales. For limit cycle behavior, for example, we expect that the amplitude of the oscillation is only a function of the slow time scale. The left side of the rate equation can now be written as,

$$\frac{\partial}{\partial \tau}\begin{pmatrix} p(\tau_0,\tau_1,\varepsilon) \\ q(\tau_0,\tau_1,\varepsilon) \end{pmatrix} \Longrightarrow \left\{ \frac{\partial}{\tau_0} + (\Delta-\Delta_c)\frac{\partial}{\tau_1} \right\} \begin{pmatrix} p(\tau_0,\tau_1,\varepsilon) \\ q(\tau_0,\tau_1,\varepsilon) \end{pmatrix}. \tag{105}$$

Since the last term in Eq.(97) represents the nonlinear term for the solution, we adopt the following expansion,

$$\begin{pmatrix} p(\tau_0,\tau_1,\varepsilon) \\ q(\tau_0,\tau_1,\varepsilon) \end{pmatrix} = \sum_{j=0}^{\infty} \varepsilon^{j+1} \begin{pmatrix} p_j(\tau_0,\tau_1) \\ q_j(\tau_0,\tau_1) \end{pmatrix}. \tag{106}$$

Therefore, any finite solution will indicate that the limit cycle occurs for values of the parameter away from the critical point, $Tr\,(M) = 0$, i.e., away from the bifurcation point. With Eq. (106), the nonlinear term in Eq. (97) acquires the following expansion in terms of the smallness parameter,

$$\begin{pmatrix} N^p \\ N^q \end{pmatrix} = \varepsilon^2 \begin{pmatrix} N_2^p \\ N_2^q \end{pmatrix} + \varepsilon^3 \begin{pmatrix} N_3^p \\ N_3^q \end{pmatrix} + O\left(\varepsilon^4\right), \tag{107}$$

where

$$\begin{pmatrix} N_2^p \\ N_2^q \end{pmatrix} = (1-\mathcal{G})^2 \begin{pmatrix} \left[p_0q_0 + (\mathcal{G}/4)\,q_0^2 + (1-\mathcal{G})\,p_0^2\right] \\ -\left[p_0q_0 + (\mathcal{G}/4)\,q_0^2\right] \end{pmatrix}, \tag{108}$$

$$\begin{pmatrix} N_3^p \\ N_3^q \end{pmatrix} = \begin{pmatrix} \begin{bmatrix} p_0q_1 + p_1q_0 + (\mathcal{G}/2)\,q_1q_0 + \frac{(1-\mathcal{G})}{4}p_0q_0^2 \\ +\frac{\Delta_1\mathcal{G}}{(1-\mathcal{G})^3}q_0^2 + \left\{\frac{\Delta_2}{(1-\mathcal{G})^3}\right\}^{\frac{1}{2}} 2p_0q_0 \\ +2(1-\mathcal{G})\,p_0p_1 - (1-\mathcal{G})^2\,p_0^3 \end{bmatrix} \\ -\begin{bmatrix} p_0q_1 + p_1q_0 + (\mathcal{G}/2)\,q_1q_0 + \frac{(1-\mathcal{G})}{4}p_0q_0^2 \\ +\frac{\Delta_1\mathcal{G}}{(1-\mathcal{G})^3}q_0^2 + \left\{\frac{\Delta_2}{(1-\mathcal{G})^3}\right\}^{\frac{1}{2}} 2p_0q_0 \end{bmatrix} \end{pmatrix}. \tag{109}$$

We did not show nonlinear terms with fractional powers of ε in Eq.(107) associated with Δ_1 in Eqs. (98) and (99), since the left hand side of the rate equation does not contain fractional powers of ε. To eliminate the occurrence of these fractional powers of ε, we have to make $\Delta_1 = 0$ in the expansion of Δ, Eq. (100), and also in Eqs.(101) and (109).

Upon substituting all the expanded quantities in the nonlinear rate equation, Eq. (97), we obtain a hierarchy of simpler equations. Those arising from the first up to the third powers of ε are given below,

$$\mathcal{L}_0 \begin{pmatrix} p_0 \\ q_0 \end{pmatrix} = 0, \tag{110}$$

$$\mathcal{L}_0 \begin{pmatrix} p_1 \\ q_1 \end{pmatrix} = \begin{pmatrix} N_2^p\left(p_0, q_0\right) \\ N_2^q\left(p_0, q_0\right) \end{pmatrix}, \tag{111}$$

$$\mathcal{L}_0 \begin{pmatrix} p_2 \\ q_2 \end{pmatrix} + \Delta_2 \mathcal{L}_1 \begin{pmatrix} p_0 \\ q_0 \end{pmatrix} = \begin{pmatrix} N_3^p\left(p_0, q_0, p_1, q_1\right) \\ N_3^q\left(p_0, q_0, p_1, q_1\right) \end{pmatrix}, \tag{112}$$

where

$$\mathcal{L}_0 = \left(\frac{\partial}{\partial \tau_0} - (M_c) \right), \tag{113}$$

$$\mathcal{L}_1 = \left(\frac{\partial}{\partial \tau_1} - (M_1) \right). \tag{114}$$

The first equation in the hierarchy is a simple eigenvalue problem, analogous to our linear-stability analysis before. The only difference is that the present eigenvalue problem has to be solved with values of the parameter at the critical point, where $Tr\,(M) = 0$, using the matrix (M_c). The solutions to Eqs.(110), (111), and (112) are detailed in the Appendix of Buot et al.[34] to second order in ε , where it is shown that limit cycle exists for $\Delta < \Delta_c$.

Thus, to second order in the smallness parameter, the limit cycle solution is given as,

$$\begin{pmatrix} \Pi \\ \mathcal{Q} \end{pmatrix} = \begin{matrix} \Pi_0 \\ \mathcal{Q}_0 \end{matrix} + \left(\left| \frac{\Delta - \Delta_c}{\Delta_2} \right| \right)^{\frac{1}{2}} \begin{pmatrix} p_0 \\ q_0 \end{pmatrix} + \left(\left| \frac{\Delta - \Delta_c}{\Delta_2} \right| \right) \begin{pmatrix} p_1 \\ q_1 \end{pmatrix} + O\left(\left| \frac{\Delta - \Delta_c}{\Delta_2} \right|^{\frac{3}{2}} \right), \tag{115}$$

where we have

$$\begin{pmatrix} p_0 \\ q_0 \end{pmatrix} = \frac{|\Theta(\infty)|}{\mathcal{G}^{\frac{1}{2}}} \begin{pmatrix} 2\ \cos \Omega\tau \\ -2\mathcal{G}^{\frac{1}{2}} \cos \Omega\tau - (1-\mathcal{G})^{\frac{1}{2}}\, 2\ \sin \Omega\tau \end{pmatrix} = \frac{2\,|\Theta(\infty)|}{\mathcal{G}^{\frac{1}{2}}} \begin{pmatrix} \cos \Omega\tau \\ -\sin(\Omega\tau + \Phi) \end{pmatrix}, \tag{116}$$

and

$$\Phi = \tan^{-1} \left\{ \frac{\mathcal{G}}{(1-\mathcal{G})} \right\}^{\frac{1}{2}}, \tag{117}$$

$$\Omega = \mathcal{G}^{\frac{1}{2}}(1-\mathcal{G})^{\frac{3}{2}} + \left\{ \mathrm{Im}\,\eta(\Delta-\Delta_c) + \frac{\mathrm{Im}\sigma}{\Delta_2}\left|\Theta(\infty)\right|^2(\Delta-\Delta_c)\right\}. \quad (118)$$

We also have,

$$\begin{pmatrix} p_1 \\ q_1 \end{pmatrix} = \left|\Theta(\infty)\right|^2 \frac{(1-\mathcal{G})}{\mathcal{G}} \begin{pmatrix} 2\mathcal{G} \\ -\frac{1}{2} \end{pmatrix} + \frac{\left|\Theta(\infty)\right|^2}{3\mathcal{G}(1-\mathcal{G})} \begin{pmatrix} \left\{ \begin{matrix} 4(1-\mathcal{G})^3 \cos 2\Omega\tau \\ -\omega(1+2\mathcal{G})\sin 2\Omega\tau \end{matrix} \right\} \\ \left\{ \begin{matrix} \left(\frac{15}{2}\mathcal{G}-\frac{9}{2}-3\mathcal{G}^2\right)(1-\mathcal{G})\cos 2\Omega\tau \\ +\omega\left(8-3\mathcal{G}-2\mathcal{G}^{-1}\right)\sin 2\Omega\tau \end{matrix} \right\} \end{pmatrix}. \quad (119)$$

We note that Eq. (119) also contains a time-independent term, indicating a higher-order shift of the center of the limit cycle from the stationary point. Therefore, the average value of is given by,

$$\begin{pmatrix} \Pi \\ \mathcal{Q} \end{pmatrix}_{average} = \begin{pmatrix} \Pi_0 \\ \mathcal{Q}_0 \end{pmatrix} + \left(\left|\frac{\Delta-\Delta_c}{\Delta_2}\right|\right)\left|\Theta(\infty)\right|^2 \frac{(1-\mathcal{G})}{\mathcal{G}} \begin{pmatrix} 2\mathcal{G} \\ -\frac{1}{2} \end{pmatrix} + O\left(\varepsilon^3\right)$$

$$= \begin{pmatrix} \Pi_0 \\ \mathcal{Q}_0 \end{pmatrix} + higher-order\ corretions, \quad (120)$$

where the leading higher-order corrections come from the time-independent terms. Thus, we have demonstrated that a unique limit cycle exists away, ($\Delta < \Delta_c$), from the bifurcation point, ($\Delta = \Delta_c$), and the average value is determined by the time-independent terms.

In examining the dependence of various quantities on the driving source,$\mathcal{G}$, we make the assumption that $(\Delta - \Delta_c)$ is approximately a constant. The physical reason for this is that Δ is a measure of the inductive delay for type I RTS as indicated in Fig. 11(c). Δ is large for small inductive delay and small for large inductance. For larger driving source, $\mathcal{G}$, which happens immediately after the current-peak bias, Δ_c is smaller and we also expect larger inductive delay at this point, and hence Δ is also smaller. For larger values of bias in the plateau, meaning weaker driving source $\mathcal{G}$, Δ is larger and we also expect the inductive delay to be smaller, meaning Δ is also larger. Therefore, $|\Delta - \Delta_c|$ is approximately constant. It is taken small enough such that the second term in Eq. (120) is only a very small correction to the first term, otherwise one has to include other higher order terms. Similar relation also holds for the type II RTS.

3.5. *Discussion*

3.5.1. *Average Value of the Current*

Because of the strong correlation between the oscillatory charges, N_e and N_w for type I RTS and N_B and P for type II RTS, the averaged measurable value of the current is determined by the sum of the stationary values of N_e and N_w or N_B and P. More specifically for type I RTS, the average current is $I_{dc} \approx \left(N_e^0 + N_w^0\right)/\tau_{eff}$,

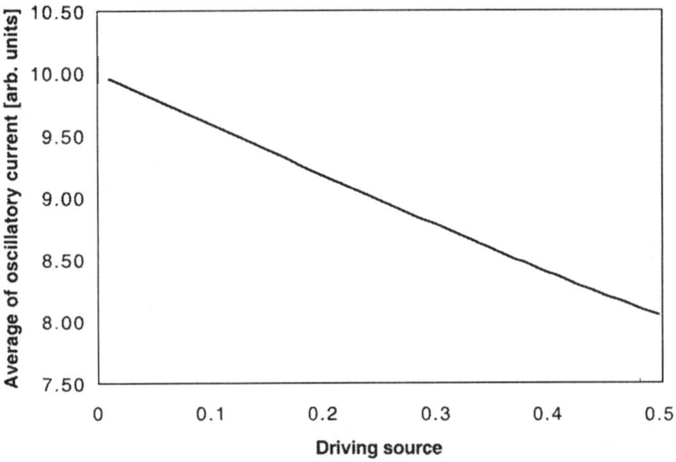

Fig. 16. Plot of average of the oscillatory current as a function of the driving source $\mathcal{G}$, for $\Delta < \Delta_C$. Increasing the drain bias goes with the decreasing values of $\mathcal{G}$ for type I RTS and with increasing values of $\mathcal{G}$ for type II RTS.

as indicated in the equivalent circuit model of Fig. 11d, where $\tau_{eff} \approx 2\tau_c$. Referring to Eqs. (89) and (91)-(92), the leading average value of this sum is determined by,

$$\Sigma^0(\mathcal{G}) = \Pi^0 + \mathcal{Q}^0 = \frac{\mathcal{G}}{(1-\mathcal{G})} + \left(\frac{(1-\mathcal{G})}{\Delta}\right)^{\frac{1}{2}}. \tag{121}$$

Therefore, we have,

$$\frac{\partial \Sigma^0(\mathcal{G})}{\partial \mathcal{G}} = \frac{1}{(1-\mathcal{G})^2} - \left(\frac{1}{4\Delta(1-\mathcal{G})}\right)^{\frac{1}{2}} \begin{cases} < 0 \ for \ \Delta < \Delta_c(\text{oscillatory}), \\ > 0 \ for \ \Delta > \Delta_c(\text{non-oscillatory}). \end{cases} \tag{122}$$

We conclude from Eq. (122) that the average of the oscillatory current increases as $\mathcal{G}$ decreases. We expect the charging rate, Q/RC , to be proportional to the energy difference (which decreases with bias) between the QW energy level and the conduction band edge of the emitter of type I RTS. On the other hand, $\mathcal{G}$, which measures the maximum Zener tunneling rate, increases with bias for type II RTS.

We estimated the range of $\mathcal{G}$ to be < 0.5. Figure 16 shows the average value of $\Sigma^0(\mathcal{G})$ for $\Delta < \Delta_c$, i.e., oscillatory condition. It is important to point out however that in the absence of intrinsic oscillatory behavior, our nonlinear model indicates that the stationary current is proportional to N_w or N_B alone, which increases with the driving source, $\mathcal{G}$, shown in Fig. 17. Thus in the absence of intrinsic oscillation, the current is expected to exhibit a decreasing behavior as a function of bias (increase of bias means decrease of $\mathcal{G}$ for type I RTS). Thus, our nonlinear model is

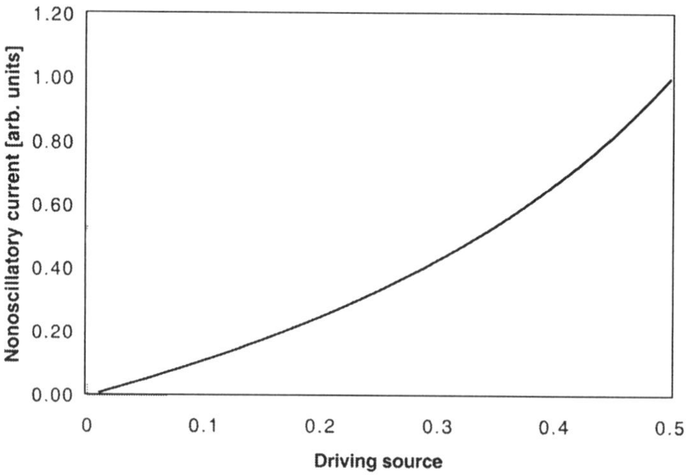

Fig. 17. Plot of the stable stationary values of plateau current as a function of $\mathcal{G}$ for $\Delta > \Delta_C$.

able to descriminate the presence or absence of oscillation in type I and type II RTS's.

3.5.2. *Relation Between Amplitude and $\mathcal{G}$*

For type I RTS, the oscillatory current is proportional to $p/\tau_c + \partial q/\partial\tau$ (refer to Fig. 11(c)) expanded to second order using Eqs. (115)-(119). We now show that the amplitude of oscillation increases with $\mathcal{G}$. From Eqs. (115) and (116), this amplitude of the fundamental frequency component is typified by the following expression,

$$\mathcal{A} = \frac{\Omega}{\beta}\left(|\Delta - \Delta_c|\right)^{\frac{1}{2}}\left[\frac{64\,\mathcal{G}}{(1-\mathcal{G})^5\left\{4 + 19\mathcal{G} - 8\mathcal{G}^2\right\}}\right]^{\frac{1}{2}}. \tag{123}$$

Figure 18 is a plot of this amplitude as a function of $\mathcal{G}$. The interference between the two cosine terms in the expression for the current further enhances this behavior in the physical range of $\mathcal{G}$. For type II RTS, the amplitude of oscillation has the same dependence on $\mathcal{G}$. The physical significance of these results is discussed in more detail by Buot and Krowne[46] and Buot et al.[34]

3.5.3. *Oscillation Frequency and its Dependence on $\mathcal{G}$*

The fundamental frequency of oscillation is given by $\omega_f = \alpha\Omega$, where the factor α arises by virtue of the conversion of τ to real time t. We have,

$$\omega_f = \alpha\sqrt{\mathcal{G}}\,(1-\mathcal{G})^{\frac{3}{2}} + \alpha\left\{\mathrm{Im}\eta\,(\Delta - \Delta_c) + \frac{\mathrm{Im}\sigma}{\Delta_2}\left|\Theta(\infty)\right|^2(\Delta - \Delta_c)\right\}. \tag{124}$$

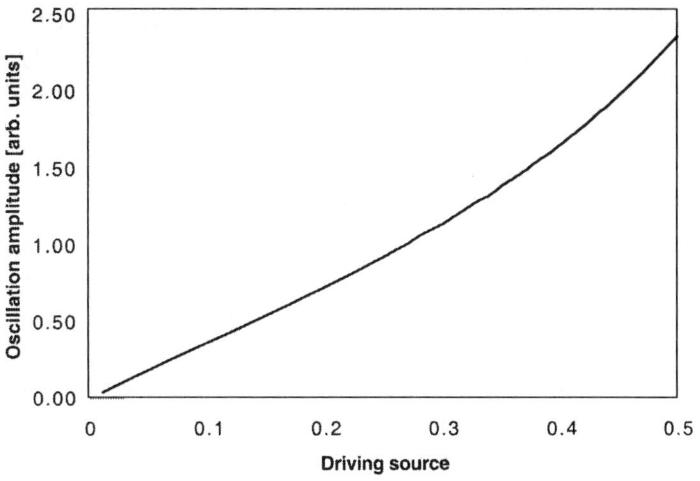

Fig. 18. Plot of the amplitude of oscillation of oscillatory current as a function of $\mathcal{G}$ for $\Delta < \Delta_C$.

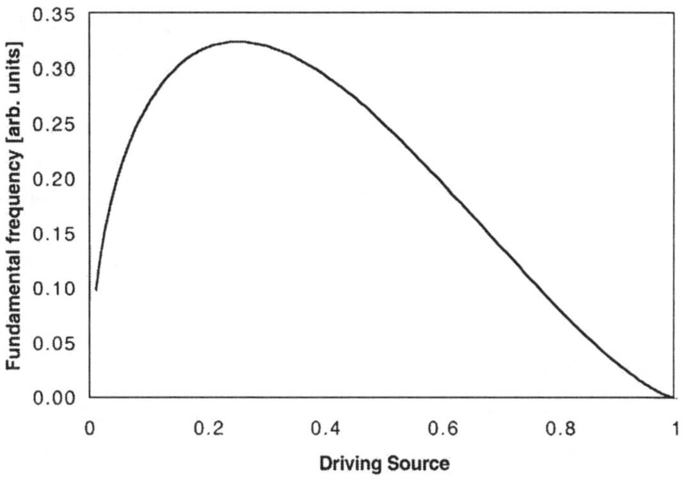

Fig. 19. Plot of $\sqrt{\mathcal{G}\,(1-\mathcal{G})^3}$ in Eq. (124) for ω_f as a function of $\mathcal{G}$ for $\Delta < \Delta_C$.

Thus, oscillation is indeed driven by the source, $\mathcal{G}$. Figure 19 shows that within the range of $\mathcal{G}$ considered in Figs. 16-18, the frequency temporarily increases as $\mathcal{G}$ decreases towards zero. For type I RTS, we see that while the amplitude of oscillation decreases, the frequency of oscillation increases at first before sharply decreasing to zero towards the end of the plateau. These behaviors were indeed noticeable in various numerical quantum transport simulations of type I RTS, and lead to the complete resolution of the long-standing controversy about the different measured I-V characteristics.[34]

3.5.4. *Equivalent-circuit Model of THz Generation*

Recently, Poltoratsky and Rychov[48] made a study of the general form of the coupled nonlinear differential equations obeyed by the equivalent-circuit model (ECM) of type I resonant-tunneling heterostructure. The equivalent circuit they used is a modification ECM originally introduced by Buot and Jensen[36] and Buot and Rajagopal[45], by adding a small capacitor across the nonlinear circuit element, i.e., in parallel with the nonlinear NDR circuit element and in series with the inductance. Indeed, the ECM they have proposed is in agreement with the theory discussed here, as can be seen from Fig. 11(c) [see also Buot, Zhao et al[34]] which suggests the presence of a small capacitor in parallel with the nonlinear circuit element $i\,(v)$ to account for the quantum-well stored charge N_w, which is out of phase with the emitter-drain capacitance stored charge N_e. The added small capacitor is intended to account for the charging of the quantum well when current passes through the inductive element, which is in parallel with the larger emitter-drain capacitance. The added small capacitor therefore describes the faster built-up of charge in the quantum well when the larger emitter-drain capacitor is discharging by virtue of the current through the inductance as a result of the realignment between the bottom of allowed energy levels in the emitter and quantum-well energy level.

The authors applied a quite general criterion in form of a theorem under which the resulting form of coupled nonlinear differential equations generate stable and unstable limit cycles. On the basis of this criterion the authors were able to explain THz generation and the presence of hysteresis and plateau-like behavior of the RTS I-V characteristics.

3.6. *Concluding Remarks*

We have demonstrated that theories of self-oscillation of type I and type II RTS's possess inherent pairing of relevant dynamical variables and parameter space, or duality. The distinguishing feature between the two theories lies in the behavior of the driving source $\mathcal{G}$. Whereas $\mathcal{G}$ decreases with bias for type I RTS, in contrast it increases with bias for type II RTS. The pairing or parallelism, referred to here as the duality theory of type I and type II RTS's, also lies on the correspondence between their physical variables, namely, $\beta N_w \Leftrightarrow \beta P$ and $\beta N_e \Leftrightarrow \beta N_B$(refer to Eq.(89). The electron current across the device always has contribution that is proportional to either N_w or P in both ac and dc situations. On the other hand, N_e and N_B act as stored capacitive charges in type I and type II, respectively, and hence are expected to contribute to the current only under oscillatory condition ($\Delta < \Delta_c$). Indeed, under this oscillatory condition, the bias dependence of N_e^0

and N_B^0 dominate the average I-V behavior leading to the decreasing current with increase in $\mathcal{G}$. For non-oscillatory situation ($\Delta > \Delta_c$), it is the bias dependence of N_w^0 and P^0 which dominate the I-V behavior leading to an increasing current with $\mathcal{G}$.

For type I RTS, the limit cycle solution supports the current oscillation in the plateau range found in various numerical quantum transport simulations, as well as in other analytical model employing the solution of many-particle Schroedinger equation. Specifically, it predicts a rising plateau current and a decreasing amplitude of current oscillation as a function of bias in the plateau range. Within this plateau, the frequency of oscillation is found to increase, reaches a maximum, and then decreases sharply to zero as a function of bias. It also predicts larger signal amplitude and the presence of higher harmonics just after the current peak. All these findings are in agreement with numerical simulations and experiments. Furthermore, in the absence of oscillation, our model predicts a decreasing current in the plateau. This happens if there is a weak quantizing field in the emitter brought about, for example, by a large barrier width.

A type I RTS device, under appropriate device parameters, has the potential for operating as a novel all solid-state THz source. This device must be operated just after the resonant current peak to maximize the output power at THz frequencies. This new intraband THz source is expected to significantly extend the domain of application of the traditional IMPATT and Gunn effect microwave solid-state sources. The results suggest further refinements of the ECM of RTS in a form of small capacitor across the nonlinear NDR circuit element.

In type II RTS, the dynamical behavior of a coupled systems of *duon* and unpaired trapped hole charge has been shown to give physical explanation of the measured I-V characteristic of AlGaSb/InAs/AlGaSb double-barrier structure. The stimulated production of *duons* and Zener tunneling of electrons lead to an autonomous control of the position of the energy level of the quantum well by the trapped charge. The self-oscillating trapped hole charge provides the physical control mechanism behind a novel interband-tunnel high-frequency-source RTS device. This oscillation is useful for applications well beyond the range of applications of the traditional IMPATT and Gunn effect devices. In practice, for a given material parameter Δ, we choose the operating bias, directly related to $\mathcal{G}$, such that $Tr\,(M) > 0$. For a range of $\mathcal{G}$ where this is satisfied, we can optimized the operating point to realize a THz source with optimum power and frequency. Since in general the *duon* formation is a higher-order process involving Zener tunneling coupled with resonant tunneling to form a polarization pair, we expect Δ to be considerably less than one (refer to Eq. (89)). Thus, in realistic device $Tr\,(M) > 0$ should be satisfied.

4. A Multi-Band Physics-Based Transport Model for Staggered-Bandgap Tunneling Structures

4.1. *Introduction*

In the past, resonant tunneling diodes (RTDs) were investigated as potential low-power sources of electromagnetic energy within the submillimeter-wave region. Here, the extremely fast response associated with the negative differential resistance (NDR) of RTDs provided an adequate gain mechanism for the generation of terahertz (THz) energy. Unfortunately, while double-barrier RTDs were successfully implemented as two-terminal oscillators up to 712 GHz,[49] the power output levels were restricted to microwatt levels primarily as a result of low-frequency design constraints (i.e., suppression of bias circuit oscillations). Furthermore, these fundamental limitations in the conventional implementation of RTDs as oscillation sources immediately motivated further investigations into "intrinsic" instability mechanisms[9] as a novel approach for circumventing the power restrictions. In the work presented here a novel family of staggered-bandgap heterostructures is considered that may provide a new avenue for the development of very high frequency solid-state oscillators. Specifically, staggered bandgap heterostructures can admit significant interband tunneling currents [put in ref]. Hence, the possibility exists that these interband currents might be engineered to develop a nanoscale feedback between adjacent valence and conduction band wells. This novel use of nanotechnology seems extremely feasible from a fundamental perspective; however, detailed theoretical studies will be required to enable a useful insight into the basic phenomenon. This paper will establish an accurate description of the electron transport within resonant tunneling diodes with staggered-bandgap heterostructures.

This paper investigates RTDs based on a type II resonant tunneling structure, i.e. a structure in which the quantum well and barrier semiconductors possess a type II band gap alignment to each other (see Fig. 20). The quantum well is formed by InAs sandwiched between two barrier regions (AlGaSb/InAs/AlGaSb double barrier structures). The left AlxGa1-xSb barriers is adjacent to a highly doped InAs emitter while the right barrier is adjacent to the undoped InAs spacer which is grown on the highly doped InAs collector region. Models are developed and utilized to study the conduction-band transport and these results are compared to existing measurement results within the scientific literature. As will be shown, electron transport within the conduction band is highly dependent on the coupling between the conduction and valence bands and an accurate estimate of current density requires the application of a multi-band model. In this work, a six-band Kane model is derived that yields very good agreement with experimental measurement. This research represents an early, and necessary step towards a fully time-dependent analysis of the transport physics and a complete understanding of nanoscale feedback within staggered-bandgap heterostructures.

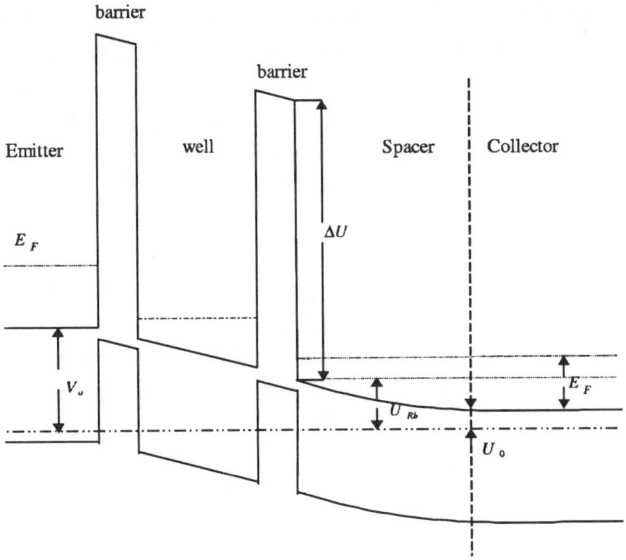

Fig. 20. AlGaSb/InAs/AlGaSb staggered-bandgap structure under study.

4.2. *The Interband Transport Model*

At lower bias voltages the current within RTDs with staggered-bandgap structures (SBSs) is dominated by conduction-band electron transport. Hence, the first step towards an accurate description of these devices is the development of an adequate physical model for the electronic motion in the conduction band. In most investigations of conduction-band tunneling processes the analysis considers single-band transport and ignores any coupling (e.g., arising out of changes in effective mass at the hetero-interfaces) between the transverse (i.e., perpendicular to the barriers) and in-plane wave-vector components. However, in SBSs it is not sufficient to apply such decoupled methods even if one is only interested in the conduction-band electrons (i.e., electrons with energies lying in the conduction band). It is very important to recall that the luxury of utilizing single-band transport models is made possible only because energy within the band of interest is relatively low-energy. Specifically, because the in-band energy (i.e., $E - E_C$) is relatively small compared to the energy gaps between the band of interest and all lower energy bands. Since the SBS under consideration contains InAs and AlGaSb layers, with relatively narrow bandgaps, one must apply the six-band Kane model with properly coupled multi-band wave functions.

The derivation of the necessary multi-band formulation for accurately describing transport within the SBSs begins with a study of Schrödinger's wave equation subject to a regular periodic potential (i.e., defined by $U_{lattice}(\vec{r})$) and the application of the Bloch condition on the basis states. The physics of electron motion will first be considered in bulk material without any applied bias. In situations where multi-band transport effects are expected, the electron wavefunction can be defined as $\Psi_{n\vec{k}}(\vec{r}) = u_{n\vec{k}}(\vec{r})\exp(i\vec{k}\cdot\vec{r})$, where $u_{n\vec{k}}(\vec{r})$ is the Bloch lattice function and n denotes the band index. The Bloch representation of the wavefunction may be utilized to derive the Schrödinger-Bloch equation[50]

$$(\hat{H}_0 + \hat{H}_{int})u_{n\vec{k}}(\vec{r}) = E'_n(\vec{k})u_{n\vec{k}}(\vec{r}), \quad E'_n(\vec{k}) = E_n(\vec{k}) - \hbar^2k^2/2m, \tag{125}$$

where $\hat{H}_0 = -(\hbar^2/2m)\nabla_r^2 + U_{lattice}(\vec{r})$ is the Hamiltonian for the semiconductor system, $U_{lattice}(\vec{r})$ is the potential of the semiconductor lattice, $\hat{H}_{int} = \hbar(\vec{k}\cdot\hat{P})/m$ is the electron-lattice interaction operator, $\hat{P}$ is the momentum operator and $E_n(\vec{k})$ are the energy eigenvalues of the Schrödinger equation. For many semiconductor transport problems $u_{n\vec{k}}(\vec{r})$ is a sufficiently weak function of $\vec{k}$ and it is possible to derive and effectively utilize a single-band effective mass equation.[51] However, as will be illustrated next, this is not the case for the SBS system under consideration.

The use of a single-band transport model (e.g., conduction band) implies that the electron energy may be accurately described by a perturbation-based expansion around the energy-band minimum (e.g., $E_c = E_c^{(0)} + E_c^{(1)} + E_c^{(2)} + \ldots\ldots$). However in the case under study here, a perturbational analysis[52] of Eq. (125) around $\vec{k} = 0$ may be used to show that the strongest contributions to transport in the direct bandgap semiconductors is derived collectively from the contributions from the lowest conduction band and highest valence bands (i.e., the degenerate light and heavy hole). Perturbation theory requires that the matrix elements for this six-band system obey the equality

$$| < n|\hat{H}_{int}|m > | \ll |E_n^{(0)} - E_m^{(0)}|, \tag{126}$$

where n and m are the conduction- and valence-band indexes and $E_n^{(0)}$ and $E_m^{(0)}$ are the corresponding band energies at $\vec{k} = 0$. For the six-band system under consideration all the extrema occur at $\vec{k} = 0$ and since the perturbing interaction term is linear in $\vec{k}$ (which means that the first order term is zero) we know that the total energy within the nth (or mth) band is given by $E_n = E_n^{(0)} + E_n^{(2)} + \ldots\ldots$ ($E_m = E_m^{(0)} + E_m^{(2)} + \ldots\ldots$). Second order perturbation theory tells us that the conduction band energy is given by[53]

$$E_c^{(2)} = | < v|\hat{H}_{int}|c > |^2/E_g,$$

where c denotes the conduction band, v denotes the valence band and $E_g = E_c^{(0)} - E_v^{(0)}$ is the bandgap, and this directly yields

$$E_c - E_c^{(0)} = \frac{| < v|\hat{H}_{int}|c > |^2}{E_g}, \tag{127}$$

which can be substituted into Eq. (127) to derive the conditions necessary for applying the effective mass approximation

$$|E_c - E_c^{(0)}| \ll E_g. \tag{128}$$

However, for the material systems and heterostructures under consideration the electron energies both in the conduction and valence bands will easily violate the condition in Eq. (128). Hence, a proper description of the electron and hole transport requires a multi-band model and due to this requirement the six-band Kane model will be developed here for this purpose.

The Schrödinger-Bloch equation given in Eq. (125) provides important guidance towards a methodology for solving the basic transport problem within a multi-band system.[50] As the Bloch lattice function must be periodic with the semiconductor lattice under consideration, Eq. (125) defines a regular boundary value problem. Furthermore, for a given value of $\vec{k} = \vec{k'}$ Eq. (125) will admit periodic solutions $u_{n\vec{k}=\vec{k'}}(\vec{r})$ only for certain energy eigenvalues E_n. However, for each value of $\vec{k'}$ there will be an infinite number of these energy eigenvalues $E_n(\vec{k'})$. The infinite set $u_{n\vec{k}=\vec{k'}}(\vec{r})$ are eigenfunctions of a Hermitian operator so they comprise an orthogonal set that can be normalized. As they also form a complete set, all the Bloch functions are expandable according to,

$$u_{n\vec{k}=\vec{k'}}(\vec{r}) = \sum_m a^n_{m\vec{k}=\vec{k'}}(\vec{r}) u_{n\vec{k}=0}(\vec{r}), \tag{129}$$

where the functions at $\vec{k} = \vec{k'}$ have been expanded in terms of the functions at $\vec{k} = 0$. Equation (129) may now be used to write the Bloch form of the electron wavefunction as

$$\Psi_{n\vec{k}}(\vec{r}) = \sum_m a^n_{m\vec{k}=\vec{k'}}(\vec{r}) u_{n\vec{k}=0}(\vec{r}) \exp(i\vec{k}\cdot\vec{r}), \tag{130}$$

for specified value of wavevector which is now written simply as $\vec{k}$. Equation (130) shows that the expansion coefficients, $a^n_{m\vec{k}=\vec{k'}}(\vec{r})$, correspond to the envelope functions for the electron wavefunction within the nth energy band when written in the $\vec{k} = 0$ Bloch basis. Equation (6) (130) provides a convenient formalism for treating the electron transport within a multi-band system once the Bloch functions at $\vec{k} = 0$ are defined. Indeed, as is well known,[51] the use of these envelope functions enables one to avoid a direct treatment of the rapidly varying functions $u_{n\vec{k}}$.

Hence, the goal is now to define the general form of the Bloch functions for the conduction and valence bands in the six-band Kane-model framework. These functions will be defined at $\vec{k} = 0$ where the interaction term is not active. We begin by recognizing that the material systems under consideration possess cubic symmetry and this implies that at $\vec{k} = 0$ the conduction band (CB) Bloch functions must have s-symmetry while the valence band (VB) Bloch functions must have p-symmetry. Here, conventional notations will be utilized where the spatial part of the CB Bloch function is denoted by S (which is symmetric under cubic transformations in real space) and the VB Bloch functions by X, Y, and Z, respectfully. If we initially ignore the influence of spin then the conduction band is non-degenerated and S can be uniquely defined. On the other hand, the valence band is threefold degenerated even before the incorporation of spin effects. Hence, the VB Bloch function may be represented by any linear superposition of the three spatial functions X, Y, Z. The VB Bloch functions are definable from the quantum mechanical representation for total angular momentum and the z-projection of the angular momentum. The basis of representation for the normalized Bloch eigenfunctions is derived using the commuting operators for orbital angular momentum squared (i.e., $\hat{L}^2$) and z-projected angular momentum (i.e., $\hat{L}_z$) as[54]

$$\hat{L}^2\psi_{lm} = \hbar^2 l(l+1)\psi_{lm}, \quad \hat{L}_z\psi_{lm} = m\hbar\psi_{lm}, \tag{131}$$

where for the p-symmetric functions $l = 1$ and m assumes the values $\{-1, 0, 1\}$. The eigenfunctions of the angular momentum are the spherical harmonics (as both angular momentum operators,

$$\hat{L}^2 = -\hbar^2[(1/\sin^2\theta)\partial^2/\partial\phi^2 + (1/\sin\theta)(\partial/\partial\theta)\sin\theta(\partial/\partial\theta)]$$

and

$$\hat{L}_z = (\hbar/i)\partial/\partial\phi,$$

only depend on angle in spherical coordinates leading to the eigenfunctions, $\psi_m = \exp(im\phi)P_m(\theta)$, $P_0 = \cos\theta$, $P_\pm = \sin\theta$) and may be related to the spatial functions according to

$$\psi_\pm = \mp\frac{1}{\sqrt{2}}|(X \pm iY)>, \quad \psi_0 = |Z>, \tag{132}$$

where the subscript for l has been dropped and the m values of $\{-1, 0, 1\}$ are denoted as $\{+, 0, -\}$.

The eigenfunctions in Eq. (132) will be modified due to the existence of electron spin. Consider first the case where $\hat{H}_0 = -(\hbar^2/2m)\nabla_r^2 + U_{lattice}(\vec{r})$ does not depend on electron spin - e.g., if the possibility of spin-orbit interaction introduced through $U_{lattice}(\vec{r})$ is ignored. In this simple case the modification is trivial and all that is required is to account for the introduced degeneracy. Specifically, the conduction band is two-fold degenerated and the valence band is six-fold degenerated. Here, all the spatial functions may be multiplied by simple spinors (i.e., denoted as $|\uparrow> = \begin{pmatrix}1\\0\end{pmatrix}$ or $|\downarrow> = \begin{pmatrix}0\\1\end{pmatrix}$) to account for the doublets introduced by spin. Hence, the CB eigenfunctions become $|S\uparrow>$ and $|S\downarrow>$, and the VB eigenfunctions become $\mp\frac{1}{\sqrt{2}}|(X \pm iY)\uparrow>$, $\mp\frac{1}{\sqrt{2}}|(X \pm iY)\downarrow>$and $|Z\uparrow>$ and $|Z\downarrow>$. Now that the general notation has been established the influence of spin-orbit interaction will be included. The total momentum at $\vec{k} = 0$ is defined through the operator $\hat{J} = \hat{L} + \hat{s}$. Since neither $\hat{L}$ or $\hat{s}$ individually commute with the Hamiltonian, the eigenstates for the VB system must be labeled in accordance with the commuting operator $\hat{J}$. The eigenfunctions must now be derived from

$$\hat{J}^2\psi_{jM} = \hbar^2 j(j+1)\psi_{jM}, \quad \hat{J}_z\psi_{jM} = M\hbar\psi_{jM}, \tag{133}$$

where there are six valence bands that split according to two possible values of the quantum number j. Here one has a four-fold degenerate VB defined by $j = 3/2$

with M values of $\{-3/2, -1/2, 1/2, 3/2\}$ and a two-fold VB defined by $j = 1/2$ with M values of $\{-1/2, 1/2\}$. While this general analysis can not predict which band is highest, experimental investigations have demonstrated that the four-fold degenerated $j = 3/2$ VB is the upper band in InAs and AlGaSb. The eigenvalues for the total spin squared and z-projected spin are definable from

$$\hat{s}^2\xi_{s\sigma} = \hbar^2 s(s+1)\xi_{s\sigma} = \frac{3}{4}\hbar^2\xi_{s\sigma}, \quad \hat{s}_z\xi_{s\sigma} = \hbar\sigma\xi_{s\sigma}, \tag{134}$$

where s assumes the value $1/2$ and σ assumes the values $\{-1/2, 1/2\}$. The goal now is to derive the eigenfunctions of the z-projected total angular momentum (i.e., defined from the operator $\hat{J}_z = \hat{L}_z + \hat{s}_z$) for the four-fold degenerate VB defined by $j = 3/2$. These eigenfunctions must be a superposition of the product of ψ_{jM} and $\xi_{s\sigma}$, and are determined by a linear sum over the allowed values of m and σ where $M = m + \sigma$. The proper eigenfunctions may be derived from[53]

$$|jM> = \sum_{m,\sigma} C^{jM}_{m\sigma}\psi_m\xi_\sigma, \tag{135}$$

where

$$C^{jM}_{m\sigma} = (-1)^{M+1/2}\sqrt{2j+1}\begin{pmatrix} l & s & j \\ m & \sigma & -M \end{pmatrix}. \tag{136}$$

The last expression in Eq.(1.4.12) is defined by the Wigner 3j-symbols and they are derivable for the system under study from[54]

$$\begin{pmatrix} l=1 & s=1/2 & j=3/2 \\ M-(\sigma=1/2) & \sigma=1/2 & -M \end{pmatrix} = \begin{pmatrix} 1 & 1/2 & 3/2 \\ M-1/2 & 1/2 & -M \end{pmatrix}$$
$$= (-1)^{M+1/2}\sqrt{\frac{3/2+M}{12}}, \tag{137}$$

and

$$\begin{pmatrix} l & s & j \\ -m & -\sigma & M \end{pmatrix} = (-1)^{l+s+j}\begin{pmatrix} l & s & j \\ m & \sigma & -M \end{pmatrix}. \tag{138}$$

Application of the previous relations leads directly to the VB Bloch functions valid at $\vec{k} = 0$. Therefore, electron transport may be described in the six-band Kane-type model by the S-like CB ($|v, j, s>$) Bloch functions[55]

$$|c, \frac{1}{2}, \frac{1}{2}> = |iS\uparrow>, \quad |c, \frac{1}{2}, \frac{1}{2}> = |iS\downarrow>, \tag{139}$$

where arbitrary phase has been introduced, and P-like valence-band ($|v, j, s>$) Bloch functions

$$|v, \frac{3}{2}, -\frac{1}{2}> = \frac{1}{\sqrt{6}}[|(X-iY)\uparrow> +2|Z\downarrow>], \quad |v, \frac{3}{2}, \frac{3}{2}> = -\frac{1}{\sqrt{2}}|(X+iY)\uparrow>, \tag{140}$$

$$|v, \frac{3}{2}, \frac{1}{2}> = -\frac{1}{\sqrt{6}}[|(X+iY)\downarrow> -2|Z\downarrow>], \quad |v, \frac{3}{2}, -\frac{3}{2}> = \frac{1}{\sqrt{2}}|(X-iY)\downarrow>, \tag{141}$$

The next task is to derive the matrix representation of the Schrödinger-Bloch equation in terms of the six-band model. The resulting matrix Hamiltonian will be the six-band Kane model we need to define the tunneling currents within the staggered-bandgap heterostructures under study. For electron conduction in a particular band (i.e., here the conduction band) and at a specified wavevector (and energy), the six-band representation of this system has the matrix form $\sum_m \hat{H}_{nm} a_m = E_n a_n$, where the a_m's are the envelope functions derived earlier in Eq. (130). This formulation is to be applied to the study of electron tunneling through heterostructures where transport occurs in the transverse or z-coordinate direction. While this is a one-dimensional analysis, it is important to note that the in-plane (i.e., $\hat{x}$ and $\hat{y}$) coordinates will play an important role in this multi-band transport process as there will be interband momentum (i.e., k_x and k_y) coupling effects. The matrix elements can be derived from the Schrödinger-Bloch equation given in Eq. (125). The interaction operator may be expressed as $\hat{H}_{int} = \hbar(\vec{k}\cdot\vec{P})/m = \hbar(k_t \cos\phi \hat{p}_x + k_t \sin\phi \hat{p}_y + k_z \hat{p}_z)/m$ where k_t is defined as the in-plane momentum vector in the direction of the electron motion.

This previous formalism can be used to construct a very efficient matrix system for deriving the envelope functions within each region of the SBS if a proper transformation of basis is applied. Specifically, it is possible to choose a transformation of the in-plane coordinates of the Bloch functions such that we obtain

$$
\begin{aligned}
|S_\pm> &= \frac{1}{\sqrt{2}}\{i\lambda|S\uparrow> \pm i\lambda|S\downarrow>\} \\
|v_{l\pm}> &= \frac{1}{\sqrt{2}}\{\lambda|v,\frac{3}{2},\frac{1}{2}> \pm\lambda^*|v,\frac{3}{2},-\frac{1}{2}>\} \\
|v_{h\pm}> &= \frac{1}{\sqrt{2}}\{\lambda e^{-i\phi}|v,\frac{3}{2},\frac{3}{2}> \mp\lambda^* e^{i\phi}|v,\frac{3}{2},-\frac{3}{2}>\}
\end{aligned}
\tag{142}
$$

where $\lambda = \exp(i\pi/4 - i\phi/2)$ and $\{S_+, v_{l+}, v_{h+}\}$ and $\{S_-, v_{l-}, v_{h-}\}$ are the decoupled bases. The introduced transformation rotates the in-plane coordinates around the z-axis by an angle ϕ and this results in an alignment of $\hat{x}$ with the k_t vector. Here, the dependence on the in-plane wave-vector is defined as $k_x = k_t\cos\phi$ and $k_y = k_t \sin\phi$ where k_t is defined as the in-plane momentum vector in the direction of the electron motion.. In the transformed basis, $|S_\pm>$ is the wave function in the 2-fold degenerate conduction-band and $|v_{l\pm}>$ and $|v_{h\pm}>$ are the wave functions in the 4-fold degenerate valence-band (light and heavy holes at $k_t = 0$). The wave function within any individual region of the SBS may now be conveniently expressed as the sum of two components (i.e., $\Psi = \Psi_+ + \Psi_-$) according to the definition

$$
\Psi_\pm = a_{c\pm}|S_\pm> + a_{l\pm}|v_{l\pm}> + a_{h\pm}|v_{h\pm}>, \tag{143}
$$

and as will next be demonstrated, they can be used to derive a partially-decoupled matrix representation for the Schrödinger equation of the general form

$$
\begin{pmatrix} H_+ & 0 \\ 0 & H_- \end{pmatrix} \begin{pmatrix} (a_+) \\ (a_-) \end{pmatrix} = E \begin{pmatrix} (a_+) \\ (a_-) \end{pmatrix} \tag{144}
$$

where E is the electron eigen-energy and ,

$$(a_+) = \begin{pmatrix} a_{c+} \\ a_{l+} \\ a_{h+} \end{pmatrix}, \quad (a_-) = \begin{pmatrix} a_{c-} \\ a_{l-} \\ a_{h-} \end{pmatrix}, \tag{145}$$

are the decoupled envelope-function vectors. At this point, a matrix system as defined in Eq. (144) is to be developed for electron transport in the SBS as shown in Fig.1. The structure considered here will be modeled as six separate spatial regions (i.e., emitter, left barrier, well, right barrier, spacer, and collector) where the device profile is defined from the conduction-band edge function $U(z)$. The SBS will be subject to an applied bias of with the following assumptions on the potential energy (PE) variations in each region. The emitter and collector regions have constant doping profiles and all accumulation/depletion effects will be ignored in emitter region for this initial study. Hence the emitter region has a constant PE profile. The left barrier, well, and right barrier are undoped and all space charging effects will be ignored in this particular analysis - hence linear PE drops will be employed. The spacer is undoped but electron screening effects from the collector region will be include, hence, the spacer region will exhibit a spatial variation in PC as shown in Fig. 20. Also, PE drop will be treated in the collector region, however, this will be a small for reasonable collector lengths. Hence, as will be discussed later, the collector length will be extended to infinity to approximate the PE variations in the collector. Furthermore, this particular study will focus on conduction band (CB) transport where the electron injection energies are $E_{cR} > 0$ which is referenced to the CB edge at the collector contact on the right.

The inclusion of the PE profile function into the analysis leads to a total Hamiltonian of the form $\hat{H}_0 + \hat{H}_{int} + U(z)$ which is substituted into the matrix system in Eq. (18) to derive the final model equations. For the InAs emitter, well, spacer and collector regions, this procedure yields

$$\begin{aligned}
&(U(z) - |E_{cR}|)a_{c\pm} + \hbar\sqrt{\frac{E_{gw}}{2m_w}}(\mp\frac{\sqrt{3}}{2}k_t a_{h\pm} + (\hat{k}_z \mp \frac{ik_t}{2})a_{l\pm}) = 0 \\
&(|E_{cR}| - U(z) + E_{gw})a_{l\pm} - \hbar\sqrt{\frac{E_{gw}}{2m_w}}(\hat{k}_z \mp \frac{ik_t}{2})a_{c\pm} = 0 \\
&(|E_{cR}| - U(z) + E_{gw})a_{h\pm} \pm \hbar\sqrt{\frac{3E_{gw}}{8m_w}}k_t a_{c\pm} = 0
\end{aligned} \tag{146}$$

where the differential, $\hat{k}_z = -i(\partial/\partial z)$, operator has been introduced. Here, m_w is the conduction-band effective-mass in the well and E_{gw} is the band gap of the InAs. Similarly, the set of equations in the barrier can be written as

$$\begin{aligned}
&(U(z) + \Delta U - |E_{cR}|)a_{c\pm} \\
&+\hbar\sqrt{\frac{E_{gw}}{2m_w}}(\mp\frac{\sqrt{3}}{2}k_t a_{h\pm} + (\hat{k}_z \mp \frac{ik_t}{2})a_{l\pm} = 0 \\
&(|E_{cR}| - U(z) + E_{gb} - \Delta U)a_{l\pm} - \hbar\sqrt{\frac{E_{gw}}{2m_w}}(\hat{k}_z \mp \frac{ik_t}{2})a_{c\pm} = 0
\end{aligned}$$

$$(|E_{cR}| - U(z) + E_{gb} - \Delta U)a_{h\pm} \pm \hbar\sqrt{\frac{3E_{gw}}{8m_w}}k_t a_{c\pm} = 0 \tag{147}$$

where E_{gb} is the band gap of the AlGaSb and ΔU is the conduction-band off-set between InAs and AlGaSb. Note that the electron masses are determined mainly by conduction-valence band coupling and that the interband momentum matrix element is the same everywhere. (Hence, the absence of a barrier mass, m_b, above). As shown in Eqs. (146) and (147), the problem has been reduced into two systems of first-order differential equations for the envelope-function vectors (a_+) and (a_-). The system of six equations can be reduced down to four - i.e., by substitution using the last Eqs. from (146) and (147). The envelope functions within the InAs emitter, well, spacer and collector regions can now be found from

$$(U(z) - |E_{cR}| + \frac{3\hbar^2 k_t^2 E_{gw}}{8m_w[|E_{cR}| - U(z) + E_{gw}]})a_{c\pm} + \hbar\sqrt{\frac{E_{gw}}{2m_w}}(\hat{k}_z \mp \frac{ik_t}{2})a_{l\pm} = 0,$$

$$(|E_{cR}| - U(z) + E_{gw})a_{l\pm} - \hbar\sqrt{\frac{E_{gw}}{2m_w}}(\hat{k}_z \mp \frac{ik_t}{2})a_{c\pm} = 0. \tag{148}$$

Similarly, the envelope functions in the barrier may be derived from the reduced system

$$(U(z) + V_0 - |E_{cR}| + \frac{3\hbar^2 k_t^2 E_{gw}}{8m_w[|E_{cR}| - U(z) - V_0 + E_{gw}]})a_{c\pm}$$
$$+\hbar\sqrt{\frac{E_{gw}}{2m_w}}(\hat{k}_z \mp \frac{ik_t}{2})a_{l\pm} = 0,$$

$$(|E_{cR}| - U(z) + E_{gb} - V_0)a_{l\pm} - \hbar\sqrt{\frac{E_{gw}}{2m_w}}(\hat{k}_z \mp \frac{ik_t}{2})a_{c\pm} = 0. \tag{149}$$

While the PE profile within the InAs collector region is nonconstant, the asymptotical solutions for the envelope functions at infinity (i.e., denoted now by $a_{Rc\pm}$, $a_{Rl\pm}$, $a_{Rh\pm}$) can be derived directly from Eq. (149) as

$$a_{Rc\pm} = \frac{t_{R\pm}}{\sqrt{N_R}}e^{ik_{Rz}z},$$

$$a_{Rl\pm} = \hbar\sqrt{\frac{E_{gw}}{2m_w}}(k_{Rz} \pm \frac{ik_t}{2})\frac{a_{Rc\pm}}{|E_{cR} + E_{gw}|} \tag{150}$$

along with $a_{Rh\pm}$ that is obtained from the last Equation in (147). Here, the energy dispersion relation is given by $|E_{cR}|(E_{gw} + |E_{cR}|) = E_{gw}(k_t^2 + k_{Rz}^2)\hbar^2/2m_w$ and $N_R = 1 + |E_{cR}|/(|E_{cR}| + E_{gw})$. Equation (152) will be important later for developing a numerical algorithm for calculating the entire device profile for the SBS. Similarly, the envelope functions throughout the AlGaSb emitter region (i.e., denoted now by $a_{Lc\pm}, a_{Ll\pm}, a_{Lh\pm}$), where a constant PE profile is assumed, are given by

$$a_{Lc\pm} = \frac{1}{\sqrt{N_L}}(e^{ik_{Lz}z} + r_{L\pm}e^{-ik_{Lz}z}), \tag{151}$$

$$a_{Ll\pm} = \frac{\hbar}{\sqrt{N_L}(E + E_{gw})}\sqrt{\frac{E_{gw}}{2m_w}}((k_{Lz} \pm \frac{ik_t}{2})(e^{ik_{Lz}z} - (k_{Lz} \mp \frac{ik_t}{2})r_{L\pm}e^{-ik_{Lz}z}),$$

along with $a_{Lh\pm}$ that is obtained from the last equation in (23). Here, $E = E_{cR} - eV_a$ is the electron energy in the emitter, r_L is the reflection coefficient and eV_a is the applied bias. The energy dispersion relation here is given by $E(E+E_{gw}) = E_{gw}(k_t^2 + k_{Lz}^2)\hbar^2/2m_w$ and $N_L = 1 + E/(E+E_{gw})$. The final relations presented in Eqs. (148) - (152) provides an efficient formalism for describing the quantum mechanics of conduction-band electrons in the SBS system. Specifically, the envelope functions can be derived from Eqs. (148) - (152) once space charge effects are included (e.g., through the incorporation of Poisson's equation). Hence, the final major theoretical consideration is specifying the definition of electron current density for the multi-band system and this will now be addressed.

At this point, it is important to note that the usual definition for current (or probability) flux given by, $\vec{J} = (-ie\hbar/2m)((\nabla\Psi)^*\Psi - \Psi^*(\nabla\Psi))$, is not applicable for multi-band transport as it is strictly valid for wavefunction solutions derived from the single-band effective-mass equation. Indeed, a rigorous definition for current flux for multi-band transport must be derived from more general principles. For an arbitrary electromagnetic vector field, where the electric field is given by $\vec{E} = -\nabla U - \partial\vec{A}/\partial t$, the fundamental operational relation between electron flux at any spatial point and the system Hamiltonian is[56]

$$\hat{\vec{J}} = \frac{\delta\hat{H}(\hbar\vec{k} - e\vec{A})}{\delta\vec{A}} = -\frac{e}{\hbar}\frac{\delta\hat{H}}{\delta\vec{k}} \tag{152}$$

where $\vec{A}$ is the vector potential and where the necessary substitution $\hbar\vec{k} \to \hbar\vec{k} - e\vec{A}$ has been made. The current density operator in Eq. (153) defines the current density at spatial point $\vec{r}$ and electron momentum value $\vec{k}$ as

$$\vec{J}(\vec{r},\vec{k}) = \Psi^+(\vec{r})\hat{\vec{J}}(\vec{r},\vec{k})\Psi(\vec{r}) = -\frac{e}{\hbar}\left[\Psi^+(\vec{r})\frac{\partial\hat{H}(\vec{r},\vec{k})}{\partial\vec{k}}\Psi(\vec{r})\right]. \tag{153}$$

Therefore, for the one-dimensional system under the consideration the z-directed electron flux of a single wavefunction state can be expressed in turns of the decoupled basis as $J_z(\vec{k}) = -e(\Psi_-^*\partial\hat{H}_-/\partial k_z\Psi_- + \Psi_+^*\partial\hat{K}_+/\partial k_z\Psi_+)/\hbar$. The total current flux through the SBS can now be derived by averaging over the ensemble of electrons entering at the emitter contact and collector contacts. Here, the averaging is performed over all electronics states at the emitter with $k_t > 0$ and $E > 0$, and all electronic states at the collector with $k_t < 0$ and $E_{cR} > 0$. This statistical averaging over all available states leads to a total current density expression of

$$J_z = \frac{e\hbar}{4\pi^2 m_w}\int_0^\infty dk_t k_t \int_0^\infty \frac{dk_{Lz}k_{Rz}}{N_R(1+|E_{cR}/E_{gw})}T(E,k_t),$$
$$T(E,k_t) = [t_{R+}(E,k_t)|^2 + t_{R-}(E,k_t)|^2][f(E) - f(E+eV_a)] \tag{154}$$

where $f(E)$ is the Fermi function and the transmission coefficients, t_{R+} and t_{R-} are obtained from solution of the system of envelope equations defined in Eqs. (148) - (152). The previously developed model can be used to determine the electron transport characteristics within the SBS through a self-consistent solution with spatial variations of the potential energy profile $U(z)$.

This investigation considers conduction-band transport through the SBS structure depicted in Fig. 20. As discussed earlier, the structure is divided into emitter, left barrier, well, right barrier, spacer and collector regions. In this preliminary study the potential variations in the heavily doped emitter region will be ignored (i.e., as was done in deriving the envelope equations given in (152)). Also, electron-charging effects in the undoped barrier-well-barrier regions will be ignored such that PE drops (and a constant electric field F_R) may be employed from the collector to spacer regions. As this formulation is being developed for the future study of interband tunneling currents, an undoped spacer region has been included following the second barrier. Hence, as shown in Fig. 1 there will be significant screening in the spacer region due to charge diffusion from the heavily doped collector. A self-consistent description of the PE variations throughout the SBS may be derived through the application of the Thomas-Fermi approximation for the electron density within the spacer and collector regions. This approach can be utilized to approximate the nonlinear variations of $U(z)$ within the spacer and collector regions. Note that the while electron screening will induce PE variations in both the spacer and collector, the change in potential in the collector will be very slowly varying. Therefore, the collector PE variations will be treating indirectly by allowing this region to be infinite in extent. Here, it is possible to develop a numerical description where potential is chosen as an independent variable for the solution of the Schrödinger equation. Specifically, Poisson's equation is used to establish the relationship between PE and the spatial coordinate and the solutions given in Eq. (152) provide the boundary conditions on $a_{Rc\pm}$, $a_{Rl\pm}$ and $a_{Rh\pm}$ at infinity. When the previously discussed approaches are combined, it is possible to derive a semi-analytical model for calculating the PE drops throughout the SBS.[55]

This PE profile model discussed above can now be solved self-consistently with the two decoupled envelope-function systems to define the SBS current density characteristics. As shown in Eqs. (148) and (152), the final multi-band electron transport problem has been conveniently reduced to two independent pairs of coupled analytical equations (i.e., in emitter and collector) and coupled differential equations (i.e., in the barriers, well and spacer) that can be used to individually calculate the (a_+) and (a_-) envelope functions. In this formulation, a complete set of envelope-function solutions may be obtained by solving the analytical equations for $\{a_{c\pm}, a_{l\pm}\}$ self-consistently with the first-order differential equations for $\{a_{c\pm}, a_{l\pm}\}$. The $a_{h\pm}$ envelope functions then follow from the last equations in (146) and (147), depending on the spatial region concerned. The device regions with analytical solutions (i.e., emitter and collector) are joined to those with differential equations by requiring the continuity of individual envelope functions at the spatial interfaces. Here, numerical methods must be employed to solve the coupled differential equations within the barrier, well and spacer regions.

The final aspect of the solution process concerns the specification of boundary conditions. While the transmission coefficients, $t_{R\pm}$, are the physical parameters needed to specify the current density in Eq. (154), they also represent (along with the reflection coefficients $r_{L\pm}$) the effective boundary conditions of the decoupled matrix systems that must be specified to find $\{a_{c\pm}, a_{l\pm}\}$. However, each of the decoupled matrix systems independently represent the quantum states Ψ_+ and Ψ_-. Hence, the total probability flux for each of these quantum states must sum to unity independently and it is possible to show using Eq. (153) that

$$|t_{R+}|^2 K(E, k_{Rz}, k_{Lz}) + |r_{L+}|^2 = 1, \quad |t_{R-}|^2 K(E, k_{Rz}, k_{Lz}) + |r_{L-}|^2 = 1, \qquad (155)$$

where

$$K(E, k_{Rz}, k_{Lz}) = \frac{k_{Rz} N_L (1 + E/E_{gw})}{k_{Lz} N_R (1 + |E_{cR}|/E_{gw})} \qquad (156)$$

explicitly involves the ratio of wavevectors in the emitter and collector, respectively. Hence, the conservation rules given in Eq. (156) allow for the closure of the decoupled matrix systems and for the generation of self-consistent solutions of the multi-band transport problem.

4.3. *Simulation Results and Conclusions*

Numerical simulation studies were performed on AlGaSb/InAs/AlGaSb double-barrier RTD structures to consider the influence of multi-band transport effects. The staggered-bandgap RTD considered in this theoretical study possessed composition and layer thickness taken from a previous experimental investigation.[40] For all these investigations, the AlGaSb barriers are 20$\mathring{A}$ thick, the InAs well is 75$\mathring{A}$ thick, and the InAs collector-spacer is 100$\mathring{A}$. Initially, a double-barrier AlxGa1-xSb/InAs system with $x = 0.4$ was considered and the InAs emitter and collector doping (which was not given in Chow and Schulman[40]) was set to $3 \times 10^{18} cm^{-3}$. First, Fig. 21 compares the transmission probability (i.e., for zero bias) results for the staggered-bandgap heterostructure that are obtained using the single-band and multi-band model. As shown, the single-band model predicts two quasi-discrete energy levels whereas the multi-band model predicts four quasi-discrete energy levels. These results alone demonstrate the general importance of the multi-band transport formalism as the number and position of the quasi-discrete levels is a crucial element in determining the current-voltage characteristics. Now consider Fig. 22 that gives the results for J obtained from a single-band transport model with (dotted line) and without (dashed line) $|t_k|^2$ dependence on k_t. These results are only included to show that effective mass differences do play a noteworthy role in these structures. Fig. 22 also gives results for obtained from the multi-band model without (dot-dashed line) and with (solid line) a collector spacer layer. As shown, the magnitude and position of the peak in J is strongly influenced by multi-band transport and by the inclusion of a collector spacer layer. Fig. 23 gives a direct comparison of the results from the multi-band tunneling model to those measured by Chow and Schulman[40] for an AlxGa1-xSb/InAs RTD with . The dashed curve gives the J

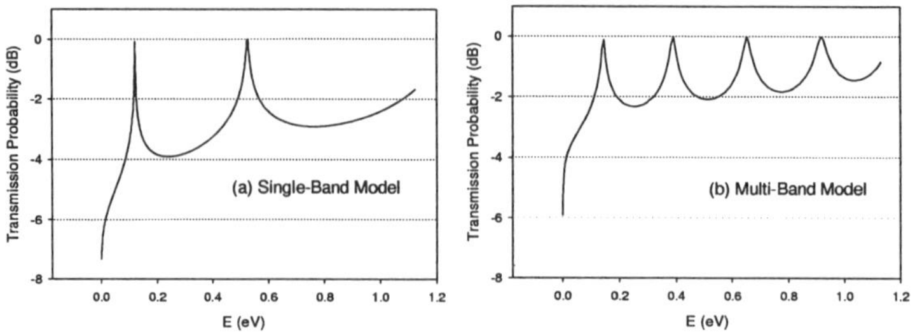

Fig. 21. Transmission Probability for, (a) Single-band and (b) Multi-band, models.

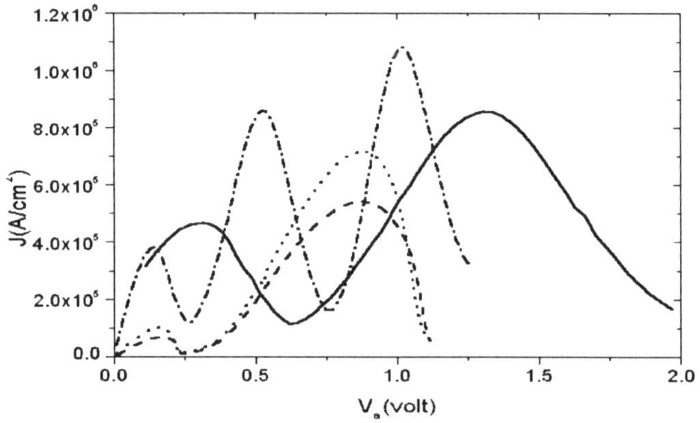

Fig. 22. Single-band and Multi-band Results.

results obtained from the model with an InAs emitter doping of . Here, a fitting of the low-bias results to that obtained experimentally was used to derive this emitter doping as it was not given in Chow and Schulman.[40]

Fig. 23 also directly compares the actual experimental results (solid curve) to the multi-band modeling results (dotted line) with an included shift potential shift of volts. This shift in potential is necessary to account for the barriers that arise in the actual experimental structure due to the emitter spacer (i.e., volts for the Fermi energy shift between the undoped well and the doped emitter) and the finite conductivity of the lightly-doped (i.e., 0.04 volt drop for the doped region) collector spacer. It is also important to note that the comparison is only made to the peak of the experimental curve, above which interband hole current comes into significance. These multi-band transport results show excellent agreement with experiment and

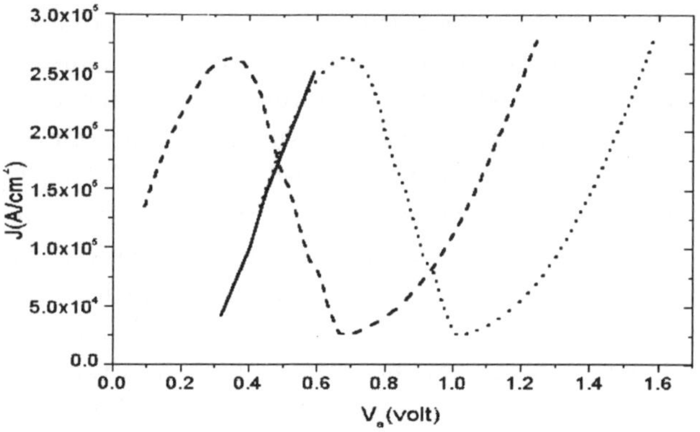

Fig. 23. Multi-band and Experimental Results.

establish an initial foundation for the accurate study of nanoscale feedback processes within staggered bandgap tunneling structures in the future. Future phases of this work will combine conduction and valence electron transport processes.

5. A Nonequilibrium Green's Function Transport Theory for Multi-band Tunneling Structures

5.1. *Introduction*

Research on charge carrier transport in semiconductors is of great interest both from the fundamental physics point of view[57] and for the prospects of device development and application.[58,59]. Device modeling is of key importance in this research process since it provides the necessary input parameters for the design and manufacture of devices. Such device modeling involves the solution of transport equations jointly with the Poisson equation self-consistently, incorporating appropriate boundary conditions. Needless to say, the accuracy of the transport equations in representing the quantum dynamics of momentum and energy transfer in nanostructures is of central importance. In semiconductor transport theories, the focus is mainly on localized conservation laws for number density, momentum, and energy of the particles, ie: balance equations which describe the physical processes involved in conduction. For example, carriers are energized by the applied field and they transfer part of the energy to the lattice by carrier-phonon interactions, leading to an increase of the lattice temperature. This affects the strain tensor of the material and in turn the motion and distribution of the carriers. Conservation laws of this type have been derived from the Boltzmann transport equation by considering its first three moments with approximations, leading to hydrodynamic equations.[60] A quantum mechanical counterpart can be derived using the Wigner function equation, leading to quantum hydrodynamic equations.[61] There is a fundamental difficulty in such hydrodynamic formulations in regard to closure of the system of equations and the attendant approximations concerning scatterings. One variant of such moment equations was derived by Lei and Ting for conditions of strong electron-electron scattering, with the inclusion of dissipative random impurity and phonon scatterings accurate to first order.[62] These equations have successfully explained a variety of transport phenomena and have been employed to analyze semiconductor devices.[63] However, the Lei-Ting balance equations cannot describe highly inhomogeneous systems, and are limited in validity by a first order description of dissipative scatterings. As such, they are of limited utility for highly nonequilibrium transport in nanostructure quantum devices having a length scale on the order of $100\AA$. While the Schrödinger equation may describe some features of carrier transport at low temperature, finite temperature and many-body effects do substantially influence transport in quantum devices.[64] To properly deal with such conduction effects in quantum devices, we need to develop nonequilibrium many-body quantum transport equations.

Nonequilibrium Green's function techniques, initiated by Schwinger[65] and Keldysh,[66] have been employed in studies relating to quantum transport by several authors.[26,58,65,67,68,66,69] Derivations of quantum transport equations have frequently been based on the first order gradient expansion suggested by Kadanoff and Baym.[68] This approximation assumes that the "center of mass" coordinates are slowly varying quantities in time and space. Hence, such an approximation precludes a full quantum description of tunneling and interference, which dominate quantum based heterostructure device characteristics. Generally speaking, quantum transport theories based on the first order gradient expansion method can not be used to describe ultrafast phenomena such as the characteristics of carriers in very high frequency devices and femtosecond optical phenomena, etc. .

The size of a mesoscopic system is about $100\AA$. Thus, even for a very small voltage across the system, the electric field is very high. Correspondingly, the carriers in many quantum devices subject to such high fields are usually in a state far from equilibrium. For this reason, there has been an effort in recent years to employ the nonequilibrium Green's function technique[58,70] to simulate quantum devices. Datta et al. used a relaxation time approximation in the self-energy function and employed a localized Boson model of carrier-phonon interactions to reduce the number of coordinates in mesoscopic devices and to facilitate the simulation of resonant tunneling devices (RTD).[71] However, a constant relaxation time approximation is relatively crude in describing scattering.

The single particle nonequilibrium Green's function, which alone determines electron transport, involves two space time coordinates, $\mathbf{r_1}, t_1$, and $\mathbf{r_2}, t_2$, which are often described in terms of Weyl-transform variables $\mathbf{R}, T$ and $\mathbf{r}, t$ according to

$$\mathbf{R} = \frac{1}{2}(\mathbf{r_1} + \mathbf{r_2}), \qquad T = \frac{1}{2}(t_1 + t_2),$$

$$\mathbf{r} = \mathbf{r_1} - \mathbf{r_2}, \qquad t = t_1 - t_2.$$

where $(\mathbf{R}, T)$ is said to represent the "*center of mass*" motion and $(\mathbf{r}, t)$ the relative "miscroscopic" motion. The recognition that $(\mathbf{r}, t)$ described microscopic motion is important for quantum devices. Such systems of about $100\mathring{A}$ are macroscopically large compared with the very fast microscopic motions of the carriers, so that Fourier transforms with respect to ($\mathbf{r}$, t) may be used to transform the space-time coordinates to energy-momentum coordinates, in particular for mesoscopic systems. Consideration of other low-dimensional device features may decrease the number of the coordinates needed to three or two. Devices having dimensions much less than $100\mathring{A}$ have only a few carriers, constituting a few-body problem rather than a many-body problem of the type contemplated here.

Recently, with the introduction of a lattice-space Weyl-Wigner formulation of quantum dynamics jointly with nonequilibrium Green's-function techniques, Buot and Jensen provided a rigorous derivation of an exact integral form of the equation for a quantum distribution function describing nanostructure transport for single band semiconductors.[72] The advantages of the integral form of this quantum transport equation are that (a) it is better suited for numerical work than the equations of gradient expansion theories, and (b) it is capable of faithfully representing quantum effects in its structure. An approximation to the resulting equation has been used to simulate a resonant tunneling device (RTD).[81] Theoretical RTD studies[22] to date fall short of successfully dealing with all of the quantum features and scattering interactions addressed in the Buot-Jensen approach.

Quantum transport in *multi-band* semiconductor systems has attracted considerable research interest over the years. However, the existing treatments are limited by single particle theory[30] or are specialized to steady-state.[82] Experimentally, Chow and Schulman found hysteresis and bistability of the I-V characteristic of a multiband resonant tunneling structure consisting of AlGaSb-InAs-AlGaSb.[40] Though some of the physics of this experimental result is qualitatively understood,[83] a detailed and quantitative explanation of the experimental data is lacking. There is a need for the development of nonequilibrium quantum many-body and non steady-state transport equations for *multi-band* systems to sort out the complicated features involved.

In this section, we develop many-body quantum transport equations for multi-band semiconductors using a nonequilibrium Green's function approach for electrons/holes weakly coupled to a phonon bath. In particular, we formulate local conservation laws for the number density, energy and momentum of carriers in *multi-band* semiconductor systems.

5.2. *First Principle Derivation of Equation of Motion of Multi-Band Nonequilibrium Green's Function*

As we have stated in the above section, the motion of carriers in quantum devices is the core of quantum transport. The carriers are in states far from equilibrium states. In the far-from-equilibrium states, the functions which describe the motion of carriers, such as correlation functions, response functions, and relation functions, should be two-time-variable functions. The reasons for this are: the inverse temperature β is itself a time-evolving function; the ensemble used in the average can be a time-evolving function, too. For describing the transport of carriers in the nonequilibrium states, the best way is to use nonequilibrium close-time-path Green's function method. Nonequilibrium Green's functions are the tools of choice for analyzing nonequilibrium transport in which the states of the system are far from equilibrium, as in transport in quantum devices. For multi-band semiconductors, the nonequilibrium Green's function may be defined by

$$G_{ab}(x_1, x_2) = -\frac{i}{\hbar} < \psi_a(x_1)\psi_b^+(x_2) >_{\mathcal{T}}, \tag{157}$$

where a and b are energy band indices, $\psi_a(x)$ and $\psi_a^+(x)$ are annihilation and creation operators of the carrier field, and $x = (\mathbf{r}, t)$. $\psi_a(x)$ and $\psi_a^+(x)$ obey the following commutation rules,

$$\{\psi_a(\mathbf{r}t), \psi_b^+(\mathbf{r}'t)\} = \delta_{ab}\delta(\mathbf{r} - \mathbf{r}'), \tag{158}$$

$$\{\psi_a(\mathbf{r}t), \psi_b(\mathbf{r}'t)\} = \delta_{ab}\delta(\mathbf{r} - \mathbf{r}'), \tag{159}$$

$$\{\psi_a^+(\mathbf{r}t), \psi_b^+(\mathbf{r}'t)\} = \delta_{ab}\delta(\mathbf{r} - \mathbf{r}'), \tag{160}$$

where { ..., ... } denotes anticommutation, that is {A, B}=AB+BA. The time t varies over a closed-time-path contour. The average $< ... >_{\mathcal{T}}$ may be written in terms of interaction picture operators as

$$< O(t) >_{\mathcal{T}} = \frac{Tr\{\rho(-\infty)\mathcal{T}[S_{\mathcal{T}}O(t)]\}}{Tr\{\rho(-\infty)S_{\mathcal{T}}\}}, \tag{161}$$

where $\rho(-\infty) = e^{-\beta H}$, with the Hamiltonian, H, augmented by a source-interaction part $H_{ext}(t)$, and

$$S_{\mathcal{T}} = \mathcal{T}exp\{-\frac{i}{\hbar}\int_{\mathcal{T}} H_{ext}(\tau)d\tau\}. \tag{162}$$

$\mathcal{T}$ denotes time ordering along the closed-time-path contour (The time arguments run from the initial time t_0 to the far future, the largest time argument of the physical quantity and then back to t_0). Once the Green's function is obtained, current density can be calculated from the nonequilibrium Green's function. The current density in the ath band can be expressed by Green's function as

$$\begin{aligned}\mathbf{J}_a &= < J_a(\mathbf{r}t) >_a \\ &= 2\sum_{bcd}\left(ReD_{bc}^{ad}\frac{\partial}{\partial x_d}G_{cb}(x, x') + \hbar B_{bc}^{d}\delta_d^a G_{cb}(x, x)\right)\end{aligned} \tag{163}$$

where $G_{cb}(x, x')$ is the multi-band Green's function defined by Eq.(157) For a multi-band semiconductor system, the Hamiltonian of the system can be written as

$$H(t) = H_e + H_p + H_{ep} + H', \tag{164}$$

where H_e, H_p, H_{ep}, and H' are carrier Hamiltonian, free phonon Hamiltonian, carrier-phonon interaction Hamiltonian, and source Hamiltonian, respectively. These terms are given as follows:

$$H_e = \sum_{ab} \int \mathbf{dr_1} \psi_a^+(\mathbf{r_1}t_1) h_{ab}(\mathbf{r_1}) \psi_b(\mathbf{r_1})$$

$$+\frac{1}{2}\sum_{ab} \int d\mathbf{r_1} d\mathbf{r_2} \psi_a^+(\mathbf{r_1}t_1)\psi_b^+(\mathbf{r_2}t_1) V(\mathbf{r_1}-\mathbf{r_2}) \psi_b(\mathbf{r_2}t_1)\psi_a(\mathbf{r_1}t_1), \tag{165}$$

$$H_p = \sum_{ilk} \frac{P_{ilk}^2}{2M_{lk}} + \frac{1}{2}\sum_{ilk}\sum_{i'l'k'} \Phi_{ii'}(lk, l'k') Q_{ilk}(t_1) Q_{i'l'k'}(t_1), \tag{166}$$

$$H_{ep} = -\sum_{ilk}\sum_{a} \gamma_{ilk}^a \int d\mathbf{r_1} Q_{ilk}(t_1) \psi_a^+(\mathbf{r_1}t_1)\psi_a(\mathbf{r_1}t_1), \tag{167}$$

$$H' = \sum_a \int d\mathbf{r_1} e_a \psi_a^+(\mathbf{r_1}t_1)\psi_a(\mathbf{r_1}t_1) U(\mathbf{r_1}t_1) + \sum_{ilk} J_{ilk}(t_1) Q_{ilk}(t_1)$$

$$+\sum_a \int d\mathbf{r_1} [\psi_a^+(\mathbf{r_1}t_1)\xi_a(1) + \xi_a^* \psi_a(1)]. \tag{168}$$

In the above equations, a and b are band indices; e_a is the charge of band carrier species a; h_{ab} is the single-particle intra-band ($a = b$) and inter-band ($a \neq b$) Hamiltonian, including external electric and magnetic fields; $\psi_a^+(\mathbf{r_1}t_1)$ ($\psi_a(\mathbf{r_1}t_1)$) creates (annihilates) a carrier in the $a'th$ band, respectively and $V(\mathbf{r_1} - \mathbf{r_2})$ is the Coulomb carrier-carrier interaction; $Q_{ilk}(t_1)$ is the phonon displacement field operator where (ijk) stands for the location of an ion and P_{ilk} is its momentum, M_{lk} its mass and $\Phi_{ii'}$ arises from ion potential coupling, and γ_{ijk}^a describes the strength of electron-phonon coupling. The summations in the above equations are with respect to energy band indices a, b. In the following derivation, we use Arabic letters to denote energy bands and Greek letters (η , etc.) for the branches of the closed time path. The Greek letters may take the values + or −, which stand for the upper time-path branch or the lower time-path branch, respectively. $U(\mathbf{r})$ is an external potential, J_{ilk} is a phonon source and ξ_a and ξ_a^* are carrier sources. The equation of motion of the average $< \cdots >$ of $\psi_a(\mathbf{r_1}t_1)$ may be written as

$$i\hbar\frac{\partial < \psi_a(1_\eta) >}{\partial t_1} = \sum_b h_{ab}(\mathbf{r_1}) < \psi_b(1_\eta) >$$

$$+e_a \sum_b \int d\mathbf{r_3} e_b V(\mathbf{r_1} - \mathbf{r_3}) < \psi_b^+(\mathbf{r_3}t_{1_\eta}^+)\psi_b(\mathbf{r_3}t_{1_\eta}^+)\psi_a(1_\eta) >$$

$$-\sum_{ilk} \gamma_{ilk}^a < Q_{ilk}(t_{1_\eta})\psi_a(1_\eta) > +\xi_a(1_\eta) + eU(1_\eta) < \psi_a(1_\eta) > . \tag{169}$$

The average in the above equation is defined by Eqs. (161) and (162). In terms of the functional derivative technique, the Green's function may be written as

$$G_{ab}(1_\eta, 2_{\eta'}) = \left[\frac{\delta}{\delta\xi_b(2_{\eta'})} < \psi_a(1_\eta) >\right]_{\xi\to 0}, \tag{170}$$

where η and η' stand for the branches of the time path. It is straightforward to prove the following formulae:

$$\sum_{b'} e_{b'} \left[\frac{\delta}{\delta \xi_b(2_{\eta'})} < \psi^+_{b'}(3_\eta)\psi_{b'}(3_\eta)\psi_a(1_\eta) > \right]_{\xi \to 0} = i\hbar\eta \frac{\delta}{\delta U(3_\eta)} G_{ab}(1_\eta, 2_{\eta'})$$

$$+ G_{ab}(1_\eta, 2_{\eta'}) \sum_{b'} e_{b'} G_{b'b'}(3_\eta, 3_\eta), \tag{171}$$

$$\frac{\delta}{\delta \xi_b(2_{\eta'})} < Q_{ilk}(t_{1\eta})\psi_\alpha(1_\eta) > = i\hbar\eta \frac{\delta}{\delta J_{ilk}(t_{1\eta})} G_{ab}(1_\eta, 2_{\eta'})$$

$$+ < Q_{ilk}(t_{1\eta}) > G_{ab}(1_\eta, 2\eta'), \tag{172}$$

$$\frac{\delta}{\delta U(3_\eta)} G_{ab}(1_\eta, 2_{\eta'}) = - \int d4d5 G_{ac}(1_\eta, 4_\xi) \frac{\delta G^{-1}_{cd}(4_\xi, 5_\zeta)}{\delta U(3_\eta)} G_{db}(5_\zeta, 2_{\eta'}), \tag{173}$$

and

$$\frac{\delta}{\delta J_{ilk}(t_{1\eta})} G_{ab}(1_\eta, 2_{\eta'}) = - \int d4d5 G_{ac}(1_\eta, 4_\xi) \frac{\delta G^{-1}_{cd}(4_\xi, 5_\zeta)}{\delta J_{ilk}(t_{1\eta})} G_{db}(5_\zeta, 2_{\eta'}). \tag{174}$$

Here, the repeated indices are understood to be summed over. Defining the electron-electron interaction vertex and the electron-phonon interaction vertex, respectively, as

$$\Gamma_{cd}(4_\xi, 5_\zeta; 3_\eta) = - \frac{\delta G^{-1}_{cd}(4_\xi, 5_\zeta)}{\delta U(3_\eta)}, \tag{175}$$

$$\tilde{\Gamma}_{cd}(4_\xi, 5_\zeta; ilk, t_{1\eta}) = - \frac{\delta G^{-1}_{cd}(4_\xi, 5_\zeta)}{\delta J_{ilk}(t_{1\eta})}, \tag{176}$$

and the carrier selfenergy as

$$\Sigma_{ad}(1_\eta, 5_\zeta) = \Bigg\{ e_a U(1_\eta)\delta(1,5) - i\hbar e_a \sum_c e_c \int d3 V(1_\eta - 3_\eta) G_{cc}(3_\eta, 3_\eta)$$

$$+ \left\{ \sum_{ilk} \gamma^a_{ilk} < Q_{ilk}(t_{1\eta}) > \right\} \delta_{ad}(1_\eta, 5_\zeta)$$

$$+ i\hbar\eta e_a \int d3 \int d4 V(1_\eta - 3_\eta) G_{ac}(1_\eta, 4_\xi) \Gamma_{cd}(4_\xi, 5_\zeta; 3_\eta)$$

$$- i\hbar \sum_{ilk} \gamma^a_{ilk} \int d4 G_{ac}(1_\eta, 4_\xi) \tilde{\Gamma}_{cd}(4_\xi, 5_\zeta; ilk, t_{1\eta}), \tag{177}$$

the equation of motion of Green's function may be written as

$$G^{-1}_{ad} G_{db} = \delta_{ab}, \tag{178}$$

where

$$G^{-1}_{ad} = G^{-1}_{0,ad} - \Sigma_{ad}, \tag{179}$$

and

$$G^{-1}_{ad,0} = \left[i\hbar\delta_{ad}\frac{\partial}{\partial t_1} - h_{ad}(\mathbf{r_1})\right]. \tag{180}$$

The above results, particularly Eq.(177), are in agreement with those of Baym,[74] Joshi and Rajagopal,[75] and Tso and Horing.[76,77] In these equations, we have used matrix expressions for Green's functions and electronic selfenergies which are defined, in the physical limit $H' \to 0$, by

$$G_{ad}(1,2) = G^r_{ad}(1,2)\begin{pmatrix} 1 & 0 \\ 0 & 1 \end{pmatrix} + \begin{pmatrix} G^<_{ad}(1,2) & -G^<_{ad}(1,2) \\ G^>_{ad}(1,2) & -G^>_{ad}(1,2) \end{pmatrix}, \tag{181}$$

where

$$G^>_{ad}(1,2) = -\frac{i}{\hbar} < \psi_a(1)\psi^+_d(2) >, \tag{182}$$

$$G^<_{ad}(1,2) = -\frac{i}{\hbar} < \psi^+_b(2)\psi_a(1) >, \tag{183}$$

$$G^r_{ad}(1,2) = \theta(t_1 - t_2)[G^>_{ad}(1,2) - G^<_{ad}(1,2)], \tag{184}$$

and

$$G^r_{ad}(1,2) - G^a_{ad}(1,2) = G^>_{ad}(1,2) - G^<_{ad}(1,2). \tag{185}$$

For the selfenergy matrix, the definitions are similar to those of the Green's functions as follows:

$$\Sigma_{ad}(1,2) = \Sigma^r_{ad}(1,2)\begin{pmatrix} 1 & 0 \\ 0 & 1 \end{pmatrix} + \begin{pmatrix} \Sigma^<_{ad}(1,2) & -\Sigma^<_{ad}(1,2) \\ \Sigma^>_{ad}(1,2) & -\Sigma^>_{ad}(1,2) \end{pmatrix}, \tag{186}$$

where

$$\Sigma^r_{ad}(1,2) = \theta(t_1 - t_2)[\Sigma^>_{ad}(1,2) - \Sigma^<_{ad}(1,2)], \tag{187}$$

$$\Sigma^r_{ad}(1,2) - \Sigma^a_{ad}(1,2) = \Sigma^>_{ad}(1,2) - \Sigma^<_{ad}(1,2). \tag{188}$$

In the above equation, $< Q_{ilk}(t_{1\eta}) >$ is the mean value of the phonon displacement in a nonequilibrium state. Using the commutation relation

$$[u_i(lk), P_{i'}(l'k')] = i\hbar\delta_{ll'}\delta_{kk'}\delta_{ii'}, \tag{189}$$

and the Heisenberg equation, we obtain the equation of motion of $< Q^\eta_{ilk}(t_1) >$ as

$$D_0^{-1} < \mathbf{Q} >= -\mathbf{J} + i\hbar\sum_a \int d\mathbf{r_1}\gamma^a G_{aa}(1,1^+), \tag{190}$$

where we have used matrix notation and

$$D_0^{-1}(x_1,x_2) = [M_{lk}\frac{\partial^2}{\partial t_1^2} + \Phi]\delta(x_1 - x_2) \tag{191}$$

is the inverse of the non-interacting phonon propagator D_0 (recall from Eq.(167) that M_{lk} are the ion masses and $\Phi \to \Phi_{ii'}$ represents the matrix of " spring constants" of the lattice). The phonon Green's function may be defined as

$$D(ilkt_{1\eta}, i'l'k't_{2\eta'}) = -i\hbar \frac{\delta}{\delta J_{i'l'k'}(t_{2\eta'})} < Q_{ilk}(t_{1\eta}) > . \tag{192}$$

From Eqs.(190) and (192) we obtain the equation of motion of the phonon Green's function as

$$\begin{aligned} D_0^{-1}D &= i\hbar I - (i\hbar)^2 \sum_a \int d\mathbf{r}_1 \gamma^a \frac{\delta}{\delta \mathbf{J}'} G_{aa}(1,1^+) \\ &= i\hbar I + (i\hbar)^2 \sum_a \int \gamma^a d\mathbf{r}_1 G_{ab}(1,2)\tilde{\Gamma}_{bc}(2,3;J')G_{ca}(3,1^+), \end{aligned} \tag{193}$$

where I is the unit matrix. It should be noted that Eqs.(190) and (192) are not closed. To close these equations, it is desirable to express the vertex functions in terms of the interaction potential and Green's functions. From Eqs.(175),(177), and (178), in terms of the matrix notation, the electron-electron interaction vertex function may be written as

$$\begin{aligned} \Gamma_{ad}(4,5;3) = &\Bigg[e_a\delta(4,3) - i\hbar e_a \sum_t e_t \int d6d7d8V(4-6)G_{tt}(6,7)\Gamma_{uv}(7,8;3)G_{vt}(8,6) \\ &+ \sum_{ilk} \gamma^a_{ilk} \frac{\delta}{\delta U(3)} < Q_{ilk}(t_4) > \Bigg] \delta(4-5) \\ &+ i\hbar\eta e_a \int d6d7V(4-6) \left[\frac{\delta G_{ac}(6,7)}{\delta U(3)} \Gamma_{cd}(7,5;3) + G_{ac}(4,6) \frac{\delta \Gamma_{cd}(7,5;3)}{\delta U(3)} \right] \\ &- i\hbar \sum_{ilk} \gamma^a_{ilk} \int d6 \left[\frac{\delta G_{ac}(4,6)}{\delta U(3)} \tilde{\Gamma}_{cd}(6,5;ilk,t_1) + G_{ac}(4,6) \frac{\delta \tilde{\Gamma}_{cd}(6,5;ilk,t_1)}{\delta U(3)} \right] . \end{aligned} \tag{194}$$

Employing Eq.(178), the functional derivative $\delta < Q > /\delta U$ may be evaluated as

$$\begin{aligned} \frac{\delta < Q_{ilk}(t_4) >}{\delta U(3)} &= i\hbar \sum_{ai'l'k'} \\ &\int d1 D_0(ilkt_4, i'l'k't_1)\gamma^a_{i'l'k'} \int d2d4G_{ac}(1,2)\Gamma_{cd}(2,4;3)G_{da}(4,1), \end{aligned} \tag{195}$$

where we have used $\delta G_{ab}(4,5)/\delta U(3)$ as given by Eq.(173). Employing Eq.(178), the functional derivative of the electron-electron interaction vertex function may be written as

$$\frac{\delta\Gamma_{ad}(1,2;3)}{\delta U(9)} = -\Gamma_{ab}(1,4;9)G^{-1}_{bc}(2,5)G_{cf}(4,6)\Gamma_{ft}(6,7;3)G_{td}(7,5)$$
$$-G^{-1}_{ab}(1,4)\Gamma_{bc}(2,5;9)G_{cf}(4,6)\Gamma_{ft}(6,7;3)G_{td}(7,5)$$
$$+G^{-1}_{ab}(1,4)G^{-1}_{bc}(2,5)\frac{\delta G_{cf}(4,6)}{\delta U(9)}\Gamma_{ft}(6,7;3)G_{td}(7,5)$$
$$+G^{-1}_{ab}(1,4)G^{-1}_{bc}(2,5)G_{cf}(4,6)\frac{\delta\Gamma_{ft}(6,7;3)}{\delta U(9)}G_{td}(7,5)$$
$$+G^{-1}_{ab}(1,4)G^{-1}_{bc}(2,5)G_{cf}(4,6)\Gamma_{ft}(6,7;3)\frac{\delta G_{td}(7,5)}{\delta U(9)}, \tag{196}$$

and

$$\frac{\delta\tilde{\Gamma}_{ad}(1,2;3)}{\delta U(9)} = -\Gamma_{ab}(1,4;9)G^{-1}_{bc}(2,5)G_{cf}(4,6)\tilde{\Gamma}_{ft}(6,7;3)G_{td}(7,5)$$
$$-G^{-1}_{ab}(1,4)\Gamma_{bc}(2,5;9)G_{cf}(4,6)\tilde{\Gamma}_{ft}(6,7;3)G_{td}(7,5)$$
$$+G^{-1}_{ab}(1,4)G^{-1}_{bc}(2,5)\frac{\delta G_{cf}(4,6)}{\delta U(9)}\tilde{\Gamma}_{ft}(6,7;3)G_{td}(7,5)$$
$$+G^{-1}_{ab}(1,4)G^{-1}_{bc}(2,5)G_{cf}(4,6)\frac{\delta\tilde{\Gamma}_{ft}(6,7;3)}{\delta U(9)}G_{td}(7,5)$$
$$+G^{-1}_{ab}(1,4)G^{-1}_{bc}(2,5)G_{cf}(4,6)\tilde{\Gamma}_{ft}(6,7;3)\frac{\delta G_{td}(7,5)}{\delta U(9)}. \tag{197}$$

Similarly, we have

$$\tilde{\Gamma}_{ad}(4,5;ilk) = \left[-i\hbar e_a \sum_t e_t \int d6V(4-6)d7d8G_{tt}(6,7)\tilde{\Gamma}_{uv}(7,8;ilk)G_{vt}(8,6)\right.$$
$$\left.+\sum_{ilk}\gamma^a_{ilk}\frac{\delta}{\delta J_{ilk}} < Q_{ilk}(t_4) >\right]$$
$$+i\hbar\eta e_a \int d6d7V(4-6)\left[\frac{\delta G_{ac}(6,7)}{\delta J_{ilk}}\Gamma_{cd}(7,5;ilk) + G_{ac}(4,6)\frac{\delta\Gamma_{cd}(7,5;ilk)}{\delta J_{ilk}}\right]$$
$$-i\hbar\sum_{ilk}\gamma^a_{ilk}\int d6\left[\frac{\delta G_{ac}(4,6)}{\delta J_{ilk}}\Gamma_{cd}(6,5;i'l'k',t_1)\right.$$
$$\left.+G_{ac}(4,6)\frac{\delta\tilde{\Gamma}_{cd}(6,5;i'l'k',t_1)}{\delta J_{ilk}}\right], \tag{198}$$

where

$$\frac{\delta\Gamma_{ad}(1,2;3)}{\delta J_{ilk}} = -\tilde{\Gamma}_{ab}(1,4;ilk)G_{bc}^{-1}(2,5)G_{cf}(4,6)\Gamma_{ft}(6,7;3)G_{td}(7,5)$$
$$-G_{ab}^{-1}(1,4)\tilde{\Gamma}_{bc}(2,5;ilk)G_{cf}(4,6)\Gamma_{ft}(6,7;3)G_{td}(7,5)$$
$$+G_{ab}^{-1}(1,4)G_{bc}^{-1}(2,5)\frac{\delta G_{cf}(4,6)}{\delta J_{ilk}}\Gamma_{ft}(6,7;3)G_{td}(7,5)$$
$$+G_{ab}^{-1}(1,4)G_{bc}^{-1}(2,5)G_{cf}(4,6)\frac{\delta\Gamma_{ft}(6,7;3)}{\delta J_{ilk}}G_{td}(7,5)$$
$$+G_{ab}^{-1}(1,4)G_{bc}^{-1}(2,5)G_{cf}(4,6)\Gamma_{ft}(6,7;3)\frac{\delta G_{td}(7,5)}{\delta J_{ilk}}, \tag{199}$$

and

$$\frac{\delta\tilde{\Gamma}_{ad}(1,2;3)}{\delta J_{ilk}} = -\tilde{\Gamma}_{ab}(1,4;ilk)G_{bc}^{-1}(2,5)G_{cf}(4,6)\tilde{\Gamma}_{ft}(6,7;3)G_{td}(7,5)$$
$$-G_{ab}^{-1}(1,4)\tilde{\Gamma}_{bc}(2,5;ilk)G_{cf}(4,6)\tilde{\Gamma}_{ft}(6,7;3)G_{td}(7,5)$$
$$+G_{ab}^{-1}(1,4)G_{bc}^{-1}(2,5)\frac{\delta G_{cf}(4,6)}{\delta J_{ilk}}\tilde{\Gamma}_{ft}(6,7;3)G_{td}(7,5)$$
$$+G_{ab}^{-1}(1,4)G_{bc}^{-1}(2,5)G_{cf}(4,6)\frac{\delta\tilde{\Gamma}_{ft}(6,7;3)}{\delta J_{ilk}}G_{td}(7,5)$$
$$+G_{ab}^{-1}(1,4)G_{bc}^{-1}(2,5)G_{cf}(4,6)\tilde{\Gamma}_{ft}(6,7;3)\frac{\delta G_{td}(7,5)}{\delta J_{ilk}}. \tag{200}$$

Thus, we have a closed set of ten coupled equations (albeit closed in terms of the variational differential equations Eqs.(168), (178), (190), (193)-(200)) in ten unknown functions, $G(1,2)$, $< Q_{ilk}(t) >$, $\Sigma(1,2)$, $\Gamma(1,2;3)$, $\tilde{\Gamma}(1,2;3)$, $\frac{\delta\tilde{\Gamma}(1,2;3)}{\delta U(4)}$, $\frac{\delta\Gamma(1,2;3)}{\delta U(4)}$, $\frac{\tilde{\Gamma}(1,2;3)}{\delta J_{ilk}}$, $\frac{\delta\Gamma(1,2;3)}{\delta J_{ilk}}$, and $D(ilk,i'l'k')$, which involve self-consistent solutions.

5.3. *Kinetic Description of Quantum Transport Equation for Multi-Band Semiconductors*

In the above section, we derived, from the first principle, the EOM of MBNGF. Carrier NGF and phonon NGF may be principally obtained from a set of closed equations. However, the calculation is too complicated to be implemented in a real problem since it needs a formidable computing time. In practice, simplification to the equations is necessary. In dealing with multi-band systems, coupling between energy bands is most difficult part that affects the implement of multi-band transport equations to a real physical system. Thus, decoupling between energy band should be the first step in simplifying the multi-band transport equations.

5.3.1. *Decoupling of EOM of MBNGF*

Two Band-Model of Semiconductor Systems

The matrix notation of equations of motion of Green's function, Eq.(178), may be written as

$$\int dx_3 G_{ad}^{-1}(x_1, x_3) G_{db}(x_3, x_2) = \delta_{ab}(x_1, x_2), \tag{201}$$

where

$$G_{ad}^{-1}(x_1, x_2) = G_{ad,0}^{-1}(x_1, x_2) - \Sigma_{ad}(x_1, x_2), \tag{202}$$

and

$$G_{ad,0}^{-1}(x_1, x_2) = \left[\pm i\hbar\delta_{ad}\frac{\partial}{\partial t_3} - H_{ad}(x_3)\right]\delta(x_1 - x_3). \tag{203}$$

For the sake of simplicity, we first consider a two-band model of semiconductors. Thus, a, b, d in Eqs.(201)-(203) may be taken as e for the conduction band and h for the valence band. Symbolically, the above equation may be written as

$$G_{ee}^{-1}G_{ee} + G_{eh}^{-1}G_{he} = 1, \tag{204}$$

$$G_{ee}^{-1}G_{eh} + G_{eh}^{-1}G_{hh} = 0, \tag{205}$$

$$G_{he}^{-1}G_{ee} + G_{hh}^{-1}G_{he} = 0, \tag{206}$$

$$G_{he}^{-1}G_{ee} + G_{hh}^{-1}G_{hh} = 1. \tag{207}$$

From Eq.(206), we have

$$G_{he} = G_{hh}\Sigma_{he}G_{ee}. \tag{208}$$

Substituting this equation into Eq.(204), we have

$$[G_{ee}^{-1} - G_{eh}^{-1}G_{hh}G_{he}^{-1}]G_{ee} = 1. \tag{209}$$

Making use of Eqs.(3.46) and (3.47) and defining Σ'_{ee} as

$$\Sigma'_{ee} = \Sigma_{ee} + \Sigma_{eh}G_{hh}\Sigma_{he}, \tag{210}$$

Eq.(209) may be written as

$$[G_{ee,0}^{-1} - \Sigma'_{ee}]G_{ee} = 1, \tag{211}$$

or, equivalently,

$$\left[i\hbar\frac{\partial}{\partial t_1} - H(x_1)\right]G_{ee}(x_1, x_2) = \delta(x_1, x_2) + \int dx_3 \Sigma'_{ee}(x_1, x_3)G_{ee}(x_3, x_2). \tag{212}$$

Dealing similarly with Eqs.(205) and (207) and defining

$$\Sigma'_{hh} = \Sigma_{hh} + \Sigma_{he}G_{ee}\Sigma_{eh}, \tag{213}$$

we obtain

$$G_{eh} = G_{ee}\Sigma_{eh}G_{hh} \tag{214}$$

and

$$\left[i\hbar\frac{\partial}{\partial t_1} - H(x_1)\right]G_{hh}(x_1, x_2) = \delta(x_1, x_2) + \int dx_3 \Sigma'_{hh}(x_1, x_3)G_{hh}(x_3, x_2). \tag{215}$$

Three Band-Model of Semiconductor Systems

Considering a semiconductor with a conduction band (e), a light hole band(l), and a heavy hole band (h), Eq.(206) may be written in the form

$$G_{ee}^{-1}G_{ee} + G_{eh}^{-1}G_{he} + G_{el}^{-1}G_{le} = 1, \tag{216}$$

$$G_{ee}^{-1}G_{eh} + G_{eh}^{-1}G_{hh} + G_{el}^{-1}G_{lh} = 0, \tag{217}$$

$$G_{ee}^{-1}G_{el} + G_{eh}^{-1}G_{hl} + G_{el}^{-1}G_{ll} = 0, \tag{218}$$

$$G_{he}^{-1}G_{ee} + G_{hh}^{-1}G_{he} + G_{hl}^{-1}G_{he} = 0, \tag{219}$$

$$G_{he}^{-1}G_{eh} + G_{hh}^{-1}G_{hh} + G_{hl}^{-1}G_{lh} = 1, \tag{220}$$

$$G_{he}^{-1}G_{el} + G_{hh}^{-1}G_{hl} + G_{hl}^{-1}G_{hl} = 0, \tag{221}$$

$$G_{le}^{-1}G_{ee} + G_{lh}^{-1}G_{he} + G_{ll}^{-1}G_{le} = 0, \tag{222}$$

$$G_{le}^{-1}G_{eh} + G_{lh}^{-1}G_{hh} + G_{ll}^{-1}G_{lh} = 0, \tag{223}$$

$$G_{le}^{-1}G_{el} + G_{lh}^{-1}G_{hl} + G_{ll}^{-1}G_{ll} = 1. \tag{224}$$

From Eqs.(218) and (221),

$$G_{el} + G_{ee}G_{eh}^{-1}G_{hl} + G_{ee}G_{el}^{-1}G_{ll} = 0, \tag{225}$$

$$G_{hh}G_{he}^{-1}G_{el} + G_{hl} + G_{hh}G_{hl}^{-1}G_{hl} = 0. \tag{226}$$

Thus we find

$$G_{hl} = -[1 - G_{hh}G_{he}^{-1}G_{ee}G_{eh}^{-1}]^{-1}G_{hh}G_{he}^{-1}[1 - G_{ee}G_{el}^{-1}]G_{ll}, \tag{227}$$

$$G_{el} = G_{ee}G_{eh}^{-1}[1 - G_{hh}G_{he}^{-1}G_{ee}G_{eh}^{-1}]^{-1}G_{hh}G_{he}^{-1}[1 - G_{ee}G_{el}^{-1}]G_{ll}. \tag{228}$$

Substituting Eqs.(227) and (228) into Eq.(224), we have

$$[G_{ll,0}^{-1} - \Sigma'_{ll}]G_{ll} = 1, \tag{229}$$

where

$$\begin{aligned}\Sigma'_{ll} = \Sigma_{ll} + \Sigma_{le}G_{ee}\Sigma_{eh}[1 - G_{hh}\Sigma_{he}G_{ee}\Sigma_{eh}]^{-1}G_{hh}\Sigma_{he}[1 - G_{ee}\Sigma_{eh}] \\ -\Sigma_{lh}[1 - G_{hh}\Sigma_{he}G_{ee}\Sigma_{eh}]^{-1}G_{hh}\Sigma_{he}[1 - G_{ee}\Sigma_{eh}].\end{aligned} \tag{230}$$

Similarly, we obtain

$$G_{lh} = [1 - G_{ll}\Sigma_{le}G_{ee}\Sigma_{el}]^{-1}G_{ll}[\Sigma_{le}G_{ee}\Sigma_{eh} - \Sigma_{lh}]G_{hh}, \tag{231}$$

$$\begin{aligned}G_{eh} = - \; & G_{ee}[\Sigma_{eh} - \Sigma_{el}[1 - G_{ll}\Sigma_{le} \\ & G_{ee}\Sigma_{el}]^{-1}G_{ll}[\Sigma_{le}G_{ee}\Sigma_{eh} - \Sigma_{lh}]G_{hh},\end{aligned} \tag{232}$$

and

$$[G_{hh,0}^{-1} - \Sigma'_{hh}]G_{hh} = 1, \tag{233}$$

where

$$\Sigma'_{hh} = \Sigma_{hh} + [\Sigma_{eh} - \Sigma_{el}[1 - G_{ll}\Sigma_{le}G_{ee}\Sigma_{el}]^{-1}G_{ll}[\Sigma_{le}G_{ee}\Sigma_{eh} - \Sigma_{lh}] \\ -[1 - G_{ll}\Sigma_{le}G_{ee}\Sigma_{el}]^{-1}G_{ll}[\Sigma_{le}G_{ee}\Sigma_{eh} - \Sigma_{lh}]. \quad (234)$$

Further, from Eqs.(216), (219), and (221), we have

$$G_{he} = [1 - G_{hh}\Sigma_{hl}G_{ll}\Sigma_{lh}]^{-1}G_{hh}[\Sigma_{he} - \Sigma_{hl}G_{ll}\Sigma_{le}]G_{ee}, \quad (235)$$

$$G_{le} = G_{ll}[\Sigma_{le} - \Sigma_{lh}[1 - G_{hh}\Sigma_{hl}G_{ll}\Sigma_{lh}]^{-1} \\ G_{hh}[G_{he} - \Sigma_{hl}G_{ll}\Sigma_{le}]G_{ee}, \quad (236)$$

and

$$[G^{-1}_{ee,0} - \Sigma'_{ee}]G_{ee} = 1, \quad (237)$$

where

$$\Sigma'_{ee} = \Sigma_{ee} + \Sigma_{eh}[1 - G_{hh}\Sigma_{hl}G_{ll}\Sigma_{lh}]^{-1}G_{hh}[\Sigma_{he} - \Sigma_{hl}G_{ll}\Sigma_{le}] \\ +\Sigma_{el}G_{ll}[\Sigma_{le} - \Sigma_{lh}[1 - G_{hh}\Sigma_{hl}G_{ll}\Sigma_{lh}]^{-1}G_{hh}[G_{he} - \Sigma_{hl}G_{ll}\Sigma_{le}]. \quad (238)$$

The above equations show that the diagonal parts of the equations of motion of Green's functions for multi-band semiconductors are the same as those for single band semiconductors *except for the selfenergies.* The coupling between electrons and holes and between heavy holes and light holes for a three band model of semiconductors are embedded in the modified selfenergies defined by Eqs. (236), (234) and (238), and by Eqs. (210) and (213) for two band model of semiconductors.

5.3.2. *Integration Form of EOM of MBNGF*

We have exhibited the equations of motion of the Green's function for a multi-band semiconductor in a decoupled form similar to that of a single-band model with the effects of the coupling among energy bands expressed in terms of modified electron self energies. However, in the interest of generality and to readily demonstrate the effects of band-band coupling on the kinematic characteristics of the carriers, we retain the form of the Green's function equation of motion as given by Eq.(178). The band electron Green's function equation of motion has the form (sum over repeated band indices a, b and $\delta_{ab}(x_1, x_2) = \delta_{ab}\delta(x_1 - x_2)$)

$$\left[i\hbar\delta_{ac}\frac{\partial}{\partial t_1} - H_{ac}(x_1)\right]G_{cb}(x_1, x_2) = \delta_{ab}(x_1, x_2) + \int dx_3\Sigma_{ac}(x_1, x_3)G_{cb}(x_3, x_2), \quad (239)$$

and the adjoint equation is given by

$$\left[-i\hbar\delta_{ac}\frac{\partial}{\partial t_2} - H_{ac}(x_2)\right]G_{cb}(x_1, x_2) = \delta_{ab}(x_1, x_2) + \int dx_3G_{ac}(x_1, x_3)\Sigma_{cb}(x_3, x_2). \quad (240)$$

Here, $\Sigma(x_1, x_2)$ is the self-energy and $H(x)$ is the one electron Hamiltonian in position-time representation. $\Sigma(x_1, x_2)$ incorporates electron interaction effects with other electrons, with phonons and with random impurities. Band-band interaction effects are represented by the off-diagonal terms of $H_{ab}\,(a \neq b)$. Furthermore, $\Sigma(x_1, x_2)$ includes band-band interaction effects through Coulomb interactions, phonon exchange, etc.

To develop quantum transport equations, we examine the equations of motion for the correlation function $G^<_{ab}$, which is related to the carrier density of the $a'th$ band for $a = b$. In the interest of generality and to readily demonstrate the effects of band-band coupling on the kinematics characteristics of the carriers, we retain the form of the Green's function equation of motion as given by Eq.(178). Setting $\mathcal{H}(x_1, x_2) = H(x_1)\delta(x_1 - x_2)$, and noting time-order on the real axis in accordance with the Langreth algebra, we have

$$i\hbar\frac{\partial}{\partial t_1}G^<_{ab}(x_1, x_2) = \int dx_3 \mathcal{H}(x_1, x_3)G^<(x_3, x_2)_{ab}$$
$$+\int dx_3 \Sigma^r(x_1, x_3)G^<(x_3, x_2)_{ab} + \int dx_3 \Sigma^<(x_1, x_3)G^a(x_3, x_2)_{ab}, \quad (241)$$

$$-i\hbar\frac{\partial}{\partial t_2}G^<_{ab}(x_1, x_2) = \int dx_3 G^<(x_1, x_3)\mathcal{H}(x_3, x_2)_{ab}$$
$$+\int dx_3 G^r(x_1, x_2)\Sigma^<(x_3, x_2)_{ab} + \int dx_3 G^<(x_1, x_2)\Sigma^a(x_3, x_2)_{ab}, \quad (242)$$

where $G^{r(a)}$ is the retarded (advanced) Green's function, and $\Sigma^{r(a)}$ is the retarded (advanced) self-energy. These functions are defined in accordance with the notation,

$$F^r(x_1, x_2) = \theta(t_1 - t_2)[F^>(x_1, x_2) - F^<(x_1, x_2)], \quad (243)$$
$$F^a(x_1, x_2) = -\theta(t_2 - t_1)[F^>(x_1, x_2) - F^<(x_1, x_2)]. \quad (244)$$

Following the procedures of Buot and Jensen, we subtract Eq.(241) from Eq.(242), obtaining

$$i\hbar\left(\frac{\partial}{\partial t_1} + \frac{\partial}{\partial t_2}\right)G^<_{ab}(x_1, x_2) = [\mathcal{H}, G^<](x_1, x_2)_{ab}$$
$$+[\Sigma^<, ReG^r](x_1, x_2)_{ab} + \frac{i}{2}\{\Sigma^<, A\}(x_1, x_2)_{ab} - \frac{i}{2}\{\Gamma, G^<\}(x_1, x_2)_{ab}, \quad (245)$$

where [,] and {,} represent commutator and anticommutator, respectively. $A = -2ImG^r$ is the carrier spectral function and $\Gamma = -2Im\Sigma^r$ is the damping function. We have used a space-time matrix notation in which multiplication stands for the following integral

$$(AB)(x_1, x_2)_{ab} = \int dx_3 A(x_1, x_3)_{ac}B(x_3, x_2)_{cb}. \quad (246)$$

Introducing the notation

$$X = \frac{1}{2}(x_1 + x_2) \qquad \text{and} \qquad x = x_1 - x_2, \quad (247)$$

where $x = (\mathbf{r}, t), X = (\mathbf{R}, T)$, we have

$$i\hbar\frac{\partial}{\partial T}G^<_{ab}(x,X) = [\mathcal{H}, G^<]_{ab}(x,X)$$
$$+[\Sigma^<, ReG^r]_{ab}(x,X) + \frac{i}{2}\{\Sigma^<, A\}_{ab}(x,X) - \frac{i}{2}\{\Gamma, G^<\}_{ab}(x,X). \tag{248}$$

As pointed out in the Introduction, the Fourier transform is useful in regard to the microscopic coordinates of carriers in mesoscopic systems. Accordingly, we define the transform as

$$G^<_{ab}(q,X) = \int dx e^{\frac{iqx}{\hbar}} G^<_{ab}(x,X), \tag{249}$$

where $q = (\mathbf{q}, E/\hbar)$, and Eq.(3.92) yields

$$i\hbar\frac{\partial}{\partial T}G^<_{ab}(q,X) = \int dx e^{\frac{iqx}{\hbar}}[\mathcal{H}, G^<](q,X)_{ab} + \int dx e^{\frac{iqx}{\hbar}}[\Sigma^<, ReG^r](x,X)_{ab}$$
$$+\frac{i}{2}\int dx e^{\frac{iqx}{\hbar}}\{\Sigma^<, A\}(x,X)_{ab} - \frac{i}{2}\int dx e^{\frac{iqx}{\hbar}}\{\Gamma, G^<\}(x,X)_{ab}. \tag{250}$$

To evaluste Eq.(250), we examine the following integral structure,

$$I(q,X) = \int dx e^{\frac{iqx}{\hbar}} \int dx_3 A(x_1,x_3)B(x_3,x_2). \tag{251}$$

Since we may write a translation operator as an exponential in accordance with

$$A(y, X + \frac{1}{2}z) = exp\left[\frac{1}{2}z\frac{\partial^{(A)}}{\partial X}\right]A(y,X),$$

and

$$exp\left[\frac{\hbar}{2i}\frac{\partial}{\partial q}\frac{\partial^{(A)}}{\partial X}\right]e^{\frac{i}{\hbar}qz} = e^{\frac{i}{\hbar}qz}exp\left[\frac{1}{2}z\frac{\partial^{(A)}}{\partial X}\right],$$

we have

$$I(q,X) = e^{i\Lambda}A(q,X)B(q,X), \tag{252}$$

where the Poisson bracket differential operator Λ is given by ($\partial^{A,B}$ acts on A, B only, to the exclusion of B, A, respectively)

$$\Lambda = \frac{\hbar}{2}\left[\frac{\partial^{(A)}}{\partial q}\frac{\partial^{(B)}}{\partial X} - \frac{\partial^{(B)}}{\partial q}\frac{\partial^{(A)}}{\partial X}\right]. \tag{253}$$

Thus, Eq.(250) may be written as

$$i\hbar\frac{\partial}{\partial T}G^<_{ab}(q,X) = e^{i\Lambda}[\mathcal{H}, G^<](q,X)_{ab} + e^{i\Lambda}[\Sigma^<, ReG^r](q,X)_{ab}$$
$$+\frac{i}{2}e^{i\Lambda}[\Sigma^<, A](q,X)_{ab} - \frac{i}{2}e^{i\Lambda}[\Gamma, G^<](q,X)_{ab}. \tag{254}$$

Making use of the "exponential" displacement operation and noting that

$$exp\left[\frac{\hbar}{2}\left(\frac{\partial^{(A)}}{\partial q}\frac{\partial^{(B)}}{\partial X}-\frac{\partial^{(B)}}{\partial q}\frac{\partial^{(A)}}{\partial X}\right)\right]=exp\left[\frac{\hbar}{2}\frac{\partial^{(A)}}{\partial q}\frac{\partial^{(B)}}{\partial X}\right]exp\left[-\frac{\hbar}{2}\frac{\partial^{(B)}}{\partial q}\frac{\partial^{(A)}}{\partial X}\right]$$
$$=exp\left[-\frac{\hbar}{2}\frac{\partial^{(B)}}{\partial q}\frac{\partial^{(A)}}{\partial X}\right]exp\left[\frac{\hbar}{2}\frac{\partial^{(A)}}{\partial q}\frac{\partial^{(B)}}{\partial X}\right], \tag{255}$$

we obtain

$$I(q,X)=\frac{1}{(h^4)^2}\int dvdudq'dX'e^{\frac{i}{\hbar}(X-X')v}e^{\frac{i}{\hbar}(q-q')u}A(q-\frac{v}{2},X+\frac{u}{2})B(q',X')$$
$$=\frac{1}{(h^4)^2}\int dvdudq'dX'e^{\frac{i}{\hbar}(X-X')v}e^{\frac{i}{\hbar}(q-q')u}A(q',X')B(q+\frac{v}{2},X-\frac{u}{2}). \tag{256}$$

Defining

$$K_A^{\pm}(q,X-X';X,q-q')=\int dvdue^{\frac{i}{\hbar}(X-X')v}e^{\frac{i}{\hbar}(q-q')u}A(q\pm\frac{v}{2},X\mp\frac{u}{2}) \tag{257}$$

and making use of Eqs.(252) and (253), Eq.(254) can be written as

$$i\hbar\frac{\partial}{\partial T}G^<_{ab}(q,X)$$
$$=\frac{1}{(h^4)^2}\int dq'dX'[K_H^-(q,X-X';X,q-q')G^<(q',X')$$
$$-G^<(q',X')K_H^+(q,X-X';X,q-q')]_{ab}$$
$$+\frac{1}{(h^4)^2}\int dq'dX'[K^-_{\Sigma^<}(q,X-X';X,q-q')ReG^<(q',X')$$
$$-ReG^<(q',X')K^+_{\Sigma^<}(q,X-X';X,q-q')]_{ab}$$
$$+\frac{i}{2(h^4)^2}\int dq'dX'[K^-_{\Sigma^<}(q,X-X';X,q-q')A(q',X')$$
$$-A(q',X')K^+_{\Sigma^<}(q,X-X';X,q-q')]_{ab}$$
$$-\frac{i}{2(h^4)^2}\int dq'dX'[K^-_{\Gamma}(q,X-X';X,q-q')G^<(q',X') \tag{258}$$
$$-G^<(q',X')K^+_{\Gamma}(q,X-X';X,q-q')]_{ab}.$$

Eq.(259) is a general formulation for the time evolution of the correlation function $G^<$ in the phase-space and energy-time domain. It provides a framework for the description of all types of quantum transport phenomena, as it is devoid of any assumption of slow variation and "gradient expansion" which have limited the validity of many early theories of many-body quantum-transport. The first two terms on the right hand side (RHS) of Eq.(259) may be viewed as describing the generalized quantized motion of particles (i.e., quantum tunneling and interference phenomena) in phase space with a more complicated energy-momentum relation due to the influence of a potential and scatterings. The last two terms on the RHS of this equation describe particle transfers in phase space due to collisions and scatterings (broadening effects) and correspond to the collision terms of the classical

Boltzmann equation. The effects of energy band coupling are also included in the last three terms on the RHS of this equation. Such effects of energy band coupling on carrier transport will be discussed below.

5.3.3. *Simplification of Transport Equations*

Single-Band Systems

For single band semiconductors or multi-band semiconductors without coupling between energy bands, we have

$$B(q',X')K_A^{\pm}(q,X-X';X,q-q') = K_A^{\pm}(q,X-X';X,q-q')B(q',X'),$$

where B may be $G^<, \Sigma^<$, etc. . Defining

$$K_A^s(q,X-X';X,q-q') = K_A^+(q,X-X';X,q-q') + K_A^-(q,X-X';X,q-q'), \quad (259)$$

and

$$K_A^c(q,X-X';X,q-q') = K_A^+(q,X-X';X,q-q') - K_A^-(q,X-X';X,q-q'), \quad (260)$$

Eq.(259) can be written as

$$\begin{aligned} i\hbar\frac{\partial}{\partial T}G^<(q,X) &= \frac{1}{(h^4)^2}\int dq'dX'K_H^c(q,X-X';X,q-q')G^r(q',X') \\ &+\frac{1}{(h^4)^2}\int dq'dX'K_{\Sigma^<}^c(q,X-X';X,q-q')ReG^r(q',X') \\ &+\frac{i}{2(h^4)^2}\int dq'dX'K_{\Sigma^<}^s(q,X-X';X,q-q')A(q',X') \\ &-\frac{i}{2(h^4)^2}\int dq'dX'K_\Gamma^s(q,X-X';X,q-q')G^<(q',X'). \end{aligned} \quad (261)$$

This equation is the same as that derived by Buot and Jensen[25] except that our result is expressed in electronic coordinates, rather than lattice vectors. If scatterings can be ignored and the single particle Hamiltonian takes the form of Eq.(259), we may employ the Wigner function defined as

$$f_w(\mathbf{q},\mathbf{R},t) = \frac{1}{2\pi\hbar}\int dE(-i)G^<(\mathbf{q},E,\mathbf{R},t) \quad (262)$$

to rewrite Eq.(261) in the well known form

$$\begin{aligned} \frac{\partial}{\partial t}f_w(\mathbf{q},\mathbf{R},t) &= -\frac{1}{m^*}\mathbf{q}\cdot\frac{\partial}{\partial\mathbf{R}}f_w(\mathbf{q},\mathbf{R},t) \\ &+\frac{1}{h^4}\int d\mathbf{q}'d\mathbf{u}\sin\frac{(\mathbf{q}-\mathbf{q}')\cdot\mathbf{u}}{\hbar}\left[V(\mathbf{R}+\frac{\mathbf{u}}{2})-V(\mathbf{R}-\frac{\mathbf{u}}{2})\right]f_w(\mathbf{q}',\mathbf{R},t). \end{aligned} \quad (263)$$

For the higher level simplification of Eq.(261), the concrete form of the self-energies must be considered. In quantum devices, the following interaction mechanisms contribute to the intra-band self-energies of carriers. Considering the lowest approximation to the carrier-phonon interaction, we retain only the zero order terms in self-energies in Eq.(261). Incorporating the definition of Winger function, i.e. Eq.(262), into our derivation and integrating Eq.(261) with respect to energy, we obtain the following equation.

$$\frac{\partial}{\partial t} f_w(\mathbf{q}, \mathbf{R}, t) = -\frac{1}{m^*}\mathbf{q} \cdot \frac{\partial}{\partial \mathbf{R}} f_w(\mathbf{q}, \mathbf{R}, t)$$
$$+\frac{1}{h^4}\int d\mathbf{q}' d\mathbf{u} \sin\frac{(\mathbf{q}-\mathbf{q}')\cdot\mathbf{u}}{\hbar}\left[V(\mathbf{R}+\frac{\mathbf{u}}{2}) - V(\mathbf{R}-\frac{\mathbf{u}}{2})\right] f_w(\mathbf{q}', \mathbf{R}, t)$$
$$\frac{-i}{h}\int dE[\Sigma^< A - \Gamma G^<]. \tag{264}$$

In the above equation, A and Γ are spectrum function and decay function, respectively. For carrier-phonon interaction, the meaningful and significant contribution to the carrier-phonon self-energy is given by the so-called "one-phonon" diagram. Since the phonon propagator $D(1,2)$ is nonlocal in time, the carrier-phonon self-energy can be expressed on the closed-time-path contour as

$$\Sigma(1,2) = iG(1,2)D(1,2) \tag{265}$$

By fixing the time order, the above equation may be expressed on real time axis as

$$\Sigma^{\gtrless}(1,2) = iG^{\gtrless}(1,2)D^{\gtrless}(1,2) \tag{266}$$

Thus, the decay function may be expressed as

$$\Gamma(p,q) = -\frac{1}{h}\int dk[G^>(p+k,q)D^>(k) - G^<(p+k,q)D^<(k)] \tag{267}$$

For phonon, if we approximate the spectral weight function by the δ-function, the Fourier transform of the phonon Green's function can be expressed as

$$D^<(\mathbf{k}, E') = -iM_p^2[(N_\mathbf{k}+1)\delta(E'-\Omega_k) + N_\mathbf{k}\delta(E'+\Omega_\mathbf{k})] \tag{268}$$

and

$$D^>(\mathbf{k}, E') = -iM_p^2[(N_\mathbf{k}+1)\delta(E'+\Omega_k) + N_\mathbf{k}\delta(E'-\Omega_\mathbf{k})] \tag{269}$$

where Ω_k is the energy-momentum relation for phonons. For uniform systems at steady-state condition, the correlation function for the carriers may be written as

$$-iG^<(\mathbf{p}, E) = f(\mathbf{p}, E)A(\mathbf{p}, E) \tag{270}$$

$$iG^>(\mathbf{p}, E) = [1 - f(\mathbf{p}, E)]A(\mathbf{p}, E) \tag{271}$$

where

$$\frac{1}{h}\int dEA(\mathbf{p}, E) = 1 \tag{272}$$

These equation are called "Kadanoff-Baym ansatz". For nonuniform systems and/or nonstationary states, these relations don't hold for nonequilibrium transport. A "new" ansatz, the " generalized Kadanoff-Baym ansatz" should be employed. Substituting Eqs.(268)-(272) into Eqs.(266) and (267), we obtain

$$\Sigma^{<}(p,q) = \frac{i}{h^3}\int d\mathbf{k}\,\{f(\mathbf{p}+\mathbf{k},E+\Omega_k)A(\mathbf{p}+\mathbf{k},E+\Omega_k)M_p^2(N_{\mathbf{k}}+1)$$
$$+f(\mathbf{p}+\mathbf{k},E-\Omega_k)A(\mathbf{p}+\mathbf{k},E-\Omega_k)M_p^2N_{\mathbf{k}}\} \tag{273}$$

$$\Gamma(p,q) = \frac{1}{h^3}\int d\mathbf{k}\,\{[N_{\mathbf{k}}+f(\mathbf{p}+\mathbf{k},E+\Omega_k)]A(\mathbf{p}+\mathbf{k},E+\Omega_k)M_p^2$$
$$+[N_{\mathbf{k}}+1-f(\mathbf{p}+\mathbf{k},E-\Omega_k)]A(\mathbf{p}+\mathbf{k},E-\Omega_k)M_p^2N_{\mathbf{k}} \tag{274}$$

Considering the case that the spectral function $A(\mathbf{p},E)$ can be approximated by a delta function, $A(\mathbf{p},E) = h\delta(E-E_P)$, the collision integration term in Eq.(108) can be written as

$$\hbar\frac{\partial f}{\partial t}\bigg|_{coll} = \frac{-i}{h}\int dE[\Sigma^{<}A-\Gamma G^{<}]$$
$$= \frac{1}{2\pi}\int_{-\infty}^{\infty} d\mathbf{k}'[w_{\mathbf{k}\leftarrow\mathbf{k}'}f(\mathbf{R},\mathbf{k}',t) - w_{\mathbf{k}'\leftarrow\mathbf{k}}f(\mathbf{R},\mathbf{k},t)] \tag{275}$$

where

$$w_{\mathbf{k}\leftarrow\mathbf{k}'} = h\delta(E_{\mathbf{k}}-E_{\mathbf{k}'}+\Omega_{\mathbf{k}'-\mathbf{k}})M_{\mathbf{k}-\mathbf{k}'}^2(N_{\mathbf{k}-\mathbf{k}'}+1)[1-f(\mathbf{k})]$$
$$+h\delta(E_{\mathbf{k}}-E_{\mathbf{k}'}-\Omega_{\mathbf{k}'-\mathbf{k}})M_{\mathbf{k}-\mathbf{k}'}^2N_{\mathbf{k}-\mathbf{k}'}[1-f(\mathbf{k})] \tag{276}$$

$$w_{\mathbf{k}'\leftarrow\mathbf{k}} = h\delta(E_{\mathbf{k}}-E_{\mathbf{k}'}+\Omega_{\mathbf{k}'-\mathbf{k}})M_{\mathbf{k}'-\mathbf{k}}^2(N_{\mathbf{k}-\mathbf{k}'}+1)[1-f(\mathbf{k}')]$$
$$+h\delta(E_{\mathbf{k}}-E_{\mathbf{k}'}-\Omega_{\mathbf{k}'-\mathbf{k}})M_{\mathbf{k}-\mathbf{k}'}^2N_{\mathbf{k}-\mathbf{k}'}[1-f(\mathbf{k}')] \tag{277}$$

In practice, the relaxation-time approximation is always used to study the behavior of carriers near equilibrium state. In relaxation time approximation,

$$\frac{1}{2\pi}\int_{-\infty}^{\infty} d\mathbf{k}'w_{\mathbf{k}'\leftarrow\mathbf{k}} = \frac{\hbar}{\tau(\mathbf{k})} \tag{278}$$

where $\tau(\mathbf{k})$ is the momentum relaxation time which is a constant for a certain material. In equilibrium state, the collision term vanishes, i.e., $\hbar\frac{\partial f}{\partial t}\big|_{coll} = 0$. Thus, the detailed balance condition suggests that

$$w_{\mathbf{k}\leftarrow\mathbf{k}'} = \hbar f_0(\mathbf{R},\mathbf{k},t)/(\tau(\mathbf{k})\rho_0(\mathbf{R}) \tag{279}$$

where $f_0(\mathbf{R},\mathbf{k},t)$ is the equilibrium Wigner function and

$$\rho_0(\mathbf{R},t) = \frac{1}{2\pi}\int_{-\infty}^{\infty} f_0(\mathbf{R},\mathbf{k}) \tag{280}$$

is the equilibrium density of carriers.

Substituting Eqs.(279) to (280) into Eq.(276), we obtain the collision integration as

$$\hbar\frac{\partial f}{\partial t}\bigg|_{coll} = \frac{\hbar}{\tau\rho_0}\left[\rho(\mathbf{R},t)f_0(\mathbf{R},\mathbf{k}) - \rho_0(\mathbf{R},t)f_(\mathbf{R},\mathbf{k})\right] \tag{281}$$

where the carrier density, ρ, is given by

$$\rho(\mathbf{R},t) = \frac{1}{2\pi}\int_{-\infty}^{\infty} f(\mathbf{R},\mathbf{k}) \tag{282}$$

Thus the exact transport equation Eq.(108) can be simplified as

$$\begin{aligned}
&\frac{\partial}{\partial t} f_w(\mathbf{q},\mathbf{R},t) = -\frac{1}{m^*}\mathbf{q}\cdot\frac{\partial}{\partial \mathbf{R}} f_w(\mathbf{q},\mathbf{R},t) \\
&+\frac{1}{h^4}\int d\mathbf{q}' d\mathbf{u} \sin\frac{(\mathbf{q}-\mathbf{q}')\cdot\mathbf{u}}{\hbar}\left[V(\mathbf{R}+\frac{\mathbf{u}}{2}) - V(\mathbf{R}-\frac{\mathbf{u}}{2})\right] f_w(\mathbf{q}',\mathbf{R},t) \\
&+\frac{\hbar}{\tau\rho_0}\left[\rho(\mathbf{R},t) f_0(\mathbf{R},\mathbf{k}) - \rho_0(\mathbf{R},t) f_(\mathbf{R},\mathbf{k})\right]
\end{aligned} \tag{283}$$

This is the Wigner function equation. We will employ this equation to study the kinetic behavior of carrier transport through a resonant tunneling structure and the distribution of the heat dissipated into the structure in terms of phonon-carrier interaction in the following chapters.

Two-Band Systems

The general form for the multi-band single particle Hamiltonian is given in Ref. 27. Often, the coupling between the heavy holes and the electrons can be neglected, and we will treat them separately. However, since the mass of light holes is approximately equal to that of electrons, the coupling between electrons and light holes can not be ignored. This coupling is well described by the two-band model Hamiltonian derived from Kane's $\mathbf{k}\cdot\mathbf{p}$ scheme:[78,79]

$$H = \begin{bmatrix} H_e(\mathbf{q},\mathbf{r}) & H_{eh}(\mathbf{q}) \\ H_{he}(\mathbf{q}) & H_h(\mathbf{q},\mathbf{r}) \end{bmatrix} = \begin{bmatrix} E_c + \frac{q^2}{2m_0} + V_c & \Pi_{cv} q \\ \Pi_{cv} q & E_v - \frac{q^2}{2m_0} + V_v \end{bmatrix}, \tag{284}$$

where $E_c(E_v)$ is the conduction (valance) band edge, m_0 is the mass of a free electron, $V_c(V_v)$ is the potential for electrons (light holes), Π_{cv} is the matrix element of the momentum operator between the conduction band and the light-hole valance band. It is convenient to employ the decoupling technique to determine decoupled equations of motion for the Green's functions for electrons and light holes, G_{ee} and G_{hh}, separately. Once G_{ee} and G_{hh} are obtained, G_{he} and G_{eh} can be easily obtained. The coupling between the conduction band and valance band can be incorporated as additional self-energy contributions given by

$$\Sigma_{ee,c} = (H_{eh} + \Sigma_{eh}) G_{hh} (H_{he} + \Sigma_{he}), \tag{285}$$

and

$$\Sigma_{hh,c} = (H_{he} + \Sigma_{he}) G_{ee} (H_{eh} + \Sigma_{eh}) \tag{286}$$

where $G_{ee}(G_{hh})$ is the electron(light hole) Green's function. The carrier interaction self-energies may be denoted as

$$\Sigma'_{ee} = \Sigma_{ee} + \Sigma_{ee,c}, \tag{287}$$

and

$$\Sigma'_{hh} = \Sigma_{hh} + \Sigma_{hh,c}, \tag{288}$$

where Σ_{ee} and Σ_{hh} are the self-energies of electrons and holes without considering coupling and transitions between energy bands, ie., due to electron-electron and hole-hole Coulomb and phonon-mediated interactions, respectively. Replacing the selfenergies Σ in Eq.(261) by the above equations, and replacing H in Eq.(261) by $H_{ee} = H_e$ for electrons, and $H_{vv} = H_v$ for holes, we obtain the equations of motion of the Green's functions for carriers involving two bands (a=e,h).

If scatterings can be ignored in the system, the equations of motion of the correlation functions may be written as

$$\begin{aligned} i\hbar\frac{\partial}{\partial T}G^<_{aa}(q,X) &= \frac{1}{(h^4)^2}\int dq'dX'K^c_{H_{aa}}(q,X-X';X,q-q')G^<_{aa}(q',X') \\ &+\frac{1}{(h^4)^2}\int dq'dX'K^c_{\Sigma^<_{aa,c}}(q,X-X';X,q-q')ReG^r_{aa}(q',X') \\ &+\frac{i}{2(h^4)^2}\int dq'dX'K^s_{\Sigma^<_{aa,c}}(q,X-X';X,q-q')A_{aa}(q',X') \\ &-\frac{i}{2(h^4)^2}\int dq'dX'K^s_{\Gamma_{aa,c}}(q,X-X';X,q-q')G^<_{aa}(q',X'),\ a=e,h. \end{aligned} \tag{289}$$

In the above equation, the first term on the RHS arises primarily from intra-band carrier dynamics and the remaining terms on the right hand side of the equation are due to the coupling between the energy bands. Substituting the two band Hamiltonian and using a parabolic approximation for each of the bands into the above equation we have,

$$\begin{aligned} i\hbar\frac{\partial}{\partial T}G^<_{aa} &= -\frac{\mathbf{q}}{m_0}\cdot\frac{\partial}{\partial \mathbf{X}}G^<_{aa}(q,X) \\ &+\frac{1}{\hbar^3}\int d\mathbf{q}'du e^{\frac{i}{\hbar}(\mathbf{q}-\mathbf{q}')\cdot\mathbf{u}}\left[V_a(\mathbf{x}-\frac{\mathbf{u}}{2})-V_a(\mathbf{x}+\frac{\mathbf{u}}{2})\right]G^<_{aa}(\mathbf{q}',e,X)+s_a, \\ a &= e,h. \end{aligned} \tag{290}$$

where $s_a(q,X)$ is the source term, given by

$$\begin{aligned} s_a(q,X) &= +\frac{1}{(h^4)^2}\int dq'dX'K^c_{\Sigma^<_{aa,c}}(q,X-X';X,q-q')ReG^<_{aa}(q',X') \\ &+\frac{i}{2(h^4)^2}\int dq'dX'K^s_{\Sigma^<_{aa,c}}(q,X-X';X,q-q')A_{aa}(q',X') \\ &-\frac{i}{2(h^4)^2}\int dq'dX'K^s_{\Gamma_{aa,c}}(q,X-X';X,q-q')G^<_{aa}(q',X'),\ a=e,h. \end{aligned} \tag{291}$$

It should be noted that both scatterings and coupling between energy bands contribute to the source terms in the equation of continuity for particle number. However, there is a difference in the mechanisms of change of particle number in the $a'th$ band between that caused by scattering and by that caused by coupling between energy bands. Scattering do cause transitions of carriers between energy bands (as well as within a band) and accordingly, they change particle number in a given band, a. On the other hand, the coupling between energy bands gives rise to a change of the distribution function of the carriers in the $a'th$ band. Thus, coupling changes the number density of carriers in the $a'th$ band by changing the distribution function. In the context of transport this coupling produces a *drag* effect in which the electrons and holes tend to drag each other against the direction in which they are normally propelled by the electric field.

If we consider the lowest order approximation to the coupling between the energy bands, the source terms can be further simplified. Dropping all scattering terms and treating band-band coupling by H_{cv} alone (using the definition of the coupling matrix elements given by Eq.(285)), Eq.(290) can be simplified as

$$\begin{aligned} i\hbar\frac{\partial}{\partial T}G^<_{aa} = & -\frac{\mathbf{q}}{m_0}\cdot\frac{\partial}{\partial \mathbf{X}}G^<_{aa}(q,X) \\ & +\frac{1}{\hbar^3}\int d\mathbf{q}'d\mathbf{u}e^{\frac{i}{\hbar}(\mathbf{q}-\mathbf{q}')\cdot\mathbf{u}}\left[V_a(\mathbf{x}-\frac{\mathbf{u}}{2})-V_a(\mathbf{x}+\frac{\mathbf{u}}{2})\right]G^<_{aa}(\mathbf{q}',e,X) \\ & +iH^2_{cv}\left[G^<_{a'a',0}(q,X)A_a(q,X)+2ImG^r_{a'a',0}(q,X)G^<_{aa}(q,X)\right], \\ & \qquad a',a=e,h \qquad \text{and} \qquad a\neq a'. \end{aligned} \tag{292}$$

(No summation over the repeated index $aa(a'a')$ here.) Employing the Langreth algebra, Eqs.(208) and (215), we may express the inter-band correlation functions as

$$\begin{aligned} & G^<_{eh}(q,X)=H_{cv}(q)\frac{1}{h^4}\int dvdudq'dX'e^{\frac{i}{\hbar}(X-X')v}e^{\frac{i}{\hbar}(q-q')u} \\ & \left[G^r_{ee,0}(q-\frac{v}{2},X+\frac{u}{2})G^<_{hh}(q',X')+G^<_{ee,0}(q-\frac{v}{2},X+\frac{u}{2})G^a_{hh}(q',X')\right], \end{aligned} \tag{293}$$

and

$$\begin{aligned} & G^<_{he}(q,X)=H_{cv}(q)\frac{1}{h^4}\int dvdudq'dX'e^{\frac{i}{\hbar}(X-X')v}e^{\frac{i}{\hbar}(q-q')u} \\ & \left[G^r_{hh,0}(q-\frac{v}{2},X+\frac{u}{2})G^<_{ee}(q',X')+G^<_{hh,0}(q-\frac{v}{2},X+\frac{u}{2})G^a_{ee}(q',X')\right]. \end{aligned} \tag{294}$$

Eqs.(292) to (294) are the equations of motion of the Green's functions for a two band system when the scattering can be completely ignored. It is important to note that the self-energies in the equations are known functions. They are determined by the nondiagonal terms of the Hamiltonian and the Green's functions in the absence of scatterings with no coupling between energy bands. It should be noted that since the band-band coupling terms in the Hamiltonian are an artifact of the perturbative determination of the band structure for a perfect crystal in a periodic potential with no lattice oscillations or impurities, one must expect that Eqs.(292)-(294) do not provide a mechanism of resistance, nor do they yield momentum relaxation. While the system is thus subject to a dc "run-away", this theory will provide interesting features in ac transport concerning resonance involving inter-band transitions.

5.4. *Dynamic Description of Quantum Transport Equations for Multi-Band Semiconductors*

Kinetic description determines the whole transport features of systems. This statement is partially true since a hypothesis, the temperature of lattice and carriers are constant, has been used when we calculate the green's function. In a real device system, the carriers get energy from the applied electric field and heated themselves. The electron-phonon interaction dissipates part of the carrier's energy into the lattice thereby leading to the increase of the lattice temperature and furthermore to the change of the distribution of phonons and then the dissipation of carrier energy into the phonon systems. This two-way influence of temperature changing can not be properly described by kinetic formation of quantum transport equations. To account for energy exchanging in the system, dynamic formation of quantum transport equations must be used in modeling quantum devices. In this chapter, we discuss quantum transport equations from the viewpoint of dynamics. We discuss particle number balance equation at first.

5.4.1. *Particle Number Balance Equation*

For single band semiconductors and multi-band semiconductors in which energy band coupling may be ignored, it is often useful to employ a parabolic approximation for each band, in the form,

$$H(q,X)_{ab} = (\frac{q^2}{2m_a^*} + V(X))\delta_{ab}, \tag{295}$$

where m_a^* is the effective band mass of the carriers in the $a'th$ band. In this approximation, Eq.(259) may be written as

$$\begin{aligned} i\hbar\frac{\partial}{\partial T}G^<_{ab}(q,X) &= -i\hbar\nabla\cdot\frac{\mathbf{q}}{m_a^*}G^<_{ab}(q,X) \\ &+\frac{1}{h^4}\int dq'du e^{\frac{i}{\hbar}(q-q')\cdot u}[V(X+\frac{u}{2})-V(X-\frac{u}{2})]_a G^<_{ab}(q',X) \\ &+\frac{1}{(h^4)^2}\int dq'dX'[K^-_{\Sigma^<}(q,X-X';X,q-q')ReG^<(q',X') \\ &-ReG^<(q',X')K^+_{\Sigma^<}(q,X-X';X,q-q')]_{ab} \\ &+\frac{i}{2(h^4)^2}\int dq'dX'[K^-_{\Sigma^<}(q,X-X';X,q-q')A(q',X') \\ &-A(q',X')K^+_{\Sigma^<}(q,X-X';X,q-q')]_{ab} \\ &-\frac{i}{2(h^4)^2}\int dq'dX'[K^-_{\Gamma}(q,X-X';X,q-q')G^<(q',X') \\ &-G^<(q',X')\ K^+_{\Gamma}(q,X-X';X,q-q')]_{ab}. \end{aligned} \tag{296}$$

The number density for carriers in the $a'th$ band is given by

$$n(X)_a = \int\frac{d\mathbf{q}}{(2\pi\hbar)^3}\frac{dE}{2\pi}(-i)G^<_{aa}(q,X), \tag{297}$$

and the current density is

$$\mathbf{j}(X)_a = \int \frac{d\mathbf{q}}{(2\pi\hbar)^3} \frac{dE}{2\pi} \frac{\mathbf{q}}{m_a^*} (-i) G_{aa}^{<}(q, X), \tag{298}$$

leading to the relation

$$\frac{\partial n(X)_a}{\partial T} + \nabla \cdot \mathbf{j}(X)_a = -s_a, \tag{299}$$

where s_a is the nonequilibrium source of the carriers in the $a'th$ band,

$$\begin{aligned} s_a = & \frac{1}{(h^4)^3} \int dq \int dq' dX' [K_{\Sigma^<}^{-}(q, X - X'; X, q - q') ReG^{<}(q', X') \\ & - ReG^{<}(q', X') K_{\Sigma^<}^{+}(q, X - X'; X, q - q')]_{aa} \\ & + \frac{i}{2(h^4)^3} \int dq \int dq' dX' [K_{\Sigma^<}^{-}(q, X - X'; X, q - q') A(q', X') \\ & - A(q', X') K_{\Sigma^<}^{+}(q, X - X'; X, q - q')]_{aa} \\ & - \frac{i}{2(h^4)^3} \int dq \int dq' dX' [K_{\Gamma}^{-}(q, X - X'; X, q - q') G^{<}(q', X') \\ & - G^{<}(q', X') K_{\Gamma}^{+}(q, X - X'; X, q - q')]_{aa}. \end{aligned} \tag{300}$$

Eq.(299) is the equation of continuity for the $a'th$ band of a multi-band system. Microscopically, scatterings cause carriers to undergo transitions from the $a'th$ band to other bands and there are also transitions from other bands to the $a'th$ band, so that there is variation of the number of particles in the $a'th$ band. Energy coupling between bands also contributes to the source term. This change of carrier number can not be described by the current term defined in Eq.(298). The existence of such a particle source term in Eq.(299) is a distinguishing feature by which the number conservation law for multi-band semiconductor systems differs from that for single band semiconductor systems. For single band semiconductors or multi-band semiconductors where the coupling between energy bands can be ignored, it is readily shown that $s_a = 0$. Defining overall number density, current density, and source terms by the following equations,

$$u(X) = \sum_a u(X)_a, \qquad \mathbf{j}(X) = \sum_a \mathbf{j}(X)_a, \qquad s(X) = \sum_a s(X)_a, \tag{301}$$

we obtain the local continuity equation of particle number as

$$\frac{\partial n(X)}{\partial T} + \nabla \cdot \mathbf{j}(X) = -s(X), \tag{302}$$

for all bands combined.

5.4.2. *Energy and Momentum Conservation Laws*

Eq.(249) shows that the correlation function varies with respect to T, the "center of mass" time. Of course, the conservation law for particle number alone can not fully reflect the physical processes which occur when an external field is applied to the system. To describe such transport dynamics it is necessary to solve the coupled localized conservation laws for particle number, energy, and momentum consistently. To formulate the localized energy and momentum conservation laws, it is helpful to consider how the Green's function changes with respect to the relative coordinate x.

Fourier transforming Eq.(249) with respect to x, differentiating with respect to x, and then setting $x = 0$, we have

$$i\hbar\frac{\partial}{\partial T}\int dqqG^{<}_{ab}(q,X) = \int dqq[\mathcal{H},G^{<}](q,X)_{ab} + \int dqq[\Sigma^{<},ReG^{r}](q,x)_{ab}$$
$$+\frac{i}{2}\int dqq\{\Sigma^{<},A\}(q,X)_{\alpha\alpha} - \frac{i}{2}\int dqq\{\Gamma,G^{<}\}(q,X)_{\alpha\alpha}. \tag{303}$$

In terms of the notation used in Eqs. (256) and (257), we obtain

$$i\hbar\frac{\partial}{\partial T}\int dqqG^{<}_{ab}(q,X)$$
$$= \frac{1}{(h^4)^2}\int dq'dX'dqq[K^{-}_{H}(q,X-X';X,q-q')G^{<}(q',X')$$
$$-G^{<}(q',X')K^{+}_{H}(q,X-X';X,q-q')]_{ab}$$
$$+\frac{1}{(h^4)^2}\int dq'dX'dqq[K^{-}_{\Sigma^<}(q,X-X';X,q-q')ReG^{<}(q',X')$$
$$-ReG^{<}(q',X')K^{+}_{\Sigma^<}(q,X-X';X,q-q')]_{ab}$$
$$+\frac{i}{2(h^4)^2}\int dq'dX'dqq[K^{-}_{\Sigma^<}(q,X-X';X,q-q')A(q',X')$$
$$-A(q',X')K^{+}_{\Sigma^<}(q,X-X';X,q-q')]_{ab}$$
$$-\frac{i}{2(h^4)^2}\int dq'dX'dqq[K^{-}_{\Gamma}(q,X-X';X,q-q')G^{<}(q',X')$$
$$-G^{<}(q',X')K^{+}_{\Gamma}(q,X-X';X,q-q')]_{ab}. \tag{304}$$

This equation is an exact expression of the energy-momentum conservation laws in phase-space and energy-time domain. We discuss the energy conservation law and momentum conservation law, respectively, in the following subsections.

Energy Conservation Law

Considering the component of Eq.(304) ($q_4 = E$), and setting $a = b$, we have

$$\frac{\partial u_a}{\partial T} = -\frac{1}{i\hbar(h^4)^3}\int dq'dX'dqE[K_{\overline{H}}^{-}(q, X - X'; X, q - q')G^{<}(q', X')$$
$$-G^{<}(q', X')K_H^{+}(q, X - X'; X, q - q')]_{aa}$$
$$-\frac{1}{i\hbar(h^4)^3}\int dq'dX'dqE[K_{\Sigma^<}^{-}(q, X - X'; X, q - q')ReG^{<}(q', X')$$
$$-ReG^{<}(q', X')K_{\Sigma^<}^{+}(q, X - X'; X, q - q')]_{aa}$$
$$-\frac{i}{2(h^4)^3}\int dq'dX'dqE[K_{\Sigma^<}^{-}(q, X - X'; X, q - q')A(q', X')$$
$$-A(q', X')K_{\Sigma^<}^{+}(q, X - X'; X, q - q')]_{aa}$$
$$+\frac{i}{2(h^4)^3}\int dq'dX'dqE[K_{\Gamma}^{-}(q, X - X'; X, q - q')G^{<}(q', X')$$
$$-G^{<}(q', X')K_{\Gamma}^{+}(q, X - X'; X, q - q')]_{aa}, \tag{305}$$

where the energy density of carriers in the $a'th$ band is defined as

$$u_a(X) = -i\int \frac{d\mathbf{q}}{(2\pi\hbar)^3}\frac{dE}{2\pi}EG_{aa}^{<}(q, X). \tag{306}$$

This is the most general form of the local energy conservation law for single band or multi-band semiconductor systems. It includes all quantum microscopic processes on the right hand side of this equation. The first two terms on the RHS of Eq.(305) are the energy variation with respect to time due primarily to band carrier dynamics in the absence of scattering. The remaining terms on the RHS of this equation are the energy change rates which are due primarily to the effects of carrier collisions and scatterings (broadening effects) and the coupling between energy bands. For a single-band system, or in the case where the coupling matrix elements of a multi-band Hamiltonian can be ignored, the Hamiltonian can be taken in the form given by Eq.(295). If we define the energy flux density in the $a'th$ band as

$$\mathbf{S}(X)_a = -i\int \frac{d\mathbf{q}}{(2\pi\hbar)^3}\frac{dE}{2\pi}E\frac{\mathbf{q}}{m_a^*}G_{aa}^{<}(q, X), \tag{307}$$

Eq.(305) can be written as

$$\frac{\partial u(X)_a}{\partial T} + \nabla\cdot\mathbf{S}(X)_a = -w_a, \tag{308}$$

where w_a is the power supplied to the system. It may be expressed as

$$w_a = w_{1a} + w_{2a}, \tag{309}$$

where

$$w_{1a}(X) = \frac{1}{(h^4)^2}\int dq'dqduEe^{\frac{i}{\hbar}(q-q')u}\left[V(X + \frac{u}{2}) - V(X - \frac{u}{2})\right]_a G_{aa}^{<}(q', X), \tag{310}$$

and

$$w_{2a}(X) = \frac{1}{(h^4)^3} \int dq'dX'dqE[K^-_{\Sigma^<}(q, X - X'; X, q - q')ReG^<(q', X')$$
$$-ReG^<(q', X')K^+_{\Sigma^<}(q, X - X'; X, q - q')]_{aa}$$
$$+\frac{i}{2(h^4)^3} \int dq'dX'dqE[K^-_{\Sigma^<}(q, X - X'; X, q - q')A(q', X')$$
$$-A(q', X')K^+_{\Sigma^<}(q, X - X'; X, q - q')]_{aa}$$
$$-\frac{i}{2(h^4)^3} \int dq'dX'dqE[K^-_{\Gamma}(q, X - X'; X, q - q')G^<(q', X')$$
$$-G^<(q', X')K^+_{\Gamma}(q, X - X'; X, q - q')]_{aa} \qquad a = e, h. \tag{311}$$

w_{1a} is the rate of local energy variation due primarily to the single particle potential, which may be related to time-dependent applied fields. If the potential is independent of time, this term vanishes, in accordance with the conservation of mechanical energy in the absence of scattering. w_{2a} includes contributions from two physical processes: The first is the dissipation of power into the host lattice by carrier-phonon interactions. This is the Joule heat generated by the electric current flowing through the system. The second is the energy transferred between energy bands, associated with band-band interaction and band-carrier drag effects. It is easy to identify this effect in two-band systems. Defining the total local energy density $u(X)$, the total local energy flux $\mathbf{S}(X)$, and the total local power $w(X)$ (for all bands combined) by

$$u(X) = \sum_a u(X)_a, \qquad \mathbf{S}(X) = \sum_a \mathbf{S}_a(X), \qquad w(X) = \sum_a w(X)_a \tag{312}$$

we obtain the localized energy conservation law for the system as

$$\frac{\partial u(X)}{\partial T} + \nabla \cdot \mathbf{S}(X) = -w(X). \tag{313}$$

For two-band systems, the single particle Hamiltonian is given by Eq.(284). As mentioned in the previous section, multi-band equations can be decoupled, with the coupling between energy bands expressed in terms of selfenergies.

Accordingly, we may obtain the local energy conservation law for two-band semiconductor systems by replacing the selfenergies in Eq.(309) by Eqs.(210) and (213). If scattering interactions can be ignored, we may set $\Sigma_{hh} = \Sigma_{ee} = 0$. Thus, considering only the lowest approximation to the coupling terms between energy bands, the local energy conservation law for carriers in the $a'th$ band can be written as

$$\frac{\partial u_\alpha}{\partial T} + \nabla \cdot \mathbf{S}_a(X) = -w_a(X), \tag{314}$$

where

$$w_a(X) = w_{1a}(X) + w_{ca}(X).$$

$w_{1a}(X)$ is given by Eq.(310) and

$$w_{ca} = \frac{i}{h^4}\int dqE\left|H_{cv}^2(q_z)\right|^2 \; [G^<_{a'a',0}(q,X)A_a(q,X) + 2ImG^r_{a'a',0}(q,X)G^<_{aa}(q,X)]$$
$$a', a = e, h \qquad \text{and} \qquad a' = a, \tag{315}$$

is the local power needed for the carriers in the *ath* band to overcome the drag due to carriers in the other band. In an electric field, the electrons move in the direction opposite to the direction of motion of the holes. Since the interaction between electrons and holes tends to resist motion of the electrons, opposing the free flow of electrons in the external field, this interaction slows the electrons and reduces their energy. Thus, part of the energy the electrons obtain from the electric field is transferred to holes by the band-carrier-drag mechanism. w_{ca} is a measure of this energy transfer. The drag mechanism also causes momentum transfer between electrons and holes, as discussed in the following subsection.

Momentum Conservation Law

Considering the first three components of Eq.(304), we have

$$\begin{aligned}
\frac{\partial \mathbf{q}_a}{\partial T} &= -\frac{1}{i\hbar(h^4)^3}\int dq'dX'dq\mathbf{q}[K_H^-(q,X-X';X,q-q')G^<(q',X') \\
&-G^<(q',X')K_H^+(q,X-X';X,q-q')]_{aa} \\
&-\frac{1}{i\hbar(h^4)^3}\int dq'dX'dq\mathbf{q}[K_{\Sigma^<}^-(q,X-X';X,q-q')ReG^<(q',X') \\
&-ReG^<(q',X')K_{\Sigma^<}^+(q,X-X';X,q-q')]_{aa} \\
&-\frac{i}{2(h^4)^3}\int dq'dX'dq\mathbf{q}[K_{\Sigma^<}^-(q,X-X';X,q-q')A(q',X') \\
&-A(q',X')K_{\Sigma^<}^+(q,X-X';X,q-q')]_{aa} \\
&+\frac{i}{2(h^4)^3}\int dq'dX'dq\mathbf{q}[K_\Gamma^-(q,X-X';X,q-q')G^<(q',X') \\
&-G^<(q',X')K_\Gamma^+(q,X-X';X,q-q')]_{aa}
\end{aligned} \tag{316}$$

This is the most general form of the local momentum conservation law for multi-band semiconductor systems. It includes all quantum microscopic processes on the right-hand-side of this equation. The first two terms on the RHS of Eq.(316) are the momentum change rate due primarily to band carrier dynamics in the absence of scattering. The remaining terms on the RHS of this equation are primarily due to carrier collisions and scattering and the coupling between energy bands. If the Hamiltonian can be taken in the form given by Eq.(284), the momentum density and the momentum current-density tensor of the *a'th* band can be expressed as

$$\mathbf{q}(X)_a = \int \frac{\mathbf{q}}{(2\pi\hbar)^3}\frac{de}{2\pi}\mathbf{q}(-i)G^<_{aa}(q,X), \tag{317}$$

and

$$\overleftrightarrow{\mathcal{P}}(X)_a = \int \frac{d\mathbf{q}}{(2\pi\hbar)^3}\frac{dE}{2\pi}\frac{\mathbf{q}\mathbf{q}}{m^*_\alpha}(-i)G^<(q,X)_{aa}, \tag{318}$$

respectively. Eq.(316) can be written as

$$\frac{\partial \mathbf{q}(X)_a}{\partial T} + \nabla \cdot \overleftrightarrow{\mathcal{P}}_a = -\mathbf{f}(X)_a. \tag{319}$$

The source term in the above equation can be expressed as

$$\mathbf{f}(X)_a = \mathbf{f}(X)_{1a} + \mathbf{f}(X)_{2a}, \tag{320}$$

where

$$\mathbf{f}(X)_{1a} = \frac{1}{(h^4)^2} \int dq dq' du \mathbf{q} e^{\frac{i}{\hbar}(q-q')u} \left[V(X + \frac{u}{2}) - V(X - \frac{u}{2}) \right]_a G^<_{aa}(q', X), \tag{321}$$

and

$$\begin{aligned}
\mathbf{f}(X)_{2a} &= \frac{1}{(h^4)^3} \int dq' dX' dq \mathbf{q} [K^-_{\Sigma^<}(q, X - X'; X, q - q') ReG^<(q', X') \\
&- ReG^<(q', X') K^+_{\Sigma^<}(q, X - X'; X, q - q')]_{aa} \\
&+ \frac{i}{2(h^4)^3} \int dq' dX' dq \mathbf{q} [K^-_{\Sigma^<}(q, X - X'; X, q - q') A(q', X') \\
&- A(q', X') K^+_{\Sigma^<}(q, X - X'; X, q - q')]_{aa} \\
&- \frac{i}{2(h^4)^3} \int dq' dX' dq \mathbf{q} [K^-_{\Gamma}(q, X - X'; X, q - q') G^<(q', X') \\
&- G^<(q', X') K^+_{\Gamma}(q, X - X'; X, q - q')]_{aa}.
\end{aligned} \tag{322}$$

In the above equation, $\mathbf{f}_{1a}$ is the local force due primarily to band carrier dynamics in the absence of scattering. $\mathbf{f}_{2a}$ is the local momentum loss rate, which consists of two parts. One is the contribution mainly by scattering in the systems, which may lead to inter-band transitions. Particles exchange momentum when they transfer from *ath* band to *a'th* band, changing the net carrier momentum in the *ath* band. Another contribution is due to the band carrier-drag effect. The mechanism of this part of the momentum transfer is quite different from the first one. Generally, the first part is identified as a frictional force experienced by the carriers due to scatterings and the second part as a drag force caused by carriers of other bands.

Defining the local momentum density $\mathbf{q}(X)$, the local momentum flux density $\overleftrightarrow{\mathcal{P}}(X)$, and the local total force density $\mathbf{f}(X)$ by

$$\mathbf{q}(X) = \sum_a \mathbf{q}(X)_a, \qquad \mathcal{P}(X) = \sum_a \overleftrightarrow{\mathcal{P}}_a(X), \qquad \mathbf{f}(X) = \sum_a \mathbf{f}(X)_a, \tag{323}$$

we obtain the local momentum conservation law for the system as follows

$$\frac{\partial \mathbf{q}(X)}{\partial T} + \nabla \cdot \overleftrightarrow{\mathcal{P}}(X) = -\mathbf{f}(X). \tag{324}$$

As indicated in the preceding subsection, if we replace the selfenergies in Eq.(322) by Eqs.(287) and (288), we may obtain the momentum conservation law for multi-band semiconductor systems in a form that is similar to that of single band semiconductor systems. However, the drag forces must be included in the RHS of the local momentum conservation law. To make this clear, we consider a two-band semiconductor system. Ignoring scatterings and considering only the lowest approximation to the coupling between energy bands, the local momentum conservation law of the carriers in the *ath* band may be written as

$$\frac{\partial \mathbf{q}_a(X)}{\partial T} + \nabla \cdot \overleftrightarrow{\mathcal{P}}_a (X) = -\mathbf{f}_a(X), \tag{325}$$

where

$$\mathbf{f}_a = \mathbf{f}_{1a} + \mathbf{f}_{2a}. \tag{326}$$

$\mathbf{f}_{1a}$ is given by Eq.(321) and

$$\mathbf{f}_{2a}(X) = \frac{i}{h^4} \int dq\mathbf{q} \left| H_{cv}(q_z) \right|^2 \left[G_{a'a',0}(q,X) A_a(q,X) + 2ImG^r_{a'a',0}(q,X) G^<_{aa}(q,X) \right]$$

$$a', a = e, h \qquad \text{and} \qquad a' \neq a. \tag{327}$$

$\mathbf{f}_{2a}$ is the drag force applied to the carriers in the *ath* band by the carriers in the *a′th* band. It is caused by coupling between energy bands and is a measure of inter-band momentum transfer of the carriers.

Thus, by the use of a two-band model for semiconductor systems, we have illustrated the coupling between energy bands as it affects the motion of carriers in terms of a band carrier-drag mechanism in three respects: the coupling between energy bands causes change in the distribution of carriers in the *ath* band, and it causes energy and momentum transfers between the energy bands.

5.5. *Summary*

We have employed closed-time-path nonequilibrium Green's functions jointly with a band-band interaction decoupling technique to derive transport equations for single- and multi-band semiconductor systems. Our results for multi-band semiconductors yield kinetic equations which are expressed in decoupled form, with the couplings between energy bands embedded in carrier self-energies. As it is extremely difficult, in terms of current computing facilities, to determine the exact nonequilibrium Green's functions, we have developed an alternative formulation in terms of balance equations based on microscopic quantum dynamics of the carriers. Defining carrier number, energy, and momentum in the *a′th* band and differentiating the kinetic transport equation with respect to the four-dimensional momentum of the carriers, we have derived band-localized number, energy, and momentum conservation laws in integral form. Our results, based on nonequilibrium microscopic carrier dynamics, describe in accurate detail the macroscopic energy and momentum transfer from carriers to the lattice. The inhomogeneous right-hand-sides of the balance equations for number (Eq.(302)), momentum (Eq.(319)), and energy (Eq.(308)) are in the nature of sources., contributed by scatterings and the band carrier drag effect resulting from coupling between the energy bands. These sources, identified in terms of microscopic nonequilibrium Green's functions and carrier self-energies (Eq.(301), (309), and (320)), provide detailed information concerning the transfer of energy(heating) and momentum which can be used to improve the simulation of quantum devices. Past analyses of the dynamics of heating and momentum transfer have often been unduly restricted to classical considerations.[62] Our treatment includes careful consideration of all quantum dynamical processes which contribute to the dissipation of heat. In part II of this series of research, we will present a numerical analysis of energy transfer and momentum transfer in heterostructure quantum dynamics based on Eq.(311) and (322).

The present formulation of quantum transport equations has several advantages. First, since the equations are decoupled with respect to inter-band interactions, we may code the simulator for n band systems in terms of that of a n-1 band system. This programming technique corresponds with the current tendency in the development of programming technology. Second, the structure of the equations is amenable to solution by modern computing techniques, in particular, parallel computing. The focus of these equations on a single band provides a more direct route for the determination of carrier motion in energy bands and to delineate the influence of inter-band interactions.

As an example of the use of the balance/transport equations, in the absence of scattering and considering only the lowest approximation to the coupling between energy bands, we have treated a two-band model of a semiconductor and derived the corresponding transport equations. Analysis of these balance equations shows that the coupling between energy bands affects the motion of carriers by the band-carrier-drag mechanism, which causes change in the quantum distribution function of the carriers the ath band and is also responsible for the transfer of energy and momentum between energy bands. However, with no scattering, band-band interaction alone does not provide a mechanism of resistance or relax momentum, since such band-band interaction refers to an approximation technique for the determination of the band structure of a perfect crystalline periodic potential. Nevertheless, our results will provide interesting information about ac transport concerning resonance involving inter-band transitions. Further application of these equations to the analysis of an inter-band RTD will be reported elsewhere.

6. Conclusions

This Chapter presented a systematic study of instability processes in nanostructures, along with a description of the new and formidable theoretical challenges. Advanced theoretical approaches are developed and applied towards the study of one-dimensional tunneling devices that possess the potential for exhibiting self-oscillations. Section II considers intrinsic oscillation processes within double-barrier quantum-well structures. Section III presents a duality theory that suggests self-oscillations may be induced within staggered-bandgap heterostructures. Section IV presents a multi-band physics-based model that can be accurately applied towards tunneling studies in staggered-bandgap heterostructures. Section V presents a Green's function formalism for rigorously incorporating quantum dissipation into highly nonequilibrium and time-dependent transport processes within single- and multi-band based devices. Taken together, these Sections serve as fundamental understanding and starting point for a serious look at the double barrier nanostructure as a potential high-frequency/high-speed device.

Appendix A. Relaxation Time Approximation

This appendix briefly presents the procedure used for calculating the relaxation time used in these studies.

According to the fundamental scattering theory, the scattering rate $W(\mathbf{k})$ is written as

$$W(\mathbf{k}) = \frac{2\pi}{\hbar}\frac{\Omega}{(2\pi)^3}\int d\mathbf{k}'| < \mathbf{k}'|H'|\mathbf{k} > |^2\delta(E_{\mathbf{k}'} - E_{\mathbf{k}} \pm \hbar\omega) \tag{A.1}$$

where the expressions of the interaction Hamiltonian H' for various interactions can be found in references.[87,86,85,15] The momentum relaxation time can be related to the scattering rate in terms of the formula $W(\mathbf{k}) = 1/\tau(\mathbf{k})$. The total effect of the interactions on the relaxation time can be expressed by using the Matthiessen's rule, i.e., $1/\tau = \sum 1/\tau_i$, where $1/\tau_i$ is the scattering rate caused by the ith interaction. It should be noted that the $\tau(\mathbf{k})$ is related to the energy of the electrons. The total relaxation time τ_i for each process can be obtained from the momentum relaxation time $\tau_i(\mathbf{k})$ by

$$\tau_i = \frac{1}{\Gamma(5/2)}\int_0^\infty x^{3/2}\tau_i(x)exp(-x)dx, \tag{A.2}$$

where Γ is the gamma function and $x = \beta E_k$. In deriving the above equation, we have assumed that the electron energy is parabolic in momentum.

References

1. Sze, S. M., (1990) *"High-Speed Semiconductor Devices"* Chapter 9 (Wiley, New York).
2. Capasso, F., Beltram, F., Sen, S., Palevski, A. and Cho, A. Y. (1994) in *High Speed Heterostructure Device*, edited by R. A. Kiehl and T. C. L. G. Sollner, Academic Press, Inc..
3. Tredieucci, A., et al, (2000) *Electronics Letters*, V36, 876.
4. Eisele, H., et al, (2000) IEEE Trans. MTT, V48,626.
5. Chanberlain, J. M., (1997) in New Directions in Terahertz Technology, edited by J. M. Chamberlain and R. E. Miles, Kluwer Academic Publishers.
6. Sollner, T. C. L. G., et al, (1984) Appl. Phys. Lett. 50, 332.
7. Brown, E. R. (1992) in Hot Carriers in Semiconductor Nanostructures (edited by J. Shah), pp469, Academic Press, Boston.
8. Lheurette, E., et al, (1992) Electronic Letters **28**, 937.
9. Woolard, D. L., et al,(1995) 53rd Annual Device Research Conference Digest, p54.
10. Woolard, D. L., et al, (1996) IEEE Trans ED, **43**,332.
11. Boric-Lubecke, O., et al, (1995) IEEE Trans. MTT., **43**, 969.
12. Kidner, C., et al, (1990) IEEE Trans. MTT. **38**,864(1990).
13. Zhao, Peiji et al, (2000a) presentation in IWCE7, Scottland; 2001, VLSI Design Vol.13, pp.413-417.

14. Ricco, B. and Azbel, M. Ya. (1984) Phys. Rev. B**29**, 1970.
15. Jensen KL, and Buot FA. (1991) Numerical simulation of intrinsic bistability and high-frequency current oscillation in resonant tunneling structures. *Physical Review Letters* **66**: 1078-1081.
16. Biegel BA & Plummer JD. (1996) Comparison of self-consistency iteration options for Wigner function method of quantum device simulation. *Physical Review* **B54**: 8070-8082.
17. Presilla, C., et al, (1991) Phys. Rev. **B**43, 5200.
18. Zhao, P., Cui, H. L. and Woolard, D., (2001) Phys. Rev. B63, 75302.
19. Woolard DL, Buot FA, Rhodes DL, Lu XL, Lux RA & Perlman BS. (1996) On the different physical roles of hysteresis and intrinsic oscillations in resonant tunneling structures. *Journal of Applied Physics* **79**: 1515-1525.
20. Schombury, E., et al, (1999) Electronics Letts. 35, 1419.
21. Woolard, D., Zhao, P. and Cui, H. L. (2001) presentation in HCIS12, Santa Fe, NM and accepted by Phys. B.
22. Frensley, W. (1987) Phys. Rev. B**36**, 1570(1987).
23. Kluksdahl et al, (1988) Phys. Rev. **B**39, 7720.
24. Wigner, E., (1932) Phys. Rev. 40,749.
25. Buot, F. A. and Jensen, K. L., (1990) Phys. Rev. **B**42, 9429.
26. Zhao, Peiji, Horing, Norman, and Cui, H. L. (2000b) Phi. Mag. 80, 1359.
27. Bordone, P., Pascoli, M., Brunetti, R., Bertoni, A. and Jacoboni, C., (1999) Phys. Rev. B**59**, 3060.
28. Frensley, W. R., (1987) Rev. Mod. Phys., vol. 36, 1570.
29. Frensley, W. R., (1992) Superlattice and Microstructures, **11**, 347.
30. Liu,Y. X., Ting, D. Z. -Y., and McGill, T. C.,(1996) Phys. Rev. B**54**, 5675.
31. Zhao, Peiji, et al, (2000c) J. Appl. Phys., **87**, 1337.
32. Yu ET, McCaldin JO & McGill TC. (1992) In: *Solid-State Physics, Advances in Research and Applications* (eds H Ehrenreich & D Turnbull) 46: 1-146. Academic Press, Boston
33. Ting DZY, Yu ET & McGill TC. (1992) Multiband treatment of quantum transport in interband tunnel devices. *Physical Review* B**45**: 3583-3592.
34. Buot FA, Zhao P, Cui HL, Woolard DL, Jensen KL & Krowne CM. (2000) Emitter quantization and double hysteresis in resonant- tunneling structures: A nonlinear model of charge oscillation and current bistability. *Physical Review* B**61**: 5644-5665.
35. Sollner TCLG. (1987) Comment on "Observation of intrinsic bistability in resonant tunneling structures". *Physical Review Letters* **59**: 1622-1623.
36. Buot FA & Jensen KL. (1991) Intrinsic high-frequency oscillations and equivalent circuit model in the negative differential resistance region of resonant tunneling devices. *International Journal for Computation & Mathematics in Electrical & Electronics Engineering*, COMPEL **10**: 241-253.
37. Goldman VJ, Tsui DC & Cunningham JE. (1987) Observation of intrinsic bistability in resonant tunneling structures. *Physical Review Letters* **58**: 1256-1259.
38. Buot FA. (1997) An interband tunnel high-frequency source. *Journal of Physics D: Applied Physics* **30**: 3016-3023.
39. Buot FA. (1998) An interband tunnel oscillator: Intrinsic bistability and hysteresis of trapped hole charge in a double-barrier structure. VLSI Design 8: 237-245
40. Chow DH & Schulman JN. (1994) Intrinsic current bistability in $InAs/Al_xGa_{1-x}Sb$ resonant tunneling devices. *Applied Physics Letters* **64**: 76-78.
41. Buot FA. (1993) Mesoscopic physics and nanoelectronics: Nanoscience and nanotechnology. *Physics Reports* 234: 73-174.

42. Abe Y. (1992) Bifurcation of resonant tunneling current due to the accumulated electrons in a well. *Semiconductor Science & Technology* **B7**: 498-501.
43. Jona-Lasino G, Presilla C & Sjöstrand J. (1995) On Schrödinger equations with concentrated nonlinearities. *Annals of Physics* **240**: 1-21.
44. Presilla C & Sjöstrand J. (1997) Nonlinear resonant tunneling in systems coupled to quantum reservoirs. *Physical Review* B**55**: 9310-9313; (1996) Transport properties in resonant tunneling heterostructures. Journal of Mathematical Physics 37: 4816-4844.
45. Buot FA & Rajagopal AK. (1993) High-frequency behavior of quantum-based devices: equivalent-circuit, nonperturbative-response, and phase-space analyses. *Physical Review* B**48**: 17217-17232 .

    The argument of this paper concerning alignment should be corrected, namely, in the absence of 2-D states in the emitter, the peak or resonant current correspond to the alignment of the quantum well energy level with the bottom of the conduction band of the emitter. In view of the present findings, oscillation occurs during passage across the boundary between allowed and forbidden 2-D states, not across the bottom of the conduction band of the emitter.
46. Buot FA & Krowne CM. (1999) Double-barrier THz source based on electrical excitation of electrons and holes. *Journal of Applied Physics* **86**: 5215-5231; 2000. Erratum. *Journal of Applied Physics* **87**: 3189.
47. Buot FA & Rajagopal AK. (1995) Theory of novel nonlinear quantum transport effects in resonant tunneling structures. *Materials Science & Engineering* B**35**: 303-317.
48. Poltoratsky EA & Rychkov GS. (2001) The dynamic nature of peculiarities of RTS static I-V characteristics. *Nanotechnology* **12**: 556-561.
49. Brown, E. R. et. al., (1991) Appl. Phys., Lett., 58(1), 2291.
50. Kroemer, H. (1994) "Quantum Mechanics for Engineering, Material Science, and Applied Physics," Chapter 15 (Prentice Hall, New Jersey)
51. Datta, S. (1989) "Quantum Phenomena," Chapter 6 (Addison-Wesley, Reading, MA).
52. Kittel, C., (1987) "Quantum Theory of Solids," Chapter 14 (Wiley, New York).
53. Landau, L. D. and Lifshitz, E. M., (1965) "Quantum Mechanics," Chapter 6 (Pergamon Press, Oxford).
54. Edmonds, A. R., (1957) "Angular Momentum in Quantum Mechanics," Chapter 3 (Princeton University Press, New Jersey).
55. Gelmont, B., Woolard, D., Zhang, W. and Globus, T. (2002) "Electron Transport within Resonant Tunneling Diodes with Staggered-Bandgap Heterostructures," accepted for publication in Solid State Electronics.
56. Akhiezer, A. I. and Berestetsky, V. B. (1953) "Quantum Electrodynamics (Part I)," Chapter 4, page 200 (State Technico-Theoretical Literature Press, Moscow).
57. Beenakker, C. W. J. and Houten, H. Van (1991) in Solid State Physics, Advances in Research and Application, edited by H. Ehrenreich and D. Turnbull (Academic, San Diego) Vol. 44.
58. Ferry, David K. and Grubin, Harold L. (1995) Solid State Physics, Vol. 49, 283.
59. See, for example, *Physics of Quantum Electron Devices*, edited by F. Cappaso, (1990) Springer Series in Electronics and Photonics Vol. 28 (Springer, Berlin).
60. Carey, G. F., Richardson, W. B., Reed, C. S. and Mulvaney, B. J., (1996) Circuit, Device and Process Simulation, John Wiley & Sons Ltd..
61. Gardner, Carl L., (1994) SIAM J. Appl. Math. **54**, 409.
62. Lei, X. L. and Ting, C. S. (1985) Phys. Rev. B**32**, 1112.
63. Lei, X. L., Horing, N. J. and Cui, H. L., (1991) Phys. Rev. Lett. **66**,3277.
64. Tanaka, Yukihiro and Akera, Hiroshi, (1996) Phys. Rev. **B**53, 3901.
65. Schwinger, J., (1961) J. Math. Phys. **2**, 407.

66. Keldysh, L. V., (1964) Zh. Eksp. Teor. Phys. **47**, 1515(1964) [Sov. Phys. – JETP **20**, 1018(1965)].
67. Mahan, G. D., (1987) Phys. Rep. **145**, 251.
68. Kadanoff, L. P. and Baym, G., (1962) Quantum Statistical Mechanics, (Benjamin, New York).
69. Jauho, A. P., (1985) Phys. Rev. B**32**, 2248.
70. Hideaki Tsuchiya and Tanroku Miyoshi, (1998) J. Appl. Phys. 83, 2574.
71. Datta, S., (1989) Phys. Rev. **B**40, 5830.
72. F. A. Buot and K. L. Jensen, Phys. Rev. B**42**, 9429(1990).
73. Lake, R. and Datta, R., (1992) Phys. Rev. **B**46, 4757.
74. Baym, G., (1961) Ann. Phys. 14,1.
75. Joshi, S. K., and Rajagopal, A. K.,(1968) in Solid State Physics **22**, 248, Ed: F. Seitz and D. Turnbull, Academic Press, New York.
76. Tso, H. C. and Horing, N. J. M., (1991) Phys. Rev. **B**44,8886; (1992), Phys. Rev. **B**46, Errata.
77. Tso, Hum Chi,(1990) Ph. D Thesis, Stevens Institute of Technology.
78. Bastard, G., (1982) Phys. Rev. B**25**, 7584.
79. White, S. R. and Sham, L. J., (1981) Phys. Rev. Letters, 47, 879.
80. Gelmont, B. L., (1987) "Three Band Kane Model and Auger Recombination," Sov. Phys. JETP, Vol 48, pp. 268-272.
81. Jensen, K. L. and Bout, F. A., (1989) Appl. Phys. Lett. **55**, 669.
82. Lake,R., G. Klimeck, G., Chris Bowen, R., and Jovanovic, D., (1997) J. Appl. Phys., **81**.
83. Buot, F. A. and Rajagopal, A. K., (1995) Materials Science and Engineering **B**35, 303.
84. Buot FA. (1999) Zener effect. In: *Wiley Encyclopedia of Electrical & Electronics Engineering* (ed JG Webster) **23**: 669-688. John Wiley & Sons, New York.
85. Nag, B. R., (1980) Electron Transport in Compound Semiconductors (Springer, Berlin)
86. Ridley, B. K., (1988) Quantum Process in Semiconductors (2nd ed.) (Oxford University, New York)
87. Tomizawa, K., (1993) Numerical Simulation of Submicron Semiconductor Devices (Artech House, Boston)

International Journal of High Speed Electronics and Systems
Vol. 13, No. 4 (2003) 1255–1286

Wigner Function Simulations of Quantum Device-Circuit Interactions

H. L. GRUBIN and R. C. BUGGELN

Scientific Research Associates, Inc.
Glastonbury, CT 06033, USA

Issues associated with modeling of quantum devices within the framework of the transient Wigner equation are addressed. Of particular importance is the structure of the Wigner function, hysteresis, and device switching time, whose value is determined by the large signal device properties.

Keywords: Wigner function, dissipation, relaxation oscillations

1. Introduction

Numerical simulation of the dc and transient behavior of semiconductor devices has been around for a long time[1]. For the most part these simulations are based on either empirical expressions or physical equations. The empirical expressions are thought to have a basis in more fundamental physical equations, but often their use in not predicated on fundamental justification; rather the issue is design efficacy. But physical modeling, while itself incomplete can open new device areas of research, lead to new device design and generate new devices, as we learned many years ago with the discovery of velocity overshoot[2]. Velocity overshoot, we recall, is a transient phenomena in which the mean carrier velocity reaches values in excess of a peak steady state velocity specific to the material under study. The phenomenon arises by incorporating acceleration into the standard, at that time, semiconductor equations. The experience gained in understanding velocity overshoot taught many of us the perils of ignoring such elementary physics concepts when examining short time scale device operation.

When dealing with quantum devices, short time phenomena is not a curiosity, it is intrinsic to understanding device physics. Indeed, the dc operation of quantum devices is probably an artificial concept arising from the need to deal with the approximations of quantum transport simulation. When dealing with transient quantum phenomena, we often begin with the Liouville equation for a Hermitian Hamiltonian:

$$i\hbar\frac{\partial\rho_{op}(t)}{dt}=\left[H(t),\rho_{op}(t)\right]. \tag{1}$$

While there is much information to be garnered from the above equation, the usual situation is to choose a particular representation and construct a differential equation from which transient device simulation algorithms are developed. A typical representation is the coordinate representation. If a two-time density matrix is introduced:

$$\rho(\mathbf{r},\mathbf{r}';t,t')=\sum_i f_i\Psi^*{}_i(\mathbf{r}',t')\Psi_i(\mathbf{r},t), \tag{2}$$

two time-evolving transport equations emerge. For the case $t=t'$, the familiar coordinate representation equation is retrieved:

$$i\hbar\frac{\partial\rho(\mathbf{r},\mathbf{r}';t)}{\partial t}+i\hbar\left(\frac{\partial\rho(\mathbf{r},\mathbf{r}';t)}{\partial t}\right)_{scattering} = -\frac{\hbar^2}{2m}\left(\frac{\partial^2}{\partial\mathbf{r}^2}-\frac{\partial^2}{\partial\mathbf{r}'^2}\right)\rho(\mathbf{r},\mathbf{r}';t)+\left(V(\mathbf{r},t)-V(\mathbf{r},t)\right)\rho(\mathbf{r},\mathbf{r}';t) \quad (3)$$

Equation (3), which includes a generic dissipation contribution, is often considered the starting point for discussing transport in nanoscale devices. It can be considered as the beginning of a discussion of transport in classical devices, as well, providing the density matrix is obtained for a suitable classical system[3]. But, to date the most extensive and successful transient device simulations are through the Wigner equation, which is related to the equation (3) via the Weyl transformation[4].

The Wigner function, first discussed in 1932[5], was introduced in terms of pure state wave functions and ignored spin contributions. In one dimension:

$$f_W(x,k)=\frac{1}{2\pi\hbar}\int_{-\infty}^{\infty}dy\psi^*\left(x+\frac{y}{2}\right)\psi\left(x-\frac{y}{2}\right)e^{iky} \quad (4)$$

The equation governing $f_W(x,k)$ was obtained by application of Schrödinger's equation. O'Connell and Wigner[6] generalized the quantum distribution function to include spin:

$$f_W(\mathbf{r},\mathbf{k},\ell)=2^3\sum_{m,m'=1,-1}\sigma^{\ell}_{m,m'}\int d\vec{\xi}\,\exp\left[2i\mathbf{k}\bullet\vec{\xi}\right]\psi^*\left(\mathbf{r}+\vec{\xi},m\right)\psi\left(\mathbf{r}-\vec{\xi},m'\right) \quad (5)$$

where the subscripts for $\sigma^{\ell}_{m,m'}$ are identified with different combinations of spin-up and spin down states:

$$\begin{pmatrix}\uparrow\uparrow & \uparrow\downarrow\\ \downarrow\uparrow & \downarrow\downarrow\end{pmatrix} \quad (6)$$

Please notice the differences between the normalizations between equations (4) and (5).

Currently, transient Wigner simulations are dominated by one-dimensional space and momentum transport, and constant effective mass. This is certain to change in the near future as algorithms are constructed for dealing with the increased size of the computation, and machine speed increases. The equation we deal with is:

$$0=\frac{\partial f_W(x,\mathbf{k})}{\partial t}+\left(\frac{\partial f_W(x,\mathbf{k})}{\partial t}\right)_{DISSIPATION}+\frac{\hbar k}{m}\frac{\partial f_W(x,\mathbf{k})}{\partial x} -\frac{1}{\pi\hbar}\lim_{L\to\infty}\int_{-L}^{L}dy\left(E_C(x+y)-E_C(x-y)\right)\int_{-\infty}^{\infty}dk'\,f_C\left(y,k_x',k_y,k_z\right)\sin\left[2\left(k_x'-k_x\right)y\right] \quad (7)$$

The first, third and fourth terms can be obtained from application of the time dependent Schrödinger equation to equation (4). Figure 1 is a representation of the term $E_C(x+y)-E_C(x-y)$ for a double barrier, where we see that only a small portion of the integrand contributes to the integral. This term when multiplied by the sine term shows significant structure is directions diagonal to the barriers. The dissipation term is the recognition that the interaction of devices with their environment involves a non-conservative exchange of energy. Most device simulations treat dissipation within the relaxation time approximation.

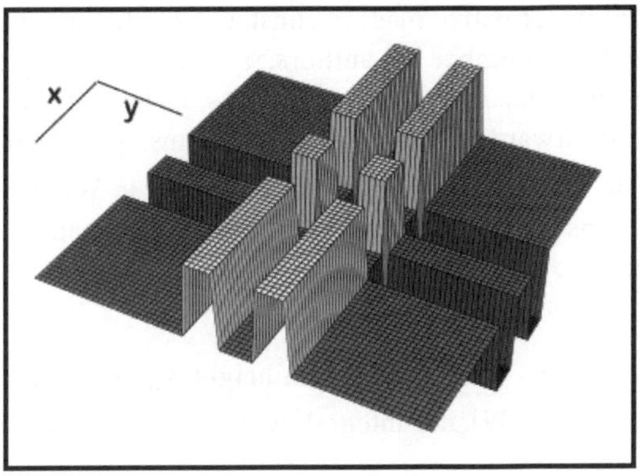

Figure 1. ***E(x)-E(y)*** **for a double barrier structure.**

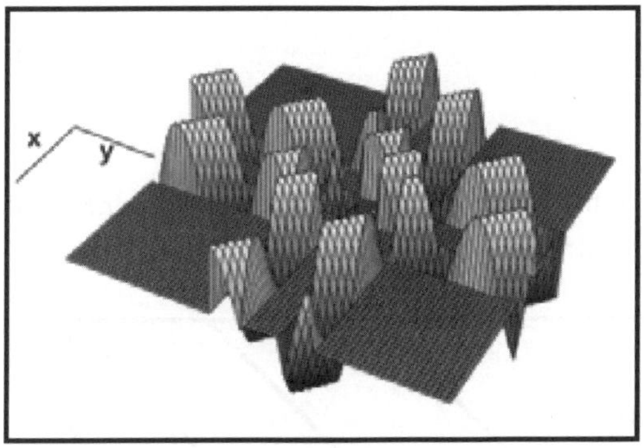

Figure 2. ***(E(x)-E(y))sin[(k'-k)(x-y)]*** **for the double barrier structure of figure 1.**

The time dependent Wigner equation provides transients and enables us to avoid the dilemmas associated with ignoring short-time effects. Indeed, because of the limited class of devices that have been studied with the Wigner equation (three-dimensional structures are not studied unless patched together with Monte Carlo procedures), the advantages of Wigner simulations emerge mainly when the envelope moves to transient device studies. While analytical studies provide insight into the tunneling times of specific quantum devices, the details associated with non-uniform fields, scattering, very short time scales, noise contributions, and large signal device- circuit and/or transmission lines interactions, require transient numerical studies.

2. The Physical Model when Spin is not a Design Parameter

The basic quantum transport equation is the Wigner equation[5], modified to include dissipation and mixed systems. The potential energy in this equation combines two contributions, the barrier/well configuration (single or multiple) and the potential energy arising from Poisson's equation. In some of the studies discussed below, the Poisson contribution was treated as $\nabla_x V_{POISSON} \cdot \nabla_k f_W(\mathbf{k},x)$. While we can imagine situations when the particle distribution is such as to yield variations in the Poisson potential that competes with spatial variations in the barrier potential, the cases examined below do not fall into this category and this contribution was treated classically.

The form of the Wigner integral is expressed as a limit. This limit is one of many ways of expressing differences with the original development of the Wigner equation. The latter we recall was for wave functions that were solutions to Schrodinger's equation with boundary conditions at $\pm\infty$. All of the cases we deal with are for systems of finite size that interact with their environment. This interaction includes dissipation. In the one space-dimension case, the interaction is expressed through boundary conditions at the end of the device, not at $\pm\infty$. Indeed, as we will see some of the most important issues in examining barrier device transport involve dissipation[7], the finite size of the domain and open boundaries[8].

In the simulations discussed below the effective mass is constant. Modifications due to variations have been considered by a number of authors, see, e.g., Barker[9], Miller[10], among others. Introducing the spatial dependent effective mass, at least in the symmetric form ($H = p(1/2m)p$) is straightforward. Keeping all of the terms associated with the effective mass has the unfortunate effect of increasing the order of the Wigner equation, requiring significant modifications to the Wigner algorithms. This can be avoided, at the loss of physics, by replacing the mass in the Wigner transport equation by a spatially dependent effective mass, i.e., $m -> m(x)$.

In all cases considered below the barriers were square with height V_0 and width Δ. This permitted an analytical integration of the Wigner integral, which was used in all the studies:

$$\lim_{L\to\infty}\int_{-L}^{+L} dy\left[E_C(x+y)-E_C(x-y)\right]\sin\left[2\left(k_x^{'}-k_x\right)y\right]$$
$$=\frac{2V_0\sin\left[2\left(k_x^{'}-k_x\right)\left(x-x_0\right)\right]\sin\left[2\left(k_x^{'}-k_x\right)\Delta\right]}{i\left(k_x^{'}-k_x\right)} \quad (8)$$

In equation (8) x_0 is the central position of the barrier. All square barriers used the same expression, with the center position, height and width of each barrier defined separately.

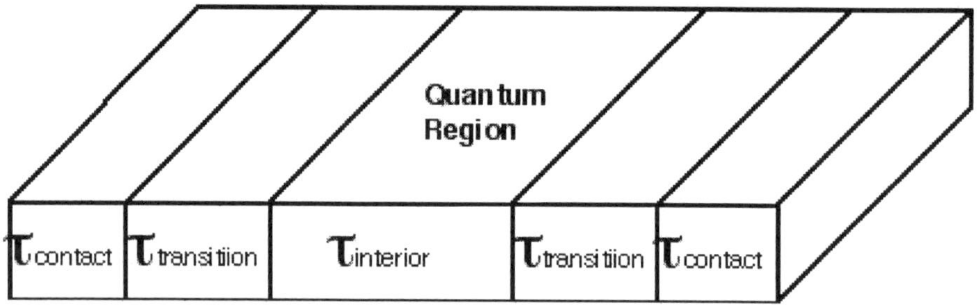

Figure 3. Five zone region used in the Wigner calculations.

Structure in equation (8) teaches that for large values of Δ and/or $x - x_0$, the Wigner integral behaves as a delta function in $k_x^{'} - k_x$ and all interactions are local. For a pair of barriers separated by a distance 2δ there is additional structure coming from a term $\cos\left[\left(k_x^{'}-k_x\right)\delta\right]$ (not shown in equation (8)).

In all simulations the computational regions are broken into classical and quantum regions, as shown in figure 3. The contact and transition regions are treated classically (the size of each region in figure 3 is not scaled to the size used in the computation), with the central region representing the quantum mechanical region. The scattering times can and do vary spatially within each of the regions shown in figure 3. In addition, the background doping which is position-dependent is varied independently of the regional designation shown in figure 3, although generally the doping levels are highest within the contact regions. The classical and quantum designation are established within the framework of the Wigner integral by multiplying equation (8) by a modulating function that is equal to unity within the 'quantum region' and zero elsewhere. For most of the calculations discussed below in which the device length is 200 *nm*, the quantum region is at least 120 *nm* long. For each of the calculations discussed below the transition between the quantum and classical regions takes place within the heavily doped region.

What about dissipation? A detailed discussion is provided by Bordone, et al[7], but the tack taken by many is to relax the Wigner function. Here several approaches have been taken. One approach, taken by Kluksdahl, et al[11], and others is to assume a relaxation time approximation where:

$$\left(\frac{\partial f_W(\mathbf{r},\mathbf{k})}{\partial t}\right)_{DISSIPATION} = -\frac{\left(f_W(\mathbf{r},\mathbf{k}) - f_0(\mathbf{r},\mathbf{k})\right)}{\tau(\mathbf{r},\mathbf{k})} \tag{9}$$

The relaxation time approximation in the above form leads to source and sink terms in the continuity equation. To avoid sink terms others[12] have multiplied the equilibrium distribution function, $f_0(\mathbf{r},\mathbf{k})$, by the ratio of the non-equilibrium carrier density to the equilibrium carrier density. We have done both, but we will confine ourselves to that represented by equation (9).

The question of interest is what is $f_0(\mathbf{r},\mathbf{k})$? We must remember that unlike many early discussions in the theory of metals[13] where it is often assumed that the electric field under equilibrium is near or approximately zero, in the case of electron devices where there are barriers, variations in background density and strong local variations in the electron charge density, the local electric field can be in the range of Mev/cm. The form of the equilibrium distribution function is dependent upon the model used to connect current at the open boundary and must represent the spatially dependent distribution associated with barriers, scattering and self-consistency. The models described below have evolved. Initially we assumed a displaced Fermi Dirac distribution function on the boundaries. While, in principle, the displaced momentum should be zero, or at least zero within five or six significant figures, it wasn't. The boundary condition was then changed to one in which the normal derivative (with respect to position) of the distribution function was set to zero. This provided the requisite zero current conditions. Further, to enhance the possibilities of flat-band open boundary conditions the relaxation time in the vicinity of the boundaries was set to values that were smaller than elsewhere in the device.

Obtaining $f_0(\mathbf{r},\mathbf{k})$ required several steps. For the first time step a Fermi-Dirac distribution function consistent with uniform fields is introduced. Then assuming fixed boundary conditions, and, of course, zero-current, the distribution function consistent with the doping variation, Poisson's equation and ***zero*** barrier height is obtained. The number of time steps required for convergence is problem specific, but usually on the order of twenty. Then with fixed boundary conditions, the barriers are gradually turned on. After convergence is reached the boundary conditions are altered to those in which the normal derivative with respect to position is zero, and the solutions again taken to convergence. This last step provides the function, $f_0(\mathbf{r},\mathbf{k})$, that is used in the subsequent calculations.

Now, consider some details in differencing. The topic most workers tend to focus on is the term containing the spatial derivative of the distribution function. In the earliest studies specific attention was given to the importance of difference schemes representing particles coming in with negative/positive values of momentum. Specifically, the spatial derivative for carriers entering with positive values of momentum is displaced by one spatial index, from particles coming in with negative values of momentum[14]. When this condition, as discussed in reference 14, was entered into our numerical schemes, the equilibrium distribution function was asymmetric with respect to momentum. This issue was avoided using the numerical scheme discussed in the appendix, where under equilibrium conditions the distribution function is symmetric. (The appendix also summarizes block matrix inversion procedures[15].) But another important issue, with respect to differencing, is continuity, a particularly important device issue. Starting from the analytical expressions for the Wigner equation, most difference techniques will not provide numerically accurate expressions of continuity unless the number of grid points used in the calculation becomes excessive. One way around this is to write a Wigner equation in difference form that supports the analytical forms of the Wigner equation in its most common integro-differential form. We as well as other[8] workers have done this; and while the results appear promising; numerically based Brillouin zone effects are introduced.

Another issue, thought by some to be based upon physical constraints, is the structure of the Wigner function. Certainly the Wigner function sustains negative values, but should the noise that appears in most numerical simulation be considered as fundamental? We cannot yet answer this question, but some of our results, although inconclusive, suggest that this noise is not necessary and is dependent on device length and the number of grid points. We will illustrate this point.

In performing the simulations both the position and momentum are written as discrete variables. The differential increment for momentum is typically taken as $\Delta k_x = 2\pi / Device\ \ Length$. Although we recognize this increment as being consistent with periodic boundary conditions on the Bloch functions; it has nothing to do with the boundary conditions imposed on the Wigner function. The smallest differential increment used in our studies is $\Delta k = \pi / Device\ \ Length$. In a typical device simulation the number of spatial grid points, N_{POS}, is 150, the number of momentum grid points N_{MOM} is 150, and $\Delta k \Delta x \cong 2\pi / (N_{POS} - 1)$. Additionally, typical simulations are for structures with a nominal length of 200 nm. The number of momentum grid points needed to reach a value of energy equal to the Fermi energy is readily calculated. (For parameters suitable for GaAs and for a Fermi energy equal to 40 meV the Fermi wave vector is reached with an index greater than '9'. Thus the computation takes place for momentum values well in excess of the Fermi energy.) The matrix we deal with was constructed from a square tridiagonal $N_{POS} \times N_{POS}$ matrix with $(N_{MOM} + 1) \times (N_{MOM} + 1)$ sub-matrices both along and straddling the diagonal. The computational matrix includes contributions from Poisson's equation and the boundaries. As indicated earlier, a discussion of the computational matrix is given in the appendix.

To examine different aspects of numerical noise and Wigner functions, several studies were performed for structures whose length was 100 nm. The 100 nm structures were not regarded as representative of any device because conditions at the boundaries did not satisfy flat band constraints. Rather they were of interest because of the insight they provided in the structure of the Wigner function. Figures 4 through figure 8 illustrate. The structure was 100 nm long with two 250 meV barriers and a 50 meV quantum well. The equilibrium distribution of potential energy, charge and background density is displayed in figure 4, which displays a non-zero slope for the density at the boundaries.

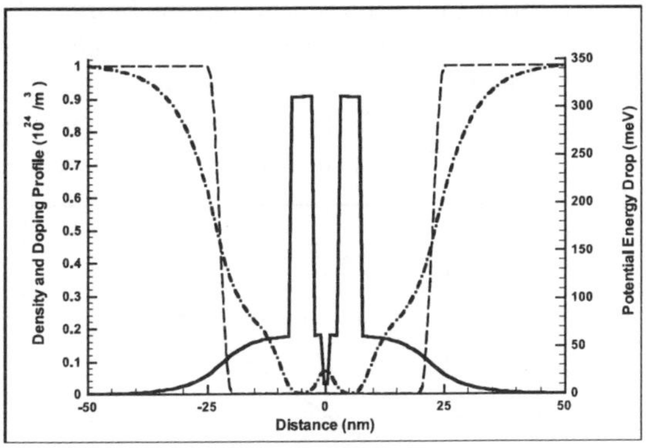

Figure 4. The equilibrium distribution of density (dot-dashed), potential energy (solid) and doping density for a 100 nm double barrier structure with a 50 meV structural quantum well.

This structure also sustains a weak amount of negative differential conductivity at potential energies between $-7k_BT$ and $-8k_BT$. The Wigner functions at equilibrium, $-7k_BT$, and $-8k_BT$ are displayed in figure 5. The equilibrium distribution displays local charge accumulation in the vicinity of the quantum well. There is a build-up of charge between the barriers, as well as an accumulation of charge on the emitter side of the first barrier, as the collector potential energy is reduced relative to the emitter. The charge build-up between the barriers reaches a peak prior to the onset of negative conductance and then decreases as shown in the last frame of figure 5. These 100nm calculations were the only ones to shown smooth distributions through the drop back in current density.

Calculations for structures 200 nm and longer, were also studied for computational noise. Several calculations were performed to isolate and perhaps determine the origin of these variations. Unfortunately, we were unable to form any concrete opinion. Instead we describe the results. The structure, for which computational noise was studied, was 200 nm long, with a nominal doping of $10^{24}/m^3$ everywhere, except in the center, where the background density was a lower $10^{21}/m^3$. For these calculations the momentum grid spacing was varied sometimes to values half of the nominal value discussed above, and the number of grid points increased. Typically, the Wigner function at bias levels below the threshold for negative differential conductivity did not display any noise. As the bias level increased, the distribution functions retained their apparently smooth structure until the bias was below/above but near threshold. For the high values of bias, the noise in Wigner function was again absent. There was some reduction of noise as the number of grid points increased, but nothing to lead to the conclusion that noise could be eliminated with increased grid points. Calculations with single barriers did not show the noise structure generated with the double barrier structures.

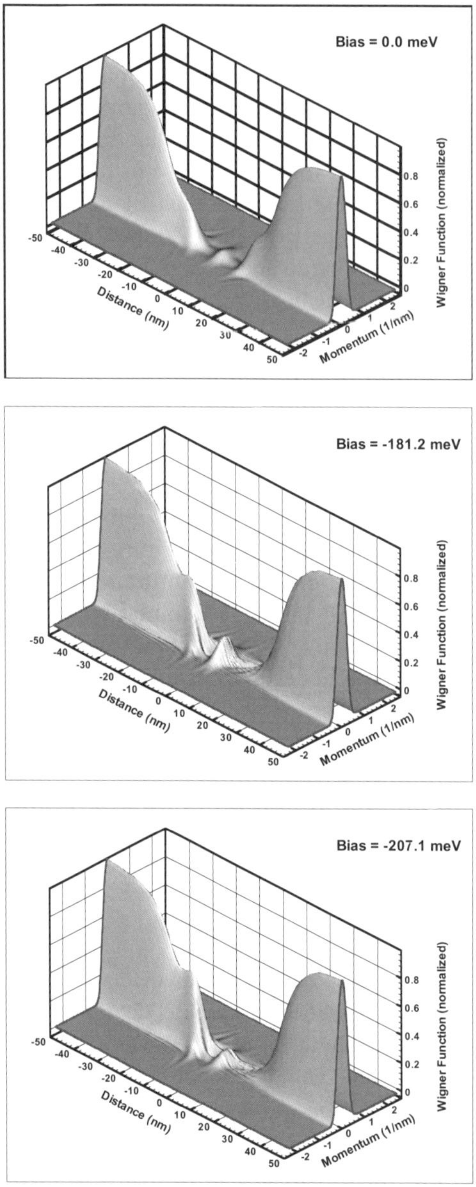

Figure 5. Wigner distribution at equilibrium and two different values of bias.

3. Calculations for Classical Structures

Wigner simulations are generally implemented for barrier problems. Rarely have techniques associated with Wigner functions dealt with classical problems, even though there is value in performing such calculations. The value of such calculations lies as a reference for the barrier calculations and as a means of determining the resistance of the structure prior to the introduction of the barrier. It also suggests that more complex circuits can be studied in conjunction with the Wigner simulations. We illustrate the use of the Wigner simulation for a classical

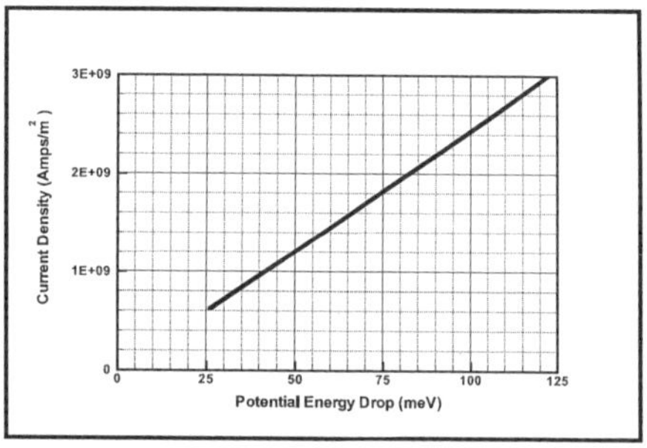

Figure 6. '*IV*' for an NIN structure.

NIN 200 nm structure with doping variations and variable scattering rates similar to those employed in the barrier problems. The current voltage characteristics are displayed in figure 6, and the equilibrium Wigner function for this classical problem is shown in figure 7.

While there will be more opportunity to discuss the structure of the Wigner function, it is important to note that the distribution function associated with the double barrier structure of figure 5 has a definitive 'k'-dependence as we move away from the center of the structure, and that this 'k' dependence is absent from the Wigner function for the 'NIN' structure displayed in figure 7. This k-dependent structure is regarded as arising from quantum interference contributions.

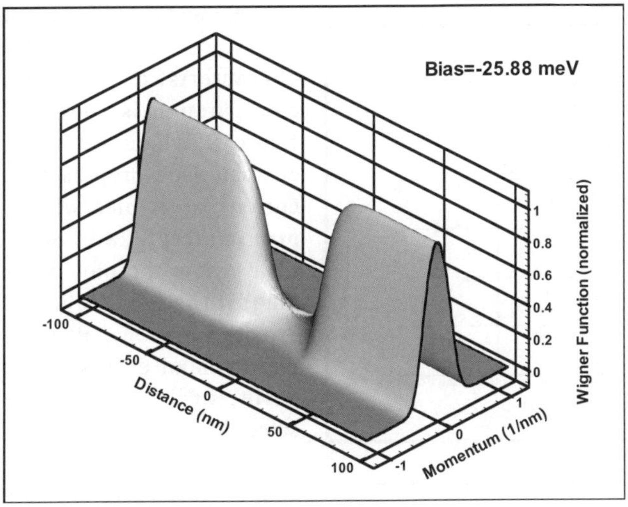

Figure 7. The Wigner distribution for an NIN structure under an applied bias.

The 'NIN' charge density and potential energy distributions at equilibrium and $-4k_BT$ are also shown. For the latter we observe a local minimum in charge density near the emitter and a local maximum near the collector. In these calculations the relaxation time, which is a function of position is smaller in the vicinity of the boundaries then elsewhere in the device. Under conditions of constant current and assuming that the forces in the vicinity of the boundaries are approximately the same, we can argue that $\partial(n\tau)/\partial x \approx 0$, or $\partial n/\partial x \approx -n\partial \ell n(\tau)/\partial x$. Then moving in from the emitter where the relaxation time increases we expect a decrease in carrier density. Similarly, within the vicinity of the collector, where the relaxation time decreases, a local initial increase in the carrier density occurs. Very similar behavior is observed when dealing with barrier structures, which is considered next.

4. Baseline DC Characteristics

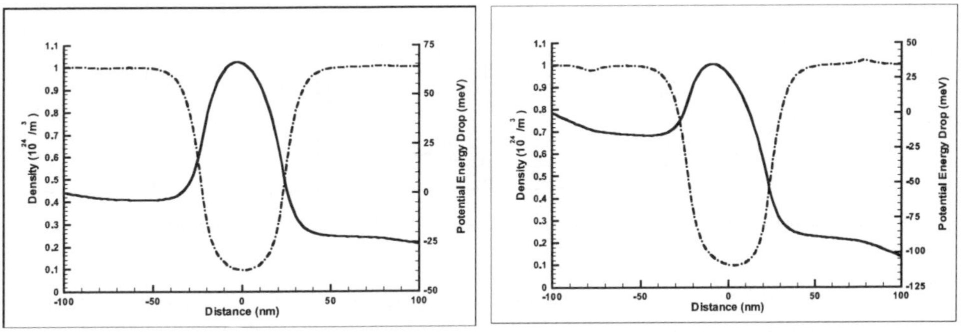

Figure 8. The charge and potential energy distribution for an NIN structure subject to two different values of bias.

The bulk of the discussion in this chapter is concerned with transients and resonant tunneling. Baseline dc calculations were performed for the NIN configuration discussed above, with two tunnel barriers centrally placed in the low-doped center regions. These barriers were each 250 meV high with a width of 5 nm, separated by 5 nm. The detail of the relaxation and doping variations are as follows: The effective mass is

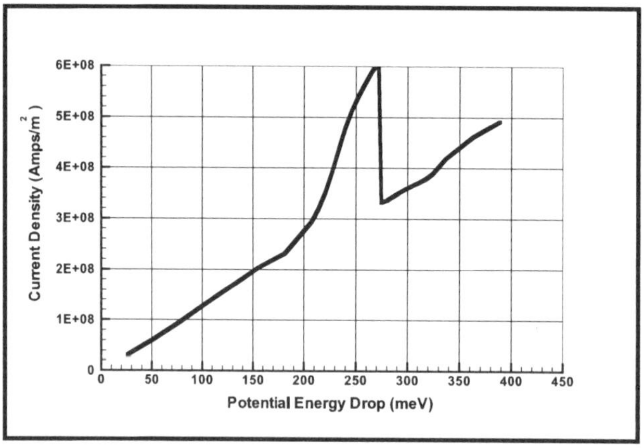

Figure 9. The dc IV characteristic for a double barrier RTD embedded within an NIN structure.

constant and equal to that of GaAs. The dc IV characteristic for this configuration is shown in figure 9 where we see a characteristic drop-back in current at a bias of approximately 280 meV. Please note the plot is for magnitude of current density versus magnitude of the potential energy at the collector. In these calculations the collector is negative with respect to the emitter. The curve in figure 9 represents two sets of data, that corresponding to increasing and decreasing applied bias. For the dc calculations, which involved very small bias increments and decrements, there wasn't any hysteresis in current versus voltage (see more recently, Shifren and Ferry[16]) . Hysteresis appeared under time dependent conditions

A comparison of the low bias dc characteristics of the NIN structure and the NIN structure with the embedded RTD displays significant differences in the apparent low field mobility of the device. Of course, mobility is not an operative concept here. Much of the potential drop falls across the barriers.

The potential energy and charge distribution at equilibrium, immediately prior

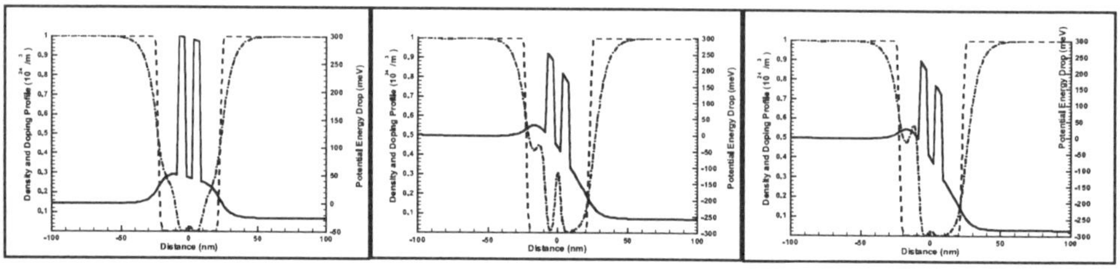

Figure 10. The distribution of potential energy and carrier density under three different conditions of dc bias.

and post threshold are displayed in figure 10. At equilibrium there is a small amount of charge accumulation in the quantum well. As the bias increases there is increased charge accumulation within the quantum well, and a local build-up of charge on the emitter side of the first barrier. This charge build-up is consistent with the formation of a 'notch' potential to the left of the emitter side barrier. There is also enhanced charge depletion on the collector side of the second barrier. Global charge neutrality is operative over the entire device. The distribution just prior to the onset of negative conductance is shown in the center frame. The final frame shows the distribution of charge immediately after the drop in current. We notice the loss of charge in the quantum well, and the build-up of charge associated with the notch potential. The potential and charge distributions on the down swing are the same as the potential and charge distribution on the up swing. It is important to point out that a number of authors have displayed results that indicate hysteresis in the dc characteristics. In our case in which the boundary conditions on the up and down swings are the same, hysteresis is absent.

5. Transient Device Simulations

The key advantage of device simulation has always been in the ability to address large signal time dependent device issues. In this case the experimental counterpart is the resonant tunneling diode as a relaxation oscillator[17]. Thus we have a quantum device that would appear to function as a very fast van der Pol oscillator, albeit one that requires study within a transmission line configuration. There are studies of such structures, although few, if any that involve space charge contributions. But such studies, particularly those involving the repetitive nature of the device response, are crucial for determining how long it takes the device to relax prior to a re-interrogation. Is this number a function of bias and of switching range? If the answer to this is positive then simple descriptions of switching times are irrelevant.

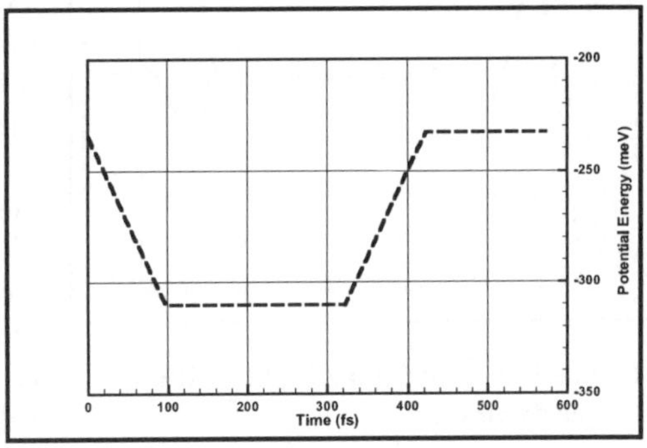

Figure 11. Voltage versus time for switching.

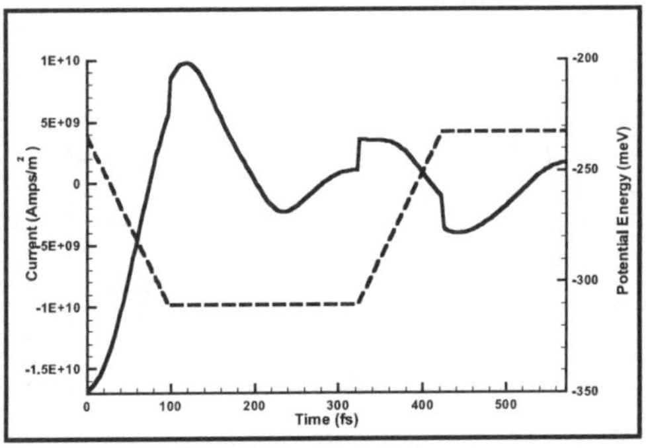

Figure 12. Voltage pulse of figure 11 and subsequent current response.

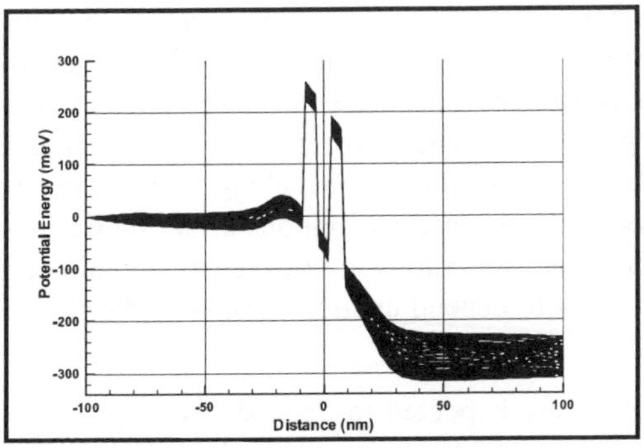

Figure 13. Superposition of potential energy distributions corresponding to the time variation of figure 12.

To study this the RTD was subjected to a controlled time dependent change in voltage (see also Zhao, et al[18]). Figure 11 shows the change in voltage. Remembering that the anode is negative with respect to the cathode and that negative values of voltage result in movement across the region of negative differential conductivity, we start from a point just below that value of voltage that results in a drop in current, to a point well beyond the region of negative conductance. The voltage is both decreased and increased at a controlled rate, and the dwell time is also controlled.

The response of the device to this change in voltage is shown in figure 12, which displays the voltage change as well as the current response at one of the contacts. Because we have dissipation within the device current continuity does not imply current that is independent of position throughout the device. The current displayed in figure 12 consists of two components, a particle current and a displacement current. The displacement current contributions arise from local changes in the electric field in the heavily doped regions at the boundaries as seen in figure 13. When the system relaxes toward equilibrium these displacement current contributions vanish. Please note, the dwell time for this calculation is approximately 200fs at which time the voltage was lowered to the value just below threshold for negative differential conductance.

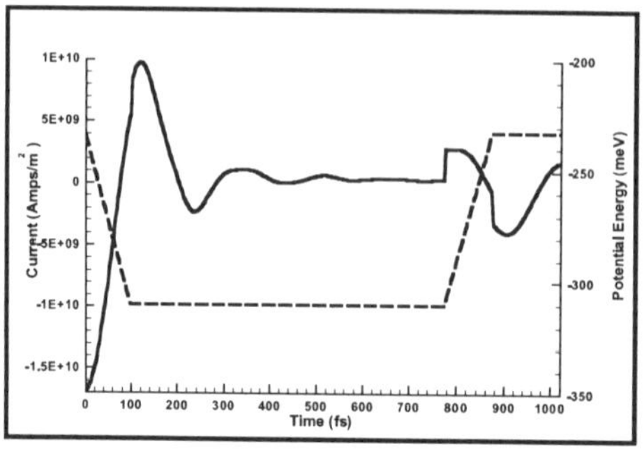

Figure 14. As in figure 12, but with a longer dwell time.

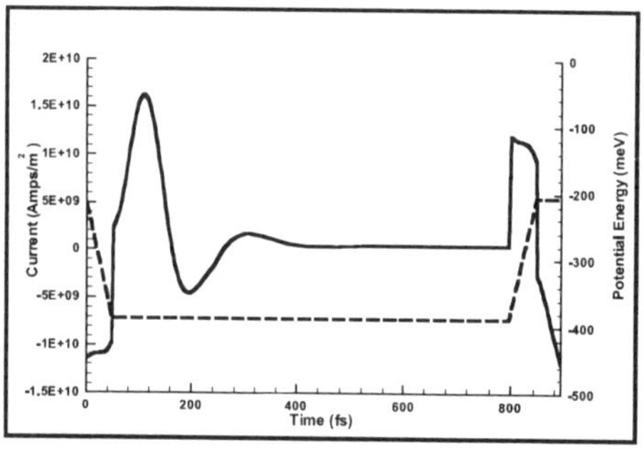

Figure 15. Transient response to a larger change in bias.

The transient calculations reveal several features. First, while we did not see dc hysteresis in the current versus voltage relation here the currents at the same values of voltage are different for increasing and decreasing voltages. This result is expected! Also note that the transient current, at least on this time scale, appears to react almost instantaneously to the change in voltage. This is likely to be an artifact of the calculation and the absence of an external circuit to influence the device response. This result teaches that for this bias change the device does not relax within 200 fs, the duration of the voltage dwell time. The relaxation time of this structure is more closely obtained from the results of figure 14, where the transient voltage pulse differs from that of figure 12 in the dwell time. It would appear from this calculation that the current relaxes in under 0.75 ps. These results speak to the switching time of the device, but not to whether this device will result in sustained oscillations at the shorter period excitation. Here, as in the case of relaxation oscillations in space charge dominated classical devices, sustained oscillations are likely to depend upon the ability of the circuit to quench any incipient space charge effects.

The switching properties of the RTD are expected to depend not only on the dwell time but also on the voltage swings. For example in another calculation (figure 15), starting from the same initial voltage level but going down to a voltage level of approximately 400 meV, the carriers appear to relax to steady state in a time close to 500 fs, which is shorter than that realized above. Also notice that the rise time for this calculation was shorter than for the figures 14 calculation, resulting in a higher displacement current contribution. This more rapid relaxation is likely to be a consequence of more rapidly removing the charge that built up in the quantum well. *These calculations indicate that the measure of a quantum device switching time is a function of the environment that the device resides in.*

While the key issue is the interaction of the device with its external circuit we have begun to study the interaction of the device with a sinusoidal input. Others have done this as well[14]. To date we have done this only for a single barrier device, and one for which there are an insufficient number of grid points. But it is worthwhile showing the transient Wigner function, because it highlights the advantages of performing simula-

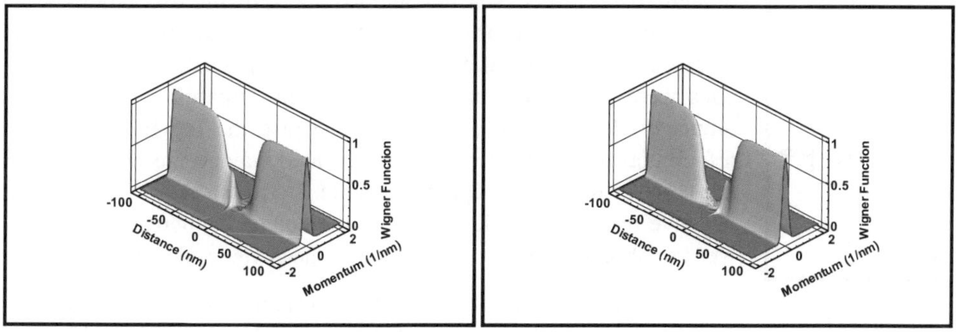

Figure 16. Response of the Wigner function for a single barrier diode subject to a sinusoidal time dependent change in voltage.

tions in which there isn't a contact preference. These results are displayed in figure 16, which displays the Wigner function at two instants of time for a structure subject to a sinusoidal change in voltage with a period of approximately 200 fs. Notice the apparent local build up of charge on either side of the barrier as the oscillation proceeds. The device appears to behave classically, with the quantum features determining the actual distribution of charge, along with quantum interference effects as represented by the momentum dependent structure in the Wigner function.

6. The RTD as a Self-Excited Oscillator

The above discussion has demonstrated that the transient characterization of an RTD is not the time it takes to go from one state to another, but rather the time it takes to recover from a switch. This feature of device behavior has been demonstrated in a wide variety of other devices including transferred electron Gunn oscillators, IMPATTs, etc. The implication for RTDs is that the maximum frequency of oscillation will be determined by a number of parameters, only one of which is the tunneling time. How do we determine the parameters for self-excited oscillation?

What is a relaxation oscillator and why is it relevant here? Several years ago Verghese and co-workers[17] demonstrated that the resonant tunnel diode could operate as a relaxation oscillator. While this was anticipated, it was the first experimental observation of such behavior. The results were expected because the resonant tunnel diode with

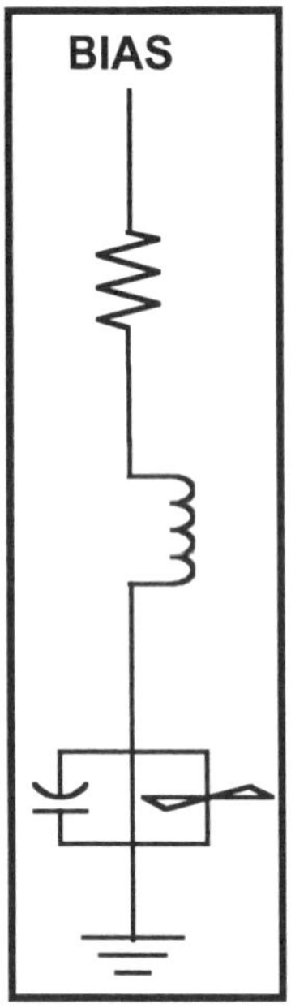

Figure 17. Circuit used for self-excited oscillation studies.

its region of negative differential conductivity, when placed in an external circuit containing reactive elements is a classic van der Pol oscillator----providing it can maintain its region of negative differential conductivity. (Note: van der Pol discussed solutions to the ordinary differential equation (ODE): $y''-\mu\left(1-y^2\right)y'+y=0$. The form of the oscillation is determined by the parameter μ, which when zero yields standard sinusoidal oscillations.) The relevance here is as follows: when a simple resistor is placed in a circuit containing an inductor and capacitor it undergoes damped oscillations. When the resistor contains a region of negative resistance, the oscillations grow. Suitable balance between the damped and growing oscillations leads to sustained oscillations. This type of oscillation characteristic received considerable attention when initial applications of tunnel diodes were proposed[19] in the sixties. The situation with resonant tunnel diodes is more complicated because the region of negative differential resistance depends in a detailed way on the way in which the carriers tunnel through the double barrier structure, and indeed negative resistance while present as the voltage is increasing, is often weak or absent when the voltage across the device is decreasing.

The first circuit dependent transient situation we examined was when the RTD was part of the circuit shown in figure 17. The RTD, which contains its own capacitive contributions as determined by Poisson's equation is in parallel with a capacitor, the combination being serially connected to an inductor and a linear resistor. The RTD is represented as a solution to the transient Wigner equation. The parameters for the simulation are obtained by first solving an ordinary differential equation with a nonlinear NDR element with parameters fit to the dc IV characteristic of the RTD under study. One of the interesting features associated with ordinary differential equations is the ability to scale. Simple scaling permits a range of identical scaled solutions with identical voltage and current swings but different frequencies. This type of scaling when applied to the RTD represented by the Wigner function permits a direct determination of those space charge features that constrain device operation.

For example: for the ODE we represent the current voltage relation of the NDR element as $I_D\left(V_D\right)$, where I_D is the dc current through the device and V_D is the voltage drop across the device. With I_P denoting the peak current at the onset of negative differential resistance, and V_P denoting the threshold voltage, or the voltage at the onset of negative differential resistance, we define a device resistance as $R_D = V_P/I_P$. Then for a reference impedance $Z_0 = \sqrt{L_{inductance}/C}$ and a reference period $T_0 = 2\pi\sqrt{L_{inductance}C}$, the relevant circuit equations are:

$$\frac{dv_D}{dt'} = \frac{2\pi Z_0}{R_D}\left(i(t') - i_D\left(v_D\right)\right) \tag{10a}$$

$$\frac{di}{dt'} = \frac{2\pi R_D}{Z_0}\left(v_{APPLIED}(t') - v_D(t') + i(t')\frac{R_{LOAD}}{R_D}\right) \tag{10b}$$

Here: $t' = t/T_0$, $v_D = V_D/V_P$, $v_{APPLIED} = V_{APPLIED}/V_P$, $i = I/I_P$. The significance of the scaling is as follows. For fixed values of Z_0/R_D and R_{LOAD}/R_D, the oscillation characteristics are independent of frequency, ***provided, the term*** $i_D(v_D)$ is independent of frequency. A frequency dependent $i_D(v_D)$ could be used to mimic hysteresis contributions in RTDs, and could provide an illustration of the role of hysteresis in determining the frequency limitations of RTDs. Rather that pursue this path, we have, as indicated above, represent the RTD as a solution to the space and time dependent Wigner equation. But we were informed by the ODE results and guided by them. Consider, then the output of the calculations.

Figure 18 displays a self-excited (relaxation) oscillation (RO) of the device represented by solutions to the Wigner equation of motion. The structure is a 200 nm device with a nominally undoped central region. The structure details will be apparent in figure 20. Density and current are obtained from the Wigner function through the following relations:

$$\begin{aligned} \rho(x,t) &= \frac{2}{(2\pi)^3}\int d^3\mathbf{k} f_w(\mathbf{k},x) \\ j(x,t) &= -\frac{2e}{(2\pi)^3}\int d^3\mathbf{k}\frac{\hbar k}{m} f_w(\mathbf{k},x) \end{aligned} \tag{11}$$

Figures 18 and 19 display the results of a ***116 GHz*** self-excited oscillation. Figure 18 shows the total current through the load, which included contributions from the circuit, and device capacitance as well as the particle current obtained from equation (12). The current in figure 19 is the particle current. In both cases we are plotting the magnitude of the current density. Figures 18 and 19 also display the magnitude of the potential energy drop as well as the magnitude of the applied bias in units of meV.

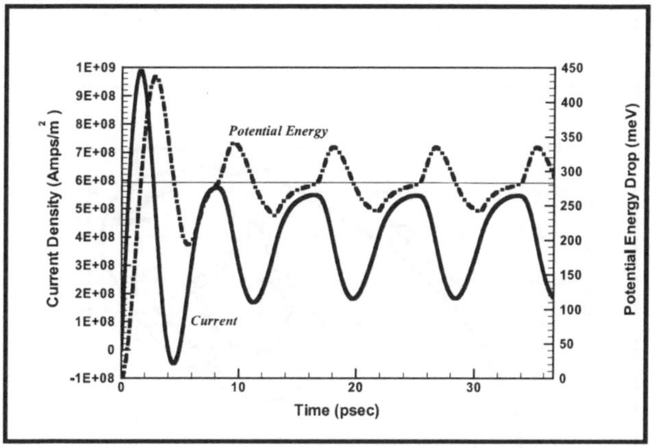

Figure 18. The time-dependent current density, potential energy and applied bias, for a 116 GHz oscillation.

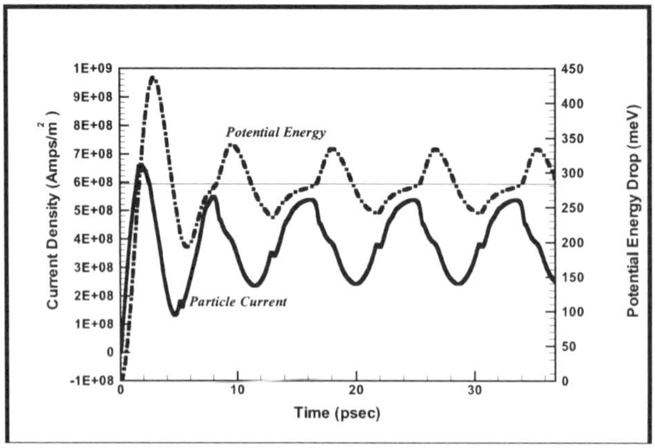

Figure 19. The time-dependent particle current, potential energy and applied bias for a 116 GHz oscillation.

The nature of the oscillation is best described by examining figure 19. Consider that point in time where the potential energy drop is at a minimum, for example $t \approx 22$ *psec*. Beyond this time with the potential energy increasing, as a result of the applied bias, there is an increase in current through the RTD. When the potential energy across the RTD reaches its threshold value there is a drop in particle current (there is also a drop in total current) and corresponding increase in potential; the increase corresponding to the presence of the NDR region. After reaching its peak value the potential energy decreases, as does the particle current. You will also notice that as the potential energy decreases in value below the threshold voltage, the particle current, rather than continuing to decrease, begins to increase; as though there was a second hysteretic NDR region.

The extent of the hysteresis is shown in figure 20. This curve is a dynamic particle current versus potential energy drop distribution; and the way to interpret this is to imagine that the curve is traced in clockwise fashion starting from the origin, and continuing until the inner loop is reached, after which it repeats ad infinitum.

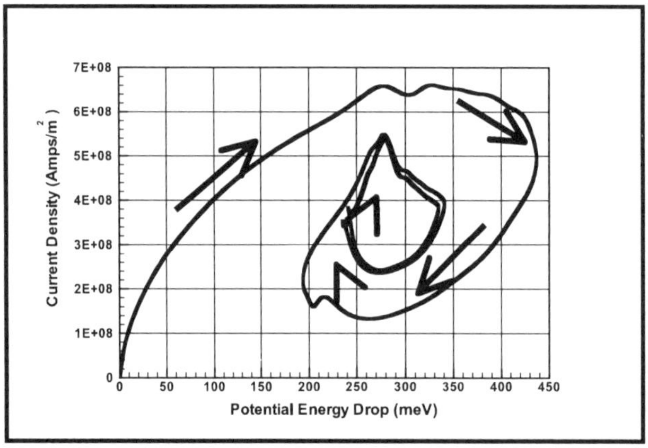

Figure 20. The particle current density versus potential energy drop as obtained from the data of figure 19.

The charge and potential distribution responsible for this time dependence is shown in the next sequence of diagrams. For example, figure 21 displays a time sequence of the potential energy as a function of distance (also shown is the doping profile for this calculation), for that portion of the cycle where the potential energy starts from a value near –240 meV and decreases to –280 meV. During this time sequence the RTD is in the low energy positive resistance portion of its cycle, and the device-plus-circuit behaves as a lossy oscillator. During this time duration, which occurs over approximately 3.5 ps there is a gradual and substantial build-up of charge in the quantum well as shown in figure 22. Accompanying this charge build-up is a movement of the downstream depletion zone toward the collector contact region.

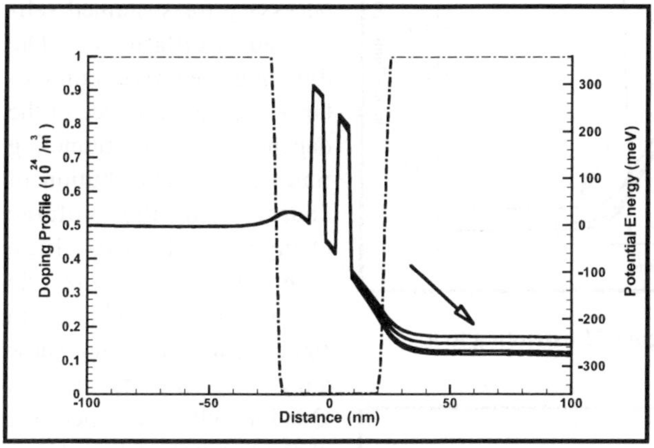

Figure 21. The potential energy and doping distribution for a duration of approximately 22-25ps, for the data of figure 18. The arrow designates the direction of the energy changes.

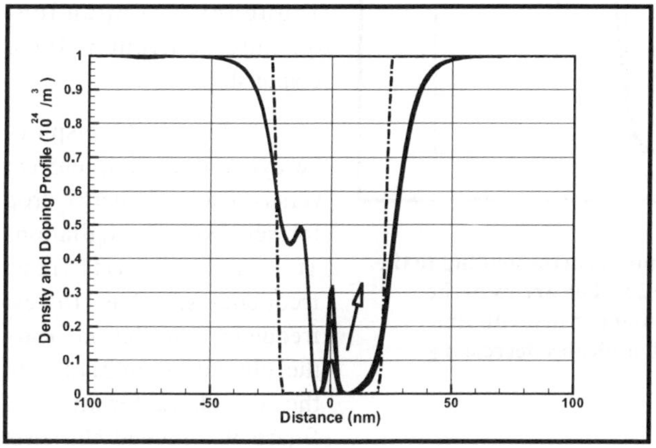

Figure 22. The carrier and doping density corresponding to the time duration associated with figure 21. The arrow designates the direction of charge build-up with increasing time.

At the threshold for negative differential resistance there is a rapid change in potential energy across the device, with the potential energy decreasing to approximately –340 meV, across a time duration of approximately 1~2 ps. This continued decrease is shown in figure 23. As displayed in figure 24 the rapid change in potential energy is accompanied by a loss of charge within the quantum well and corresponding increases in charge in the 'notch' region on the emitter side of the barrier. This change in the distribution of charge does not occur instantaneously, but occurs over an approximately 1~2 ps time duration.

Recovery, with the potential energy changing from ~-340 meV to ~-240 meV occurs over a time duration of ~2-3 ps. The change in the potential distribution is shown in figure 25 The change in the charge distribution is shown in figure 26 where we see the reduction in charge within the 'notch' potential region and a corresponding increase in charge within the quantum well. The time required for the charge to build-up within the quantum well appears to be longer than the time required to the loss of charge within the well.

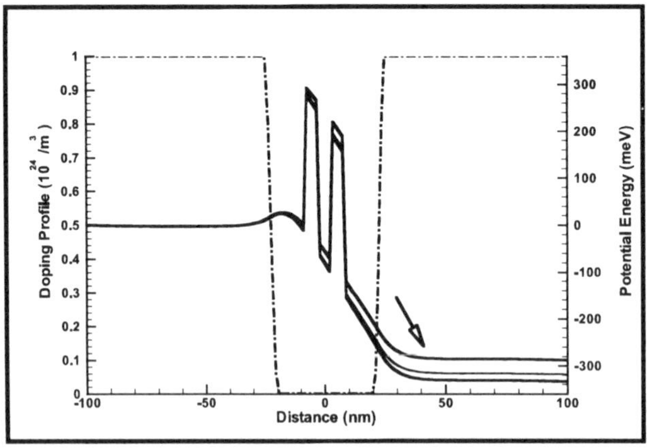

Figure 23. A continuation of figure 21 extending to a time ~ 26.5 ps.,

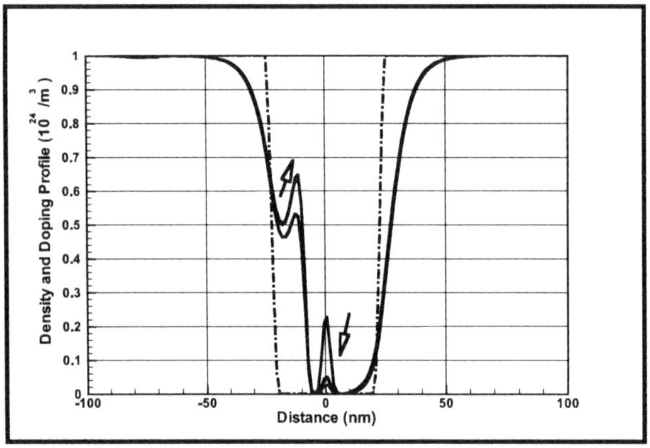

Figure 24. The density distribution corresponding to the potential distribution of figure 23. The arrow at the 'notch' region indicates increasing density with time. The arrow in the quantum well indicates decreasing charge with time.

The presence of the oscillation speaks to the issue of negative conductance, a feature necessary for sustained self-excited oscillations. The dynamic negative conductance is a consequence of the dependence of the tunneling times on the distribution of charge and the voltage changes, and bears little resemblance to the dc characteristic displayed in figure 9, which did not show any hysteresis. Qualitatively, we expect the characteristics of the charge distribution to be relatively insensitive to frequency. There should be a quantitative dependence and this should manifest itself in frequency dependent dynamic current-voltage relationships.

Figure 27 displays the dynamic particle current versus potential energy drop for self-excited oscillations at four different frequencies. The lowest frequency oscillation bears the closest resemblance to the dc characteristics of figure 9. Indeed the first panel displays some interesting structure on the downswing, where we see an increase in current for decreasing values of the potential energy below the threshold value-followed by decreases in particle current with further decreases in potential energy. Very similar behavior is displayed for the 11.5 GHz, with three differences. First there is a decrease in the magnitude of the negative differential resistance; second there is a broadening of the hysteresis loop; third there is a decrease in the dynamic peak-to-valley current ratio. Both of these developments are present in the 42.2 GHz and 71.6 GHz, and also in the 116 GHz oscillation of figure 19, with an even further decrease in the magnitude of the NDR and a broadening of the hysteresis. Indeed from the perspective of circuit analysis, the upper frequency of the self-excited oscillation may be due in part to the reduction in

NDR and the reduction in the peak-to-valley current ratio. What is needed is a careful study of the role of tunneling times under dynamic circuit controlled oscillations, in controlling these two circuit parameters.

As discussed earlier there is interesting structure in the dynamic current-voltage characteristics, particularly with the looping of the current on the downswing of the potential drop. Figure 28 provides the details, where, starting at approximately 250 ps, we follow the particle current as the potential energy drop is increased from its minimum value. At first there is an increase in particle current, followed, at threshold by a sudden (on this time scale) drop in current and a corresponding increase in the potential energy drop. As the potential energy drop decreases there is a reduction in the particle current. Just before the potential energy drop reaches a minimum, there is a sudden and rapid increase in particle current, followed by a decrease in current density and then a repeat of the cycle. This is the nature of the hysteretic structure seen in figure 27.

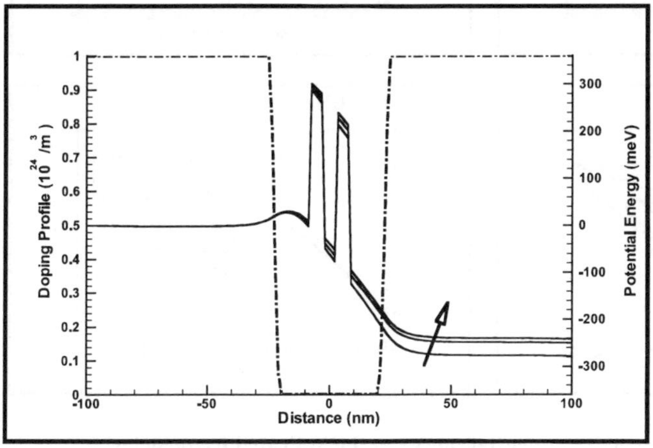

Figure 25. The potential distribution during the recovery phase of the oscillation.

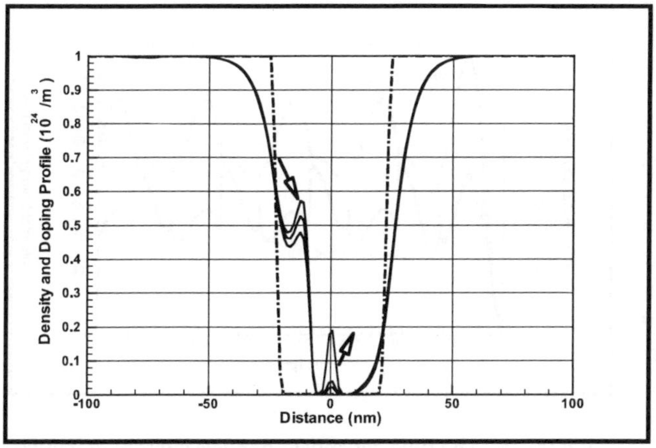

Figure 26. The charge distribution associated with the data of figure 25.

The computational results suggest that hysteresis is a time dependent phenomena and that dc hysteresis is absent. How then can we account for the experimental results that show dc hysteresis? We can only speculate. Our computations assume an energy, momentum and history independent scattering rate. Long time scattering rates could introduce hysteretic IV curves on a time scale that would be classified as dc. But this is speculative. And the simple answer is that we do not know the origin of low frequency hysteresis.

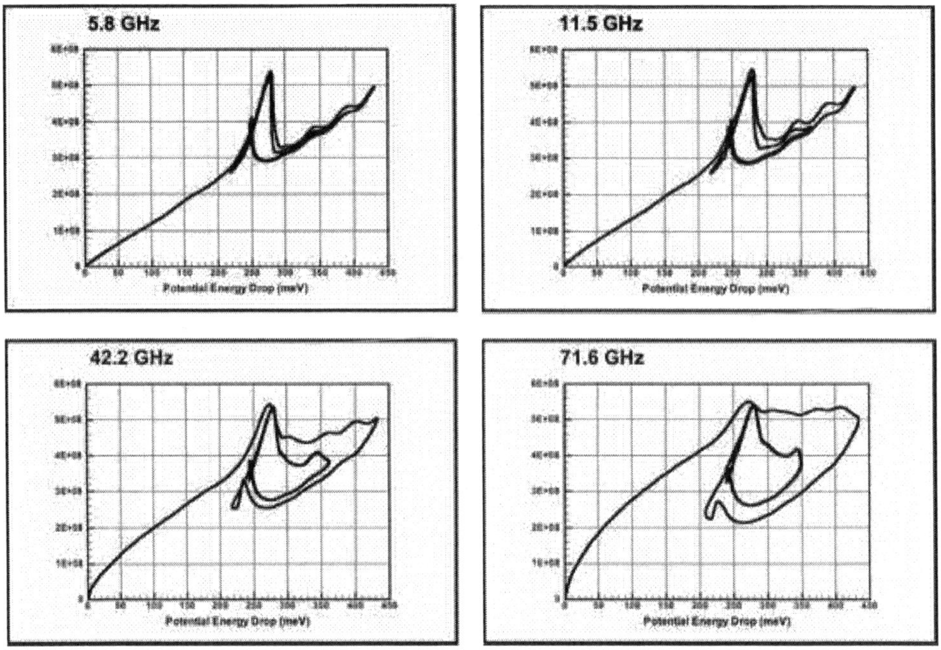

Figure 27. Particle current density vs. potential energy at the indicated frequencies.

7. Summary

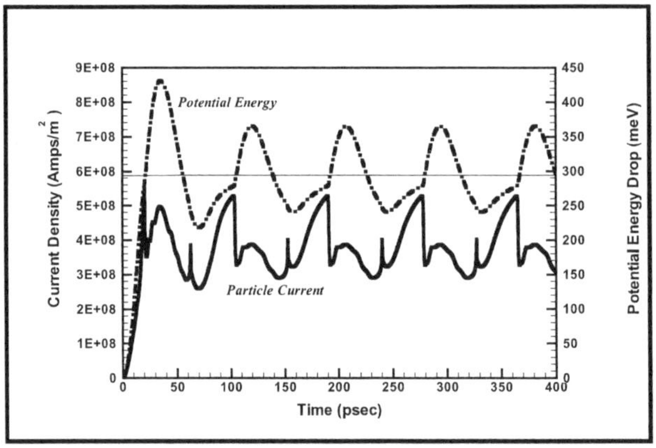

Figure 28. The time dependent particle current, potential energy drop, and the applied bias, for an 11.5 GHz self-excited oscillation.

As we have known, since the early '60s when simulations tracked the control of space charge contributions on semiconductor structures, self-sustained oscillations are possible and desired in devices sustaining negative differential conductivity. For the case of barrier devices it is, of course, quantum transport that governs the ways the carriers respond to external excitations. But there are conceptual difficulties with the picture we are using. After all we are still dealing with a 'jellium' model of transport, which appears to work better then we thought. We still have not included an appropriate transient contact model, with discrete atoms in the Wigner formulation; and detailed device circuit interactions have not been simulated. Furthermore, multiple bands, variable spatial dependent effective mass, multiple carrier species, multidimensional transient transport, remains work that will be addressed in the future. But it is clear that transient studies elucidate the physics of transport for barrier devices within the range of approximations we have accepted.

Acknowledgements

The authors are grateful for discussions with D. K. Ferry and J. P. Kreskovsky. This work was supported by the Office of Naval Research.

Appendix-Discrete Equations

A1. The Discrete Governing Equations

The relevant dimensionless Wigner equation is:

$$\frac{\partial f_w(\hat{\tau},\chi,\kappa)}{\partial \hat{\tau}} + \kappa \frac{\partial f_w(\hat{\tau},\chi,\kappa)}{\partial \chi} - \frac{\partial \phi}{\partial \chi}\frac{\partial f_w(\hat{\tau},\chi,\kappa)}{\partial \kappa} + \frac{f_w(\hat{\tau},\chi,\kappa) - \hat{f}(\hat{\tau},\chi,\kappa)}{\tau}$$
$$+\frac{i}{\pi}\underset{L->\infty}{Lim}\int_{L/\lambda}^{L/\lambda} d\xi\left[\psi(\chi+\xi)-\psi(\chi-\xi)\right]\int dk' f_w(\hat{\tau},\chi,\kappa')exp2i(\kappa'-\kappa)\xi = 0 \tag{A1}$$

The time, space and momentum increments are:

$$\hat{\tau}->n\Delta T$$
$$\chi->\left(r-\frac{N_{POS}+1}{2}\right)\Delta_\chi \tag{A2}$$
$$\kappa->\left(\ell-\frac{N_{MOM}+1}{2}\right)\Delta_\kappa$$

With the indices beginning at the lower left corner of a coordinate system, the spatial index varies in the vertical direction from $r=1$ to $r=N_{POS}$ and the momentum index varies in the horizontal direction from $\ell=1$ to $\ell=N_{MOM}$. The discrete form of the Wigner function is identified at specific values of ℓ, and at spatial points r. The discrete form of the Wigner equation is defined at points halfway between the spatial grid points. Wigner function boundary conditions are specified at $r=1$, $\ell>(N_{MOM}+1)/2$, and at $r=N_{POS}$, $\ell<(N_{MOM}+1)/2$.

The potential energy in equation (A1) is obtained from Poisson's equation, which in dimensionless form is:

$$\frac{\partial^2\phi}{\partial\chi^2} = -\hat{C}(\rho-\rho_0) \tag{A3}$$

Here $\hat{C}$ is a normalizing constant.

Time linearization is identified as:

$$f_w(\hat{\tau},\chi,\kappa) = f_{r,\ell}^{n+1} = f_{r,\ell}^{n} + \Delta f_{r,\ell}$$
$$\phi(\hat{\tau},\chi) = \phi_r^{n+1} = \phi_r^n + \Delta\phi_r \tag{A4}$$

The linearized (where we ignore the products $\Delta\phi\times\Delta f$) and discrete form of the Wigner equation, displayed next, depends upon whether ℓ is less than or greater than $N_{MOM}/2$. "Pairs" of position subscripts identify this. The upper/lower subscripts correspond to $\ell<N_{MOM}/2$, $\ell>N_{MOM}/2$, respectively.

$$\left(\frac{\Delta f_{\substack{r,\ell\\ r-1,\ell}} + \Delta f_{\substack{r+1,\ell\\ r,\ell}}}{2\Delta T} + \frac{\Delta f_{\substack{r,\ell\\ r-1,\ell}} + \Delta f_{\substack{r+1,\ell\\ r,\ell}}}{2\tau} \right) + \frac{\Delta_\kappa}{\Delta_\chi}\left(\ell - \frac{N_{MOM}+1}{2}\right)\left(\Delta f_{\substack{r+1,\ell\\ r,\ell}} - \Delta f_{\substack{r,\ell\\ r-1,\ell}}\right)$$

$$- \frac{\phi^n_{\substack{r+1\\ r}} - \phi^n_{\substack{r\\ r-1}}}{\Delta_\chi \Delta_\kappa} \left(\begin{array}{l} \dfrac{\left(\Delta f_{\substack{r+1,\ell+1\\ r,\ell+1}} - \Delta f_{\substack{r+1,\ell-1\\ r,\ell-1}} + \Delta f_{\substack{r,\ell+1\\ r-1,\ell+1}} - \Delta f_{\substack{r,\ell-1\\ r-1,\ell-1}}\right)\left(1-\delta_{\ell,1}-\delta_{\ell,N_{MOM}}\right)}{4} \\ + \dfrac{\left(\Delta f_{r+1,2} - \Delta f_{r+1,1} + \Delta f_{r,2} - \Delta f_{r,1}\right)\delta_{\ell,1}}{2} \\ + \dfrac{\left(\Delta f_{r,N_{MOM}} - \Delta f_{r,N_{MOM}-1} + \Delta f_{r-1,N_{MOM}} - \Delta f_{r-1,N_{MOM}-1}\right)\delta_{\ell,N_{MOM}}}{2} \end{array} \right)$$

$$- \frac{\Delta\phi_{\substack{r+1\\ r}} - \Delta\phi_{\substack{r\\ r-1}}}{\Delta_\chi \Delta_\kappa} \left(\begin{array}{l} \dfrac{\left(f^n_{\substack{r+1,\ell+1\\ r,\ell+1}} - f^n_{\substack{r+1,\ell-1\\ r,\ell-1}} + f^n_{\substack{r,\ell+1\\ r-1,\ell+1}} - f^n_{\substack{r,\ell-1\\ r-1,\ell-1}}\right)\left(1-\delta_{\ell,1}-\delta_{\ell,N_{MOM}}\right)}{4} \\ + \dfrac{\left(f_{r+1,2} - f_{r+1,1} + f_{r,2} - f_{r,1}\right)\delta_{\ell,1}}{2} \\ + \dfrac{\left(f_{r,N_{MOM}} - f_{r,N_{MOM}-1} + f_{r-1,N_{MOM}} - f_{r-1,N_{MOM}-1}\right)\delta_{\ell,N_{MOM}}}{2} \end{array} \right)$$

$$+ \frac{\Delta_\kappa}{4} \sum_{\ell'=1}^{N_{MOM}} \left\{ G_{\substack{r+1,\ell,\ell'\\ r,\ell,\ell'}} \Delta f_{\substack{r+1,\ell'\\ r,\ell'}} + G_{\substack{r,\ell,\ell'\\ r-1,\ell,\ell'}} \Delta f_{\substack{r,\ell'\\ r-1,\ell'}} \right\} \tag{A5}$$

$$= -\left(\frac{f^n_{\substack{r,\ell\\ r-1,\ell}} - \hat{f}_{\substack{r,\ell\\ r-1,\ell}}}{2\tau} + \frac{f^n_{\substack{r+1,\ell\\ r,\ell}} - \hat{f}_{\substack{r+1,\ell\\ r,\ell}}}{2\tau} \right) - \frac{\Delta_\kappa}{\Delta_\chi}\left(\ell - \frac{N_{MOM}+1}{2}\right)\left(f^n_{\substack{r+1,\ell\\ r,\ell}} - f^n_{\substack{r,\ell\\ r-1,\ell}}\right)$$

$$+ \frac{\phi^n_{\substack{r+1\\ r}} - \phi^n_{\substack{r\\ r-1}}}{\Delta_\chi \Delta_\kappa} \left(\begin{array}{l} \dfrac{\left(f^n_{\substack{r+1,\ell+1\\ r,\ell+1}} - f^n_{\substack{r+1,\ell-1\\ r,\ell-1}} + f^n_{\substack{r,\ell+1\\ r-1,\ell+1}} - f^n_{\substack{r,\ell-1\\ r-1,\ell-1}}\right)\left(1-\delta_{\ell,1}-\delta_{\ell,N_{MOM}}\right)}{4} \\ + \dfrac{\left(f^n_{r+1,2} - f^n_{r+1,1} + f^n_{r,2} - f^n_{r,1}\right)\delta_{\ell,1}}{2} \\ + \dfrac{\left(f^n_{r,N_{MOM}} - f^n_{r,N_{MOM}-1} + f^n_{r-1,N_{MOM}} - f^n_{r-1,N_{MOM}-1}\right)\delta_{\ell,N_{MOM}}}{2} \end{array} \right)$$

$$- \frac{\Delta_\kappa}{4} \sum_{\ell'=1}^{N_{MOM}} \left\{ G_{\substack{r+1,\ell,\ell'\\ r,\ell,\ell'}} f^n_{\substack{r+1,\ell'\\ r,\ell'}} + G_{\substack{r,\ell,\ell'\\ r-1,\ell,\ell'}} f^n_{\substack{r,\ell'\\ r-1,\ell'}} \right\}$$

Here, $G_{r,\ell,\ell'}$ is the discrete from of the Wigner integral:

$$\frac{i}{\pi}\underset{L->\infty}{Lim}\int_{L/\lambda}^{L/\lambda} d\xi\left[\psi(\chi+\xi)-\psi(\chi-\xi)\right]exp2i(\kappa'-\kappa)\xi->G_{r,\ell,\ell'} \tag{A6}$$

For square barriers an exact integration is possible, as is carried out in the discussion portion of this chapter. The delta function in equation (A5) distinguishes the way the derivative is taken with respect to momentum at the momentum boundaries and interior of the computational domain.

On the spatial boundaries:

$$\begin{aligned}
&\alpha\left(f^{n+1}_{1,\ell>\frac{N_{MOM}+1}{2}}-h^{n+1}_{1,\ell}\right)+\beta\left(f^{n+1}_{2,\ell>\frac{N_{MOM}+1}{2}}-f^{n+1}_{1,\ell>\frac{N_{MOM}+1}{2}}\right)=0\\
&\alpha\left(f^{n+1}_{N_{POS},\ell<\frac{N_{MOM}+1}{2}}-h^{n+1}_{N_{POS},\ell}\right)+\gamma\left(f^{n+1}_{N_{POS},\ell<\frac{N_{MOM}+1}{2}}-f^{n+1}_{N_{POS}-1,\ell<\frac{N_{MOM}+1}{2}}\right)=0
\end{aligned} \tag{A7}$$

For function boundary conditions: $\alpha=1,\ \beta=\gamma=0$. For derivative boundary conditions: $\alpha=0,\ \beta=\gamma=1$. Mixed boundary conditions are combinations of these conditions. The functions $h^{n+1}_{1,\ell},\ h^{n+1}_{N_{POS},\ell}$ are either specified, or are solutions to another equation.

The discrete form of Poisson's equation, which is written at discrete grid points, rather than half grid points, is:

$$\frac{\Delta\phi_r}{\Delta T_{Poisson}}+\frac{\Delta\phi_{r+1}-2\Delta\phi_r+\Delta\phi_{r-1}}{\Delta\chi^2}+C\Delta\rho_r=-C\left(\rho_r^n-\rho_{0r}\right)-\frac{\phi^n_{r+1}-2\phi^n_r+\phi^n_{r-1}}{\Delta\chi^2} \tag{A8}$$

In the above equation C is a constant and $\Delta T_{Poisson}$ is a numerical time constant that is made large enough to have no effect on the transient or steady state physics. Using trapezoidal integration:

$$\Delta\rho_r=\frac{1}{2}\sum_{\ell=1}^{N_{MOM}}\left(\Delta f_{r,\ell}+\Delta f_{r,\ell+1}\right) \tag{A9}$$

The potential energy associated with Poisson's equation is solved directly with the Wigner distribution function and so a structure consisting of N_{MOM} momentum grid points and N_{POS} position grid points, consists of $\left(N_{MOM}+1\right)^2\times N_{POS}{}^2$ elements.

Boundary conditions to the discrete from of Poisson's equation are:

$$\begin{aligned}
&\alpha_{POISSON}\left(\phi_1^{n+1}-p_1^{n+1}\right)+\beta_{POISSON}\left(\phi_2^{n+1}-\phi_1^{n+1}-\partial p_1^{n+1}\right)=0;\\
&\alpha_{POISSON}\left(\phi^{n+1}_{N_{POS}}-p^{n+1}_{N_{POS}}\right)+\beta_{POISSON}\left(\phi^{n+1}_{N_{POS}}-\phi^{n+1}_{N_{POS}-1}-\partial p^{n+1}_{N_{POS}}\right)=0;
\end{aligned} \tag{A10}$$

For function boundary conditions: $\alpha_{POISSON}=1,\ \beta_{POISSON}=\gamma_{POISSON}=0$. The functions $p_1^{n+1},\ p^{n+1}_{N_{POS}},\ \partial p_1^{n+1},\ \partial p^{n+1}_{N_{POS}}$ are either specified, or are solutions to another equation. For all of the calculations discussed in this chapter, function boundary conditions were imposed, with the potential energy at the emitter boundary set to zero. For the collector boundary the potential energy, was set to either: (a) a fixed time independent value, as in the case of the dc *IV* characteristic computation; (b) a specified time dependent value, as in the switching calculations, discussed in an earlier section; or (c) was the output of a coupled circuit equation. An illustration of the latter is the operation of the RTD as a relaxation oscillator.

A2. The Wigner Matrix

The problem them reduces to inverting the matrix:

$$\mathbf{W}(n)\bullet\Delta\mathbf{V}=\mathbf{R}(n) \tag{A11}$$

Here n represents the time step and the iteration continues until the vector $\Delta\mathbf{V}$ is smaller than a predetermined value. The right hand side vector $\mathbf{R}(n)$ is discussed in section **A4**.

The matrix $\mathbf{W}(n)$ has the structure:

$$\mathbf{W}(n)=\begin{pmatrix} \mathbf{B}(1) & \mathbf{C}(1) & 0 & \bullet & \bullet & 0 \\ \mathbf{A}(2) & \mathbf{B}(2) & \mathbf{C}(2) & 0 & \bullet & 0 \\ \bullet & \bullet & \bullet & \bullet & \bullet & \bullet \\ \bullet & \bullet & \bullet & \bullet & \bullet & \bullet \\ 0 & \bullet & 0 & \mathbf{A}(N_{POS}-1) & \mathbf{B}(N_{POS}-1) & \mathbf{C}(N_{POS}-1) \\ 0 & \bullet & \bullet & 0 & \mathbf{A}(N_{POS}) & \mathbf{B}(N_{POS}) \end{pmatrix} \tag{A12}$$

And

$$\Delta\mathbf{V}=\left(\Delta\mathbf{V}(1)\quad \Delta\mathbf{V}(2)\quad \bullet\quad \bullet\quad \Delta\mathbf{V}(N_{POS}-1)\quad \Delta\mathbf{V}(N_{POS})\right)^{T} \tag{A13}$$

Here:

$$\begin{aligned} \Delta\mathbf{V}(1)&=\left(\Delta f_{1,1}\quad \Delta f_{1,2}\quad \bullet\quad \bullet\quad \Delta f_{1,N_{MOM}}\quad \Delta\phi_1\right)^{T} \\ \Delta\mathbf{V}(2)&=\left(\Delta f_{2,1}\quad \Delta f_{2,2}\quad \bullet\quad \bullet\quad \Delta f_{2,N_{MOM}}\quad \Delta\phi_2\right)^{T} \\ \Delta\mathbf{V}(N_{POS})&=\left(\Delta f_{N_{POS},1}\quad \Delta f_{N_{POS},2}\quad \bullet\quad \bullet\quad \Delta f_{N_{POS},N_{MOM}}\quad \Delta\phi_{N_{POS}}\right)^{T} \end{aligned} \tag{A14}$$

For single species, single band transport the sub-matrices, $\mathbf{A}(i)$, $\mathbf{B}(i)$, $\mathbf{C}(i)$ are of dimension $(N_{MOM}+1)\times(N_{MOM}+1)$. The entries: $i,j\ =\ 1,2,\ldots N_{MOM}$, are obtained from the Wigner, equation. The last row and column are obtained from Poisson's equation. The sub-matrices $\mathbf{B}(1)$, $\mathbf{C}(1)$, contain the Wigner function conditions at the emitter boundary, while the sub-matrices $\mathbf{A}(N_{POS})$, $\mathbf{B}(N_{POS})$ contain Wigner function conditions at the collector boundary. The structure of all sub-matrices is determined from the discrete form of the Wigner equation at points halfway between the grid points, and Poisson's equation at the grid points. Because the Wigner function is defined at points halfway between the spatial grid points, in the absence of the Poisson contribution, the top halves of the matrices $\mathbf{B}(i)$, $\mathbf{C}(i)$ are coupled, and the bottom half of the matrices $\mathbf{A}(i)$, $\mathbf{B}(i)$ are coupled. The slanted lines in equation (A15), below, emphasize this coupling for $N_{POS}=4$. For $\ell \neq 1,\ N_{MOM}$, Poisson's equation in the discrete form of equation (A8) couples $\mathbf{A}(i),\ \mathbf{B}(i),\ \mathbf{C}(i)$.

$$\begin{pmatrix} \begin{pmatrix} \ddots\ \mathbf{B}(1)\ \ddots \\ BC \end{pmatrix} & \begin{pmatrix} \ddots\ \mathbf{C}(1)\ \ddots \\ BC \end{pmatrix} & (0) & (0) \\ \begin{pmatrix} \mathbf{A}(2) \\ \ddots\ \ddots \end{pmatrix} & \begin{pmatrix} \ddots\ \mathbf{B}(2)\ \ddots \\ \ddots\ \ddots \end{pmatrix} & \begin{pmatrix} \ddots\ \mathbf{C}(2)\ \ddots \\ \end{pmatrix} & (0) \\ (0) & \begin{pmatrix} \mathbf{A}(3) \\ \ddots\ \ddots \end{pmatrix} & \begin{pmatrix} \ddots\ \mathbf{B}(3)\ \ddots \\ \ddots\ \ddots \end{pmatrix} & \begin{pmatrix} \ddots\ \mathbf{C}(3)\ \ddots \\ \end{pmatrix} \\ (0) & (0) & \begin{pmatrix} BC \\ \ddots\ \mathbf{A}(4)\ \ddots \end{pmatrix} & \begin{pmatrix} BC \\ \ddots\ \mathbf{B}(4)\ \ddots \end{pmatrix} \end{pmatrix} \tag{A15}$$

A.3 The Matrix Elements

As an example, we set $N_{MOM}=4,\ N_{POS}=4$. Carriers with positive values of momentum are represented by the right hand part of the matrix $\mathbf{W}(n),\ \ell=3,4$; whereas carriers with negative values of momentum are represented by the left hand part of the matrix, $\ell=1,2$. Boundary conditions for positive values of momentum are represented through submatrices with the index "1". For this case, the interior matrices, $\mathbf{B}(2),\ \mathbf{C}(2)$, each 5×5, contain the coupling of $\Delta f_{2,1},\ \Delta f_{3,1}$ and $\Delta f_{2,2},\ \Delta f_{3,2}$; whereas the matrices, $\mathbf{A}(2),\ \mathbf{B}(2)$, also 5×5, contain the coupling of $\Delta f_{1,3},\ \Delta f_{2,3}$ and $\Delta f_{1,4},\ \Delta f_{2,4}$. The first four rows of $\mathbf{B}(2)$ contain non- zero elements; the first two rows of $\mathbf{A}(2)$ and the second two rows of $\mathbf{C}(2)$ contain *only* zero elements. Similar comments apply to the sub-matrices $\mathbf{A}(3),\ \mathbf{B}(3),\ \mathbf{C}(3)$. To see how entries are placed into the sub-matrices, consider the case $\ell=1,2$. The entries are:

$$\mathbf{A}(m\neq4)_{1,j}=0;$$

$$\begin{aligned}\mathbf{B}(m\neq4)_{1,j} &= \left\{\frac{\Delta_\kappa}{4}G_{m,1,j}+\frac{1}{2}\left(\frac{1}{\Delta T}+\frac{1}{\tau}\right)\delta_{1,j}+\frac{3}{2}\frac{\Delta_\kappa}{\Delta_\chi}\delta_{1,j}+\frac{\phi^n_{m+1}-\phi^n_m}{2\Delta_\chi\Delta_\kappa}\left(\delta_{1,j}-\delta_{2,j}\right)\right\}\left\{1-\delta_{j,5}\right\} \\ &\quad+\left\{\frac{f^n_{m+1,2}-f^n_{m+1,1}+f^n_{m,2}-f^n_{m,1}}{2\Delta_\chi\Delta_\kappa}\right\}\delta_{j,5};\end{aligned} \tag{A16}$$

$$\begin{aligned}\mathbf{C}(m)_{1,j} &= \left\{\frac{\Delta_\kappa}{4}G_{m+1,1,j}+\frac{1}{2}\left(\frac{1}{\Delta T}+\frac{1}{\tau}\right)\delta_{1,j}-\frac{3}{2}\frac{\Delta_\kappa}{\Delta_\chi}\delta_{1,j}+\frac{\phi^n_{m+1}-\phi^n_m}{2\Delta_\chi\Delta_\kappa}\left(\delta_{1,j}-\delta_{2,j}\right)\right\}\left\{1-\delta_{j,5}\right\} \\ &\quad-\left\{\frac{f^n_{m+1,2}-f^n_{m+1,1}+f^n_{m,2}-f^n_{m,1}}{2\Delta_\chi\Delta_\kappa}\right\}\delta_{j,5};\end{aligned}$$

$$\mathbf{A}(m \neq 4)_{2,j} = 0;$$

$$\mathbf{B}(m \neq 4)_{2j}$$

$$= \left\{ \frac{\Delta_\kappa}{4} G_{m,2,j} + \frac{\phi_{m+1}^n - \phi_m^n}{4\Delta_\chi \Delta_\kappa} \left(\delta_{1,j} - \delta_{3,j} \right) + \frac{1}{2} \left(\frac{1}{\Delta T} + \frac{1}{\tau} \right) \delta_{2.j} + \frac{1}{2} \frac{\Delta_\kappa}{\Delta_\chi} \delta_{2.j} \right\} \{1 - \delta_{1,j}\}$$

$$+ \left\{ \frac{f_{m+1,3}^n - f_{m+1,1}^n + f_{m,3}^n - f_{m,1}^n}{4\Delta_\chi \Delta_\kappa} \right\} \delta_{1,j}; \qquad \text{(A17)}$$

$$\mathbf{C}(m)_{2j}$$

$$= \left\{ \frac{\Delta_\kappa}{4} G_{m+1,2,j} + \frac{\phi_{m+1}^n - \phi_m^n}{4\Delta_\chi \Delta_\kappa} \left(\delta_{1,j} - \delta_{3,j} \right) + \frac{1}{2} \left(\frac{1}{\Delta T} + \frac{1}{\tau} \right) \delta_{2.j} - \frac{1}{2} \frac{\Delta_\kappa}{\Delta_\chi} \delta_{2.j} \right\} \{1 - \delta_{1,j}\}$$

$$- \left\{ \frac{f_{m+1,3}^n - f_{m+1,1}^n + f_{m,3}^n - f_{m,1}^n}{4\Delta_\chi \Delta_\kappa} \right\} \delta_{1,j};$$

The symmetry of the problem is dictated by the off-diagonal elements of the sub-matrices. In the absence of any field variation associated with Poisson's equation, symmetry is determined by the properties of $G_{r,\ell,\ell'}$. There is asymmetry whenever there is a local finite field.

The boundary conditions for the top boundary yield the sub-matrices $\mathbf{A}(N_{POS})$, $\mathbf{B}(N_{POS})$. For $\ell = 1,2$, we have:

$$\mathbf{A}(4)_{1,j} = -\gamma \delta_{j,1}, \ \mathbf{A}(4)_{2,j} = -\gamma \delta_{j,2};$$
$$\mathbf{B}(4)_{1,j} = (\alpha + \gamma)\delta_{j,1}, \ \mathbf{B}(4)_{2,j} = (\alpha + \gamma)\delta_{j,2} \qquad \text{(A18)}$$

We consider next the entries for $\ell = 3,4$. The boundary conditions for $r = 1$ yield the following entries:

$$\mathbf{B}(1)_{3,j} = (\alpha - \beta)\delta_{3,j}, \ \mathbf{B}(1)_{4,j} = (\alpha - \beta)\delta_{4,j};$$
$$\mathbf{C}(1)_{3,j} = \beta \delta_{3,j}, \ \mathbf{C}(1)_{4,j} = \beta \delta_{4,j};$$
(A19)

The coupling of $\ell = 3,4$ yield the entries for $\mathbf{A}(m)$, $\mathbf{B}(m)$: $m = 2,3,4$ yield:

$$\mathbf{A}(m \neq 1)_{3j}$$
$$= \left\{ \frac{\Delta_\kappa}{4} G_{m-1,3,j} + \frac{1}{2}\left(\frac{1}{\Delta T} + \frac{1}{\tau} \right)\delta_{3,j} - \frac{1}{2}\frac{\Delta_\kappa}{\Delta_\chi}\delta_{3,j} + \frac{\phi_m^n - \phi_{m-1}^n}{4\Delta_\chi \Delta_\kappa}\left(\delta_{2,j} - \delta_{4,j}\right) \right\}\left\{1 - \delta_{j,5}\right\}$$
$$+ \left\{ \frac{f_{m,4}^n - f_{m,2}^n + f_{m-1,4}^n - f_{m-1,2}^n}{4\Delta_\chi \Delta_\kappa} \right\}\delta_{j,5};$$

$$\mathbf{B}(m \neq 1)_{3,j} \tag{A20}$$
$$= \left\{ \frac{\Delta_\kappa}{4} G_{m,3,j} + \frac{1}{2}\left(\frac{1}{\Delta T} + \frac{1}{\tau} \right)\delta_{3,j} + \frac{1}{2}\frac{\Delta_\kappa}{\Delta_\chi}\delta_{3,j} + \frac{\phi_m^n - \phi_{m-1}^n}{4\Delta_\chi \Delta_\kappa}\left(\delta_{2,j} - \delta_{4,j}\right) \right\}\left\{1 - \delta_{j,5}\right\}$$
$$- \left\{ \frac{f_{m,4}^n - f_{m,2}^n + f_{m-1,4}^n - f_{m-1,2}^n}{4\Delta_\chi \Delta_\kappa} \right\}\delta_{j,5};$$

$$\mathbf{C}(m \neq 1)_{3,j} = 0;$$

$$\mathbf{A}(m \neq 1)_{4j}$$
$$= \left\{ \frac{\Delta_\kappa}{4} G_{m-1,4,j} + \frac{1}{2}\left(\frac{1}{\Delta T} + \frac{1}{\tau} \right)\delta_{4,j} - \frac{3}{2}\frac{\Delta_\kappa}{\Delta_\chi}\delta_{4,j} + \frac{\phi_m^n - \phi_{m-1}^n}{2\Delta_\chi \Delta_\kappa}\left(\delta_{3,j} - \delta_{4,j}\right) \right\}\left\{1 - \delta_{5,j}\right\}$$
$$+ \left\{ \frac{f_{m,4}^n - f_{m,3}^n + f_{m-1,4}^n - f_{m-1,3}^n}{2\Delta_\chi \Delta_\kappa} \right\}\delta_{5,j};$$

$$\mathbf{B}(m \neq 1)_{4j} \tag{A21}$$
$$= \left\{ \frac{\Delta_\kappa}{4} G_{m,4,j} + \frac{1}{2}\left(\frac{1}{\Delta T} + \frac{1}{\tau} \right)\delta_{4,j} + \frac{3}{2}\frac{\Delta_\kappa}{\Delta_\chi}\delta_{4,j} + \frac{\phi_m^n - \phi_{m-1}^n}{2\Delta_\chi \Delta_\kappa}\left(\delta_{3,j} - \delta_{4,j}\right) \right\}\left\{1 - \delta_{5,j}\right\}$$
$$- \left\{ \frac{f_{m,4}^n - f_{m,3}^n + f_{m-1,4}^n - f_{m-1,3}^n}{2\Delta_\chi \Delta_\kappa} \right\}\delta_{5,j};$$

$$\mathbf{C}(m \neq 1)_{4,j} = 0;$$

For the $N_{MOM} + 1$ row entry:

$$\mathbf{B}(1)_{5,j} = (\alpha_{POISSON} - \beta_{POISSON})\delta_{5j};\ \mathbf{C}(1)_{5,j} = \beta_{POISSON}\delta_{5j};\ \mathbf{A}(m \neq 4)_{5,j} = \frac{1}{\Delta\chi^2}\delta_{5,j};$$
$$\mathbf{B}(m \neq 1,4)_{5,j} = C\left(1 - \frac{\delta_{1,j} + \delta_{4,j}}{2}\right)\left(1 - \delta_{5,j}\right) + \left(\frac{1}{\Delta T_{Poisson}} - \frac{2}{\Delta\chi^2}\right)\delta_{5,j}; \tag{A22}$$
$$\mathbf{C}(m \neq 1)_{5,j} = \frac{1}{\Delta\chi^2}\delta_{5,j};\ \mathbf{A}(4)_{5,j} = -\gamma_{POISSON}\delta_{5,j};\ \mathbf{B}(4)_{5,j} = (\alpha_{POISSON} + \gamma_{POISSON})\delta_{5,j}$$

Notice that while there is coupling between the top halves and bottom halves of each pair of matrices, the symmetry discussed earlier is retained here.

A4. The Right Hand Side of the Matrix

The above discussion identifies the elements of the matrix $\mathbf{W}(n)$, see equation (A12). The right hand side of equation (A10) for the case we have been considering is:

$$\mathbf{R}(n)=\begin{pmatrix}\mathbf{RB}(1) & \mathbf{RC}(1) & \mathbf{0} & \mathbf{0}\\ \mathbf{RA}(2) & \mathbf{RB}(2) & \mathbf{RC}(2) & \mathbf{0}\\ \mathbf{0} & \mathbf{RA}(3) & \mathbf{RB}(3) & \mathbf{RC}(3)\\ \mathbf{0} & \mathbf{0} & \mathbf{RA}(3) & \mathbf{RB}(4)\end{pmatrix}\begin{pmatrix}\mathbf{V}(1)\\ \mathbf{V}(2)\\ \mathbf{V}(3)\\ \mathbf{V}(4)\end{pmatrix}$$
$$+\frac{1}{2\tau}\begin{pmatrix}Upper[\mathbf{V}_0(1)]+Upper[\mathbf{V}_0(2)]\\ Lower[\mathbf{V}_0(1)]+\mathbf{V}_0(2)+Upper[\mathbf{V}_0(3)]\\ Lower[\mathbf{V}_0(2)]+\mathbf{V}_0(3)+Upper[\mathbf{V}_0(4)]\\ Lower[\mathbf{V}_0(3)]+Lower[\mathbf{V}_0(3)]\end{pmatrix}+\begin{pmatrix}\mathbf{U}(n,1)\\ \mathbf{U}(n,2)\\ \mathbf{U}(n,3)\\ \mathbf{U}(n,4)\end{pmatrix} \tag{A23}$$

Here:

$$\mathbf{V}(i)=\begin{pmatrix}f_{i,1}^n & f_{i,2}^n & f_{i,3}^n & f_{i,4}^n & \phi_i^n\end{pmatrix}^T,\ \mathbf{V}_0(i)=\begin{pmatrix}\hat{f}_{i,1} & \hat{f}_{i,2} & \hat{f}_{i,3} & \hat{f}_{i,4} & 0\end{pmatrix}^T$$
$$Upper[\mathbf{V}_0(i)]=\begin{pmatrix}\hat{f}_{i,1} & \hat{f}_{i,2} & 0 & 0 & 0\end{pmatrix}^T,$$
$$Lower[\mathbf{V}_0(i)]=\begin{pmatrix}0 & 0 & \hat{f}_{i,3} & \hat{f}_{i,4} & 0\end{pmatrix}^T, \tag{A24}$$
$$\mathbf{U}(n,1)=\begin{pmatrix}0 & 0 & h_{1,3}^n+\Delta h(n)_{1,3}, & h_{1,4}^n+\Delta h(n)_{1,4} & \hat{\phi}_1^{n+1}\end{pmatrix}^T$$
$$\mathbf{U}(n,2)=\begin{pmatrix}0 & 0 & 0 & 0 & C\rho_{0,2}\end{pmatrix}^T,\ \mathbf{U}(n,3)=\begin{pmatrix}0 & 0 & 0 & 0 & C\rho_{0,3}\end{pmatrix}^T$$
$$\mathbf{U}(n,4)=\begin{pmatrix}h_{4,1}^n+\Delta h(n)_{4,1} & h_{4,2}^n+\Delta h(n)_{4,2} & 0 & 0 & \hat{\phi}_4^{n+1}\end{pmatrix}^T$$

The entries for $\mathbf{U}(n,i)$ are function boundary conditions. The generalization to larger matrices is direct.

The elements of each sub-matrix are determined by the right hand side of the difference equations and their associated boundary conditions, and are:

$$\mathbf{RA}(m)_{3,j}=-\left\{\frac{\Delta_\kappa}{4}G_{m-1,3,j}+\frac{1}{2\tau}\delta_{3,j}-\frac{1}{2}\frac{\Delta_\kappa}{\Delta_\chi}\delta_{3,j}\right\}\{1-\delta_{j,5}\}$$
$$-\left\{\frac{f_{m,4}^n-f_{m,2}^n+f_{m-1,4}^n-f_{m-1,2}^n}{4\Delta_\chi\Delta_\kappa}\right\}\delta_{j,5};$$
$$\mathbf{RA}(m)_{4,j}=-\left\{\frac{\Delta_\kappa}{4}G_{m-1,4,j}+\frac{1}{2\tau}\delta_{4,j}-\frac{3}{2}\frac{\Delta_\kappa}{\Delta_\chi}\delta_{4,j}\right\}\{1-\delta_{5,j}\}$$
$$-\left\{\frac{f_{m,4}^n-f_{m,3}^n+f_{m-1,4}^n-f_{m-1,3}^n}{2\Delta_\chi\Delta_\kappa}\right\}\delta_{5,j}; \tag{A25}$$
$$\mathbf{RA}(m)_{1,j}=-\mathbf{A}(m)_{1,j},\ \mathbf{RA}(m)_{2,j}=-\mathbf{A}(m)_{2,j},\ \mathbf{RA}(m)_{5,j}=-\mathbf{A}(m)_{5,j}$$

$$
\begin{aligned}
\mathbf{RB}(m\neq 4)_{1,j} &= -\left\{\frac{\Delta_\kappa}{4}G_{m,1,j}+\frac{1}{2\tau}\delta_{1,j}+\frac{3}{2}\frac{\Delta_\kappa}{\Delta_\chi}\delta_{1,j}\right\}\{1-\delta_{j,5}\} \\
&\quad -\left\{\frac{f^n_{m+1,2}-f^n_{m+1,1}+f^n_{m,2}-f^n_{m,1}}{2\Delta_\chi\Delta_\kappa}\right\}\delta_{j,5}; \\
\mathbf{RB}(m\neq 4)_{2,j} &= -\left\{\frac{\Delta_\kappa}{4}G_{m,2,j}+\frac{1}{2\tau}\delta_{2.j}+\frac{1}{2}\frac{\Delta_\kappa}{\Delta_\chi}\delta_{2.j}\right\}\{1-\delta_{j,5}\} \\
&\quad -\left\{\frac{f^n_{m+1,3}-f^n_{m+1,1}+f^n_{m,3}-f^n_{m,1}}{4\Delta_\chi\Delta_\kappa}\right\}\delta_{j,5}; \\
\mathbf{RB}(m\neq 1)_{3,j} &= -\left\{\frac{\Delta_\kappa}{4}G_{m,3,j}+\frac{1}{2\tau}\delta_{3,j}+\frac{1}{2}\frac{\Delta_\kappa}{\Delta_\chi}\delta_{3,j}\right\}\{1-\delta_{j,5}\} \\
&\quad +\left\{\frac{f^n_{m,4}-f^n_{m,2}+f^n_{m-1,4}-f^n_{m-1,2}}{4\Delta_\chi\Delta_\kappa}\right\}\delta_{j,5}; \\
\mathbf{RB}(m\neq 1)_{4,j} &= -\left\{\frac{\Delta_\kappa}{4}G_{m,4,j}+\frac{1}{2\tau}\delta_{4,j}+\frac{3}{2}\frac{\Delta_\kappa}{\Delta_\chi}\delta_{4,j}\right\}\{1-\delta_{5,j}\} \\
&\quad +\left\{\frac{f^n_{m,4}-f^n_{m,3}+f^n_{m-1,4}-f^n_{m-1,3}}{2\Delta_\chi\Delta_\kappa}\right\}\delta_{5,j}; \\
\mathbf{RB}(1)_{3,j} &= -\mathbf{B}(1)_{3,j};\ \ \mathbf{RB}(1)_{4,j}=-\mathbf{B}(1)_{4,j}; \\
\mathbf{RB}(4)_{1,j} &= -\mathbf{B}(4)_{1,j};\ \ \mathbf{RB}(4)_{2,j}=-\mathbf{B}(4)_{2,j}; \\
\mathbf{RB}(m)_{5,j} &= -\mathbf{B}(m)_{5,j};
\end{aligned}
\tag{A26}
$$

$$
\begin{aligned}
\mathbf{RC}(m)_{1,j} &= -\left\{\frac{\Delta_\kappa}{4}G_{m+1,1,j}+\frac{1}{2\tau}\delta_{1,j}-\frac{3}{2}\frac{\Delta_\kappa}{\Delta_\chi}\delta_{1,j}\right\}\{1-\delta_{j,5}\} \\
&\quad +\left\{\frac{f^n_{m+1,2}-f^n_{m+1,1}+f^n_{m,2}-f^n_{m,1}}{2\Delta_\chi\Delta_\kappa}\right\}\delta_{j,5}; \\
\mathbf{RC}(m)_{2,j} &= -\left\{\frac{\Delta_\kappa}{4}G_{m+1,2,j}+\frac{1}{2\tau}\delta_{2.j}-\frac{1}{2}\frac{\Delta_\kappa}{\Delta_\chi}\delta_{2.j}\right\}\{1-\delta_{j,5}\} \\
&\quad +\left\{\frac{f^n_{m+1,3}-f^n_{m+1,1}+f^n_{m,3}-f^n_{m,1}}{4\Delta_\chi\Delta_\kappa}\right\}\delta_{j,5}; \\
\mathbf{RC}(1)_{3,j} &= -\mathbf{C}(1)_{3,j};\ \ \mathbf{RC}(1)_{4,j}=-\mathbf{C}(1)_{4,j}; \\
\mathbf{RC}(m)_{5,j} &= -\mathbf{C}(m)_{5,j}.
\end{aligned}
\tag{A27}
$$

A.5 The Inversion

Matrix inversion involves inverting the sub-matrices in a procedure discussed by Varga[15]. The procedure is illustrated below. We consider the matrix equation (A12), and define the forward generation of coefficients for the vectors $\mathbf{Q}(i)$, $\mathbf{G}(i)$, defined below

$$
\begin{aligned}
&\mathbf{Q}(1) \equiv \mathbf{B}(1)^{-1} \bullet \mathbf{C}(1) \\
&\mathbf{G}(1) \equiv \mathbf{B}(1)^{-1} \bullet \mathbf{R}(n)(1) \\
&\mathbf{Q}(2) \equiv \left(\mathbf{B}(2) - \mathbf{A}(2) \bullet \mathbf{Q}(1)\right)^{-1} \bullet \mathbf{C}(2) \\
&\mathbf{G}(2) \equiv \left(\mathbf{B}(2) - \mathbf{A}(2) \bullet \mathbf{Q}(1)\right)^{-1} \bullet \left(\mathbf{R}(n)(2) - \mathbf{A}(2) \bullet \mathbf{G}(1)\right) \\
&* \\
&* \\
&\mathbf{Q}(i) \equiv \left(\mathbf{B}(i) - \mathbf{A}(i) \bullet \mathbf{Q}(i-1)\right)^{-1} \bullet \mathbf{C}(i) \\
&\mathbf{G}(i) \equiv \left(\mathbf{B}(i) - \mathbf{A}(i) \bullet \mathbf{Q}(i-1)\right)^{-1} \bullet \left(\mathbf{R}(n)(i) - \mathbf{A}(i) \bullet \mathbf{G}(i-1)\right)
\end{aligned}
\tag{A28}
$$

Then the following backward substitution yields the required vectors:

$$
\begin{aligned}
&\Delta\mathbf{V}(N_{POS}) = \mathbf{G}(N_{POS}) \\
&\Delta\mathbf{V}(i) = \mathbf{G}(i) - \mathbf{Q}(i) \bullet \Delta\mathbf{V}(i+1)
\end{aligned}
\tag{A29}
$$

Each inversion is at time step n. For those situations where the boundaries are governed by ordinary time dependent differential equations, we store appropriate time steps to satisfy the next level of the computation.

References

1. H. L. Grubin, D. K. Ferry, G. J. Iafrate, and J. R. Barker, "The Numerical Physics of Micron-Length and Submicron Length Semiconductor Devices," in *VLSI Electronics Microstructure Science*, vol. 3, N. G. Einspruch, Ed. New York: Academic Press, 1982, pp. 198-300.

2. J. G. Ruch, *IEEE Transactions on Electron Devices*, vol. ED-19, pp. 652, 1972.

3. H. L. Grubin, "Density Matrix Simulations of Semiconductor Devices," in *Quantum Transport in Ultrasmall Devices*, vol. 342, *NATO ASI Series*, A. Jauho, Ed. New York: Plenum Press, 1995, pp. 241-280.

4. H. Weyl, *Z. Physik*, vol. 46, pp. 1, 1927.

5. E. Wigner, "On the Quantum Correction for Thermodynamic Equilibrium," *Physical Review*, vol. 40, pp. 749-759, 1932.

6. R. F. O'Connell and E. P. Wigner, "Manifestations of Bose and Fermi statistics on the quantum distribution function for systems of spin-0 and spin-1/2 particles," *Physical Review A*, vol. 30, pp. 2613-2618, 1984.

7. P. Bordone, A. Bertonni, R. Brunetti, and C. Jacoboni, "Wigner Paths Method in Quantum Transport with Dissipation," *VLSI Design*, vol. 13, pp. 211-220, 2001.

8. W. R. Frensley, "Boundary conditions for open systems driven far from equilibrium," *Reviews of Modern Physics*, vol. 62, pp. 745-791, 1990.

9. J. R. Barker, D. W. Lowe, and S. Murray, "A Wigner Function Approach to Transport and Switching in Sub-micron Heterostructures," in *The Physics of Submicron Structures*, H. L. Grubin, K. Hess, G. J. Iafrate, and D. K. Ferry, Eds. New York and London: Plenum Press, 1984, pp. 277-286.

10. D. R. Miller and D. P. Neikirk, "Simulation of Intervalley Mixing in Double-barrier Diodes using the Lattice Wigner Function," *Applied Physics Letters*, vol. 58, pp. 2803, 1991.

11. N. C. Kluksdahl, A. M. Kriman, D. K. Ferry, and C. Ringhofer, "Self-consistent study of the resonant-tunneling diode," *Physical Review*, vol. B39, pp. 7720-7735, 1989.

12. K. L. Jensen and F. A. Buot, "The effects of scattering on current-voltage characteristics, transient response, and particle trajectories in the numerical simulation of resonant tunneling diodes," *Journal of Applied Physics*, vol. 67, pp. 7602-7607, 1990.

13. E. I. Blount, *Physical Review*, vol. 114, pp. 418, 1959.

14. B. A. Biegel, "Quantum Electronic Device Simulation," *Ph. D. Dissertation Submitted to Department of Electrical Engineering at Stanford University*, 1997.

15. R. C. Varga, *Matrix Iterative Analysis*: Prentice Hall, 1962.

16. L. Shifren and D. K. Ferry, "Inclusion of Non-local Scattering in Quantum Transport," *Submiitted for publication*, 2002.

17. S. Verghese, C. D. Parker, and E. R. Brown, "Phase noise of a resonant-tunnelng relaxation oscillator," *Applied Physics Letters*, vol. 72, pp. 2550-2552, 1998.

18. P. Zhao, H. L.Cui, and D. L. Woolard, "Dynamical Instabilities and I-V characteristics in resonant tunneling through double-barrier quantum well systems," *Phys. Rev. B*, vol. 63, pp. 075302-1 through -75302-14, 2001.

19. W. F. Chow, *Principles of Tunnel Diode Circuits*. New York: John Wiley and Sons, Inc, 1964.

International Journal of High Speed Electronics and Systems
Vol. 13, No. 4 (2003) 1287–1306

Continuous-Wave Terahertz Spectroscopy of Plasmas and Biomolecules

D. F. Plusquellic, T. M. Korter, G. T. Fraser, R. J. Lavrich, E. C. Benck, C. R. Bucher, A. R. Hight Walker and J. L. Domenech

Optical Technology Division
National Institute of Standards and Technology
Gaithersburg, MD 20899-8441, USA

Continuous-wave linear-absorption spectroscopy based on THz radiation generated by solid-state photomixers has been applied to the investigation of the dynamics of biomolecules in polyethylene matrices and to line shape studies of HF for diagnostics of semiconductor etching plasmas. The THz spectra of biotin and myoglobin have been obtained using a variable-temperature, cryogenic sampling system. The spectrum of biotin displays a small number of discrete absorptions over the temperature range from 4.2 K to room temperature while the spectrum of myoglobin has no obvious resonance structure at the >10% fractional absorption level. Spectral predictions from the lowest energy *ab initio* conformations of biotin are in poor agreement with experiment, suggesting the need to include condensed-phase environmental interactions for qualitative predictions of the THz spectrum. Vibrational anharmonicity is used to model the line shapes that result from drastic changes in vibrational sequence level populations of biotin over this temperature range. Anharmonicity factors ($\chi_e\omega_e/\omega_e$) at the levels of 0.1 % to 0.8 % are obtained from non-linear least squares fits of the observed resonances and illustrate their important for refining model predictions. Application of the photomixer system to line shape studies in etching plasmas has been used to study the formation efficiency and translational temperature of HF at 1.2 THz under different operating conditions. These results will aid in understanding the chemistry of industry-standard fluorocarbon and oxygenated fluorocarbon etching plasmas.

Keywords: CW THz spectroscopy, THz spectra of biomolecules, Semiconductor etching plasma diagnostics. Vibrational anharmonicity, cryogenic samples, biotin, myoglobin, far-infrared.

1. Introduction

Terahertz (THz) linear-absorption spectroscopy based on radiation produced by pumping solid-state photomixers with continuous-wave, near-infrared laser sources has been applied to the investigation of the dynamics of biomolecules and to the diagnostics of semiconductor etching plasmas. The continuous-wave photomixers offer several advantages over blackbody and pulsed-laser THz sources, including a narrow spectral bandwidth, limited by the two pump laser line widths, and excellent frequency control and intensity stability important for signal averaging and for monitoring line intensities. Moreover, the bulky pump lasers can be easily isolated from the compact THz source by the use of optical fibers for the propagation of the near-infrared pump laser beams to the photomixer. Such remote operation is convenient for performing THz measurements at sites removed from the laser facility, for coupling to certain sample systems, and for minimizing the propagation of the THz radiation through the atmosphere where it is strongly absorbed by water-vapor.

The spectrum of THz radiation is well matched to the low-frequency molecular motions important for the flexibility and thus the function of many biological molecules. These low-frequency modes are ideally investigated on small model systems, such as simple peptide mimetics, where the individual resonances can be spectrally resolved and potentially assigned to a nuclear motion with the aid of molecular modeling. In proteins and other large biomolecules, the plethora of such low-frequency motions, which increase in number approximately linearly with the number of atoms, prevents the resolution of individual lines in the spectrum necessary for obtaining any detailed dynamical information. To effectively improve the spectral resolution and thus allow the identification of individual resonant absorptions, a liquid-He-cooled, variable-temperature, cryogenic sampling system has been constructed. The initial expectation was that the cryogenic conditions would simplify the spectra through the reduction of resolved vibrational sequence structure of each embedded conformer. Instead, we found (for biotin) little change in the number of vibrational states and attribute most of the temperature dependent features to the effects of anharmonicity on the sequence structure within each vibrational state and to a lesser degree to small crystal shifts. The intrinsic widths may result from either homogeneous broadening associated with the coupling of the vibrational and phonon modes and/or inhomogeneous broadening due to local crystal defects and ultimately limit the resolution gains from sample cooling in polyethylene matrix environments.

A second application of the photomixer system is to the diagnostics of semiconductor etching plasmas. This effort complements an ongoing program on developing submillimeter-wavelength backward-wave oscillators (BWO) for plasma diagnostics. The frequencies available from these BWO sources are typically limited to < 1 THz. The photomixer produces radiation above 1 THz and thus allows the direct measurement of the expected intense rotational lines of HF and OH, important for understanding the chemistry of industry-standard fluorocarbon and oxygenated fluorocarbon plasmas. The high frequency of the rotational transitions of HF and OH, coupled with the light mass of these diatomic hydrides, gives a significant Doppler width to the resonances, which can be used for determining the translational temperature of the plasma.

Below we discuss more fully our applications of continuous-wave photomixer-based THz spectroscopy to the investigation of biological molecules and plasmas. We start with a description of the THz spectrometer and then discuss the two applications.

2. Continuous-Wave THz Spectrometer

A diagram of the continuous-wave THz laser spectrometer is shown in Fig. 1. Continuous-wave (CW) THz radiation is generated by the difference-frequency mixing of two nominally 850 nm near-infrared laser beams at the surface of a solid-state photomixer. The photomixer, manufactured at Lincoln Laboratories,[1] consists of epitaxial low-temperature-grown GaAs (LT-GaAs) on a GaAs substrate. The LT-GaAs layer is

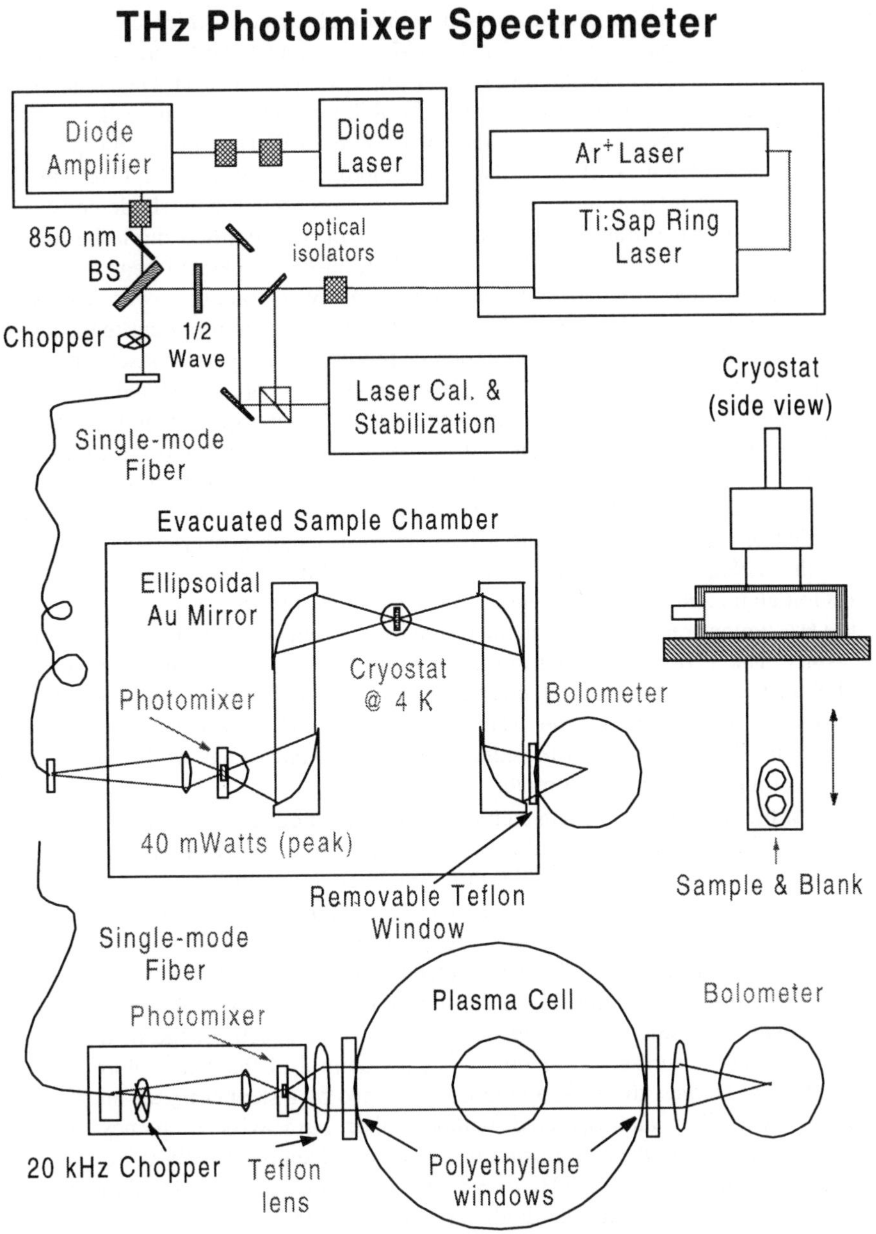

Fig. 1. A schematic diagram of the 0.1 THz to 4 THz spectrometer. The solid-matrix biomolecular sample system is shown at the top of the figure. The THz beam propagation optics used for studies of HF in plasma reactors are shown in the lower part.

topped by submicron interdigitated gold-on-titanium electrodes, which are coupled to a self-complementary spiral antenna for broad-band output.[2] The photomixer has an 8 μm × 8 μm central region containing eight 8 μm x 0.2 μm electrodes separated by 0.8 μm. The front surface is antireflection coated for 850 nm. The photomixer is operated at room temperature and biased with a voltage of 15 V to 20 V. The physical properties and performance characteristics of these types of photomixers have been given in detail elsewhere[3] and their application to THz spectroscopy has been previously demonstrated.[4]

The two near-infrared pump lasers consist of an amplified diode laser and an Ar^+-pumped CW Ti:Sapphire ring laser. The diode laser is operated near 850 nm and seeds an amplifier that produces approximately 500 mW of output power. Two optical isolators (>60 dB isolation) between the diode and amplifier limit optical feedback, thus aiding in the greater than 1000:1 suppression of power in higher-order laser modes. The Ti:Sapphire laser generates >500 mW of power. Both lasers are attenuated to 20 mW to prevent damage to the photomixer, combined on a 50 % beamsplitter, and mechanically chopped at 300 Hz prior to coupling into a single-mode fiber for propagation to the photomixer, which could be in another laboratory. The output light from the fiber is free-space propagated for approximately 25 cm and tightly focused onto the 8 μm × 8 μm active region of the photomixer using a 1.25 cm focal length aspheric lens. A $\lambda/2$ wave plate in the Ti:Sapphire beam prior to the beam splitter allows optimization of the polarization alignment at the photomixer. Dispersion in the fiber at the two colors over the 4 THz tuning range is sufficiently small that polarization preserving fiber is unnecessary. The THz power spectrum is shown as a semi-log plot in Fig 2. Above 3.5 THz, bolometric detection of the thermal radiation from pump laser heating of the photomixer is comparable to the THz power generated. Cryogenic cooling of the photomixer to 77 K is expected to reduce this background signal while improving the conversion efficiency of the photomixer.[5]

For the studies of HF in an etching plasma, the diode laser is actively locked (≈3 MHz) near 11757 cm^{-1} to an optical cavity referenced to a polarization-stabilized HeNe laser.[6] The optical layout of the calibration system is shown in Fig. 3. The Ti:Sapphire laser is operated as a single-mode source ($\Delta\nu_{laser}$ < 500 kHz) and frequency control is achieved by double-passing a small portion of the laser though an acoustic-optic modulator (AOM). The doubly-diffracted sideband is combined with and separated from the diode laser using polarizing beam splitting cubes and actively locked to the same reference cavity. The sideband is tunable over a full free spectra range of the reference cavity (250 MHz), ensuring continuous frequency control of the Ti:Sapphire laser.[7] For condensed phase studies, the Ti:Sapphire laser is tuned by a stepper motor interface to an intracavity Lyot filter and operated without other intracavity frequency narrowing elements. Consequently, the step resolution is limited to approximately 0.2 cm^{-1}. Spectra are typically acquired from 0.2 THz to 3.5 THz by scanning the Ti:Sapphire laser from 11763 cm^{-1} to 11874 cm^{-1} with the diode laser locked to the reference cavity near 11757 cm^{-1}. The frequency calibration of the THz specta is performed by recording wavemeter readings of the Ti:Sapphire laser at 0.1 cm^{-1} resolution during the course of a scan.

A silicon hyper-hemispherical lens is optically contacted to the backside of the photomixer to generate a quasi-Gaussian THz beam[8] No additional transmissive optics are used to minimize absorption loss and interference effects at the detector from surface reflections. The THz beam is propagated by means of four parabolic gold mirrors each

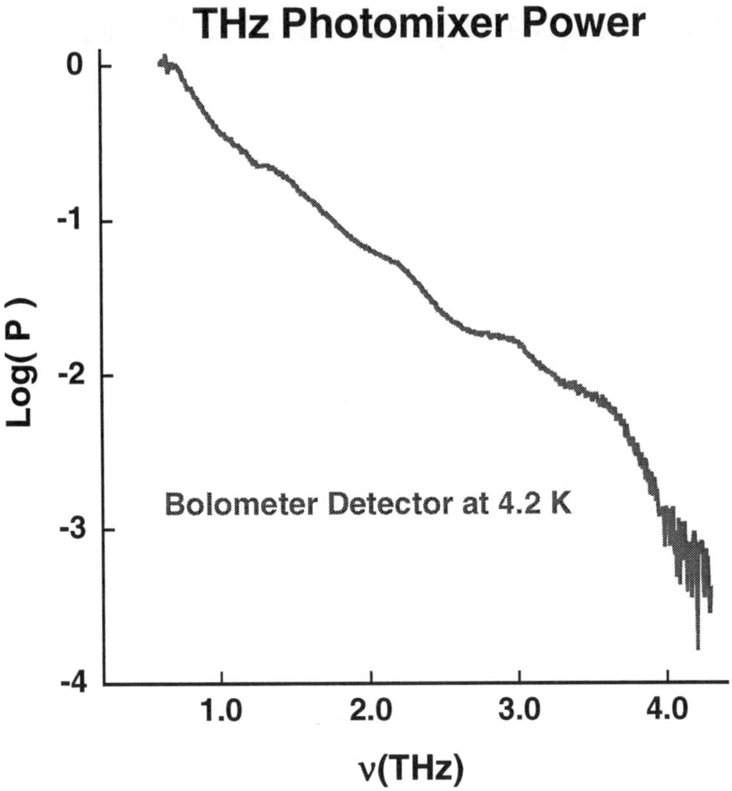

Fig. 2. The THz photomixer power spectrum measured on a 4.2 K bolometer.

having a 7.6 cm effective focal length. The first two mirrors collect, collimate, and refocus the THz beam onto the cryogenic sample to a beam waist of approximately 1 mm at 1 THz. The second two mirrors image this spot onto a liquid-He-cooled silicon composite bolometer detector fitted with a 20 μm low-pass optical filter. The detector with preamplifier has a 3 dB bandpass at 1 kHz. A 5 mm thick Teflon flat serves as the entrance window to the bolometer during sample alignment and is removed from beam path when operating the sample chamber under cryogenic conditions, allowing both the detector and sample chambers to be pumped by the liquid-He in the detector dewar.

Evacuation of the entire optical system serves two purposes. Cyrogenic cooling to 4.2 K essentially eliminates spectral broadening from sequence bands and increases the thermal population of the lowest vibrational (and perhaps conformational states) of the molecules. Also, water vapor is eliminated from the optical path. Water vapor

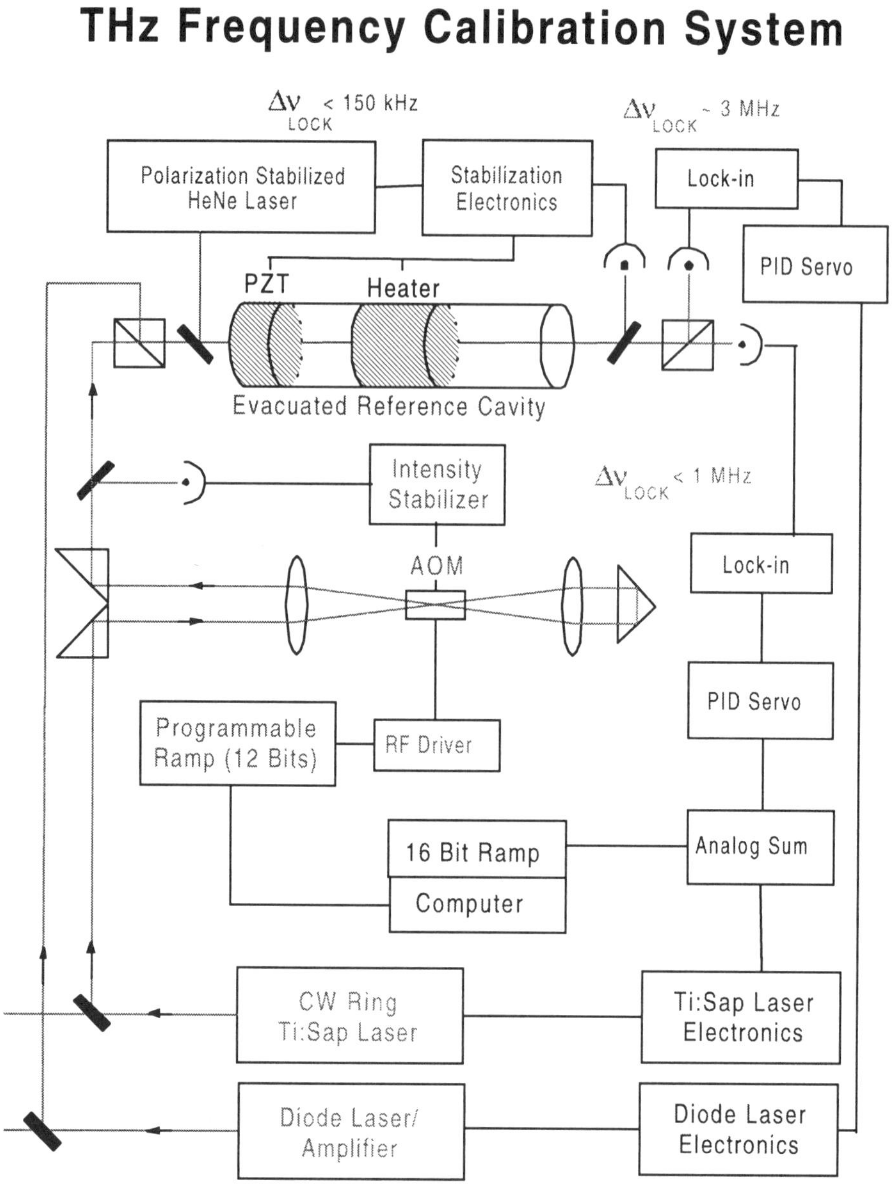

Fig. 3. The THz frequency calibration system used for lineshape studies of HF. For the solid matrix studies, the Ti:Sapphire laser is operated as a multi-mode source ($\Delta\upsilon \approx 0.2$ cm^{-1}) and the frequency is monitored using a wavemeter (not shown).

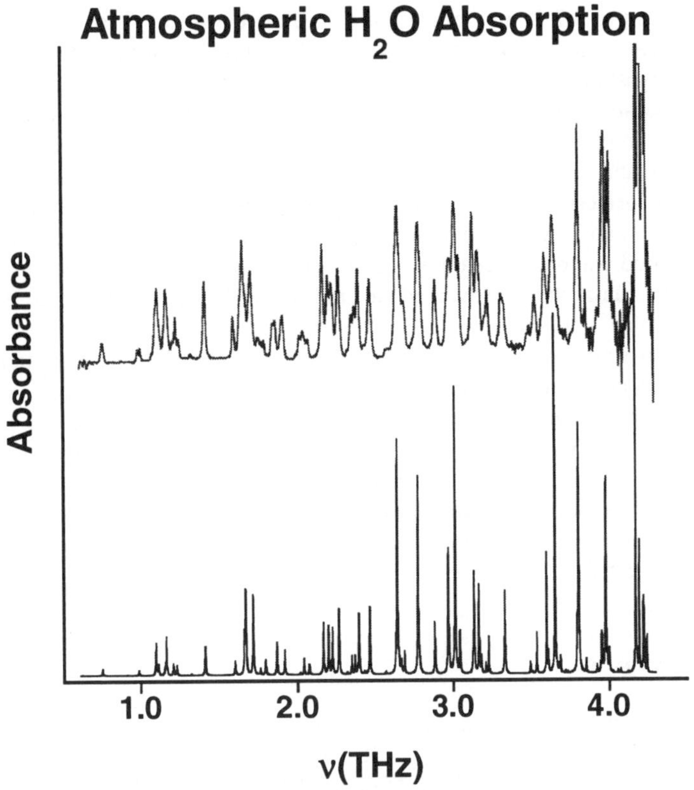

Fig. 4. The observed (top) and calculated (bottom) THz spectrum of atmospheric water vapor. The observed spectrum is offset for clarity.

absorption of THz radiation is particularly severe above 0.5 THz, as shown in Fig. 4, recorded at atmospheric pressure and laboratory humidity levels. The calculated spectrum from the HITRAN database[9] is shown in the lower panel and further serves as a test of the frequency calibration of the spectrometer. The experimental line widths of approximately 0.3 cm^{-1} are somewhat broader than the calculated pressure-broadened widths since the Ti:Sapphire laser was intentionally operating under multimode conditions. Signal saturation from the limited dynamic range of the detection system ($<10^4$) also contributes to the apparent broadening, particularly for the stronger lines having absorbances greater than 40.

Pressed disks of polyethylene with and without dry samples of biomolecules are prepared using a 1.26 cm pellet press. High-density polyethylene is used as a sample matrix due to its transparency in the THz region. The two polyethylene pellets of equal thickness, one with sample and one without sample, are inserted side-by-side into a holder and cooled to 4.2 K in a temperature regulated cryostat. THz scans of the sample and the polyethylene blank are done consecutively after interchanging the two samples by elevating the cryostat, to minimize the effects of interference produced by sample surface reflections. Lock-in signals of the bolometer and photocurrent (measured as a voltage across a 100 Ω shunt) are digitized concurrently at 100 Hz with a time constant of 30 ms. Both the sample and blank are normalized to the photocurrent signal (after subtracting off the DC current offset) to correct for laser-power variations. The normalized data are signal averaged and mapped onto a scale of constant wavenumber spacing using computer acquired wavemeter readings of the Ti:Sapphire laser frequency for calibration. Finally, sample data are normalized to the polyethylene blank and converted to absorbance.

Density functional calculations of the conformational energies and vibrational frequencies were performed using the GAUSSIAN 98 software package[10] with Becke's three-parameter exchange functional (B3) and the correlation functional of Lee, Yang, and Parr (LYP). A 6-31+G(d,p) split-valence basis set with polarization and diffuse functions was used. This level of theory has been shown to accurately reproduce experimentally observed vibrational frequencies.[11] Contributions from basis-set superposition errors (BSSE)[12] have been ignored in the reporting of the relative conformational energies.

3. Plasma Diagnostics

High-resolution THz linear-absorption spectroscopy has been applied to the detection and quantification of HF in a reactive-ion etching plasma. THz spectroscopy allows the direct observation of HF near 1.23 THz, not possible using typical backward-wave oscillators with fundamental frequency outputs less than 1 THz. In hydrofluorocarbon plasmas, such as with CF_3H as a progenitor, HF is produced in large quantities. The fundamental $J = 1 - 0$ rotational transition is an extremely sensitive probe of the presence of HF. This sensitivity comes from the large THz absorption cross section of HF arising from its favorable vibrational and rotational partition functions of 1.05 and 13.9, respectively, at 400 K, and its large electric dipole moment of 1.83 D. This high sensitivity provides the opportunity to explore the potential of using HF as a probe of the status of an etching process in the reactor or as a signature of hydrogen contamination in fluorocarbon plasmas.

HF also could potentially provide information on the temperature of the plasma, particularly for higher-temperature, inductively coupled plasmas. The relatively small mass of HF imparts a significant Doppler width to the rotational lines of approximately 4.0 MHz full width half maximum (FWHM) at 400 K and 1.25 THz. This width is proportional to $T^{1/2}$ where T is the plasma translational temperature, generally assumed to be in equilibrium with the plasma rotational temperature. As the technique requires propagation of the THz beam through the reactor, the resulting temperatures and number densities are averaged over the optical path. In principle, a systematic variation of this optical path could be combined with tomographic inversion techniques to provide spatial number-density and temperature maps.

To perform the measurements, the near-infrared laser beams are propagated from the laser laboratory to the plasma laboratory using approximately 115 m of single-mode optical fiber. The near-infrared radiation is coupled to the photomixer in the plasma laboratory, as discussed previously. The plasma reactor, shown schematically in the bottom panel of Fig. 1, is a slightly modified inductively coupled reactor based on the Gaseous Electronics Conference (GEC) reference-cell standard design. The THz beam is singly passed through the cell through high-density polyethylene windows and detected with a liquid-He-cooled composite bolometer.

For validation purposes, a series of HF calibration scans were first performed in the reactor chamber to arrive at an accurate assessment of the linewidth of the THz source. Measurements were performed using between 0.067 Pa (0.5 mTorr) to 0.67 Pa (5 mTorr) HF in Ar gas at a total pressure of 1.33 Pa (10 mTorr). An example of the line shape obtained using 0.060 Pa (0.45 mTorr) of HF is shown in the top panel of Fig. 5. All of the observed line shapes have been fit to Voigt profiles having principally Gaussian and small Lorentzian components. The calculated Doppler width of HF at room temperature is 3.4 MHz, which is smaller than the average Gaussian component of 4.7(1) MHz fit to the absorption profiles of 5 different gas mixes. We attribute this excess width to frequency noise of the diode laser that is greater than the servo-loop bandwidth of 4 kHz. Since Gaussian widths add in quadrature, the bandwidth of the THz source is calculated to be 3.1 MHz. The small Lorentzian component of 0.53(3) MHz may arise from power and/or pressure broadening. Power broadening is estimated near a Rabi frequency of 0.5 MHz assuming 100 nW of laser power focused at the center of the reactor. Pressure broadening is expected to be much less. The $J = 1 - 0$ transition of HF has a FWHM pressure broadening coefficient of approximately 0.047 MHz/Pa (6.3 MHz/Torr) at room temperature for an Ar collision partner, which corresponds to 0.063 MHz.

The bottom panel of Fig. 5 shows the absorption spectrum for HF obtained for 0.067 Pa (0.5 mTorr) of trifluoromethane (CF_3H) in Ar gas at 1.33 Pa in a radio frequency (RF) discharge operating at 300 W. An average Doppler width of 3.4(1) MHz is calculated for this profile as well as other line shapes obtained using RF powers of 200 W and 250 W. These results indicate that HF is produced with an average translational energy near room temperature regardless of plasma conditions. Boltzmann temperatures near 300 K have also been found in other direct absorption studies performed here using BWOs. These low temperatures seem unlikely and may be attributable to an artifact of the chamber design where 2/3 of the propagation path of the THz beam samples the colder gas outside of the central plasma core. Redesigns of the chamber are currently under way to reduce the size of this region.

The integrated intensities provide information on the average number density of HF molecules along the optical path through application of Beer's Law:

$$\int \ln[I_0(\nu)/I(\nu)]d\nu = \sigma n \ell_{total}$$

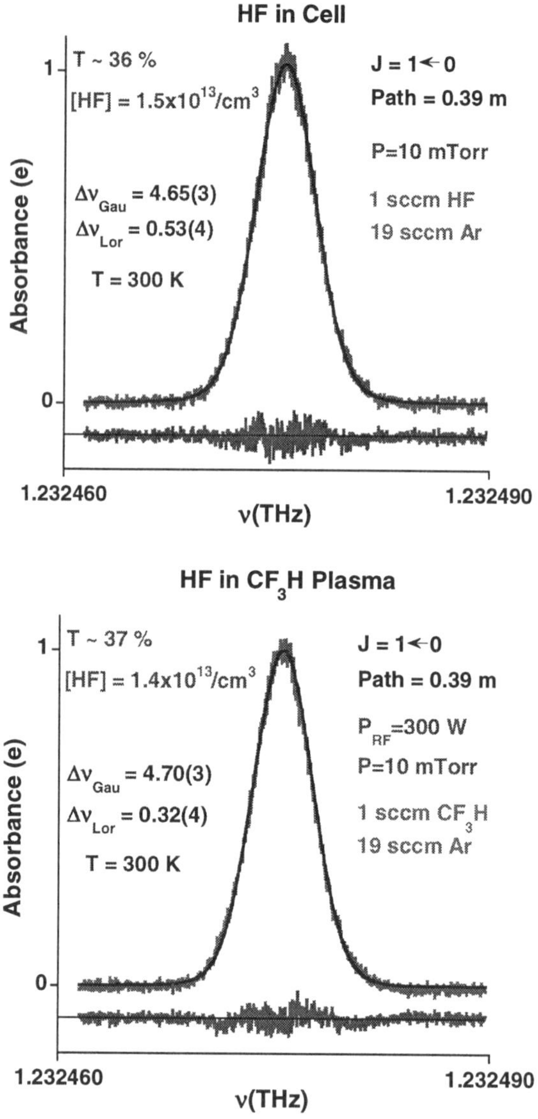

Fig. 5. The observed line shapes of the J=1-0 transition of HF in a reactive-ion etching plasma at two different pressures. Residuals from the fit are shown below each line.

Here, ν is the frequency, I_0 is the THz intensity incident on the reactor gas, I is the transmitted intensity through the reactor, σ is the integrated absorption cross section, n is the number density of HF molecules, and ℓ_{total} = 39 cm is the optical pathlength through the reactor. The bottom trace, which corresponds to a mean HF translational temperature of 300 K and a peak fractional absorption of approximately 67 % of the incident radiation, implies an HF average pathlength number density of 9.3×10^{12} cm^{-3}, or at 10 mTorr approaches 3 % of the plasma gas being present as HF. Given the relative fraction of CF_3H in Ar is 5 %, this implies a CF_3H to HF conversion efficiency of 58 %. We note that with the backward-wave oscillators, the detection sensitivity at frequencies less than 1 THz is significantly higher than with the photomixer, approaching, for instance, 10^7 cm^{-3} for SiO. The backward-wave oscillator sensitivity is also illustrated by our ability to detect DF in natural abundance (0.015 %) in similar CF_3H plasmas, with a comparable signal-to-noise ratio, although it should be noted that the DF concentration could be enriched by a kinetic-isotope effect.

4. THz Spectra of Biomolecules

THz linear-absorption spectroscopy has also been applied to the investigation of the large-amplitude, low-frequency vibrational modes of biological molecules in solid, cryogenic matrices, typically polyethylene. Direct study of the molecules in their natural aqueous environment is difficult because of the strong THz absorption of liquid water (absorption coefficient≈150 cm^{-1}). The large-amplitude modes studied are progenitors for the more complex motions critical to protein folding and function. The unique sensitivity of these modes to the molecular conformation and environment, areas of much current interest, provides much of the motivation for their study by THz methods.

The low excitation energy of the THz vibrational modes relative to the room-temperature thermal energy of kT = 210 cm^{-1} = 6.2 THz leads to significant thermal population of their excited vibrational states. Consequently, room temperature spectra measure not only the fundamental band (v = 1 - 0), but also sequence bands, v = 2 – 1, 3 – 2, etc. A vibrational mode with a fundamental frequency of 33 cm^{-1} (1 THz), for instance, will have, ignoring anharmonicity, thermal population in excess of 1 % in levels up to v = 18. The anharmonicity of the vibrational potential leads to a decreased vibrational level spacing with increasing energy, and thus a broadening and asymmetric distortion of the line shapes towards lower frequencies. Cryogenic cooling to liquid He temperatures will have a drastic effect on sequence level broadening. At 4.2 K, for example, greater than 99 % of the population will exist in the vibrational zero-point level for a 33 cm^{-1} vibrational fundamental. Therefore, absorption features obtained at these temperatures represent fundamental vibrational modes, and comparisons of spectra taken at room and liquid-He temperature quantify the anharmonicity of the potential energy surface.

Below, we discuss our investigations of the THz spectra of two biomolecules, biotin and myoglobin, which sample the range of molecular complexity present in biological systems. The observed absorption frequencies and intensities provide information about the nuclear rigidity of these molecules. The temperature dependence of these spectra has also been investigated to gain insight into the possible mechanisms of line broadening and into the spectral congestion that occurs with increasing molecular complexity.

4.1. *Biotin*

Large-amplitude motions play a critical role in the function of biotin, a vitamin coenzyme that catalyzes carboxylation and decarboxylation reactions[13] by aiding the

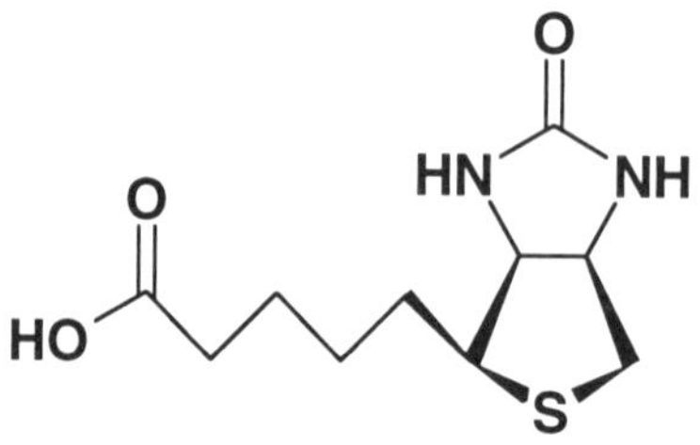

transport of intermediates between the two active sites of pyruvate carboxylase. THz spectra of biotin were acquired to obtain information about the low-frequency modes which form the basis for this transport motion.

Absorption spectra of a 5% by weight biotin in a polyethylene matrix are shown in Fig. 6 from 0.2 to 3.5 THz. The top and bottom spectra correspond to 4.2 K and 298 K, respectively. A gradual increase in the baseline absorption with frequency appears in both spectra, and is attributed, in part, to increased efficiency of Mie scattering from microcrystalline domains of the sample over the 10-fold change in THz wavelengths. The spectrum contains 12 identifiable features as discussed below with line widths that range from ≈ 1 cm^{-1} at 18 cm^{-1} to ≈ 5 cm^{-1} at 100 cm^{-1}. The same resonances appear in the room temperature spectrum, but are somewhat broader and often red-shifted.

Molecular modeling has been employed to aid in the interpretation and understanding of the observed spectral measurements. Five conformations of biotin were investigated using density functional methods. The first conformer has a configuration similar to that observed in the crystal by x-ray diffraction[14] and has been named *cis-trans*-biotin (CT) in reference to the conformation of the valeryl chain. The remaining conformers are derived from the *ab initio* work of Strzelczyk and coworkers.[15] All-*trans*-biotin (AT) has the valeryl chain fully extended and is calculated to be lower in energy than CT. Finally, the last three biotin conformers involve intramolecular hydrogen bonds between the valeryl chain and the ureido ring of the bicyclic moiety. The single hydrogen-bonded folded conformations is denoted 1HB and the two doubly hydrogen-bonded conformations are denoted 2HB-1 and 2HB-2. The conformation of the bicyclic moiety is *endo* for all the conformers, with the exception of 2HB-1 and 2HB-2 where the ring adopts an *exo* conformation.

CT was found to be the highest energy conformation, despite its resemblance to the conformation observed in the crystal.[14] The high energy of the CT conformation is presumably due to the neglect in the calculations of the stabilizing hydrogen-bonding interactions present between neighboring biotin molecules in the crystal. The AT conformer is more stable than CT by 12.0 kJ/mol, while the folded conformers are stabilized to the greatest extent, 13.8 kJ/mol, 15.1 kJ/mol, and 19.5 kJ/mol for 2HB-1,

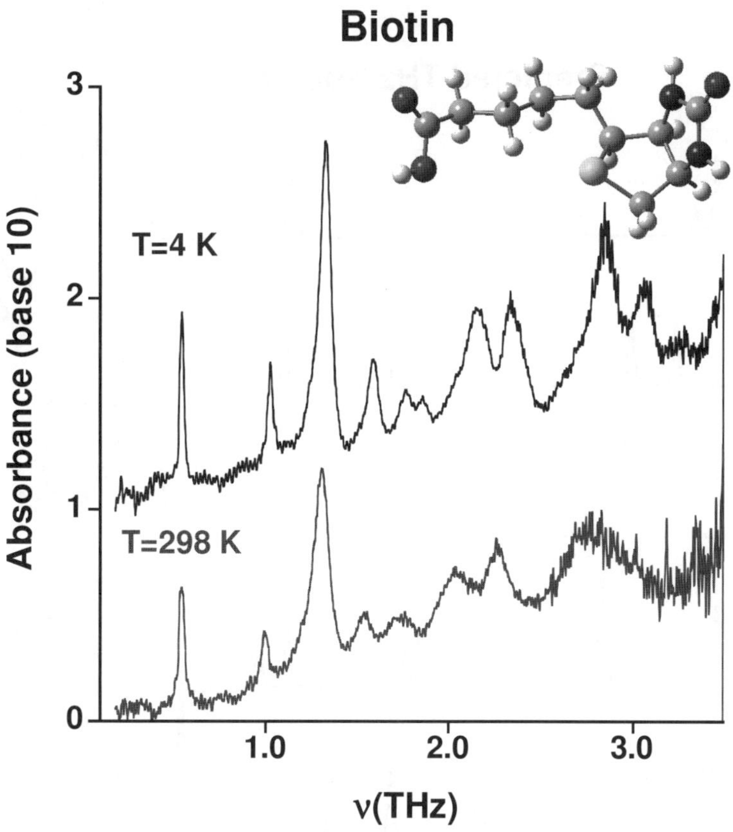

Fig. 6. The THz spectrum of biotin in a polyethylene matrix obtained at room temperature (lower trace) and at 4 K (upper trace). The upper spectrum is offset for clarity.

1HB, and 2HB-2 respectively. Our calculated values for the relative gas-phase energies are consistent with those reported previously.[15]

The calculated THz spectrum for each conformer is presented in Fig. 7. The stick spectra derived from the frequency and intensity data from the density functional calculations were convoluted with a 1 cm^{-1} (FWHM) Lorentzian line shape to approximate the experimentally observed line widths. The theoretical and experimental spectra resemble each other in overall simplicity, however, no assignments can be made with certainty, until a more complete vibrational model is presented which accounts for intermolecular hydrogen-bonding interactions and intramolecular anharmonicity.

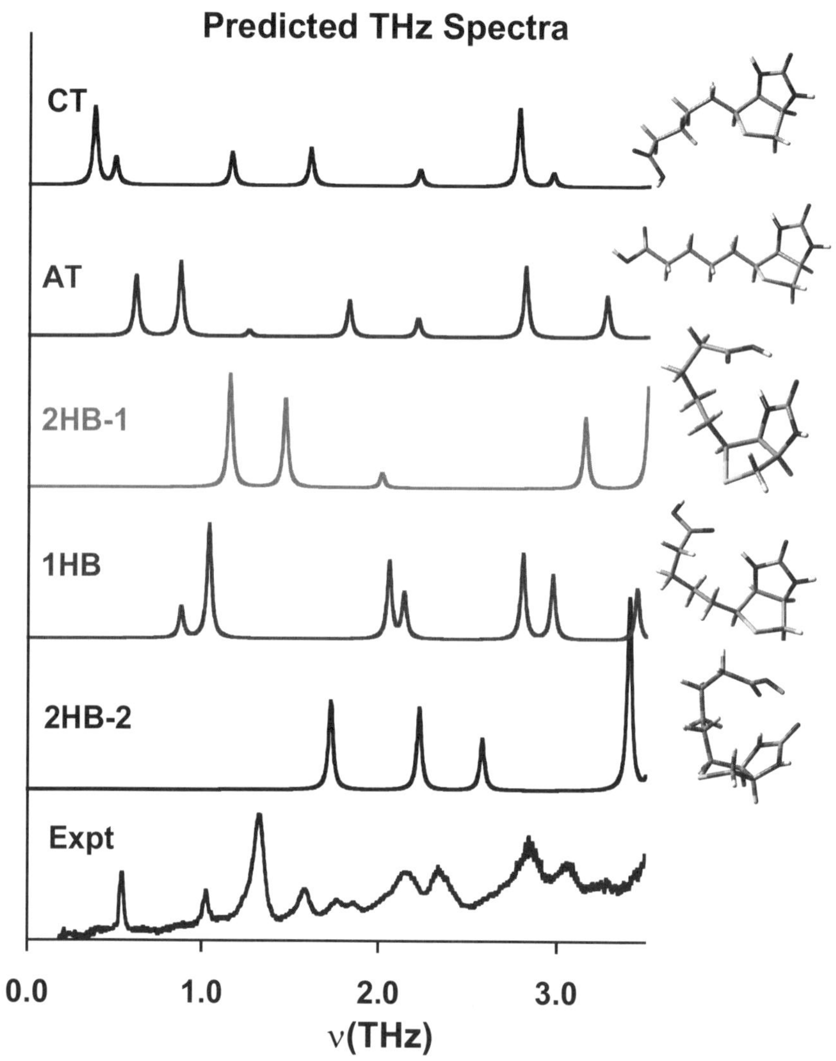

Fig. 7. The calculated vibrational spectra of the lowest energy conformations of biotin at the B3LYP/6-31+G(d,p) level of theory.

Many factors contribute to the vibrational absorption widths of samples embedded in a polyethylene matrix, including inhomogeneous broadening from differences in the local matrix environment and homogeneous broadening from vibrational energy exchange between molecules (both biotin and polyethylene) and with phonon modes. The temperature dependence of the line widths and line positions arise from structural changes in the matrix cage and from anharmonicity in the vibrational force field.

As evident from Fig. 6, many of room temperature lines are broadened and red-shifted relative to those at 4.2 K. The magnitudes of the line widths and shifts increase approximately linearly with frequency. The temperature dependence of the line broadening is consistent with the effects of vibrational anharmonicity.[16] A simple model that accounts for the observed behavior is based on the following equations:

$$I(\nu) = N\sum_{V=0}^{\infty} \Delta N_{V,V+1}(V+1)C\exp\left(\frac{-(\nu-\Delta E_V)^2}{\Delta\nu^2{}_{FWHM}}\right)$$

$$\Delta E_V = \omega_e - 2\chi_e\omega_e(V+1)$$

$$\Delta N_{V,V+1} = \frac{N_V - N_{V+1}}{\Theta}$$

$$\Theta = \sum_{V=0}^{\infty} N_V = \sum_{V=0}^{\infty}\exp\left(\frac{-(\Delta E_V - \Delta E_0)}{kT}\right) \approx \left[1-\exp\left(\frac{-\omega_e}{kT}\right)\right]^{-1}$$

Here, ω_e is the harmonic frequency of the vibration, $\chi_e\omega_e$ is the vibrational anharmonicity, $\Delta\nu_{FWHM}$ is the full-width at half maximum of each absorption line, V is the vibrational quantum number, C is the normalization factor for a Gaussian line shape and Θ is the vibrational partition function. The absorption strength or intensity, $I(\nu)$, is defined in terms of the number density, N, the normalized population difference, $\Delta N_{V,V+1}$ and the parameter, $V+1$, used to model the dependence of the square of the transition dipole moment on vibrational quanta, which is assumed to increase linearly with V. Each line-shape function is normalized to unity and therefore, population shifts between vibrational modes with changing temperature are not taken into account in this model. However, this shift will primarily affect the overall absorption strengths at the two temperatures.

Applications of this model to the absorption spectra of biotin are illustrated in Fig. 8. In the top panel are shown the minimum number of Gaussian peaks required to account for all of the observed features at 4.2 K. Three parameters of each Gaussian function (center frequency, ω_e, intrinsic width, $\Delta\omega_{FWHM}$ and intensity) were optimized in a non-linear least squares fit to this data. The anharmonicity factors, $\chi_e=\chi_e\omega_e/\omega_e$, were then fit to the room temperature spectrum using fixed values for each Gaussian function. This analysis was repeated until self-consistent values were obtained for both spectra.

Table I. Frequencies, intensities, full widths, shifts and anharmonicity constants of Biotin obtained from non-linear least squares fits of spectra obtained at 4.2 K and 298 K using Gaussian lineshape functions. Type A standard uncertainties (i.e., k=1 or 1σ) are given, as determined from the least-squares fit.

	4 K			298 K	298 K	
	ω_e (cm^{-1})	Intensity	$\Delta\omega$/cm^{-1}	χ_e	δ / cm^{-1}	χ_e
0	18.14(5)	0.78(1)	0.93(1)	0.0002(2)	-0.59(5)	0.0016(1)
1	34.34(5)	0.40(1)	1.4(1)	0.0017(5)	-1.0(1)	0.0019(1)
2	41.82(5)	0.31(1)	2.3[a]	0.0023(1)	0.06(1)	0.0034(1)
3	44.45(5)	1.40(2)	2.4(1)	0.0014(1)	-0.58(5)	0.0022(1)
4	53.24(5)	0.35(2)	2.4(1)	0.0042(1)	0.15(5)	0.0035(1)
5	59.27(7)	0.18(2)	2.6(2)	0.0038(2)	0.27(5)	0.0032(3)
6	62.2(1)	0.12(1)	2.2(2)	0.003[a]	0.2(1)	0.003[a]
7	67.8(1)	0.13(1)	3.0[a]	0.0011(1)	-0.5(1)	0.0015(1)
8	72.82(5)	0.50(2)	4.5(1)	0.0066(1)	0.22(5)	0.0067(1)
9	78.64(5)	0.47(1)	3.9(1)	0.0037(1)	1.12(5)	0.0025(1)
10	91.2(4)	0.18(1)	5.0(5)	0.005[a]	0.0(1)	0.005[a]
11	96.45(7)	0.69(1)	4.6(1)	0.0072(1)	-0.65(5)	0.0083(1)
12	103.0(1)	0.41(1)	4.9(1)	0.0066(1)	-0.55(5)	0.0070(1)

[a]Fixed in fit.

Two additional parameters were included to account for the linear increase in baseline absorption in each spectrum and an overall scaling factor was varied in the 298 K data to account for the net population shift to higher energy sequence bands outside of our scan interval.

The best fit parameters are given in Table I and the model predictions and residuals at each temperature are illustrated in the top two panels of Fig. 8. The unstructured baseline has been subtracted to aid in the visual comparison with these predictions. The center frequencies and intensities used in this model are shown as sticks in the figure. Although these fits account for the vast majority of the temperature dependent features, small discrepancies are seen in the residuals near the shoulders of some lines. It is reasonable to expect that along with such a drastic increase in temperature, small shifts in center frequencies also might occur as a result of the thermal expansion of the matrix cage. The lower panel in Fig. 8 shows the results of a similar analysis applied to the 298 K spectra except that both χ_e and ω_e were varied in the fit. The frequencies are reported as shifts relative to the 4.2 K data in Table I together with the anharmonicity factors. The overall quality of the fit is similar to that obtained at 4.2 K. Although the shifts are typically smaller than 1 cm^{-1}, the anharmonicity factors are all in excess of 0.1 % and illustrate their importance for accurately predicting vibrational frequencies.

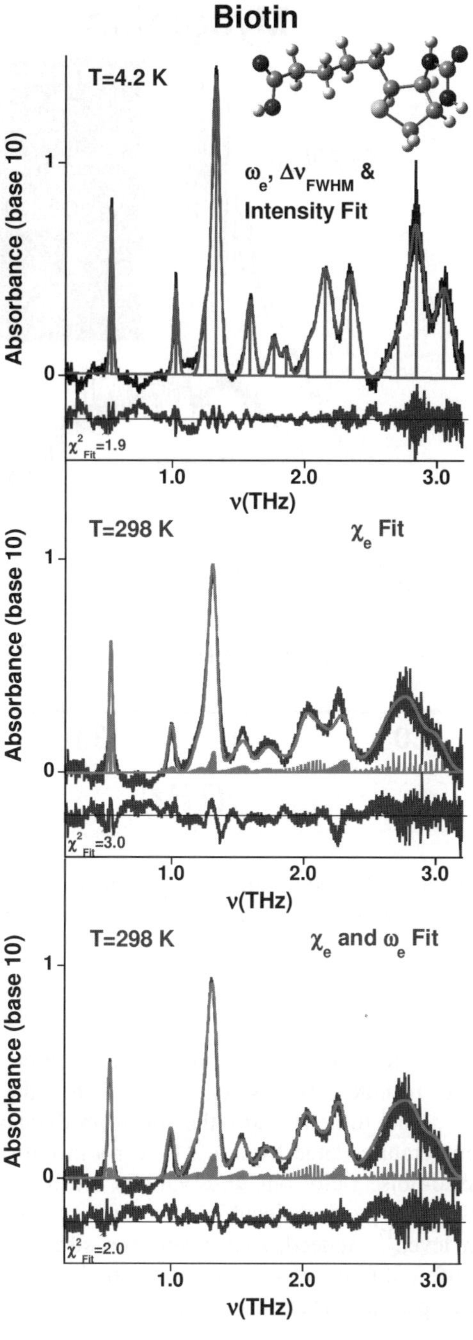

Fig. 8. Results from non-linear least squares fits that include vibrational anharmonicity.

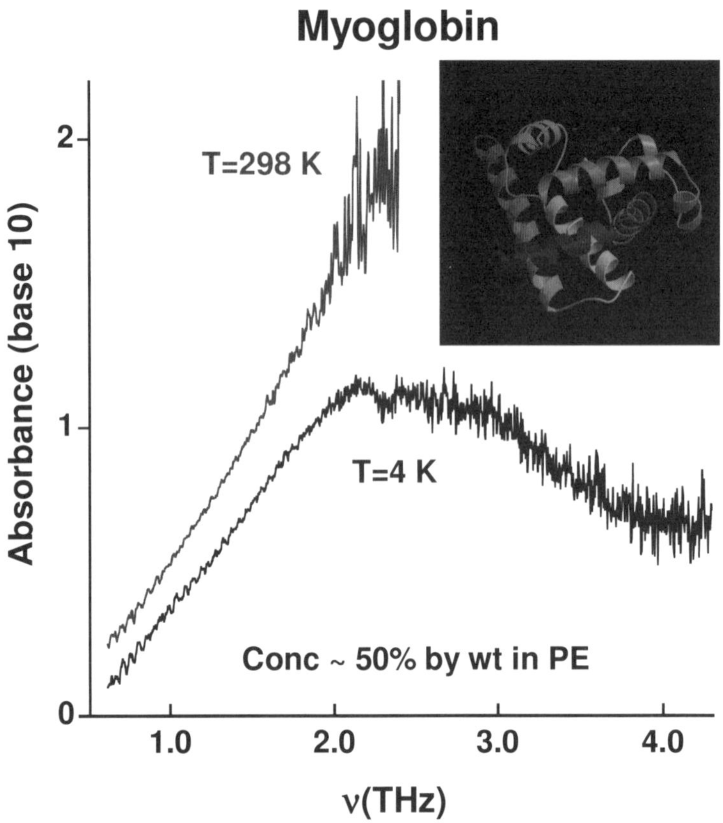

Fig. 9. The THz spectra of myoglobin in a polyethylene matrix obtained at room temperature and 4 K. Figure of structure taken from Ref. 19.

4.2. *Myoglobin*

The THz spectrum of the protein myoglobin is shown in Fig. 9. Myoglobin, with 250 amino acid residues, serves as a model for the THz spectra of large biomolecules. At our signal-to-noise ratio, the spectrum displays no obvious resonance structure over the temperature range from 4.2 K to room temperature. Because of our relatively low signal-to-noise ratio, our observations are not in disagreement with the conclusions of others that large biomolecules show resonance structure at the 1 % to 2 % fractional absorption level.[17] Indeed, such resonance structure may be expected from observations at mid-infrared frequencies where large biological molecules display highly structured vibrational spectra. That this structure would abruptly and totally vanish below 100 cm^{-1} would be surprising, particularly under cryogenic conditions where dielectric relaxation and its associated diffuse absorption features vanish

Indeed, we attribute the increased absorption observed at room temperature as due to the dielectric absorption of liquid water known to be present in our samples. The decreased absorption at 4 K is attributed to the change in the liquid hydration pockets to ice crystals with the concomitant quenching of water librational motion. Indeed, it is the quenching of the rotational motion of water that is responsible for the transparency of ice in the THz region.[18] Additional temperature studies near 273 K will directly address the importance of this effect.

5. Conclusions

To further simplify the interpretation of THz spectra, we are currently developing matrix methods based on rare gases and the quantum solid, H_2. The goal of these studies is to reduce the homogeneous and inhomogeneous factors responsible for line broadening. The inert rare-gas matrix and the removal of the vibrational degrees of freedom present in polyethylene should reduce or eliminate many of the contributors to the frequency shifts and line broadening. The quantum solid, p-H_2, provides a unique approach to eliminate broadening arising from facile energy transfer to the phonon modes of the matrix. As a result of the small size of H_2, the phonon density available for vibrational energy transfer is low. For example, the lowest rotational (J=2-0) energy transition of p-H_2 is at 355 cm^{-1} (10.6 THz),[20] which is well above the vibrational energies observed in the present study. Investigations of biomolecules trapped in Ar and H_2 matrices are currently underway.

References

1. Certain commercial products are identified in this paper in order to specify adequately the experimental or theoretical procedures. In no case does such identification imply recommendation or endorsement by the National Institute of Standards and Technology, nor does it imply that the products are necessarily the best available for the purpose.
2. K. A. McIntosh, E. R. Brown, K. B. Nichols, O. B. McMahon, W. F. DiNatale, T. M. Lyszczarz, Appl. Phys. Lett., 67, 3844 (1995); S. Verghese, K. A. McIntosh, E. R. Brown, Appl. Phys. Lett., 71, 2743, (1997).
3. E. R. Brown, Appl. Phys. Lett., 75, 769-771 (1999); S. M. Duffy, S. Verghese, K. A. McIntosh, A. Jackson, A. C. Gossard, S. Matsuura, IEEE Trans. Microwave Theory and Tech., 49, 1032 (2001).
4. A. S. Pine, R. D. Suenram, E. R. Brown, K. A. McIntosh, J. Mol. Spect., 175, 37-47 (1996); s. Matsuura, P. Chen, G. A. Blake, J. C. Pearson, H. M. Pickett, IEEE Trans. Microwave Theory and Tech., 48, 380 (2000).
5. S. Verghese, K. A. McIntosh, and E. R. Brown, Appl. Phys. Lett., 71, 2743 (1997).
6. E. Riedle, S. H. Ashworth, J. T. Farrell, Jr., and D. J. Nesbitt, Rev. Sci. Instrum. 65, 42 (1994).
7. D. F. Plusquellic, S. R. Davis and F. Jahanmir, J. Chem. Phys. 115, 225 (2001).

8. A. Gurtler, C. Winnewisser, H. Helm, and P.U Jepsen, J. Opt. Soc. Am. A., 17, 74, (2000).
9. L. S. Rothman, C. P. Rinsland, A. Goldman, S. T. Massie, D. P. Edwards, J. M. Flaud, A. Perrin, C. Camy-Peyret, V. Dana, J. Y. Mandin, J. Schroeder, A. McCann, R. R. Gamache, R. B. Wattson, K. Yoshino, K. V. Chance, K. W. Jucks, L. R. Brown, V. Nemtchinov, P. Varanasi, HITRAN Database, JQSRT, 60, 665 (1998).
10. M. J. Frisch, G. W. Trucks, H. B. Schlegel, G. E. Scuseria, M. A. Robb, J. R. Cheeseman, V. G. Zakrzewski, J. A. Montgomery, R. E. Stratmann, J. C. Burant, S. Dapprich, J. M. Millam, A. D. Daniels, K. N. Kudin, M. C. Strain, O. Farkas, J. Tomasi, V. Barone, M. Cossi, R. Cammi, B. Mennucci, C. Pomelli, C. Adamo, S. Clifford, J. Ochterski, G. A. Petersson, P. Y. Ayala, Q. Cui, K. Morokuma, D. K. Malick, A. D. Rabuck, K. Raghavachari, J. B. Foresman, J. Cioslowski, J. V. Ortiz, B. B. Stefanov, G. Liu, A. Liashenko, P. Piskorz, I. Komaromi, R. Gomperts, R. L. Martin, D. J. Fox, T. Keith, M. A. Al-Laham, C. Y. Peng, A. Nanayakkara, C. Gonzalez, M. Challacombe, P. M. W. Gill, B. G. Johnson, W. Chen, M. W. Wong, J. L. Andres, M. Head-Gordon, E. S. Replogle and J. A. Pople, Gaussian 98 (Revision A.9), Gaussian, Inc., Pittsburgh, PA, 1998.
11. M.D. Halls and H.B. Schlegel, J. Chem. Phys. 109, 10587, (1998).
12. A.K. Rappé and E.R. Bernstein J. Phys. Chem. A 104, 6117 (2000).
13. L. Stryer, Biochemistry, Fourth ed., W. H. Freeman, New York, 1995 Chapter 9.
14. G.T. DeTitta, J.W. Edmonds, W. Stallings, J. Donohue, J. Am. Chem. Soc. 98, 1920 (1976).
15. A.A. Strzelczyk, J.Cz. Dobrowolski, A.P. Mazurek, J. Mol. Struct. (Theochem) 541, 283 (2001).
16. M. Walther, P. Plochocka, B. Fischer, H. Helm, P. Uhd Jepsen, Biopolymers, 67, 310-313, (2002).
17. T. R.Globus, D. L. Woolard, A. C. Samuels, B. L. Gelmont, J. Hesler, and T. W. Crowe, and M. Bykhovskaia, J. Appl. Phys., 91, 6105, (2002).
18. C. Zhang, K-S. Lee, X.-C. Zhang, X. Wei, Y. R. Shen, Appl. Phys. Lett., 79, 491 (2001).
19. H. M. Berman, J. Westbrook, Z. Feng, G. Gilliland, T. N. Bhat, H. Weissig, I. N. Shindyalov, P. E. Bourne, The Protein Data Bank, Nucleic Acids Research, 28, 235 (2000). http://www.pdb.org.
20. T. Oka, Annu. Rev. Phys. Chem., 44, 299-333 (1993); D. P. Weliky, T. J. Byers, K. E. Kerr, T. Momose, R. M., Dickson and T. Oka, Appl. Phys. B, 59, 265 (1994).

Author Index

Volume 13 (2003)